S0-BLM-012

# HETEROLYTIC FRAGMENTATION OF ORGANIC MOLECULES

# HETEROLYTIC FRAGMENTATION OF ORGANIC MOLECULES

**TSE-LOK HO**
Department of Chemistry,
National Taiwan University

**A Wiley–Interscience Publication**
**JOHN WILEY & SONS, INC.**
New York • Chichester • Brisbane • Toronto • Singapore

This text is printed on acid-free paper.



***Library of Congress Cataloging in Publication Data:***

Ho, Tse-Lok,
Heterolytic fragmentation of organic molecules / Tse-Lok Ho.
p. cm.
"A Wiley-Interscience publication."
Includes bibliographical references (p. ) and index.
ISBN 0-471-58101-1
1. Fragmentation reactions. 2. Organic compounds. I. Title.
QD281.F7H6 1993
547.2—dc20 92-18337

Printed in the United States of America

10 9 8 7 6 5 4 3 2 1

# CONTENTS

# PREFACE

The study of heterolytic fragmentation reactions contributes to our understanding of stereoelectronic effects in organic chemistry. It is interesting to note the relationship of such reactions with retrosynthetic analysis, although the latter is purely a mental or computer-aided exercise and it does not necessarily find an exact replica in the fragmentation reaction repertoire.

We have witnessed numerous occasions in which the most expedient way to construct a molecular framework or an array of functional groups employs a fragmentation pathway. In other cases, fragmentation becomes a mandatory step to unravel the required structure.

Fragmentation reactions, whether occurring unintentionally or by design, have been reported frequently in the chemical literature. However, an extensive survey of this class of organic reactions is lacking. Considering their importance and also the implications in future synthetic adventures, a systematic muster of the fragmentation reactions seems worthwhile. Thus, this monograph is committed to a categorization of the many fragmentation possibilities according to the nature and distribution of polar functionalities in molecules or reactive intermediates. This categorization also shows, by examples, how a molecule which is undisposed to fragment may be activated to do so.

Because of the sporadic appearance of fragmentation reactions in the literature and the scarcity of cross-references and reviews, this book attempts to remedy the situation by providing a convenient source, although its comprehensiveness is not claimed. The chief criterion for inclusion of a reaction in this monograph is the presence in the molecular system of distinct donor and acceptor groups in a continuous circuit, either be they preexistent or emerging as transient species in reactions. Consequently, most pericyclic reactions are excluded. The acid-catalysed cleavage of monofunctional small ring compounds is also outside the scope of discussion.

It has been a luxury to find time and a quiet environment to complete this monograph. As always, my gratitude goes to members of my family, Honor, Jocelyn, and Daphne, for their sacrifices.

TSE-LOK HO

# 1

# INTRODUCTION

Fragmentation in organic chemistry denotes a controlled bond fission process. The majority of these reactions are heterolytic and therefore the determining elements are polar substituents. The fragmentability of an organic species is directly related to the distribution of its polar substituents and it is often quite easy to evaluate fragmentability in terms of the polarity alternation rule.

## 1.1. POLARITY ALTERNATION RULE

Polarity alternation was considered at the end of last century by a number of chemists [Lapworth, 1898] as having a profound effect on reactivity profiles of organic compounds. Despite its capability and effectiveness in dealing with diverse chemical phenomena, this empirical rule has been virtually ignored since the days modern organic theories such as resonance were developed. The polarity alternation phenomenon has solid theoretical foundations, and it is supported by calculations [Coulson, 1978].

The far-reaching consequences [Ho, 1988, 1989, 1991] that polarity alternation of molecular segments also encompass the fragmentability of organic molecules. In fact, a classification of various fragmentations adopted here is based on the distribution of polar functionalities along a short chain of atoms that engages in the process. In order to acquant the reader with the polarity alternation rule a short discussion is presented below.

While a hydrocarbon chain of an organic molecule is essentially homopolar, introduction of polar substituents tends to affect the reactivity of the member atoms profoundly. Two kinds of polar substituents exist: the donor group (**d**) including those with n- and $\pi$-electron pairs {OH, OR, OCOR, NRR′, NR(COR), SH, SR, F, Cl, Br, I} and the $+I$ substituents such as alkyl groups;

the acceptor groups (**a**) which contain empty orbitals {CHO, COR, COOH, COOR, CN, $SO_2R$, $NO_2$, $SiR_3$} and the $-I$ substituents such as $^+NR_3$. A donor group imparts acceptor characteristics to its adjoining carbon and an acceptor group demands a donor neighbor. The effect is transmitted along a chain of atoms in a gradually diminishing degree.

The presence of two polar substituents in close proximity in a carbon skeleton leads to a situation that polarity alternation is either obeyed or violated. The ground state energies of the segments conforming to polarity alternations are lower, the violating cases must be thermodynamically less stable.

It is now easy to predict the stability of α-oxy- and α-amino carbenium ions on the basis of excellent *d–a* pairings. The enormous gain in stability of carbanions flanked by a carbonyl, a cyano, a nitro, a sulfonyl, or a silyl group has a common cause in that all these substituents are acceptors. The attachment of more than one substituent of the same polarity to an atom accentuates the donor or acceptor characteristics of that center, as attested by the ionizing susceptibility of ortho esters and 1,3-dicarbonyl compounds. Of course the polarity alternation mechanism operates in conjugate systems.

There are two ways to activate an organic molecule for reaction: (i) accentuation of polarity alternation of a reaction center, which is particularly relevant in fragmentation of systems such as 1,3-diols in that the formation of the manosulfonates accelerates the bond cleavage process; and (ii) disruption of polarity alternation. The latter status provides the driving force to many 1,2-rearrangements.

The two strategies of activation are mutually exclusive. The proper employment of one of them should depend on existing functionalities.

As a relevant example in the present book one might mention the $10^3-10^4$ fold rate enhancement in the solvolysis of 3-oxobicyclo[2.2.2]oct-1-yl triflates comparing to the deoxo compounds [K. Takeuchi, 1989]. A small amount of fragmentation product has also been observed in the 5,5-dimethyl-3-oxo derivative.

## 1.2. CONTRAPOLARIZATION

Another concept needed to be introduced is called contrapolarization [Ho, 1989]. Essentially it is umpolung in situ. This donor/acceptor role change is best recognized in the reactivity of a sulfone. Thus, a sulfonyl group acidifies the neighboring C—H by virtue of its acceptor nature. On the other hand, the sulfonyl group is capable of departure from a carbon framework in substitution and elimination reactions. In these reactions the sulfonyl substituent behaves as a donor. (Note that donor and acceptor indicate their capacity for donating and accepting electrons, respectively. One must not be confused by their meanings without properly referring to the direction of the bond forming reaction, even if the bond formation process is imaginary.)

To illustrate the contrapolarization phenomenon we may consider the Mukaiyama amide synthesis [Mukaiyama, 1976]. The active reagents for this coupling of a carboxylic acid and an amine is a phosphine and a disulfide. The combination of the latter compounds leads to a phosphonium salt or a pentacovalent phosphorane. The donor trivalent phosphorus atom is thereby converted into an acceptor before it can activate the carboxylic acid. In another parlance the reaction is an oxidoreduction, valence changes occurring in the phosphorus and sulfur atoms. Actually all redox processes involve contrapolarization.

A relatively simple bifunctional molecule may be conjoint or disjoint. Conjointment means the two polar substituents are connected in a polarity alternation manner, whereas disjointment does not conform to such connection. Contrapolarization at one of the polar sites of a disjoint species would convert it into a temporarily conjoint species.

The significance of contrapolarization in certain fragmentation reactions will be described.

## 1.3. FRAGMENTABILITY

Heteropolar reactions are strongly influenced by other polar substituents close to the reaction sites. These effects are even more prominent in multicentered reactions such as fragmentation. Accordingly, fragmentability of an organic network depends on the smoothness of electron flow along the atomic chain and, excepting the terminal groups, each member of the chain engages in exchanging of an electron pair. Bond scission occurs or it can be induced readily when two or more conjoint polar substituents are present in the network which contains at least two contiguous single bonds. Thus, the appearance of a fragmentation product from solvolysis of 5,5-dimethyl-3-oxobicyclo[2.2.2]oct-1-yl triflate and the rate enhancement in the order of $10^3$–$10^4$ fold for the desmethyl compound in comparison with the parent hydrocarbyl triflate [K. Takeuchi, 1989] attest to the importance of a distal conjoint substituent. Conjoint systems in which the polar substituents are separated by conjugate $\pi$-bonds cannot fragment, they exhibit heightened ionic characters due to resonance.

Disjoint systems are resistant to fragmentation under such reaction conditions as acid or base treatment. However, contrapolarizing manipulation of disjoint molecules changes them into fragmentable states. Redox reactions are mostly commonly employed to bring about such changes, as exemplified by glycol cleavage with lead tetraacetate.

## 1.4. MECHANISTIC INFORMATION OF FRAGMENTATIONS

The defining of fragmentation as a distinct family of organic reactions was due to Grob [1955, 1957]* and it was reviewed in the late 1960s [Grob, 1967b,

* Fragmentation and the concept of $\sigma$-conjugation as proposed by A.N. Nesmeyanov was discussed by Maksimov [1965].

1969]. In this monograph the scope of fragmentation is extended further while maintaining the criterion of systems which contain at least two polar groups. Only a few of the reactions have been scrutinized in mechanistic and stereochemical contexts.

As fragmentation of the system Y—C—C—C—X encompasses the most examples and it being most thoroughly studied in terms of structural and electronic requirements, we shall outline mechanistic implications based on this system here. There are three mechanisms: two stepwise pathways, and a synchronous process.

One of the two-step mechanisms has its rate-determining step in the departure of the nucleofugal group X. Thus this mechanism is favored by substrates in which stable cations (e.g. tertiary carbenium ions) can be formed. Fragmentation of the $\beta$-donor substituted cations follows.

The adoption of this mechanism is shown by the rate increase in substrates with better leaving groups ($X = I > Br > Cl$) and the relative insensitivity of product ratios arising from fragmentation against other competing reactions including solvent capture of the cationic intermediates and cyclization.

Further evidence for nonparticipation of Y in the rate-determining step may be gained by comparing the rate of ionization of the substrate with that of the homomorphous system ($Y = C$). Usually a heteroatom Y is more electronegative than carbon and the inductive effect would suppress the ionization [Enderle, 1967].

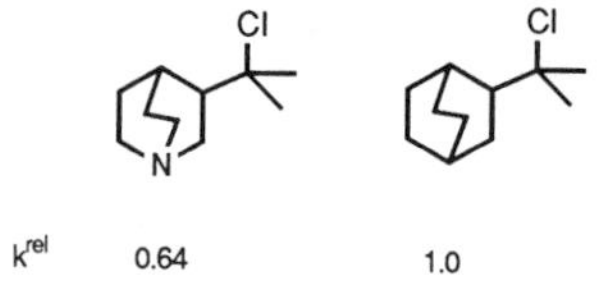

Cl Cl

N

$k^{rel}$ 0.64 1.0

The second stepwise mechanism involves formation of an ion pair by fission of an internal bond. Two conditions must be met for this mechanism to operate: low nucleofugality of group X, and stability of the anion. The first order rate constants for a series of compounds with various X would differ only slightly. On this basis, esters of 3-dimethylamino-2,2-bis(4′-nitrophenyl)propanol have been determined [Grob, 1972] to react by this mechanism.

The synchronous mechanism for fragmentation is perhaps the most interesting. The involvement of five atomic centers in the transition state necessarily puts much constraint in the structural features of the substrates. Smooth electron flow should be favored by antiperiplanarity [Klyne, 1960] of a lone pair on Y, the internal bond, and the bond with the departing X group. Such arrangement permits maximum overlap of the developing $\pi$-orbitals in the internal atoms.

This mechanism entails contrasting kinetic consequences to the cationic mechanism. The dispersion of the positive charge over four atoms results in stabilization of the transition state which is manifested in an acceleration of the reaction in comparison with the homomorphous substrate.

The steric requirements for concerted fragmentation may be satisfied by a zig-zag conformation or an eclipsed Ω conformation. These are respectively represented by the 3-bromoadamantylamines [Grob, 1964] and 4-bromoquinuclidine [Brenneisen, 1965; Grob, 1970].

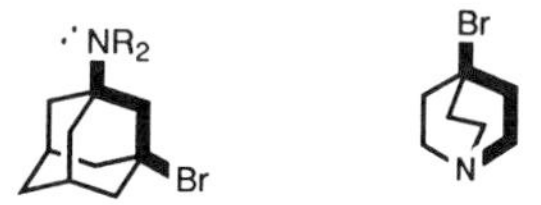


The crystal structure of the iminium ion intermediate from fragmentation of 1-bromo-3-dimethylamino-5,7-dimethyladamantane has been determined [Hollenstein, 1990]. It is very similar to the bicyclic ketone.

Further studies of the adamantanyl system [Fischer, 1974; Schmitz, 1974] have revealed that 3-chloroadamantanyl ethers and sulfides undergo $S_N1$ substitution only. However, the conjugate bases of the corresponding alcohol and thiol fragment with enormously enhanced rates.

Azabicyclooctane systems are most useful for defining the geometric stringency of the concerted fragmentation pathway. While evidence strongly indicates its existence for *N*-methyl-*endo*-6-tosyloxymethyl-isoquinuclidine, no fragmentation was observed during solvolysis of *N*-methyl-*endo*-5-tosyloxy-2-azabicyclo[2.2.2]-octane [Martin, 1973]. In *endo*-1-azabicyclo[3.2.1]oct-6-yl brosylate, the lone pair electrons of the nitrogen atom and the leaving group are deviated from antiperiplanarity by about 20°. This deviation is large enough to suppress almost completely the fragmentation [Siegfried, 1975].

4-Chloropiperidine and its *N*-alkyl derivatives undergo quantitiatve fragmentation and they show a rate enhancement of about 100-fold in comparison with chlorocyclohexane [D'Arcy, 1966]. The active conformational isomers for the fragmentation are those with axial *N*-substituents.

A remarkable result pertains to an internal fragmentation of *endo*-4-(*p*-toluenesulfonyloxy)-1-azabicyclo[3.2.1]octane and the [3.3.1]nonane homolog to the exclusion of the peripheral pathway leading to the iminium ions of 3-vinylpyrrolidine and 3-vinylpiperidine, respectively [Grob, 1975a]. These observations point to a transition state in which the ionization of the C—OTs bond is somewhat more advanced than the cleavage of the C—C bond, with the extra alkyl or cycloalkyl substituent dictating the site of fragmentation by providing some stabilization of the incipient cationic center.

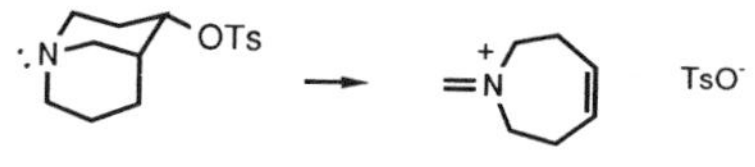

Simpler cases showing the same trend are the 1,3-dimethyl-4-piperidinyl tosylates. Production of the more highly substituted olefin is favored by a factor of 49 in the one-step fragmentation [Grob, 1975a]. It should be noted that the isomer with *cis* C-methyl and tosyloxy groups cannot undergo synchronous fragmentation without passing through a highly energetic transition state in which the two methyl groups are 1,3-diaxially disposed. The incursion of an elimination pathway and fragmentation by a stepwise process is evident.

Further examples attesting to the crucial stereoelectronic arrangement for concerted fragmentation include the 3-chlorotropanes [Bottini, 1966] and the diastereoisomeric 3-dimethylaminocyclohexyl tosylates [Burckhardt, 1967].

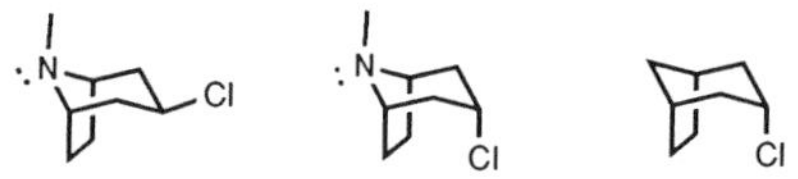

$k^{rel}$ 135,000 1 1

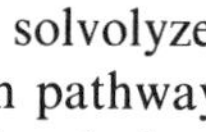

$k^{rel}$ 1

Studies of the perhydroquinoline and perhydroisoquinoline ring systems have given much insights into the mechanistic definition. *cis*-10-Chloro-*N*-methyl-decahydroisoquinoline solvolyzes 128 times as fast as *cis*-9-chlorodecalin, by exclusive fragmentation pathway. On the other hand, the trans isomer reacts 4.5 times more slowly than its homomorph, affording products of fragmentation and elimination [Geisel, 1969].

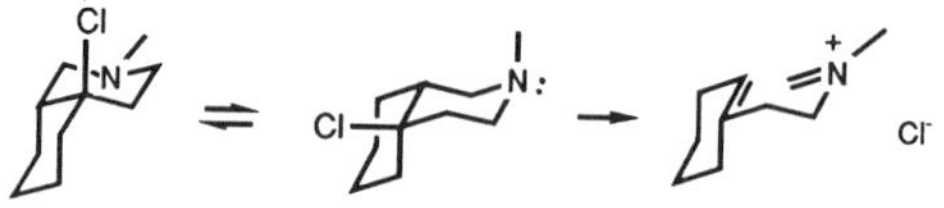

The behavior of the decahydroquinol-5-yl tosylates meets total expectation with respect to reaction rates and nature of the products [Grob, 1967a]. The isomeric *N*-methyldecahydroquinol-7-yl tosylates undergo only substitution and elmination. The preclusion of the fragmentation pathways in these cases is due to the impossibility for the nitrogen lone pair to assume a participatory orientation, that position having been permanently occupied by the carbon atom C-2.

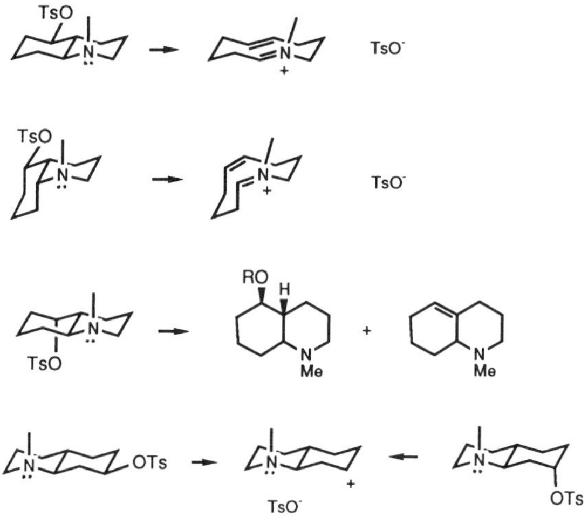

This blockade of reactive conformations cannot happen in acyclic substrates. However, it has been witnessed [Grob, 1969] that entropy is generally against the accumulation of fragmentation-prone conformations, unless special substitution patterns are introduced into the molecule. Tetramethylation of dimethylaminopropyl chloride at the $\alpha$ and $\beta$ or $\alpha$ and $\gamma$ positions appears to increase the chance of fragmentation. The nitrogen atom might be acting by internal solvation of the incipient carbenium ions.

The rigid bicyclo[2.2.1]heptane skeleton is helpful for determining stereoelectronic requirements of various reaction pathways. It has been shown that the *endo*-tosylate of a benzocyclobuteno-fused system undergoes fragmentation exclusively with rate enhancement during acetolysis [Baker, 1970]. The *exo* isomer behaves differently as concerted fragmentation is geometrically precluded.

Inhibition of fragmentation by steric constraint is also witnessed in a 1,3-diol monotosylate system that is embedded in a linear 5:4:6-fused ring skeleton [Tatsuta, 1979]. The $C_\alpha$—$C_\beta$ bond of the hydroxyl is gauche to the tosylate.

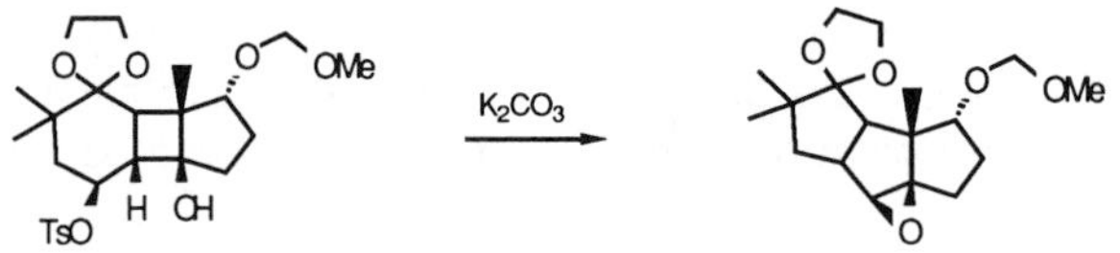

With respect to a molecular orbital assessment of the heterolytic fragmentation of X—C—C—C—N or $^{+}$C—C—C—N system [Gleiter, 1972] the electronic requirements have been determined such that the asymmetric level must be below the symmetric level. This is set by maximal through-bond coupling of the empty cation orbital and the nitrogen lone-pair. The theoretically derived conformational dependence of the through-bond coupling, that is parallel orientation of the cation orbital (or C—X bond), the C—C bond next to the nitrogen atom, and the nitrogen lone-pair must be attained.

A new observation concerning the fragmentability behavior of a rigid iodolactone with respect to its stereochemistry [Wender, 1990] is in apparent contrast to expectation. Either a mechanism involving single electron transfer operates, or the fragmentation is preceded by an enolate-assisted ionization.

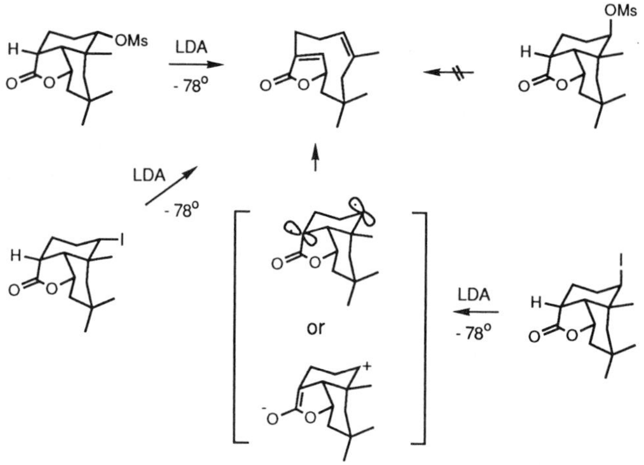

## 1.5. FRANGOMERIC EFFECT

The increase in reactivity of a system toward fragmentation due to synchronicity is called the fragomeric effect. It is defined as the ratio of the fragmentation and ionization rate constants. Frangomeric effects can be roughly estimated from the known rate constants for the homomorphous compounds with inclusion of the inductive factor.

Synchronous fragmentation is always associated with a frangomeric effect. It has been concluded, from a comparison of the activation energies and activation entropies with those of the homomorphs, that the enhanced reactivity is mainly due to a more positive activation entropy. Accordingly, a rigid system usually displays a large frangomeric effect when it undergoes a concerted fragmentation. The effect of a 6-*exo* donor substituent on the fragmentation rate increase of *exo*-2-norbornyl tosylate has been scrutinized [W. Fischer, 1979], and among these the dimethyl amino group exerts the most profound influence.

R, OTs $\xrightarrow[\text{dioxane}]{H_2O}$ CHO

| R = | H | NHAc | SMe | OMe | OH | $NH_2$ | $NMe_2$ |
|---|---|---|---|---|---|---|---|
| $k^{rel}$ | 1 | 22 | 39 | 84 | 97 | 160 | 1200 |

# 2

# a—a AND d—d FRAGMENTATIONS

The simple scheme for classifying various classes of heterolytic fragmentation reactions employed in this monograph is based on the nature of two polar substituents and the number of atoms joining them. As mentioned before, a polar substituent may be a donor (**d**) or an acceptor (**a**).

Here we begin our survey of fragmentation reactions from the **a—a** and **d—d** systems in which the polar functions are directly bonded. More extended systems carrying such designations as **a1a**, **d3d** show the number of intervening atoms at 1 and 3, respectively. It may sometimes be desirable to emphasize the bonding properties of the polar groups by adding another letter at the end. For example, retro-aldol fission falls into the category of **d2a=d** where the **a=d** segment is the carbonyl group C=O.

Autogenous or assisted ionization of the **a—a** and **d—d** systems derives its driving force from unfavorable electrostatic–electronic interactions. It must be emphasized that in the ionization process, one of the groups must undergo contrapolarization. The following examples belong to this category.

The reaction of a trimethylsilyltropenium ion with a fluoride ion gives rise to cycloheptatrienylidene which can be intercepted by fumaronitrile [Hoffmann, 1981]. In this instance the Si—C bond represents the **a—a** segment.

a a $SiMe_3$ $BF_4^-$ $F^-$ NC CN CN CN

The thermal elimination of dimethylaminocarbene from 7-dimethylamino-cycloheptatriene probably proceeds via the norcaradiene tautomer [ter Borg, 1965]. Facilitation of the decomposition by the amino group is evident as cycloelimination of methylene from cycloheptatriene requires a much higher temperature (470°C versus 110°C) [Nefedov, 1967].

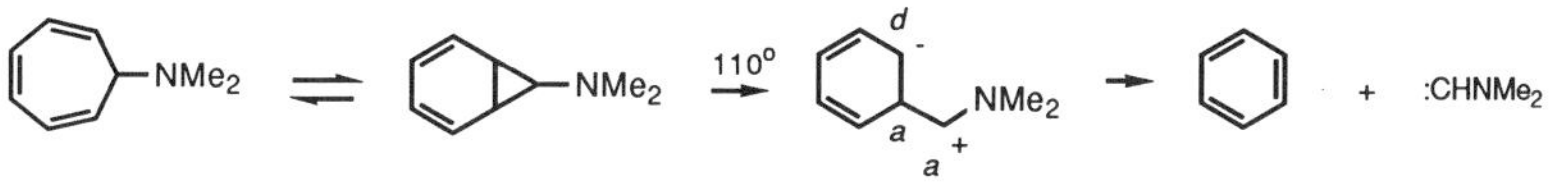

The 1,3-ditholium ion, in the canonical form of the carbocation, can be regarded as an **a—a** system. The proton at C-2 is very labile and its detachment gives rise to stabilized carbenes [Prinzbach, 1965].

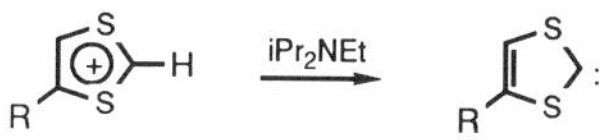

Thermal decarboxylation of pyridinecarboxylic acids is a well-known phenomenon. The decarboxylation may be considered as proceeding via the oxycarbonylcarbenium ion. The resulting carbene is a canonical form of the ylide intermediate which can be detected.

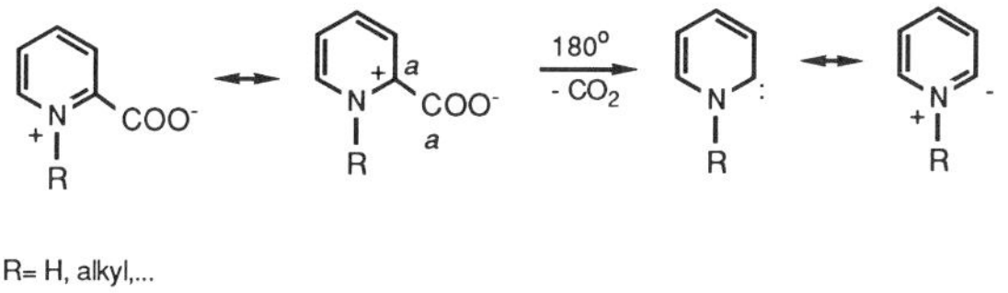

R= H, alkyl,...

The reaction of 3-bromobicyclo[3.2.1]octa-2,6-diene with potassium *t*-butoxide in dimethyl sulfoxide affords a tetracyclic carbene [Bergman, 1970]. The expulsion of the bromide ion from the carbanion is a **d—d** dissociation.

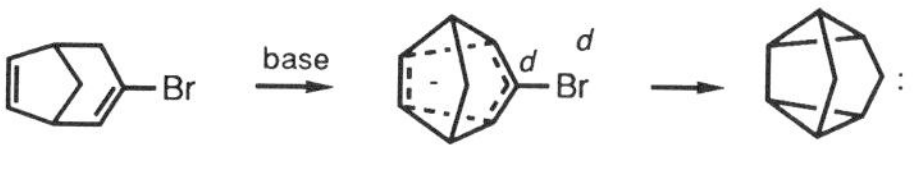

Dehydrochlorination of 3-chloro-3-methyl-1-butyne [Bramwell, 1965] may be considered as a conjugate analog.

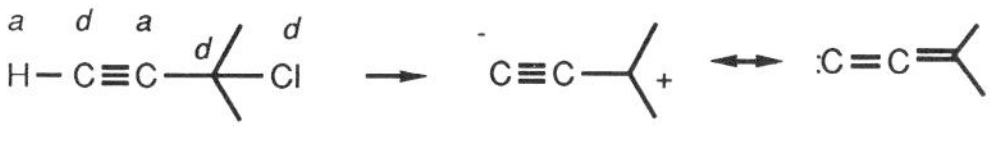

Of course, the simple alpha-elimination of HX from haloforms and the deformylative decomposition of chloral [Nerdel, 1969] by a strong base follows the same pattern.

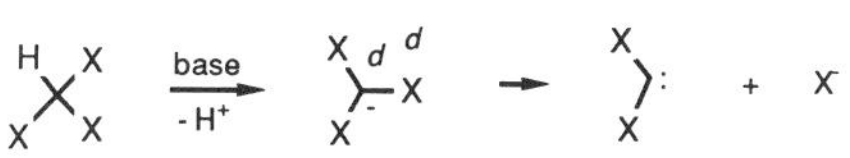

Many other methods for carbene generation are based on the same principle. Thus the decomposition of organometallic derivatives [Seyferth, 1967, 1969] and the pyrolysis of trihaloacetate salts are well-known.

$$Me_3SnCF_3 + NaI \rightarrow Me_3SnI + NaCF_3 \xrightarrow{-NaF} :CF_2 \rightarrow$$

$$PhHgCF_3 + NaI \rightarrow PhHgI + NaCF_3 \xrightarrow{-NaF} :CF_2$$

Diazoalkanes are excellent carbene precursors as they can be induced to disengage dinitrogen photolytically or with heavy metal catalysts. Nitrene formation from chloramine-T may be considered in the same terms.

$$R_2C^{-}\text{–}N_2^{+} \; (M) \xrightarrow{-N_2} R_2C:$$

An example of oxycarbene generation is shown now. The degree of facility in the thermolysis of the lactone tosylhydrazone [Agosta, 1971] is apparently influenced by the special electronic make-up.

$$\xrightarrow[0.1\ torr]{310^{\circ}}$$

N–Ts; $Na^+$

# 3

# a1a FRAGMENTATIONS

## 3.1. RETRO-CLAISEN AND RETRO-DIECKMANN REACTIONS

Retro-Claisen reactions comprise cleavage of 1,3-dicarbonyl compounds which are commonly effected by base treatment. It should be emphasized that retro-Claisen, retro-Dieckmann, and the retro-aldol fissions which will be discussed later, share the same general mechanism. The fragmentation step proceeds from the disintegration of an O—C—C—C=O segment in all these reactions. The different categorization of retro-Claisen and retro-aldol reactions is based on the variation in preexisting functional groups. The same can be said about the ethanologous and vinylogous systems in that while a normal retro-Claisen fission belongs to the **a1a** confine, the vinylogous Claisen fission would be placed in the **a3a** or **a3a=d** subfamily.

### 3.1.1. Cleavage of Three- and Four-Membered Ketones

Cyclopropanones are very reactive entities. They tend to decompose thermally or by reaction with nucleophiles. Acylcyclopropanones are even more labile, as decarbonylation (perhaps nonconcerted) occurs readily.

Based on this reactivity, an effective ring contraction procedure consisting of a dehydrochlorination-decarbonylation sequence on 2-chloro-1,3-dicarbonyl substances has been developed [Büchi, 1973b]. The decarbonylation process involves cleavage of the small ring initially at the bond which is part of the **a1a** array. However, this method is not suitable for the synthesis of cyclobutenones.

O R Cl O — $Na_2CO_3$ / xylene → O R O → -CO → R O

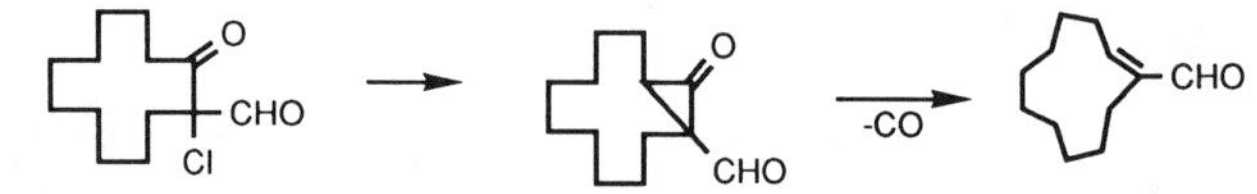

The formation of ethyl 3,3-dimethylacrylate from reaction of dimethylketene with ethyl diazoacetate [Kende, 1956] is the result of a decomposition of the cyclopropanone intermediate.

An unusual synthesis of *cis*-jasmone [Berkowitz, 1972] consisted of pyrolytic formation of an acylketene and a bicyclo[3.1.0]hexane-2,6-dione. The latter species, on losing carbon monoxide, afforded the product.

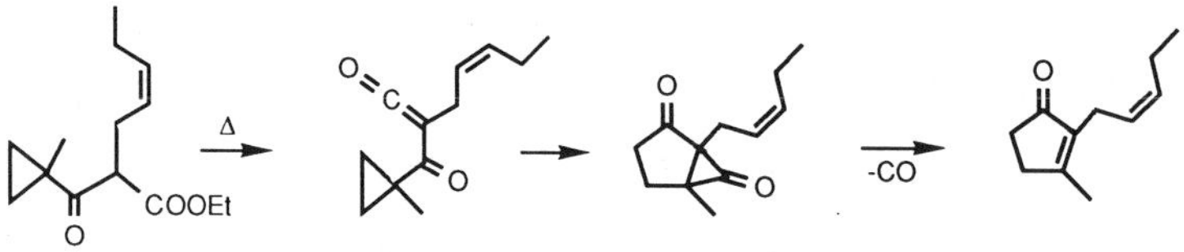

Pentacovalent bismuth species are excellent arylating agents. The reaction of phloroglucinol with triphenylbismuth carbonate [Barton, 1985] gives 2,2,4,5-tetraphenylcyclopent-4-ene-1,3-dione along with perphenylated phloroglucinol. The cyclopentenedione is derived from the tetraphenylcyclohexatrione via the bicyclic intermediate.

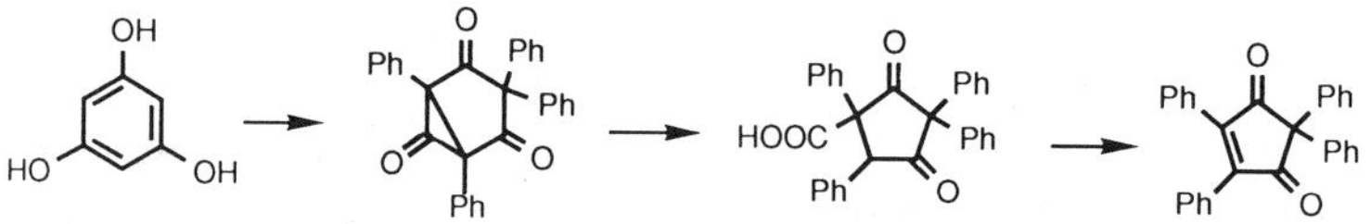

Cyclobutane-1,3-diones are also highly strained and susceptible to cleavage by various nucleophiles. Thus, Wittig and Corey reagents cause ring opening of the 2,2,4,4-tetramethyl derivative [Krapcho, 1972].

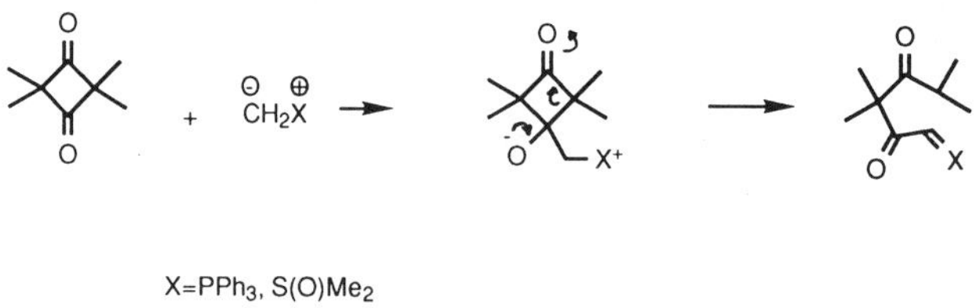

X=$PPh_3$, S(O)$Me_2$

2,2,4,4-Tetramethylcyclobutane-1,3-dione is isomerized to the $\beta$-lactone by tri-*n*-butylphosphine [Bentrude, 1972]. Opening of the four-membered dione by a retro-aldol reaction of the adduct is the expected mode of action. The best reaction pathway available to the resulting zwitterion is lactonization.

An analogous transformation has also been reported [Yoneda, 1978] of the corresponding dithione.

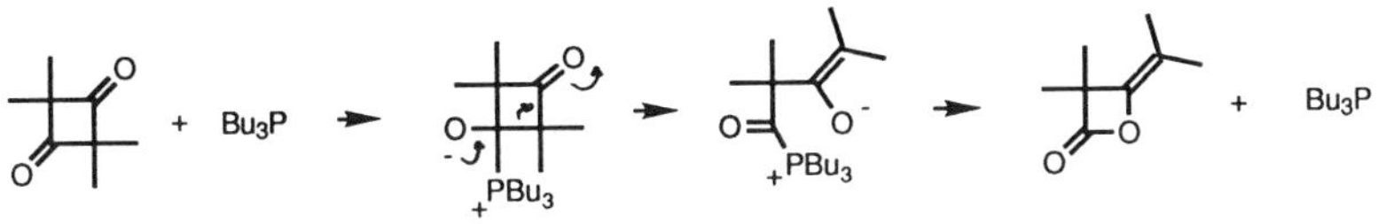

Acylcyclobutanones undergo facile ring opening to relieve strain. This reaction constitutes an integral part of the often stereoselective $\beta$-carboxylation of an $\alpha,\beta$-unsaturated ketone via photocycloaddition with allene. The latter process is regioselective [Eaton, 1964].

The reaction sequence has been successfully developed in the context of natural product synthesis. Thus, construction of the D-ring of the diterpene alkaloid veatchine [Weisner, 1968b] called for oxidative removal from the photocycloadduct, after ketalization, the exocyclic methylene group and treatment with protic acid to uncover a keto acid intermediate.

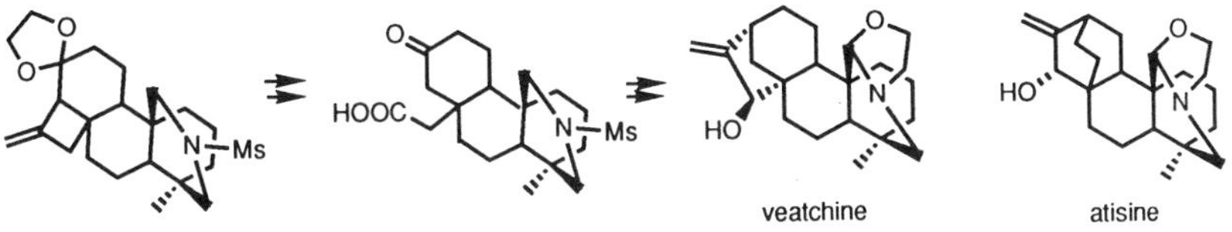

veatchine atisine

Variation of the ketone fission protocol leads to other useful intermediates. For example, reduction followed by acid treatment of an analogous compound furnishes a bicyclo[2.2.2]octanolone which is suitable for the elaboration of atisine [Guthrie, 1966].

In a reconstitution of forskolin from a desvinyl degradation product, a similar reaction sequence was employed as part of the effort in adding the vinyl group [Ziegler, 1987].

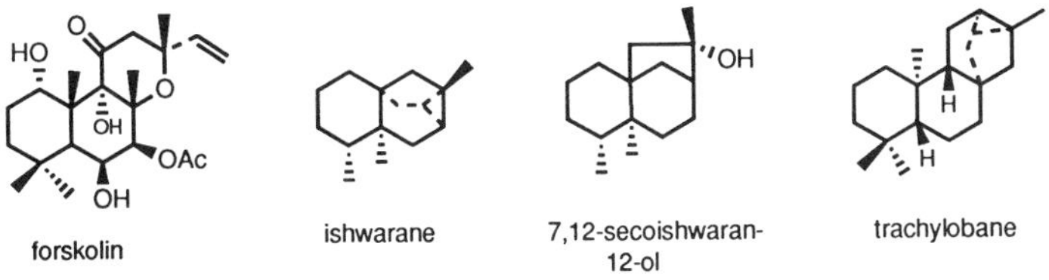

forskolin ishwarane 7,12-secoishwaran-12-ol trachylobane

Other applications to synthesis include skeletal construction of ishwarane [R.B. Kelly, 1970], 7,12-secoishwaran-12-ol [Funk, 1983], trachylobane [R.B. Kelly, 1973], and more recently, gibberellic acid [Nagaoka, 1989].

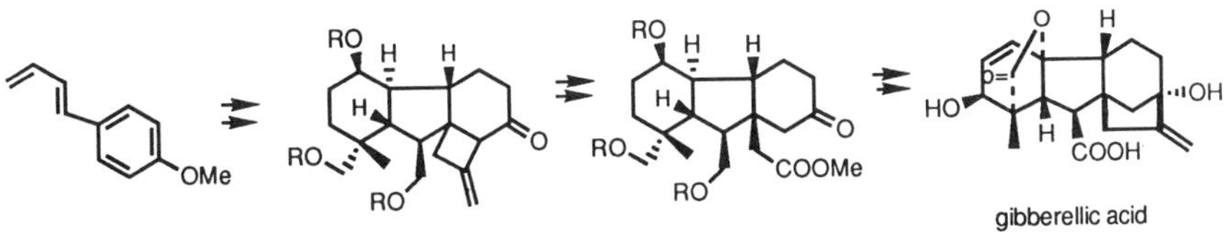

gibberellic acid

A proposed conversion of testosterone to quassin must address the problem of introducing an angular methyl group at C-8. The photocycloaddition route has been used to accomplish the task [Pfenninger, 1980], although an extensive degradation of the adduct is required.

It should be noted, however, the method fails to provide the required precursor for synthesis of eremophilone [Ziegler, 1977] because the photocycloaddition takes place from the face of 3,4-dimethyl-2-cyclohexeneone syn to the methyl group at C-4.

A remarkable retro-Claisen/Claisen/retro-Claisen transformation occurs during treatment of a tetracarbocyclic diketone intermediate en route to stemodin [Piers, 1982]. The unusual cleavage of the cyclopentanone in the presence of a fused and fragmentable cyclobutanone may be attributed to its more exposed environment.

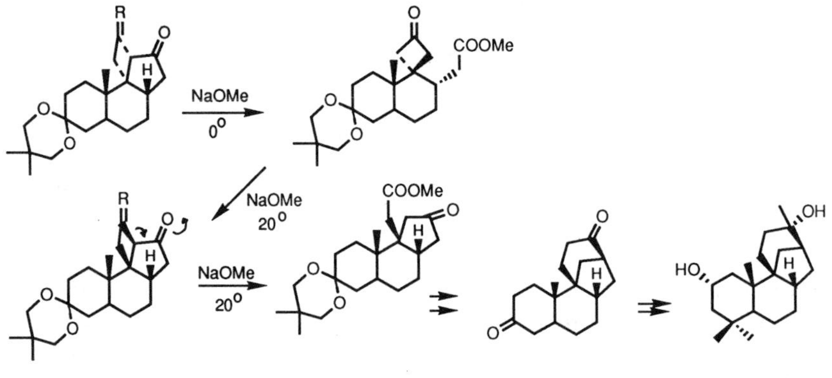

Certain spirocycles and propellanes have been assembled by way of intramolecular photocycloaddition of enones which have a tethered allene sidechain [Becker, 1975, 1980].

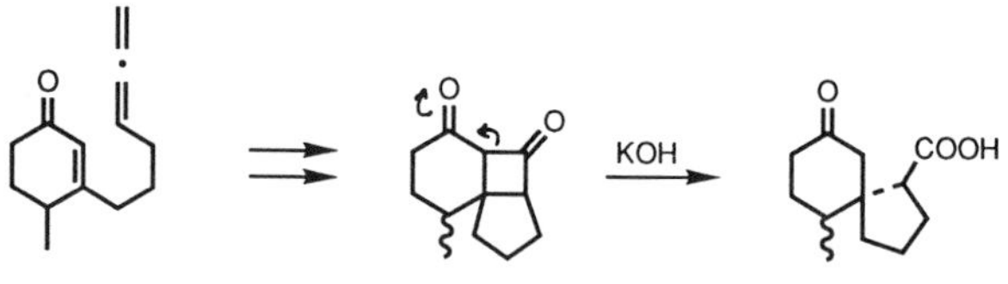

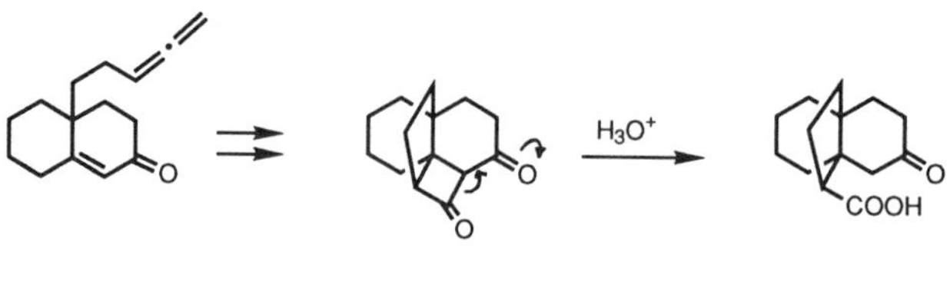

Photolysis of 6-hydroxy-2,3-benzotropone in methanol led to an indanoneacetic ester [Yoshioka, 1971]. The electrocyclized product is very susceptible to solvent attack and its fragmentation would be spontaneous.

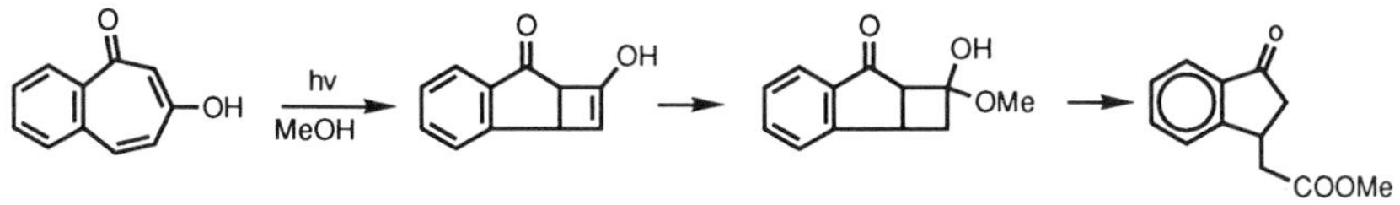

The photoisomer of α-tropolone methyl ether is a masked 1,3-diketone. It has served admirably as the building block for brefeldin-A [Baudouy, 1977]. The four-membered ring exerts stereocontrol during introduction of the alkenyl chain and it also provides an acetic ester residue which is to be fashioned into the unsaturated carbonyl segment. Accordingly, after attachment of the "lower" chain to the photoisomer, hydrolysis of the strained enol ether triggers a fragmentation.

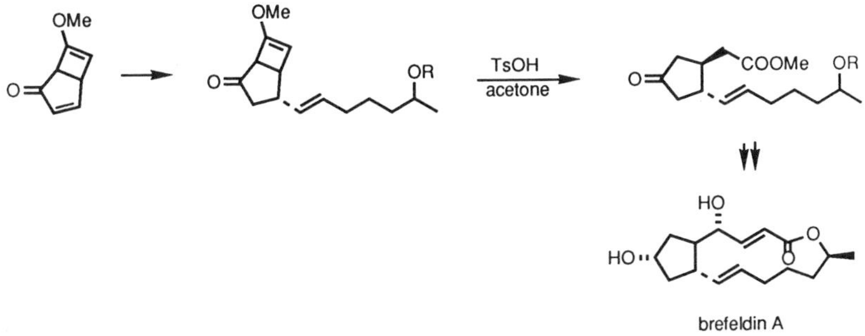

brefeldin A

Photochemical rearrangement of 8-formyl-7-hydroxy-1,2-naphthoquinone gave a γ-lactone [Ferreira, 1980]. It is conjectured that a cyclobutanolone was the intermediate.

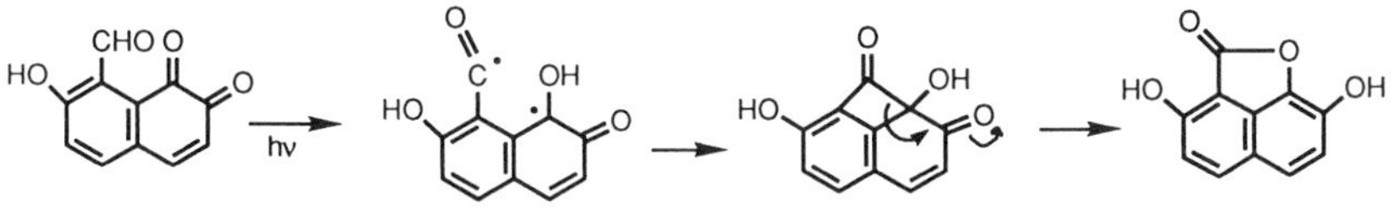

Cyclobutanones are also available by a thermal [2 + 2]cycloaddition involving ketene. A synthesis of methyl dehydrojasmonate [Lee, 1988] demonstrates an interesting ring construction plan which also meets the strict demand of functional group distribution by the fragrant material. In this particular synthesis the cyclopentanone is latent in a bicyclic intermediate, and its unraveling is required for the ultimate scission of the cyclobutanone moiety.

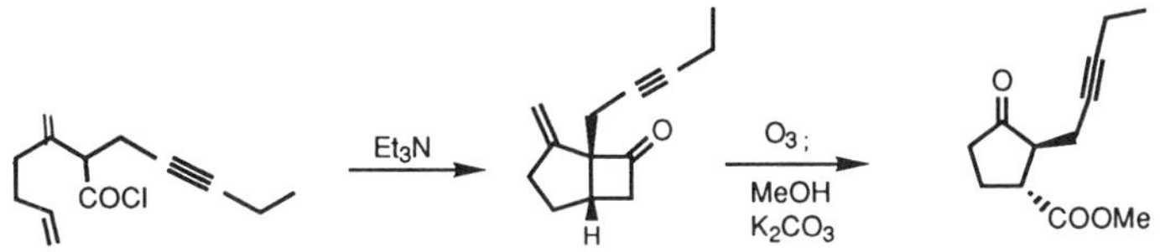

Thermal cycloadducts derived from ynamines are versatile intermediates for synthesis. Several important applications to the assembly of natural products have been realized. Among them are *erythro*-juvabione [Ficini, 1974], iso-dehydronepetalactone [Ficini, 1976], and dihydroantirhine [Ficini, 1979].

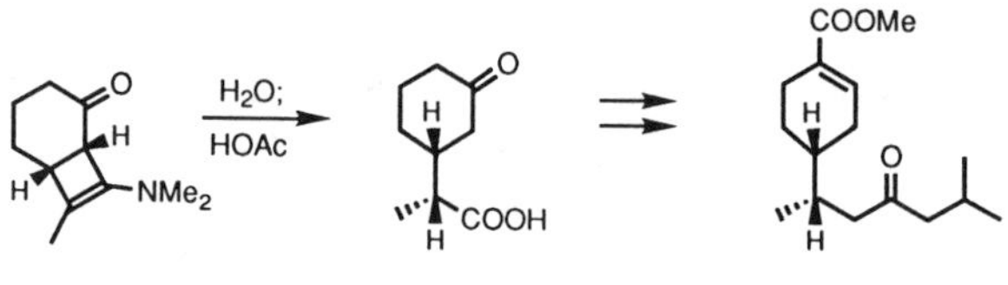

$H_2O$; HOAc

erythro-juvabione

In these synthetic schemes the hydrolysis of the cycloadducts is essential to the stereocontrol of the two contiguous asymmetric centers to be created while the sidechain is released. *exo*-Protonation of the aminocyclobutene subunit with water has been established, and if this is desirable, a more vigorous acid treatment of the resulting enaminoketone would lead to the $\beta$-diketone and then the keto acid of well-defined stereochemistry [Ficini, 1972].

Thus it is not surprising that a stereocontrolled assembly of the D-ring synthon for steroids with special reference to the correct relative configurations at C-17 and C-20 can be delegated to this same method [Desmaele, 1983a]. This building block has been elaborated into vitamin-D [Desmaele, 1983b].

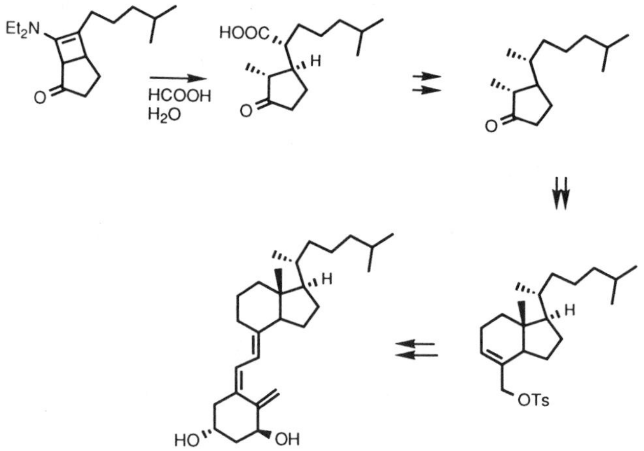

HCOOH $H_2O$

3a,4,5,6-Tetrahydroindan-1-one-4-carboxylic acids which are epimeric at C-4 are accessible from an ynamine cycloadduct by acid cleavage under regulated conditions. Photochemical addition of methanol across the enone double bond has been performed. Isomers thus available are conceivable precursors of certain yohimbe alkaloids [d'Angelo, 1983].

A reaction sequence including formation and cleavage of a benzocyclobutenone system has been proposed [Guyot, 1973] to account for the generation of homophthalimide from the reaction of bromobenzene with diethyl malonate in the presence of sodamide.

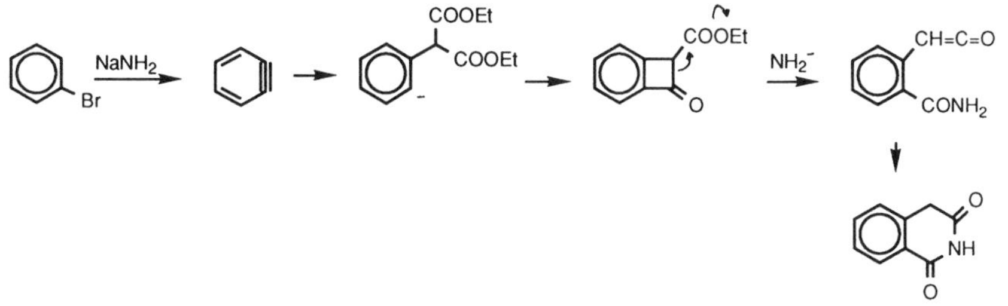

An important side-product from brominative oxidation of methyl 2-dimethylamino-3,3-dimethylcyclobutanecarboxylate is a $\gamma$-lactone ester [Agosta, 1969]. Its formation is due to ring strain.

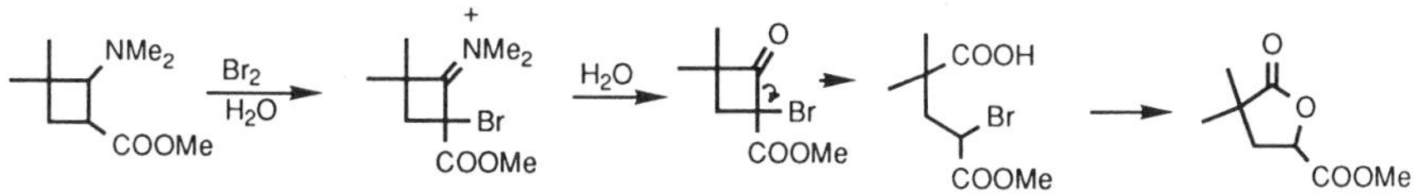

2-Azabicyclo[2.2.0]hexane-3,5-dione is readily available from a 4-alkoxy-2-pyridone via photocyclization and hydrolysis. The bicyclic compound underwent ring opening on exposure to alcohols or thiols [Katagiri, 1985].

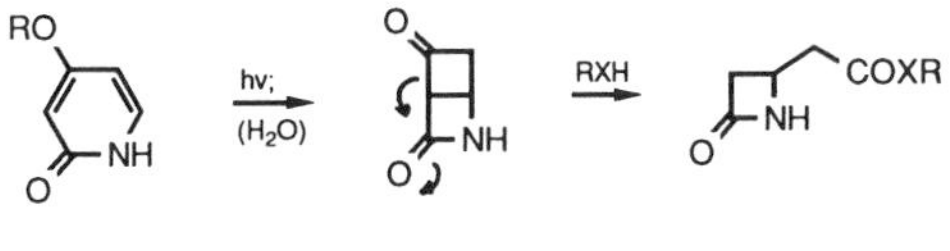

### 3.1.2. Cleavage of Five-Membered Rings

A crucial reaction sequence used in the structural elucidation of apollan-11-ol ($\alpha$-caryophyllene alcohol) [Nickon, 1970] involves the Barton reaction twice and alkaline fission of a triketone. The presence of a five-membered ring containing a *gem*-dimethyl group was established. Interestingly, the nitrite ester (actually the oxy radical) underwent epimerization during photolysis.

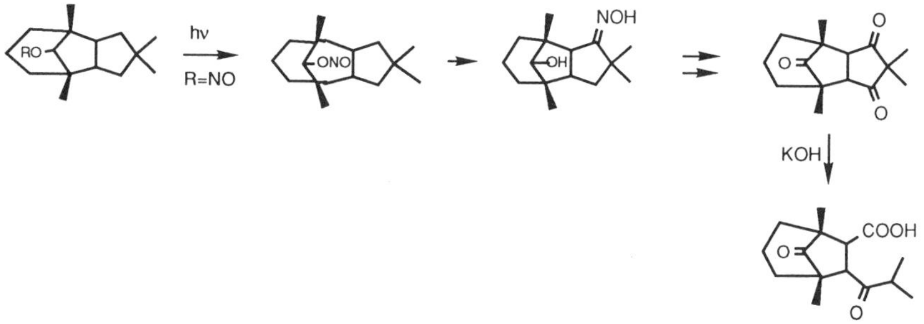

2,2′-Diphenyl-2,2′-biindan-1,1′,3,3′-tetraone is cleaved by aqueous base to afford a naphthoquinone and phthalic acid [Beringer, 1963]. This reaction is related to the reduction of 2,2′-biindan-1,1′3,3′-tetraone to form a dihydroxynaphthacenequinone [Dufraisse, 1936; Wanag, 1937; Schönberg, 1949].

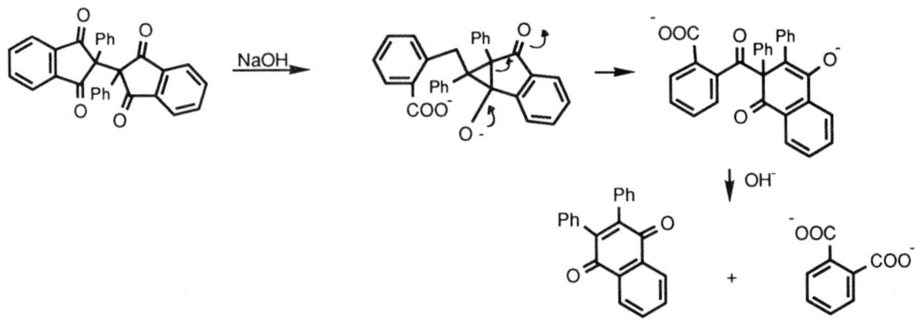

A purple indolizinoquinone was obtained from a reaction of 2,3-dihydro-2-[2(1*H*)-pyridylidene]-3-thioxoinden-1-one with a 1,4-benzoquinone [Buggle, 1980]. Michael addition resulted in a fully substituted 1,3-oxothione which is prone to ring opening by the alcoholic solvent. The attack on the ketone group instead of the thione is due to HSAB (hard and soft acids and bases principle) discrimination. Various hydroquinone intermediates were oxidized by the excess benzoquinone present.

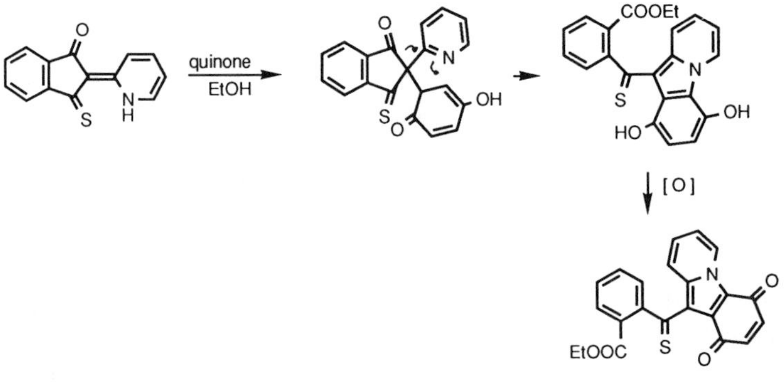

A synthetic pathway to cordrastine-I and -II [Smula, 1973] embodies formation of a spiroisoquinoline intermediate which was then submitted to a retro-Claisen fission. The cleavage is nonregioselective due to the fact that each carbonyl group is influenced by an activating (*o* or *p*-) methoxy substituent in the D-ring.

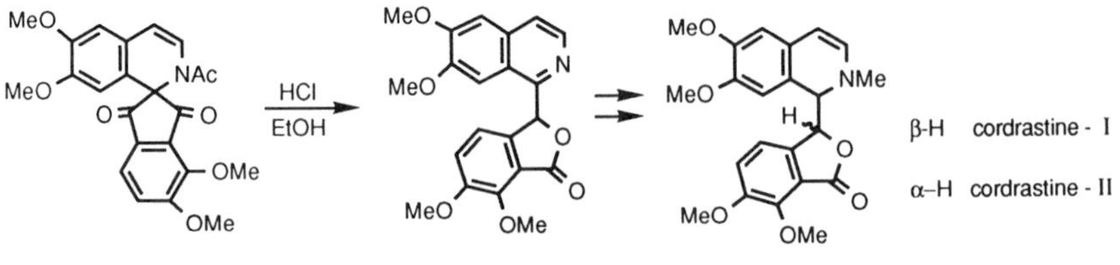

β-H cordrastine - I

α-H cordrastine - II

The behavior of 2-(*o*-cyanobenzyl)-α-tetralone toward alkali is perhaps a little devious. Instead of undergoing simple saponification, it gives 2-(*o*-

carboxylphenyl)indan-1-one [Herz, 1957]. A spirocyclic diketone must be the prefragmentation intermediate.

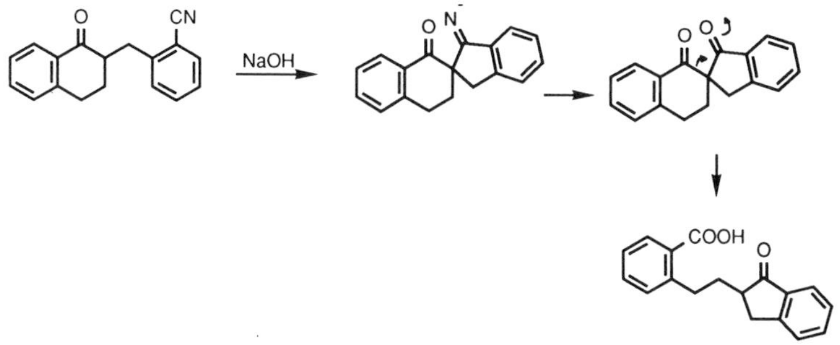

NaOH

COOH

Prolonged exposure of the antiinflammatory agent "oxepinac" to sunlight in an alkaline solution gave a diphenyl ether and a diphenylmethane derivative [Tagawa, 1984]. Formation of the latter compound can be envisaged to arise from C—C radical coupling followed by a retro-Claisen scission.

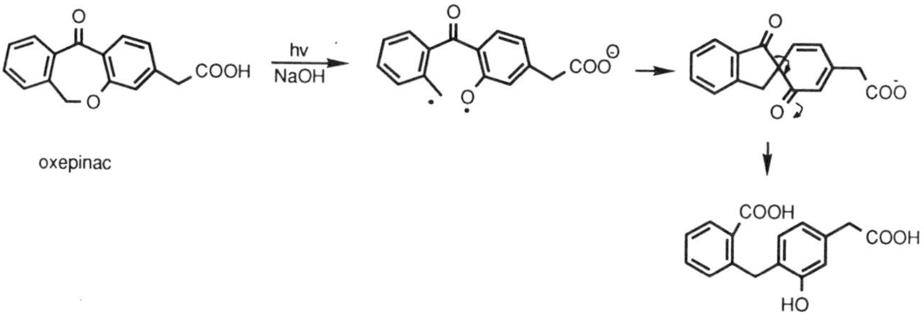

COOH hv NaOH COO

oxepinac

COOH COOH HO

The structure of a stress metabolite of sweet potato has been identified as 7-hydroxymyoporone by synthesis of its degradation product, a $C_{11}$-keto acid [Burka, 1974]. This synthesis involves acylation of a carbomethoxycyclopentanone and hydrolytic cleavage.

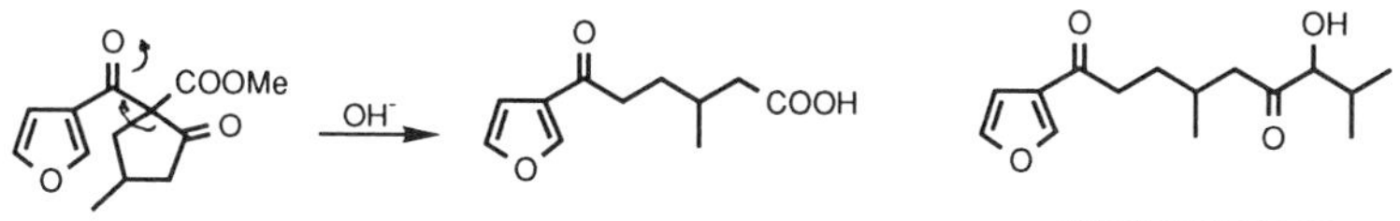

COOMe OH⁻ COOH OH

7-hydroxymyoporone

Jones oxidation of tetrahydroshikoccin [Fuji, 1985] accomplished also a retro-Claisen ring fission and aldolization to give an abietane derivative. It is possible that the aldol condensation precedes the fragmentation.

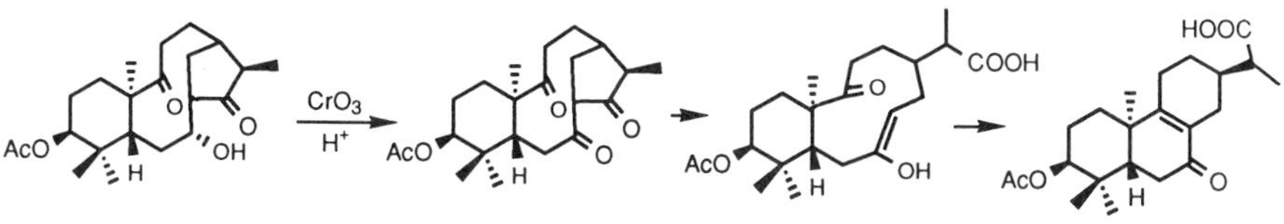

AcO OH $CrO_3$ $H^+$ AcO COOH AcO OH HOOC AcO

The diketolactone ester obtained from enmein reacts with enthanedithiol with ring opening [Fujita, 1966], apparently owing to the strain of the bridged ring system.

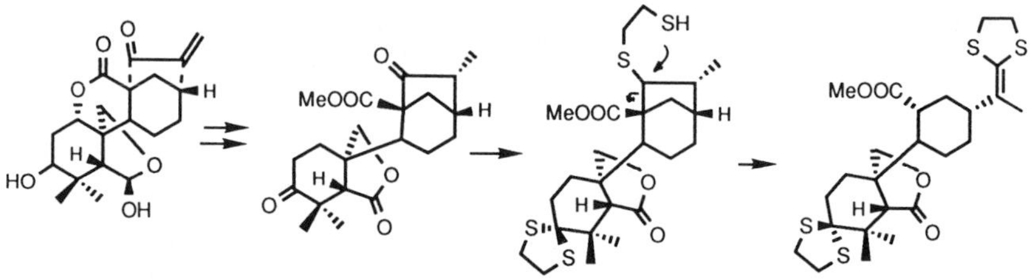

α-Alkylation of 2-cyclohexenone and 2-cyclopentenone can be conducted via bicyclic diketones which are easily obtained by condensation of the enones with thioacetic acid derivatives (Claisen–Michael condensation). After the alkylation of the diketone, a reverse Dieckmann–Michael reaction tandem is then performed [Baraldi, 1984].

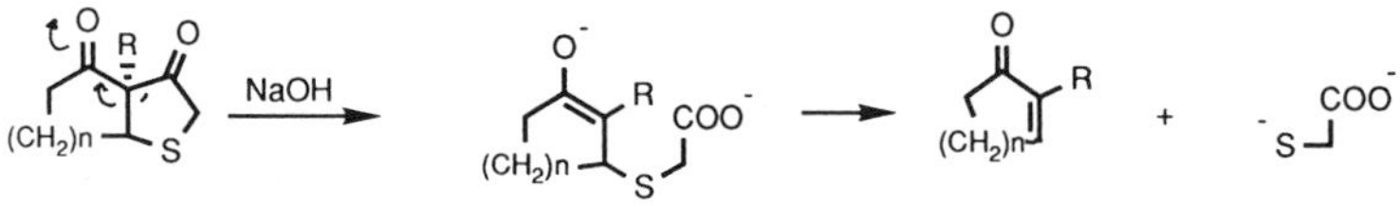

Two factors favor the selective attack on the ketone group of the heterocycle in the unraveling of the enone. First, a cyclopentanone is slightly more strained than a cyclohexanone and there is a greater tendency for cleavage of the former. Second, the influence of the sulfur atom by imparting an acceptor character to the proximate carbonyl group via the polarity alternation mechanism must be considered.

Stauranetetraone undergoes stepwise retro-Claisen cleavages to afford a diquinanedione [Han, 1982]. The tricyclic trione is more reactive and hence it cannot be isolated. The cleavage is also chemoselective, as the spirocyclic isomer is not produced.

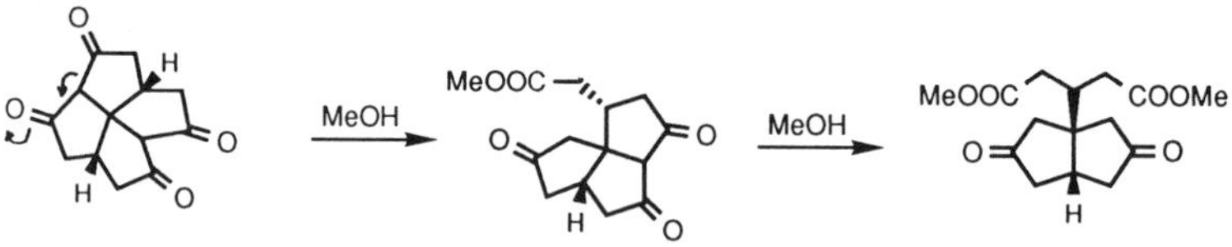

An interesting synthesis of 1,4-cycloheptanedicarboxylic acids is based on rearrangement and retro-Claisen fission of the bicyclo[4.2.0]octan-2-ones [Ranu, 1988].

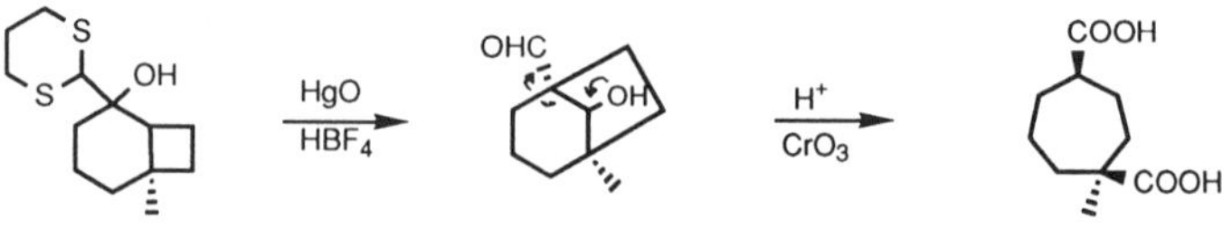

For the following transformation it has been suggested [Carruthers, 1973] that a retro-Claisen fission follows the elimination of *p*-toluenesulfonic acid, because concerted fragmentation is not unfavorable in geometric terms.

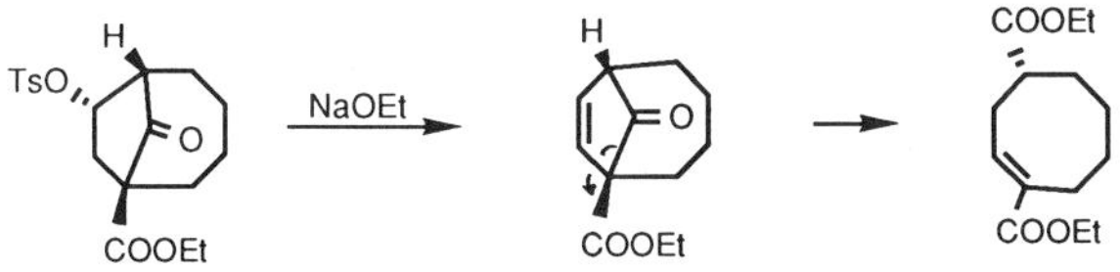

A minor product from treatment of 3-(2-methyl-1,3-dioxocyclopentan-1-yl)propanoic acid with a sodium acetate–acetic acid mixture is 7-methyl-1,4-cycloheptanedione [Bondon, 1976], The genesis of the latter compound must have involved an acidic version of the Claisen condensation to give a bridged triketone which could then undergo a retro-Claisen process at an alternative site. Hydration, carboxylation, and ketonization are the unexceptional events that follow.

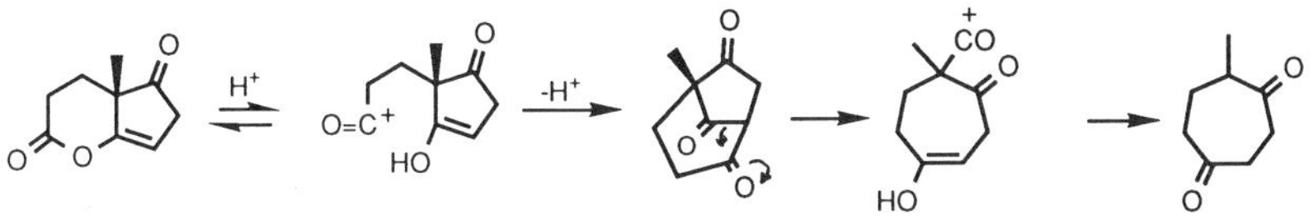

Oxidation of pyrogallol leading to purpurogallin proceeds via 3-hydroxy-1,2-benzoquinone by a Michael–aldol reaction sequence [Horner, 1961]. The strained cyclopentanone of the rearomatized tricycle is hydrated and fragmented, first by a retro-Claisen reaction and subsequently decarboxylation, prior to the final autoxidation step.

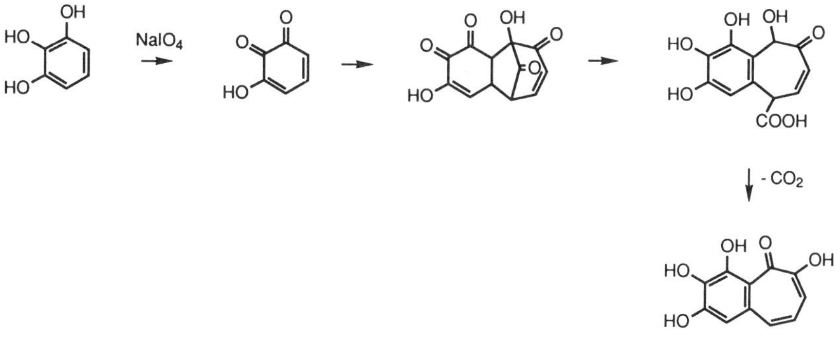

purpurogallin

3,3-Dimethylbicyclo[2.2.1]heptane-2,6-dione gives a monocyclic carboxylic acid on treatment with mild base [Beerboom, 1963]. The retro-Claisen cleavage is regioselective because one of the carbonyl groups is much more hindered (neopentyl situation).

Hydrolysis of a bicyclo[3.1.0]hexanone ester was complicated by partial rearrangement [Hudlicky, 1989b]. The intervention of a tricyclic diketone was the cause.

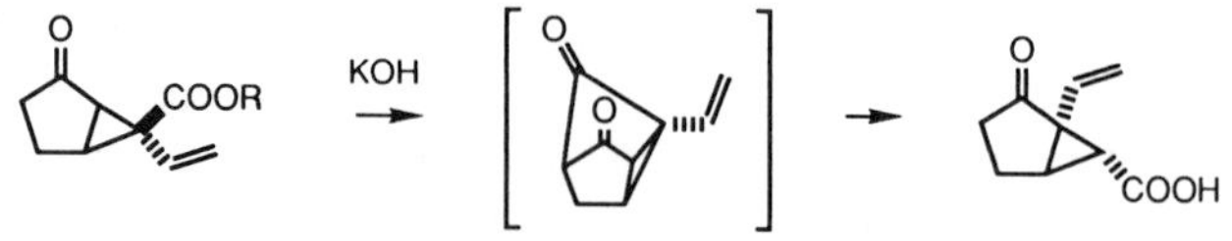

A rather unexpected result emerged from the dithioalkylation of an α-hydroxymethylene ketone in a bicyclo[2.2.1]heptane framework [Saksena, 1981]. Deformylation that is generally observed did not occur in this case presumably because the *endo* formyl group of the intermediate is hindered; also the relief of the system strain lowers the activation energy for the direct ring fission pathway.

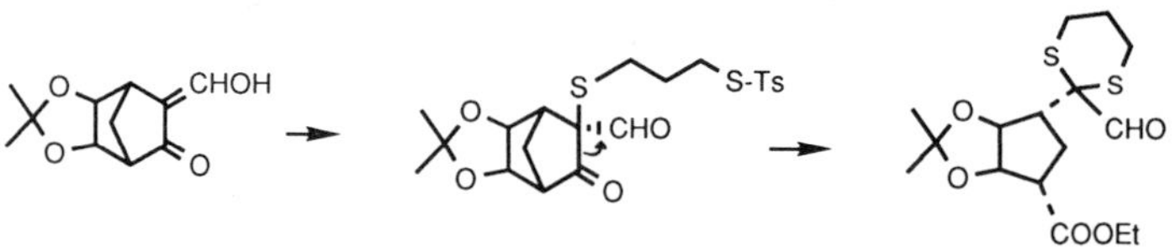

Consistent with this behaviour is the observation on the retro-Dieckmann process of the bicyclo[2.2.1]heptane and -heptene systems [Chamberlain, 1979]. The bicyclo[2.2.2]octane–octene homologs do not undergo fragmentation under identical reaction conditions.

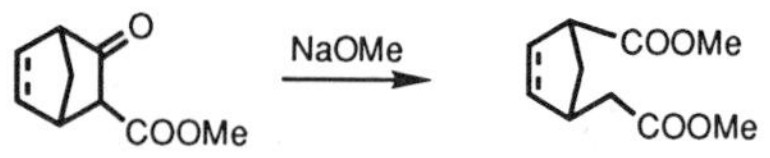

Showdomycin has been acquired by a route involving Diels–Alder and retro-Dieckmann reactions [Kozikowski, 1981].

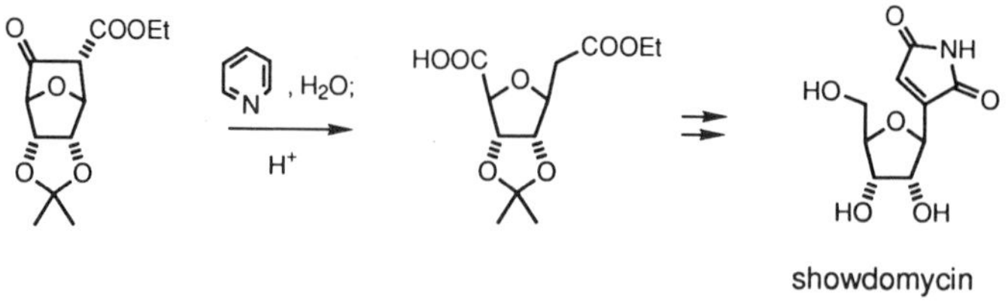

showdomycin

An approach to kopsijasmine [Magnus, 1988] based on a heptacyclic intermediate was formulated within the context of a general scheme by which successful elaboration of kopsanone had been achieved. The regio- and stereocontrol associated with the intramolecular Diels–Alder reaction is well suited for modification as required in the new endeavor. The β-dicarbonyl system derived from an α-sulfinyl lactam in two steps is readily fragmented to establish the carboxyl function.

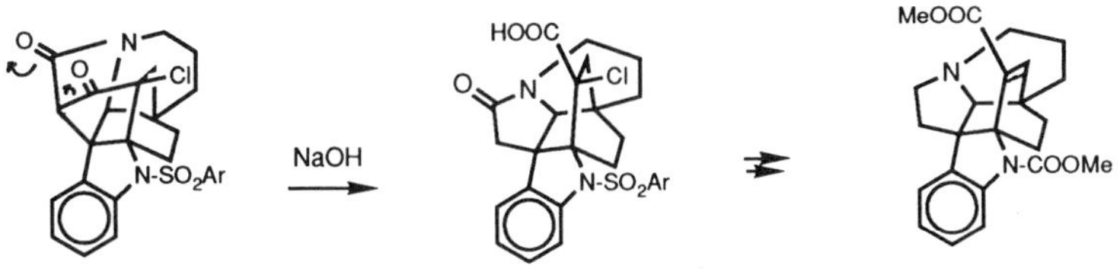

A delightful synthesis of eburnamonine [Klatte, 1977] used a cyclopropane aldehydo ester to condense with tryptamine in a stereoselective manner. Attack on the activated cyclopropane by a nucleophile, formation of a cyclopentanone thereafter, and stepwise release of two ester chains paved the way to closure of the two lactam rings. It is likely that the α-oxocyclopentanecarboxylic ester intermediate recyclized to a δ-lactam which was followed by fragmentation of the five-membered ring.

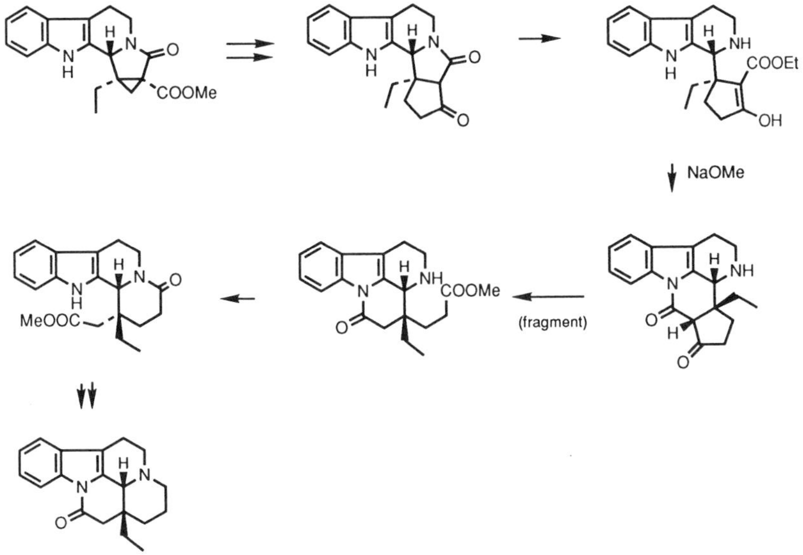

The Dieckmann condensation of *N*-carbethoxypropyl-2-pyrrolidinone furnished a dehydropyrrolizidine after acidic work-up [Arata, 1979]. In the bicyclic lactam product, n,π-overlap can no longer be enjoyed and the carbonyl bridge behaves as a strained ketone. Subsequent hydration and decarboxylation would lead to a mesocyclic amino ketone which can undergo transannular cyclodehydration.

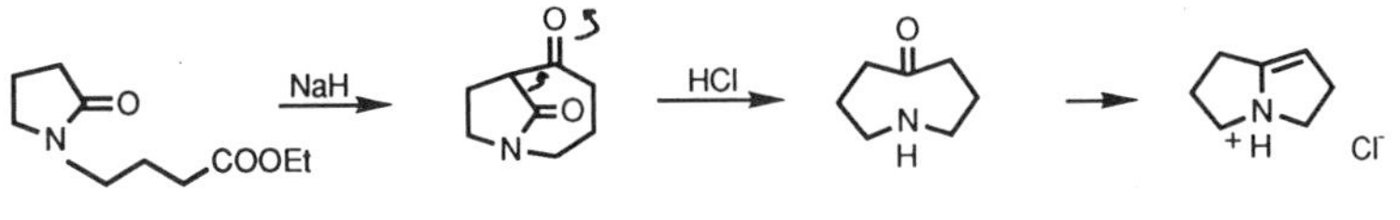

### 3.1.3. Cleavage of Six-Membered Rings

Most enolizable 1,3-diketones form stable conjugate bases on treatment with proton abstractors, but the cleavage of nonenolizable analogs occurs readily [Desai, 1932]. The "dimer" of 1,3-cyclohexanedione is a diketo acid [Conrow, 1966] which is derived from a retro-Claisen fission.

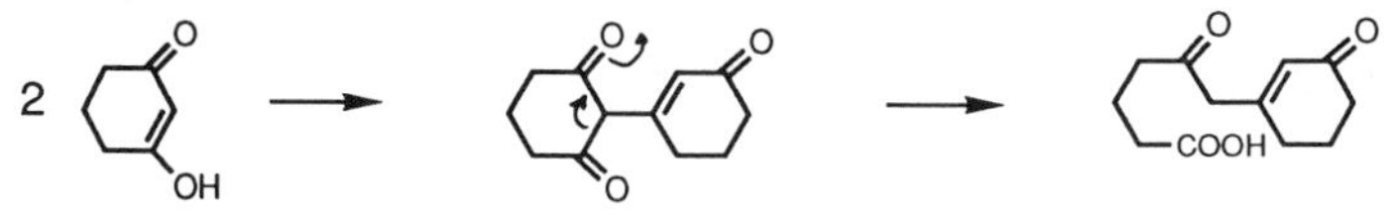

Swietenine derivatives undergo base-induced rearrangements [J.D. Connolly, 1973] which are initiated by $\beta$-dicarbonyl fission. The most fascinating aspect is the dependence of the subsequent aldol condensation on minor structural alterations. It is also interesting to note that mexicanolide, which has the same skeleton but with the double bond in the C-ring, suffers cleavage in the retro-aldol sense [J.D. Connolly, 1968].

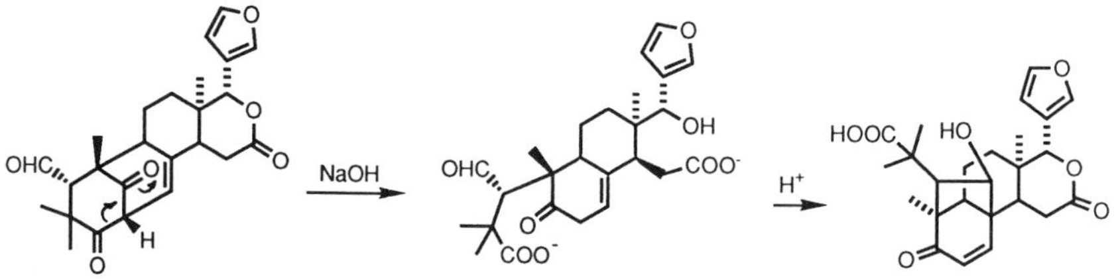

The reaction product of 3-methyl-2-cyclohexenone with methyl cyanoacetate does not come directly from the initial Michael adduct by a cyano transfer [Hill, 1975]. The correct mechanism specifies a bicyclo[2.2.2]octanedione intermediate. The selective fragmentation results in a more stable $\alpha$-cyano ketone which is stabilized as the enolate.

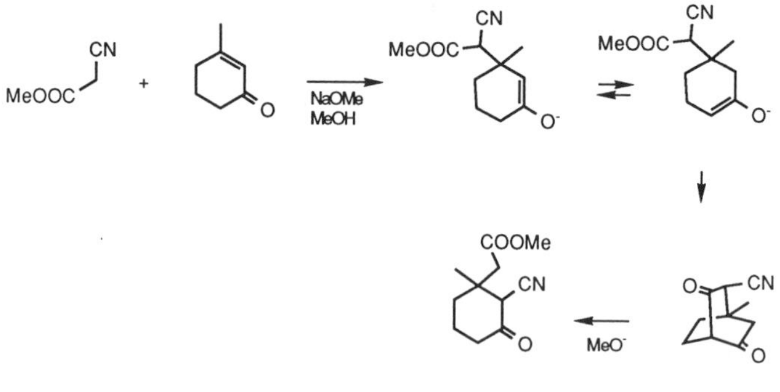

The acid hydrolysis of bicyclo[3.3.1]nonane-2.4.7-trione 7-ethyleneketal causes a retro-Claisen fission also. The resulting diketo acid is well disposed toward cyclization by the aldol process [Yamazaki, 1987].

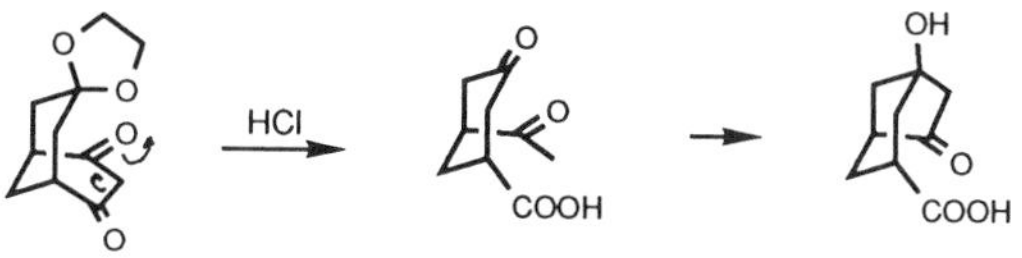

Formation of 6-hydroxy-α-tetralone by ferricyanide oxidation of 2-[3-(3-hydroxyphenyl)propyl]-1,3-cyclohexanedione [Leboff, 1987] is the result of a retro-Claisen fragmentation of the spirocyclic diketone and oxidative degradation of the keto acid.

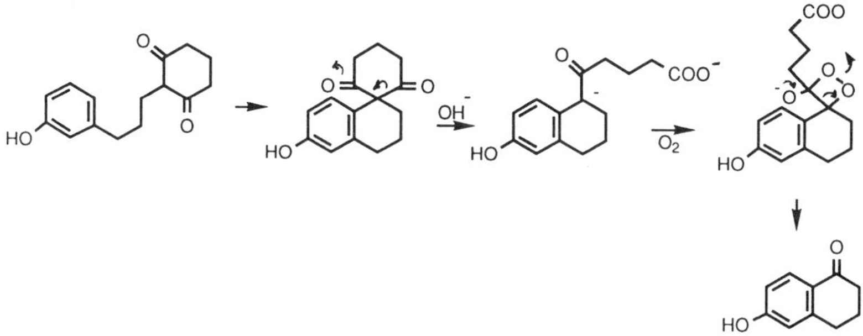

The dehydration of 2-hydroxy-2-carbethoxymethyldimedone with concentrated sulfuric acid gives a lactone [Takeuchi, 1984]. Ring opening of the 1,3-diketone system appears to be the favored first step.

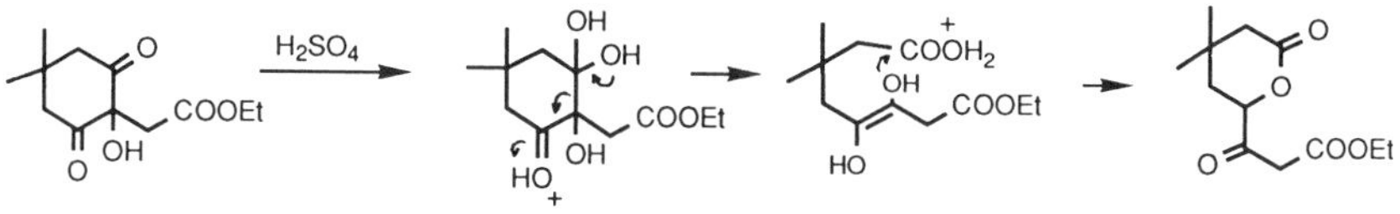

Similar transformations are described in the following equations [Chalmers, 1974].

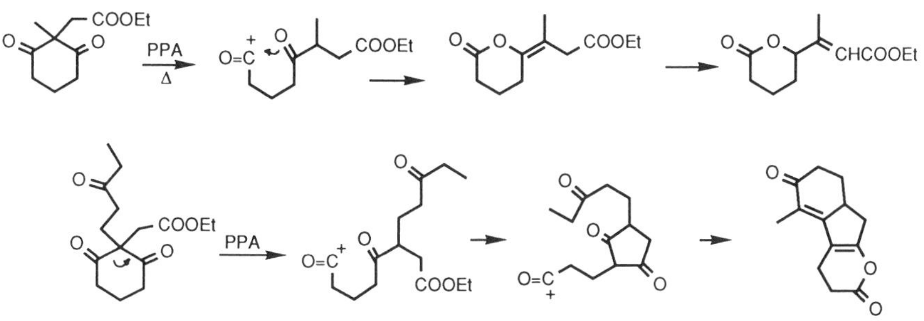

*ent*-Kaurene possesses the same carbon skeleton as phyllocladene. The difference between the two diterpenes is in the bridging of the D-ring and the absolute configuration. Regarding settlement of these stereochemical issues the conversion of both hydrocarbons to a tricyclic keto ester was an essential piece of evidence: enantiomeric keto esters were produced. A Claisen/retro-Claisen reaction sequence is involved in the transformation of *ent*-kaurene [Cross, 1963].

ent-kaurene

phyllocladene

COOMe

When the 1,3-cyclohexanedione is tethered with a proper nucleophile, intramolecular attack can be induced. The resulting ketol would then fragment to furnish a ring-enlarged product [Mahajan, 1976]. A particularly interesting application of this process to synthesis of strained macrolides [Schore, 1987] entails reaction of a substrate in which a propargyl alcohol sidechain is coordinated with a dicobalt hexacarbonyl residue. The complexation effectively changes the linear geometry of the triple bond and allows an intramolecular approach of the hydroxyl group to the ketone function.

NaH

$Co_2(CO)_6$

Carbon nucleophiles are also useful in the ring expansion described above. The following example is a retro-Dieckmann fragmentation [Xie, 1988].

tBuOK / DMSO

COOR

A retro-Dieckmann reaction is the general course for $\beta$-keto esters in which the ester group is located at a bridgehead. A homologous example of a bicyclo system described in the previous section is the following [Cope, 1954].

NaOEt

COOEt

During synthesis of huperzine-A [Qian, 1989] an attempted saponification of the angular ester group led to the destruction of the ring system. The retro-Dieckmann degradation was avoided by conducting a Wittig olefination before the hydrolysis.

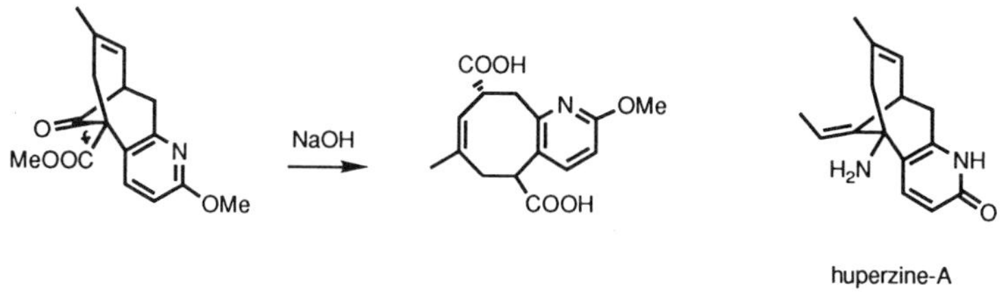

huperzine-A

Still more cases of retro-Claisen reactions are the following. These involve either the generation of the fragmentable species in situ or fragmentation of an existing 1,3-dicarbonyl system only after other reactions are completed. Thus, cyclopropanation resulted when 4-benzyloxycyclohexanone and -cycloheptanone were heated with potassium *t*-butoxide in *t*-butanol [P. Yates, 1963]. This reaction was discovered during an attempted Stobbe condensation.

A reasonable mechanism for the isomerization involves O → C transacylation, lactonization which also triggers the fragmentation, and finally an intramolecular alkylation.

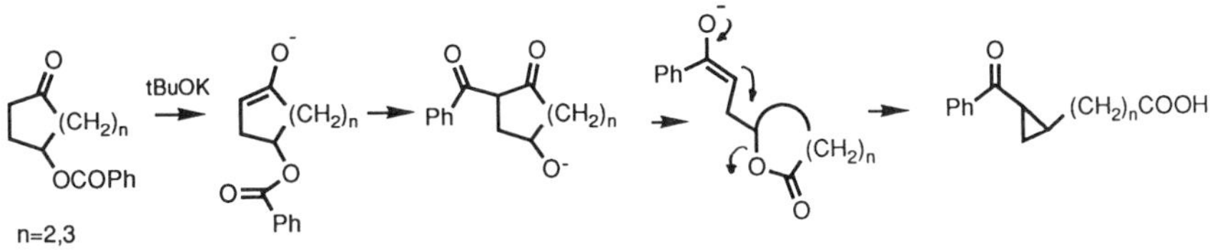

The Claisen/retro-Claisen sequence which transforms 2-methyl-5-carbethoxycycloheptanone into a cyclopentanoneacetic ester [Marshall, 1968b] is thermodynamically controlled.

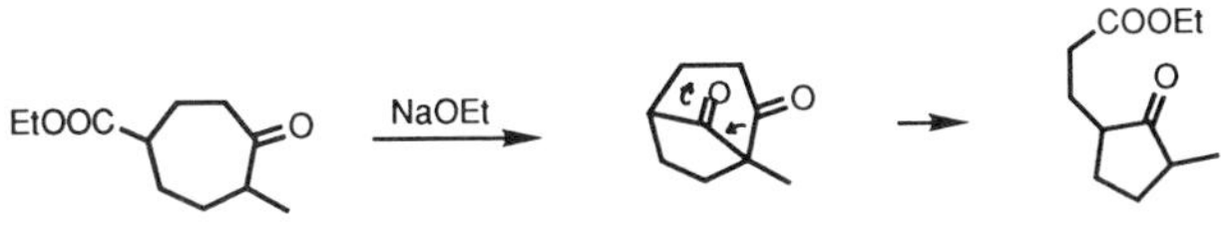

The aldolization–retro-Claisen cleavage tandem also occurs in the following transformations. The aldol step deviates from the normal Robinson annulation initiative because there is a strong 1,3-diaxial methyl : methyl interaction in the linear aldol product in one case [Baisted, 1965]. In the other case [Spencer, 1963] certain subtle electronic effects might be responsible for the favorable formation of the bridged aldol.

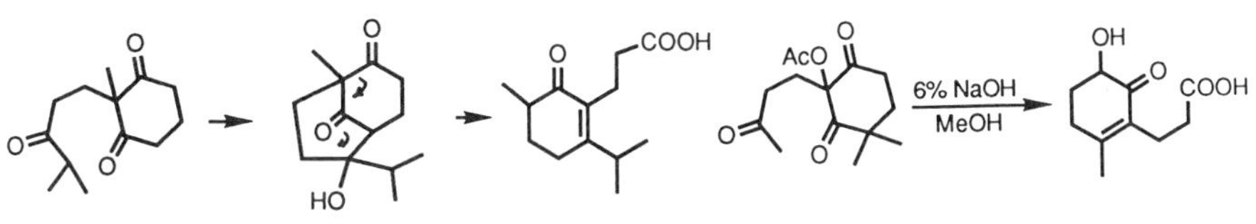

### 3.1.4. Cleavage of Other 1,3-Dicarbonyl Compounds

A very unusual route to methyl geranate starts with fragmentation of a bicyclo[3.3.0]octanone epoxide [Tsuzuki, 1977a].

Na$_2$CO$_3$ MeOH COOMe COOMe

An intramolecular deacetylation is a crucial operation in an approach to *trans*-chrysanthemic acid from α-terpineol [Ho, 1982a]. Ordinarily, a 2-acetyl-2-alkylcyclopentanone undergoes ring opening preferentially on base treatment.

H O O O O O- HO O O O H COOH H O H H OAc

A synthesis of terramycin [Muxfeldt, 1979] was based on the bisannulation of a 4-alkylidenethiazole-5-one with methyl 3-oxoglutaramate. The aldehyde portion for the preparation of the alkylidenethiazolone was ultimately derived from a Diels–Alder reaction of 1-acetoxy-1,3-butadiene with juglone acetate. After proper adjustment and protection of the functional groups the cyclohexene moiety was degraded in two stages. Disengagement of the —CH$_2$—CH= group actually involved a retro-Claisen reaction in the final step.

AcO O H AcO O CHO O$_3$ CHO AcO O CHO Na$_2$CO$_3$ CHO AcO O

OH OH NMe$_2$ CONH$_2$ OH O HO OH O Ph N H O AcO O

terramycin

The successful elaboration of dodecahedrane [Ternansky, 1982] from a heptacyclic precursor depends on a tactic enforcing a properly functionalized one-carbon chain inside the molecular cavity. A germinal alkoxymethyl group is the device for achieving the goal, and its departure was arranged after the needed carbon atom was safely locked into place.

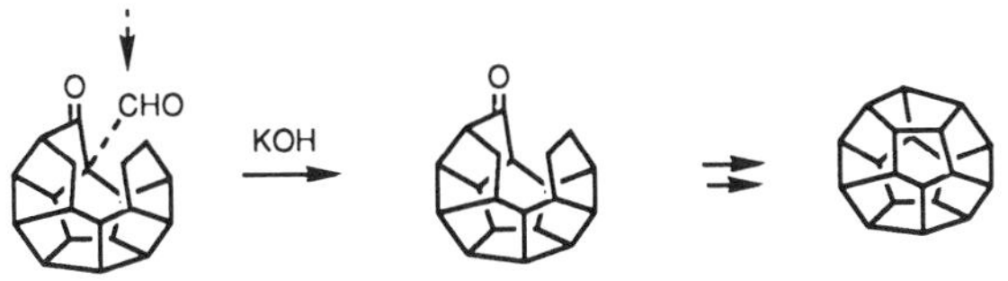

1,3-Diketones undergo Payne rearrangement on treatment with alkaline hydrogen peroxide. Fragmentation accompanies 1,2-rearrangement of the 3,5-dihydroxy-1,2-dioxolane intermediate, as illustrated in an example describing ring fission of a spirodiketone [Hawkins, 1973].

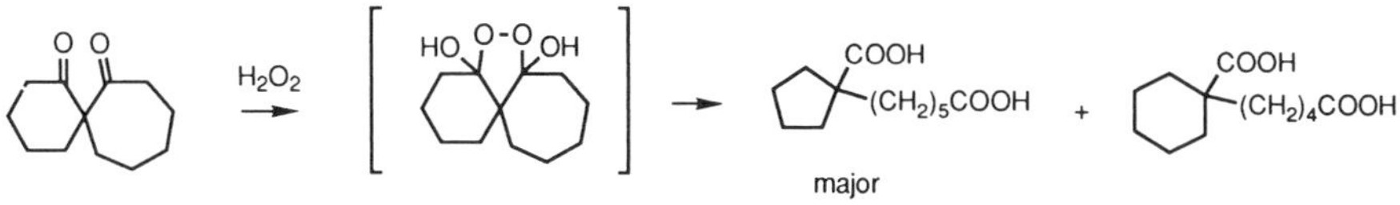

During the alkaline hydroperoxide reaction of the isoflavone elongatin [Smalberger, 1975] the loss of a skeletal carbon was actually accompanied by reduction at the benzylic position. A cyclopropanone may be the transient intermediate and the process is reminiscent of a Favorskii rearrangement.

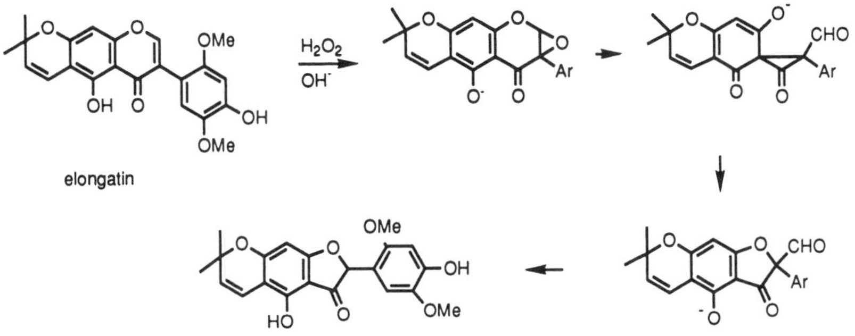

By way of contrast, the conversion of a 2-formylcyclobutanone into an enollactone during a synthesis of acorenone-B [Trost, 1975a] shows the dominance of the strain factor in regulating reactivity. Less strained and strain-free ketoaldehydes invariably lose the formyl residue under hydrolytic conditions.

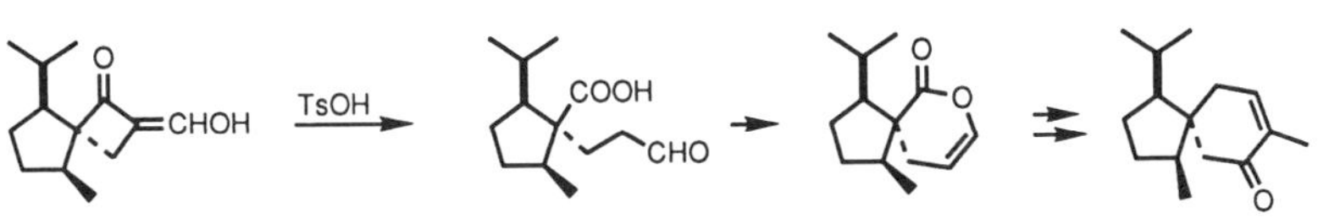

Flavonoids undergo extensive degradation on treatment with hot alkali. For example, chrysin gives benzoic acid, acetic acid and phloroglucinol under such conditions [Piccard, 1873]. The xanthone natural product mangostin also suffers an aromatic ring rupture via a series of retro-Claisen reactions [P. Yates, 1958].

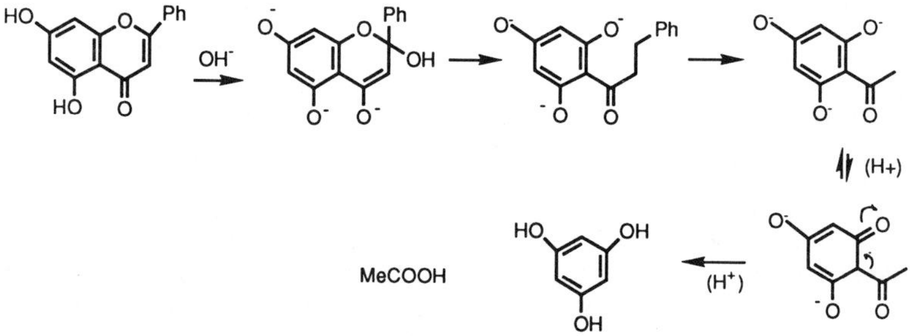

α-Functionalization of β-dicarbonyl compounds can be achieved under hyperiodine oxidation in the presence of a nucleophile [Moriarty, 1988]. An exception to this general transformation is the reaction of benzoylacetone, a mixture of α-methoxyacetophenone (63%) and methyl phenylacetate (24%) being produced. Retro-Claisen fission intervenes at the final stages of both pathways.

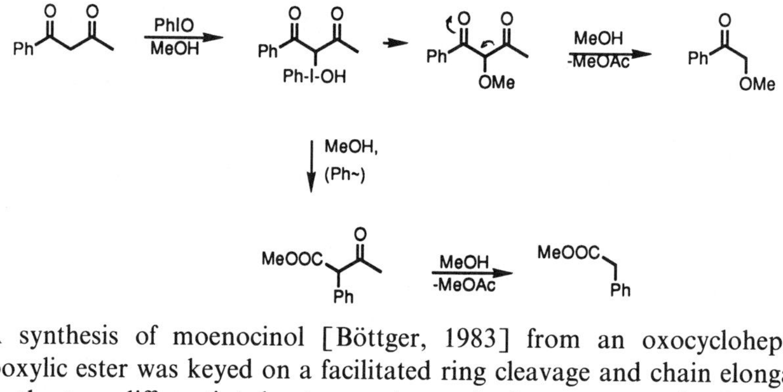

A synthesis of moenocinol [Böttger, 1983] from an oxocycloheptanecarboxylic ester was keyed on a facilitated ring cleavage and chain elongation from the two differentiated ester pendants. A shorter route via Grob fragmentation failed.

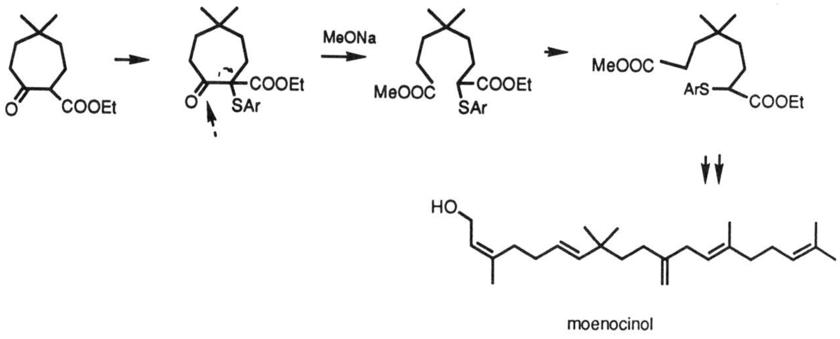

moenocinol

A newer and more convenient method for removing the ester group from malonic, β-keto, and α-cyano esters consists of heating these substrates in a dipolar aprotic solvent in the presence of an alkali metal salt [Krapcho, 1982]. The incipient carbanionic species can be intercepted intramolecularly by an electrophile [Arlt, 1981; Eilerman, 1981], enabling annulation under essentially neutral conditions.

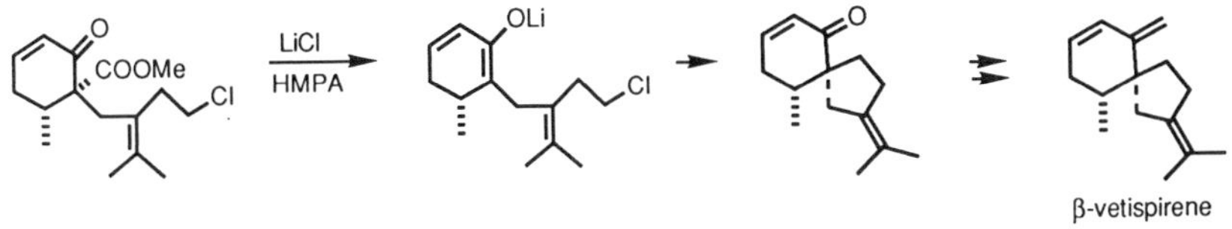

β-vetispirene

Steganone has been secured via intramolecular Perkin reaction [E. Brown, 1979]. Oxidation of the resulting β-hydroxy lactone afforded the β-ketone lactone in a mixture of the enol and keto forms. The lactonic carbonyl was removed by hydrolysis.

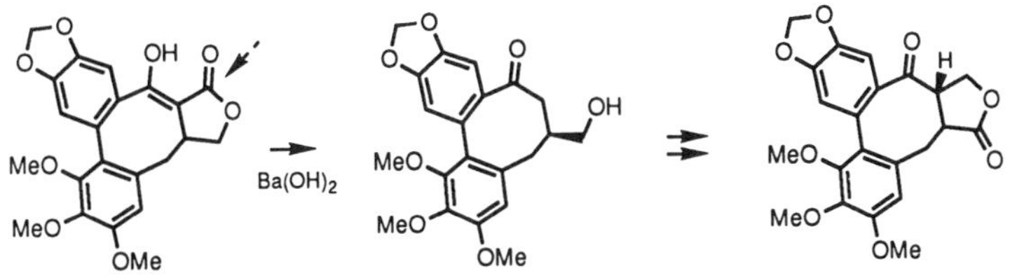

The regioselective carbethoxylation of 2-(3-oxocyclohexyl)one-2-cyclohexen on reaction with diethyl methylmalonate [Risinger, 1973] indicates the occurrence of a Michael, Claisen, retro-Claisen, and retro-Michael reaction sequence.

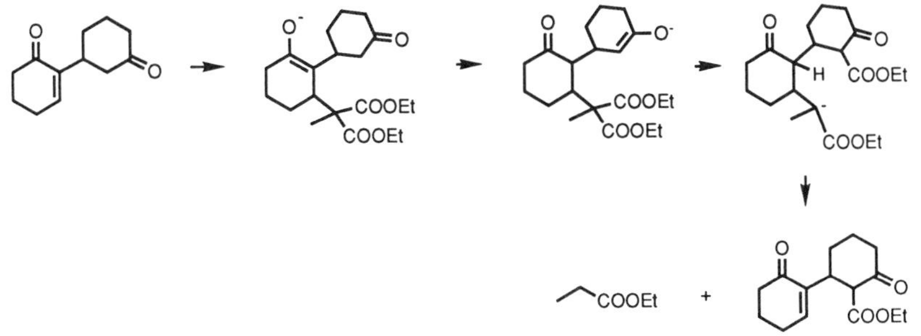

## 3.2. CLEAVAGE OF 1,3-IMINOCARBONYL COMPOUNDS

Only a few of these substances have been investigated. However, it is of interest to observe that an α-dichloro-β-iminocarbonyl compound always suffers acyl fission on exposure to nucleophiles [DeKimpe, 1985]. Contrary to the corresponding β-ketoesters, decarbalkoxylation of β-imino esters can be effected selectively and this method allows preparation of 1,1-dichloro-2-alkanones from β-keto esters in four steps.

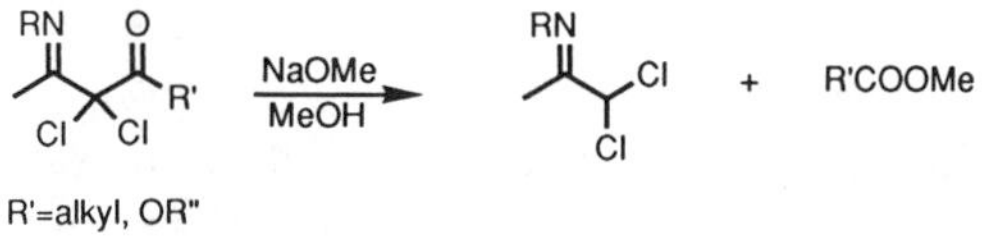

RN O RN
NaOMe / MeOH → Cl + R'COOMe
R'=alkyl, OR"

$\beta$-Amino $\alpha,\beta$-unsaturated esters lose the ester function when heated with boric acid [Bacos, 1987]. Such reactions most likely proceed via the $\beta$-imino esters.

The photoisomerization of *N*-acylindoles to 3-acylindolenines is an extremely versatile protocol for gaining access to the latter class of substances. With reference to indole alkaloid synthesis [Ban, 1983] a single intermediate obtained from the photoreaction of 6-oxo(2-aminoethyl)-6,7,8,9-tetrahydropyrido[1,2-*a*]indole has served as an intermediate for members of the eburna, aspidosperma, and strychnos families.

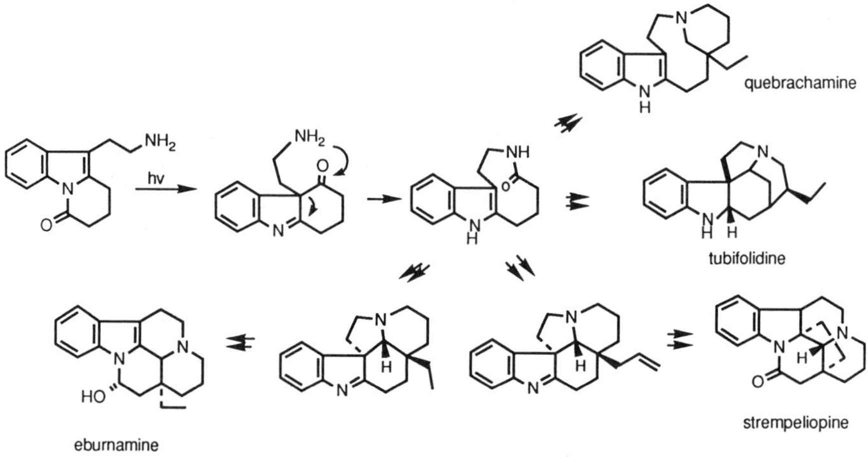

quebrachamine

hν

tubifolidine

eburnamine

strempeliopine

The formation of an indolenine from a spiroannulated oxindole [Cartier, 1989] is interesting in that a formal methyl group transfer and loss of carbon dioxide are involved. It is conceivable that a pentacyclic intermediate was formed and degraded. The degradation via two fragmentation processes was succeeded by a Pictet–Spengler cyclization.

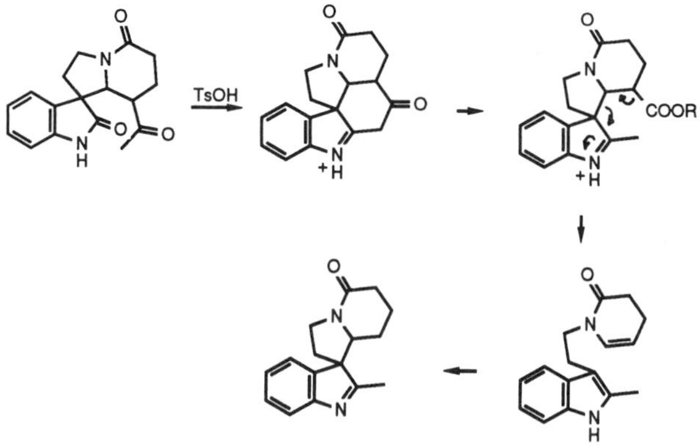

TsOH

COOR

Lead tetraacetate oxidation of deacetylgeissovelline and pyrolysis of the product unraveled the bridged framework [R.E. Moore, 1973]. The oxidation generated a fragmentable species. An unusual pathway has been postulated for the transformation of the intermediate to a tetrahydro-$\beta$-carboline derivative.

Cyclization of a thiolactam by treatment with trimethyl phosphite [Fang, 1990] was accompanied by rearrangement whereby the sulfur atom was retained. The final step of the reaction sequence is a fragmentation induced by C → S transacylation.

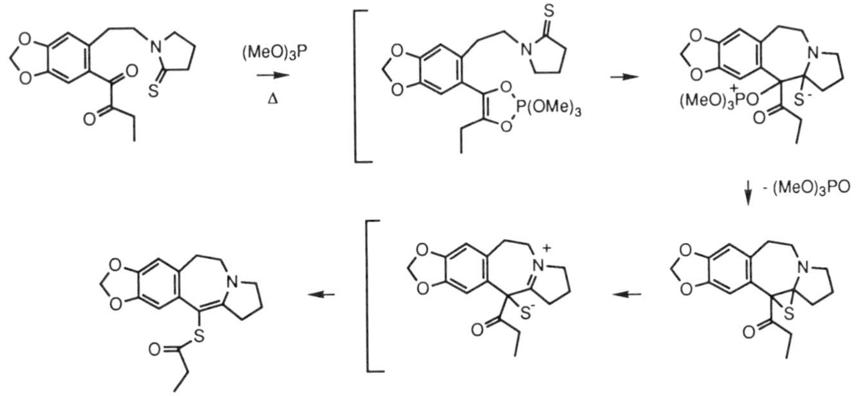

$(MeO)_3P$, Δ; $- (MeO)_3PO$

A fascinating reaction of certain 2-(2′-nitrobenzoyl)-1,3-diketones is the formation of 2-acyl-3-hydroxyquinolines [Bayne, 1975] on base treatment. This transformation must be initiated by a Michael addition onto the aromatic nucleus. The spirocyclic intermediate is cleaved to form a nitroso derivative. It is followed by cyclization and aromatization of the ensuing heterocycle.

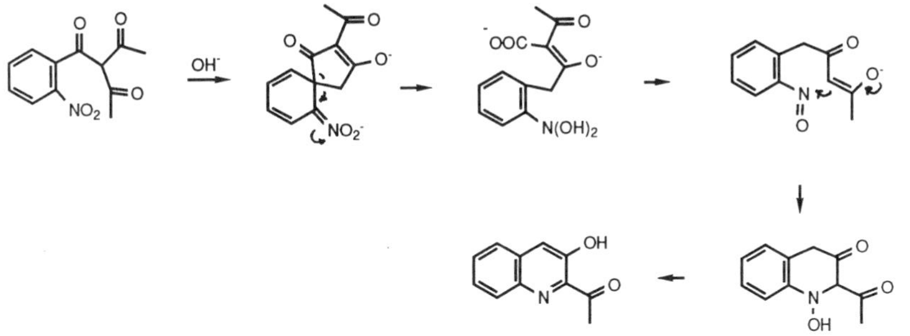

$OH^-$

The monopotassium salt of 2,5-dinitrocyclopentanone undergoes decarbonylation in the presence of aqueous sulfuric acid [Feuer, 1961] by a double **a1a** fragmentation.

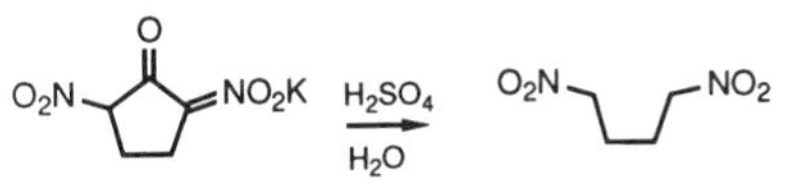

$O_2N$ ... $NO_2K$ $\xrightarrow[H_2O]{H_2SO_4}$ $O_2N$ ... $NO_2$

Although not strictly proper, a few examples concerning fragmentation of $\alpha$-cyano ketones are described here. As with 1,3-dicarbonyl compounds, fragmentation occurs when enolization of the substrate is precluded. Thus, A-nor-5$\beta$-cyano-3-keto steroids which are formed by photodecomposition of the 4-azido-4,6-dien-3-ones undergo very facile cleavage on treatment with base [E.M. Smith, 1974].

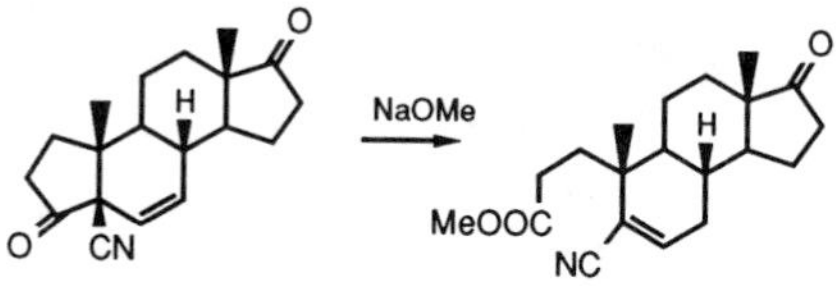

A remarkable dechlorinative cage opening of a pentacyclic diketone occurs on reaction with sodium methodoxide [T. Liu, 1979]. The two cyano groups are crucial as they facilitate the semibenzilic acid rearrangement and the final step of **d3d** fragmentation.

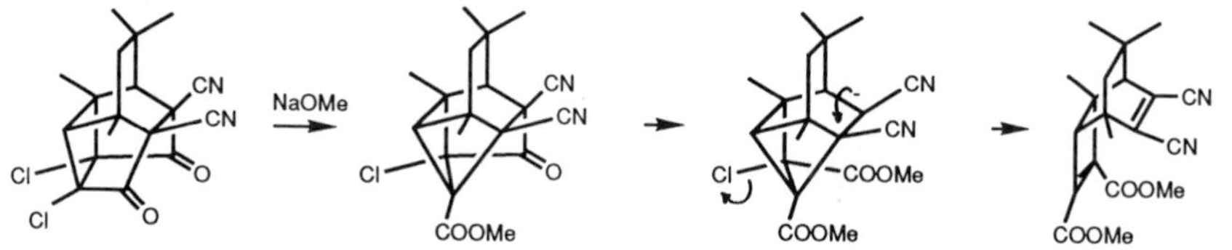

## 3.3. CLEAVAGE OF $\beta$-OXOCARBENIUM IONS

A $\beta$-oxocarbenium ion O=C—C—C$^+$ or its incipient form is extremely prone to fragmentation because the process generates a more stable acylium species and a double bond. Accordingly, fragmentation occurs when aquayamycin is treated with hydrogen chloride in methanol [Sezaki, 1970]. Ionization of the tertiary alcohol instigates the ring cleavage which leads to an anthraquinone and therefore gain in energy. The dihydro derivative of the antibiotic undergoes double dehydration without skeletal degradation.

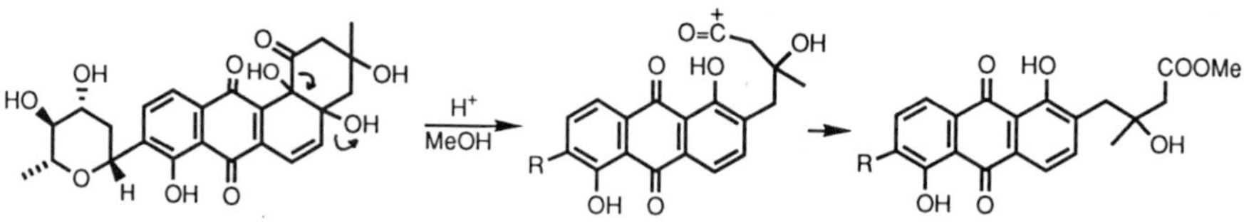

$\alpha$-Acylcarbenium ions are thermodynamically less stable than the deoxy species, consequently, rapid rearrangement of the incipient ions is expected. The formation of various lactones from *exo*-3-bromocamphor on treatment with a silver salt [Bégué, 1978] can be rationalized on the basis of a mechanism involving Wagner–Meerwein rearrangement and fragmentation.

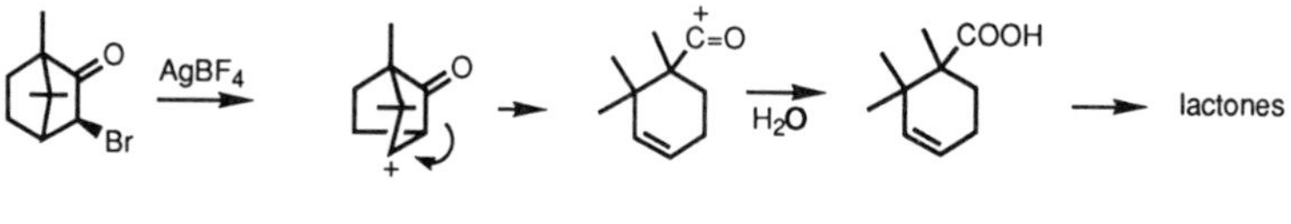

*endo*-2-Phenyl-*exo*-2-hydroxyepicamphor reacts with concentrated sulfuric acid at 0°C to afford the lactone of 3-hydroxy-4-phenyl-2,2,3-trimethylcyclohexanecarboxylic acid [Wilder, 1975]. Ionization of the hydroxyl is followed by a Wagner–Meerwein rearrangement. This bond migration removes the unfavorable electronic interactions between two adjacent acceptors and creates a fragmentable species. The fragmentation relieves ring strain.

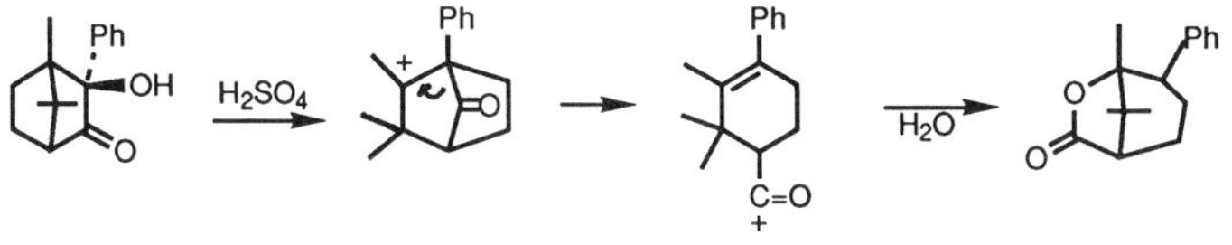

Acid catalyzed decomposition of 3-diazoisofenchone gives 1,4,4-trimethylcyclohex-3-enecarboxylic acid as one of the major products [Blattel, 1972a]. It has been postulated that *exo*-protonation precedes an anchimerically assisted ionization of the diazonium ion (here indicated by a classical structure for the sake of clarity) and fragmentation. The ensuing monocyclic acylium ion is then hydrated. Solvent effects seem to support this view.

The sole production of acidic material from 4.6.6-trimethyl-3-diazo-2-norbornanone [Blattel, 1972b] is easily explained. The Wagner–Meerwein rearrangement is facilitated by the formation of a teriary cation.

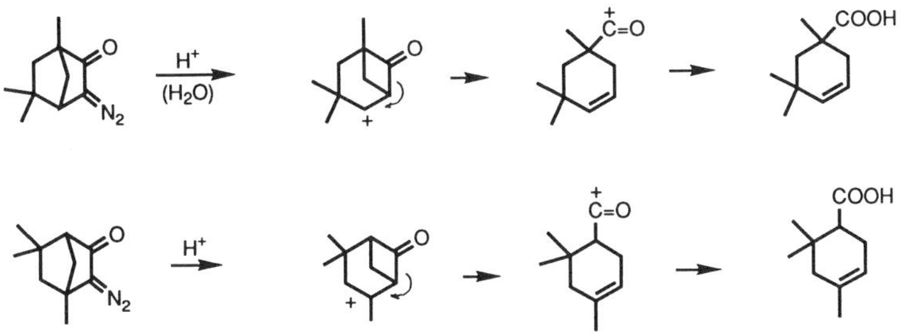

In situ creation of an **a1a** array from camphorquinone by a Wagner–Meerwein rearrangement triggers skeletal fragmentation [Gibson, 1925]. Slight modification of this process has provided a building block for the C-ring of vitamin-$B_{12}$ in a monumental synthesis [Woodward, 1968, 1971, 1973].

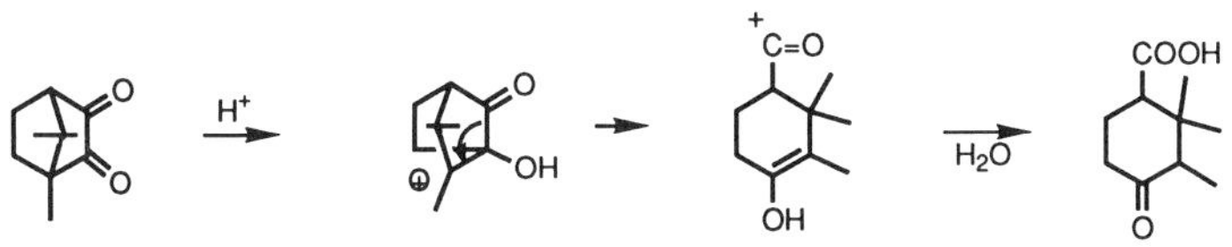

A minor product of the Wichterle reaction of ethyl 2-(3-chlorobut-2-enyl)cyclopentanone-2-carboxylate is a cycloheptenecarboxylic ester [Dauben, 1960]. Fragmentation of a bridged oxo cation is a most reasonable process in the reaction pathway.

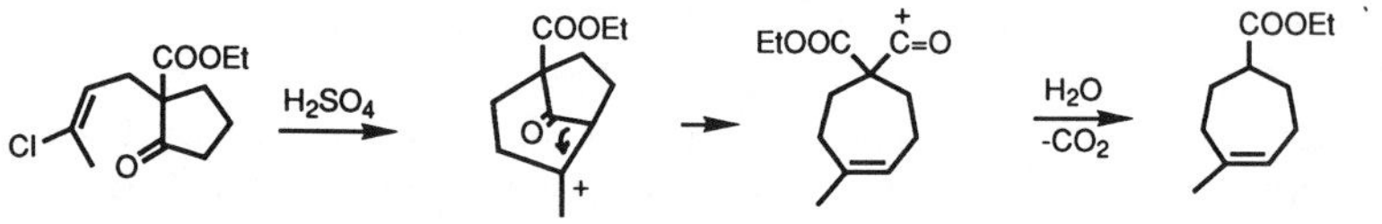

A fascinating reaction occurs when a phenylbicyclo[3.3.1]nonan-3,9-dione was submitted to the Wichterle alkylation–cyclization protocol [G.L. Buchanan, 1966a]. The expected cyclization proceeded only after a Lansbury condensation involving the nonenolizable ketone group and the resulting tetracyclic keto diol underwent ionization, fragmentation, and finally lactonization.

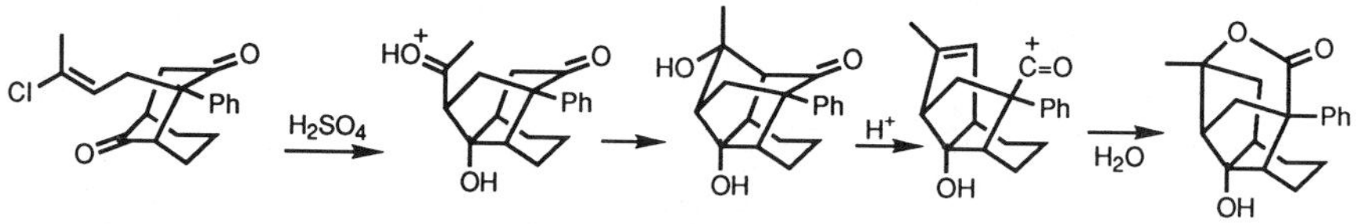

Acid treatment of 2-carboalkoxy-2-(3-oxobutyl)cyclopentanone affords a bridged lactone [Evans, 1982]. This compound results from the aldol via fragmentation and lactonization.

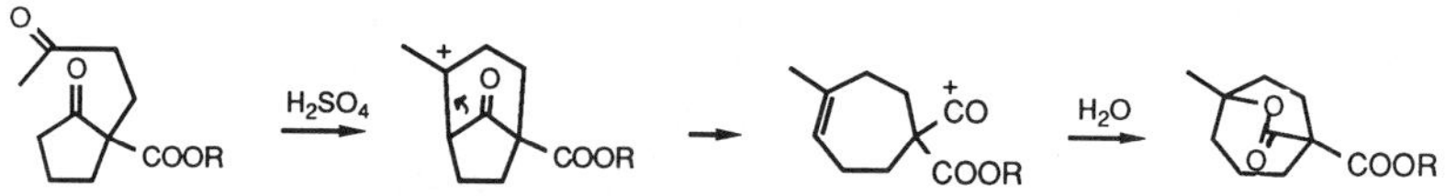

The aldolization–fragmentation sequence constitutes the crux of a synthesis of bulnesol [Tanaka, 1988]. The fragmentation may have been proceeded as indicated, or it could occur after the ketone group was derivatized.

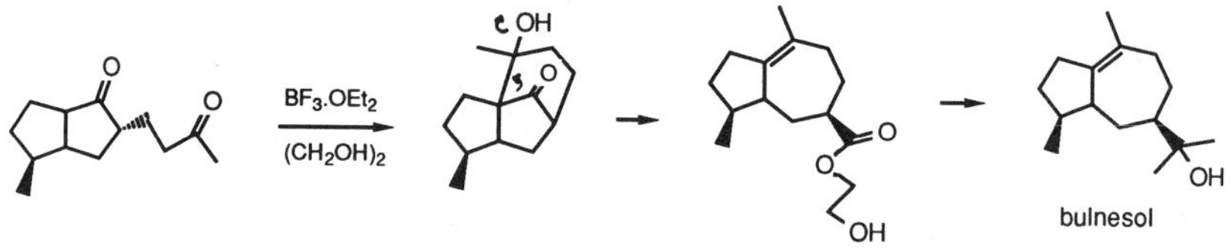

*p*-Anisyl (3-oxocycloalkyl)methyl ketones are converted into 3-anisylcycloalk-3-en-1-ylacetic acids [Dave, 1976]. This isomerization consists of aldol condensation to the ketols and a deannulation which is triggered by ionization of the very reactive benzylic hydroxyl group.

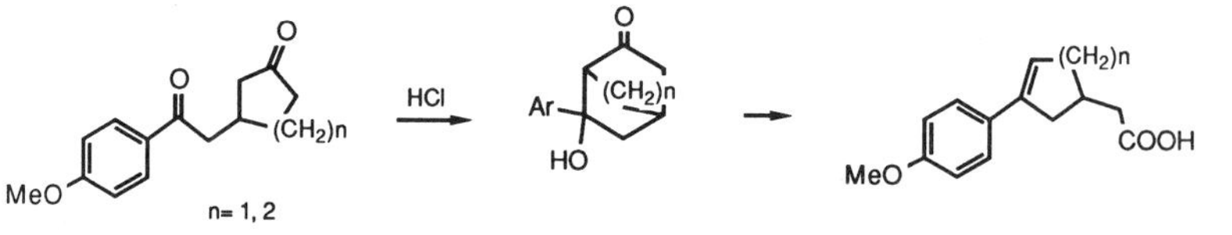

Oxidation of 20-oxoaspidospermine with *m*-chloroperbenzoic acid results in a 19-hydroxy-10-oxo derivative [Gebreyesus, 1972]. Loss of an element of acetic acid has been observed on further treatment of the latter compound with sulfuric acid–acetic anhydride.

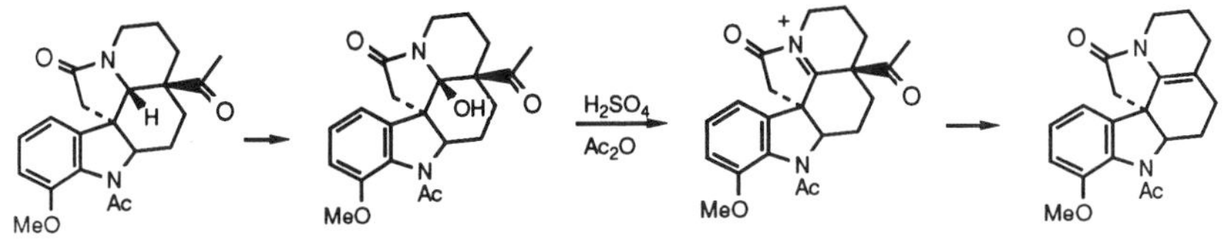

The tosyloxy ketone shown below cannot fragment according to the relationship of its functional groups. However, the inherent strain of the system forces it to rearrange, and a fragmentable intermediate apparently results [Marschall, 1972].

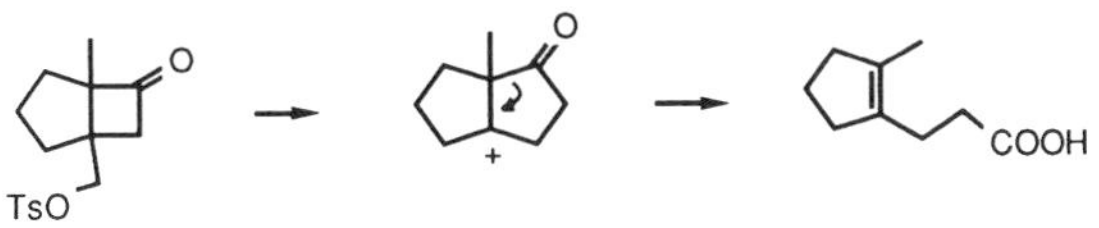

The following is a reaction whose crux is a fragmentation of a formal β-oxocarbenium ion. It is related to an attempted hydrolysis of a thioester of the carbapenem [Gordon, 1981]. The aberrant reaction apparently proceeded by mercuronium ion formation even in the presence of a sulfenyl group. The fragmentation was followed by an $S_N2'$ displacement of the mercurous acetate.

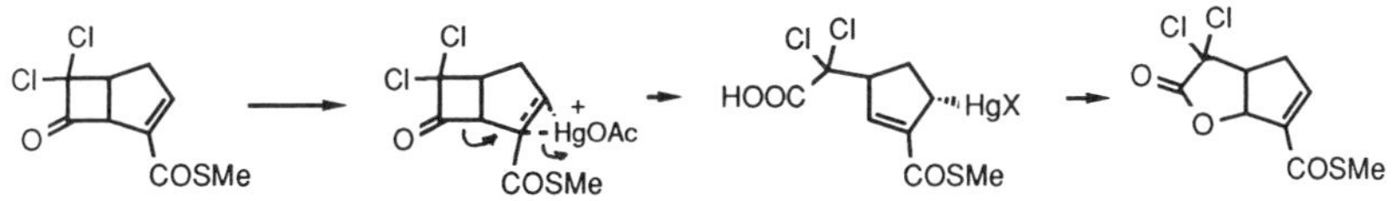

## 3.4. CLEAVAGE OF α-DIKETONE MONOTHIOKETALS

The fragmentation of α-diketone monothioketals is synthetically valuable as it leads to two functionalized residues. From cyclic ketones many aldehydo acids or esters may be prepared using this method. The protected aldehyde function is also a convenient source of a methyl group and ketone units, especially when the thioketal is a 1,3-dithiane unit. By virtue of sulfur atom's capability to assume an acceptor role, that is to stabilize an α-carbanion, the Haller–Bauer reaction is greatly facilitated.

In the 1970s the anticancer sesquiterpenes vernolepin and vernomenin aroused enormous interest in their synthetic studies. Several approaches focused on fission of a *trans*-β-decalone which possesses an oxymethyl substituent at an angular position, so that a δ-lactone can be formed has the proper *cis*-fusion to the cyclohexane ring. Such plans are advantageous because the rigid *trans*-decalin system is easy to construct and its chemical behavior, especially with regard to stereo- and regiocontrol, is fully predictable.

A prototype of this bicyclic skeleton has been prepared by a reaction sequence embodying bisalkylsulfenylation and Haller–Bauer fragmentation as key steps [Marshall, 1974b, 1975a]. It should be noted that the reagent (KOH-*t*BuOH) is rather special.

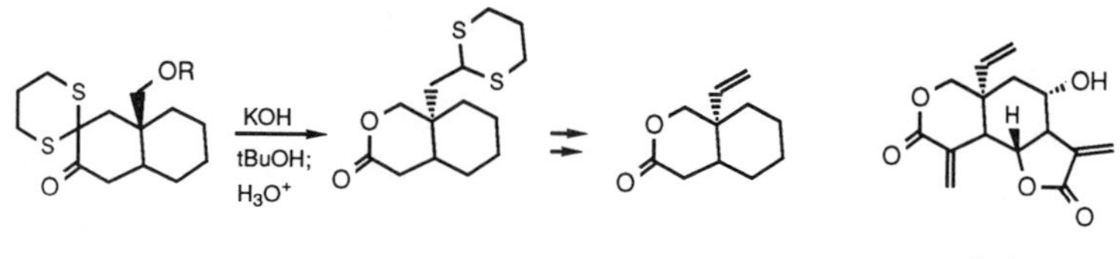

KOH tBuOH; $H_3O^+$

vernolepin

Other ingenious exploitations of this method include a synthesis of estradiol [Kametani, 1978] from a perhydroindanone, and many alkaloids. The work in the latter area is due mainly to Takano who has identified a number of ketones, especially norcamphor as suitable building blocks.

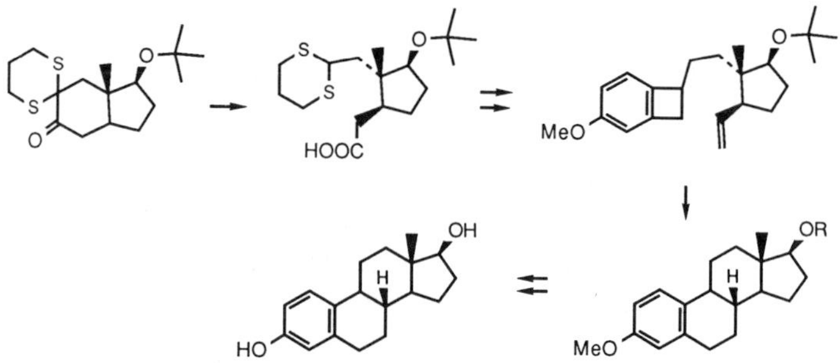

Conversion of norcamphor to various 3-substituted cyclopentanones via the bicyclic lactones, and subjecting the ketones to a bisalkylsulfenylation-fragmentation sequence, constitute the core operation of several novel syntheses. Configurational control at the α-carbon of the bicyclic lactone is straightforward. *exo*-Alkylation is always observed. Equilibration of the α-alkyllactone transforms it into the *endo* isomer.

These chemists have contemplated synthesis of the natural enantiomers of alkaloids. The synthesis of emetine requires the (+)-isomer of norcamphor [Takano, 1978b]. The critical steps include coupling of the ethylated lactone derived therefrom with 3,4-dimethoxyphenethylamine, elaboration of the resulting cyclopentanol to a α-diketone monothioketal, Haller–Bauer fragmentation, and Pictet–Spengler cyclization.

The ethyl epimer of the bicyclic lactone is ideal for synthesis of corynantheidol [Takano, 1980b].

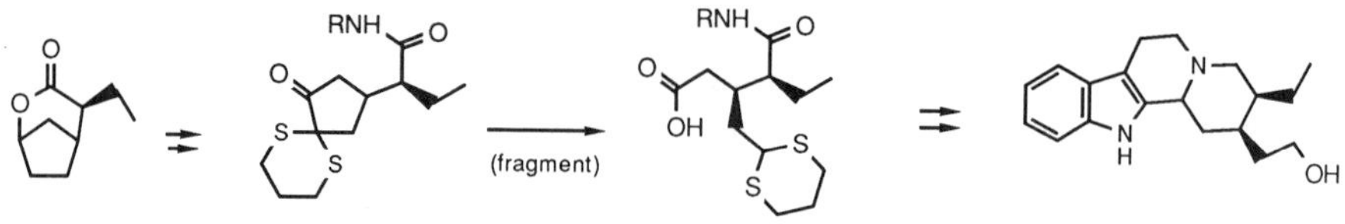

(fragment)

Norcamphor must be resolved to give the two enantiomers and it would be wasteful to reject (−)-isomer. This logistic problem has been circumvented by the development of a complementary reaction sequence [Takano, 1979c]. Thus, instead of performing a Pictet–Spengler cyclization with the latent aldehyde, the carboxylic acid chain was used to form the C-ring. It has been found that epimerization of the ethyl group and the quinolizidine ring junction of the undesirable diastereomer can be accomplished during hydride reduction of the tricyclic lactams. seco-BC intermediates must be involved.

The α-allyl lactone has been used in a synthesis of meroquinene aldehyde [Takano, 1979a]. The same lactone and its epimer are expedient precursors of semburin and isosemburin [Takano, 1982b].

From sources other than norcamphor another cyclopentanedione monothioketal has been secured and processed in a similar fashion to arrive at (−)dihydrocorynantheol [Suzuki, 1985a] and (−)-antirhine [Suzuki, 1985b]. It was very important to recognize the structural correlation of the two alkaloids with a common intermediate.

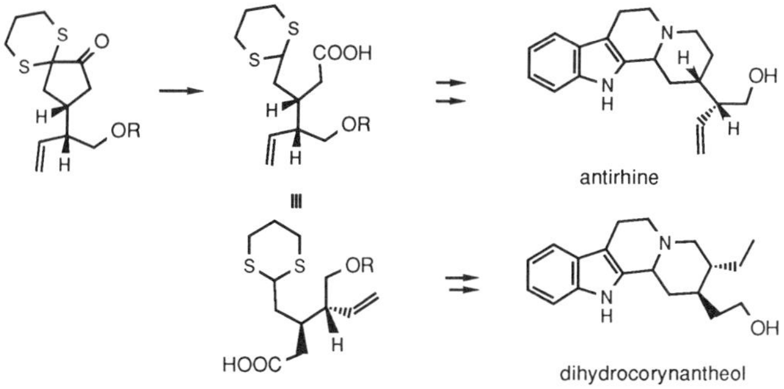

antirhine

dihydrocorynantheol

Synthesis of indole alkaloids such as quebrachamine, tabersonine [Takano, 1979a] and eburnamenine [Takano, 1977] can also utilize a cycloalkanedione monothioketal. Here the required cycloalkanone is symmetrical (a 4,4-disubstituted cyclohexanone), therefore the functionalization does not have to rely on steric effects. Of course, regioselective functionalization of the ketone is never a problem during a synthesis of the dihydrocleavamines [Takano, 1981].

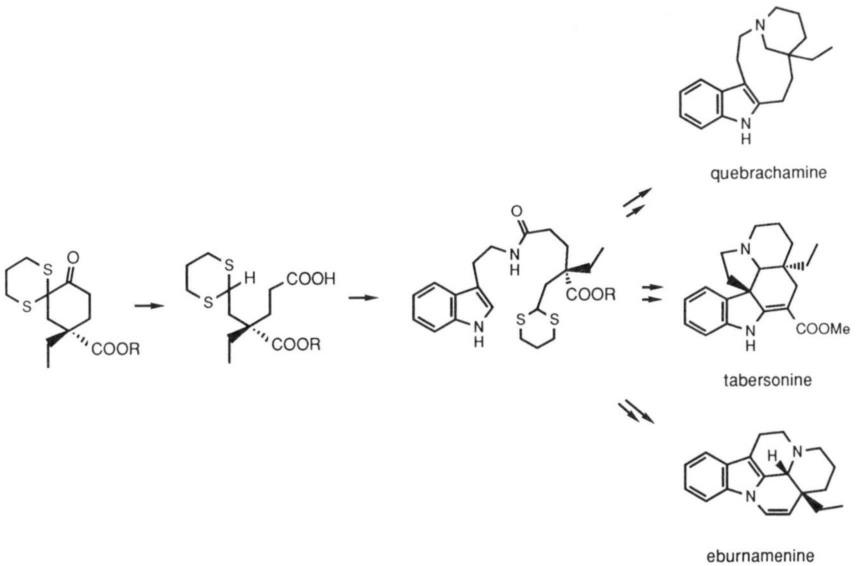

quebrachamine

tabersonine

eburnamenine

Earlier, an approach to emetine was adopted which was based on fragmentation of a tetracyclic substrate [Takano, 1978b].

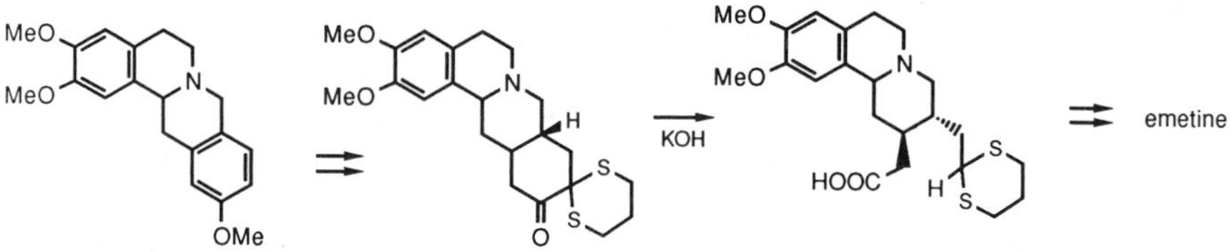

Secoalkylation is a method consisting of reductive oxocyclobutanation and subsequent cleavage of the four-membered ring. The cyclobutanone is formed by reaction of a carbonyl compound with a cyclopropylsulfonium ylide and rearrangement of the resulting oxaspirane with a Lewis acid catalyst.

The ring cleavage step is readily performed by base treatment of the bissulfenylated ketones. The versatility of such substances have been amply demonstrated by their conversion into a broad array of other ring systems and open-chain products. A synthesis of methyl desoxypodocarpate [Trost, 1973] may be mentioned.

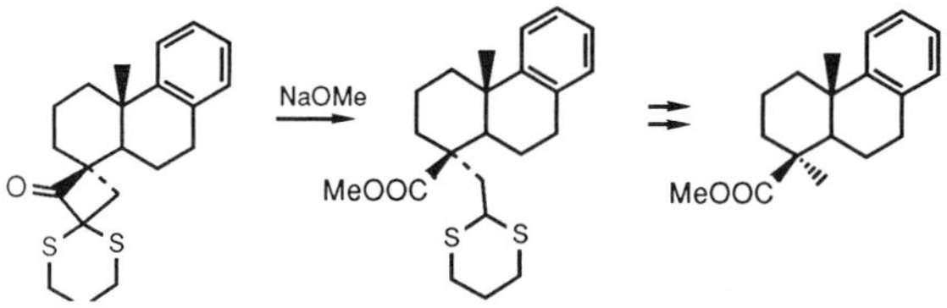

A particularly facile fragmentation occurs during an attempted preparation of the symmetrical monothioketal of 1,2,3-cyclohexanetrione [Bryant, 1975]. The incipient carbanion is stabilized by more than one acceptor group.

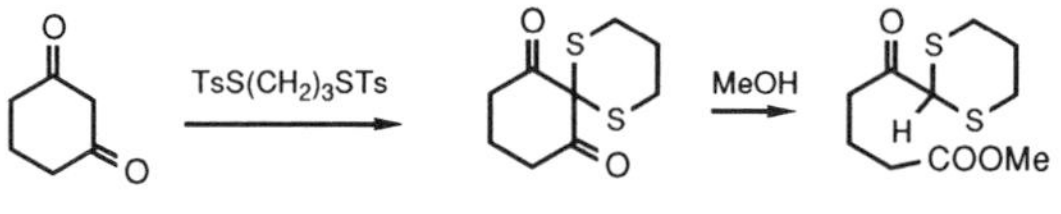

## 3.5. CLEAVAGE OF OTHER α-HETEROSUBSTITUTED KETONES

The sulfonyl group is a better carbanion stabilizer than the sulfenyl group, therefore, only one sulfonyl substituent at the α-position of a ketone is necessary to allow deacylation. A synthesis of 4-alkenoic acids takes advantage of this property and it is based on a sequence of alkylation of 3-thiacyclohexanone 3,3-dioxide, brominative ring cleavage, and Ramberg–Bäcklund rearrangement [Scholz, 1981].

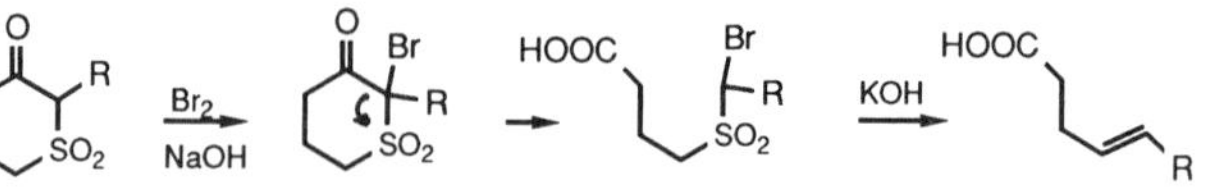

In an attempt to close the bridged system of the epiallogibberic acid skeleton by an aldol condensation it was observed [House, 1973] that alkali also induced ring cleavage of the tetracyclic product.

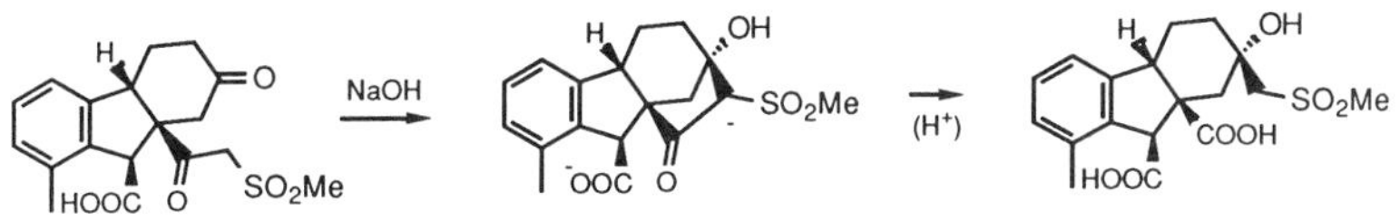

Another example pertains to the attempted synthesis of 3-sulfonylchromenes [Balasubramanian, 1980] from Mannich bases of phenols and α-sulfonyl ketones. Fragmentation supersedes dehydration of the intermediates.

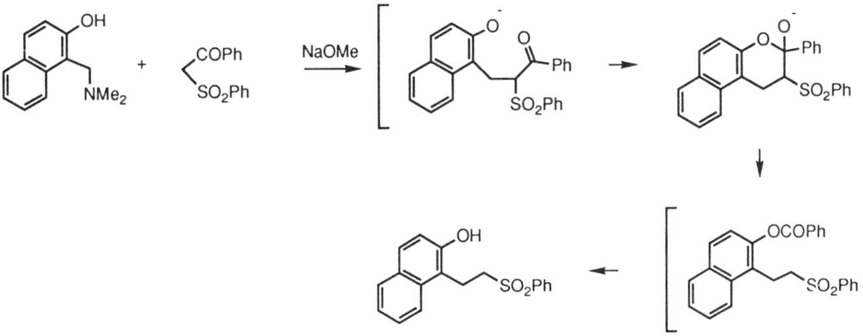

Intramolecular attack on the carbonyl group of an α-sulfonyl ketone by a nucleophile leads to C—C bond cleavage. Of α-sulfonylcycloalkanones a ring expansion may then be effected. Certain carbocyclic [Trost, 1980b; Hajicek, unpublished] and heterocyclic (e.g. lactones) systems [Bhat, 1981] have been prepared.

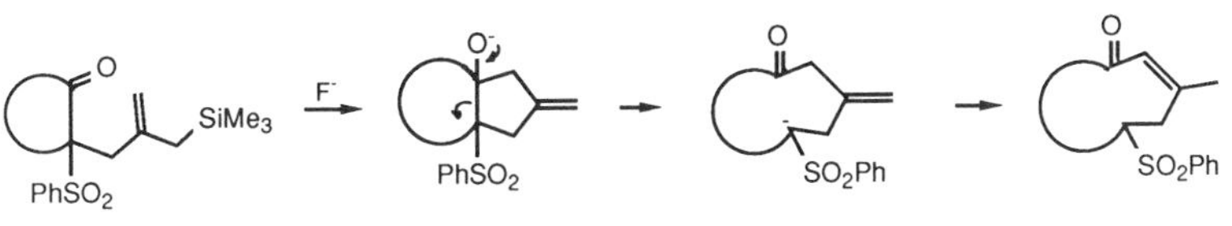

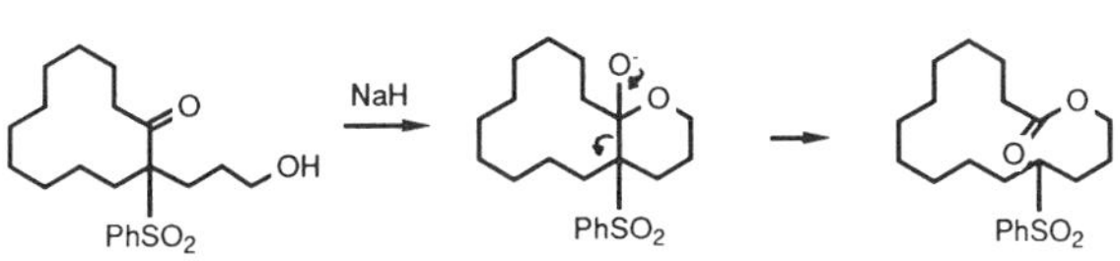

A variant is the analogous reaction with α-nitro ketones [Stach, 1988]. The nitro group is a more powerful acceptor to facilitate the fragmentation. It is also a more versatile functionality as methods for converting $CHNO_2$ to carbonyl, methylene, in addition to $CHNH_2$, are available.

As an example of the synthetic utility of this ring expansion procedure one may cite an expedient synthesis of phoracantholide-I [Ono, 1984].

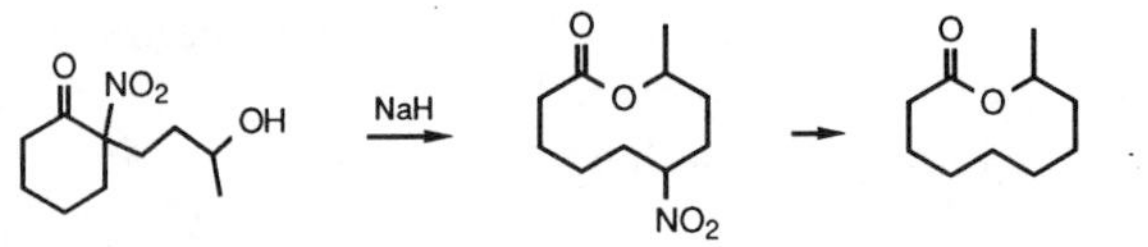

There are two potential pathways for 5-nitrobicyclo[3.3.1]nonane-2,9-dione to fragment on exposure to base. The better stabilizing ability of the nitro group causes the generation of the $\beta$-keto ester [Lorenzi-Riatsch, 1984].

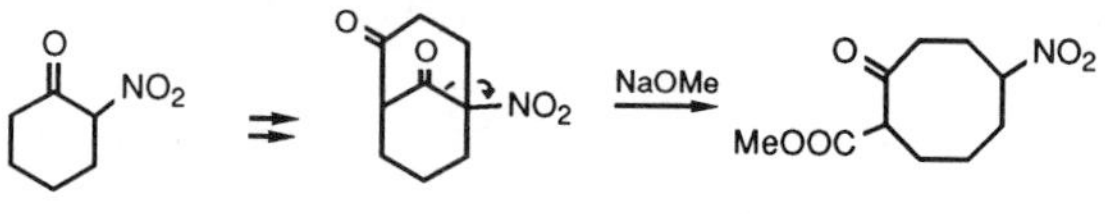

The Mannich reaction of the reduced 2,4-dinitrophenol (with borohydride) gave a bispidine, albeit in low yield [Wall, 1970]. Only half of the carbon atoms from the phenol is preserved in the product. A retro-Michael degradation occurs after fragmentation of the bicyclic nitroketone.

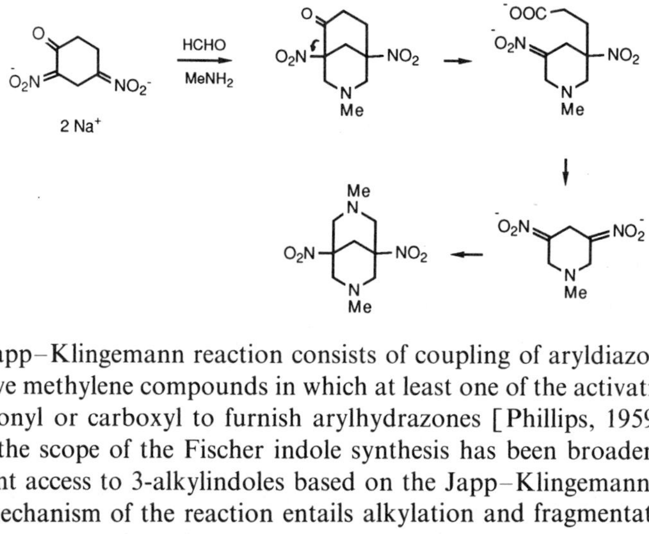

The Japp–Klingemann reaction consists of coupling of aryldiazonium salts with active methylene compounds in which at least one of the activating groups is a carbonyl or carboxyl to furnish arylhydrazones [Phillips, 1959]. Consequently, the scope of the Fischer indole synthesis has been broadened by the convenient access to 3-alkylindoles based on the Japp–Klingemann reaction.

The mechanism of the reaction entails alkylation and fragmentation of the azo derivative. Confirmation of the structure of paniculidine-A by synthesis started from a Japp–Klingemann reaction of 2-carbethoxy-6-methylcyclohexanone [Kinoshita, 1985].

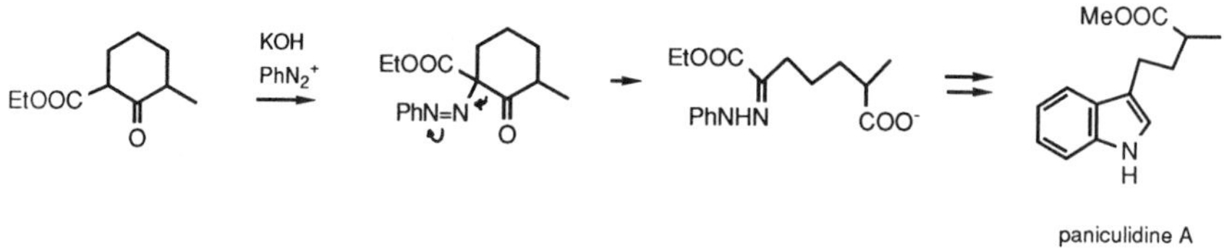

paniculidine A

A synthetic approach to the mitomycins embodied an oxidative cleavage of indole derivatives to reveal the benzazocine ring system which may then be elaborated further [Kametani, 1979a]. The arylhydrazones that carry proper function groups for cyclization of the initially formed indoles were expediently prepared by the Japp–Klingemann reaction.

It has been demonstrated that modification of the Japp–Klingemann reaction by azoation in the presence of an non-nucleophilic base (e.g. lithium diisopropylamide) and treatment of the diazo adduct with a reducing agent ($NaBH_4$) enabled synthesis of *C*-β-ribofuranosyl derivatives [Kozikowski, 1978] which should be useful for elaboration into various *C*-nucleosides.

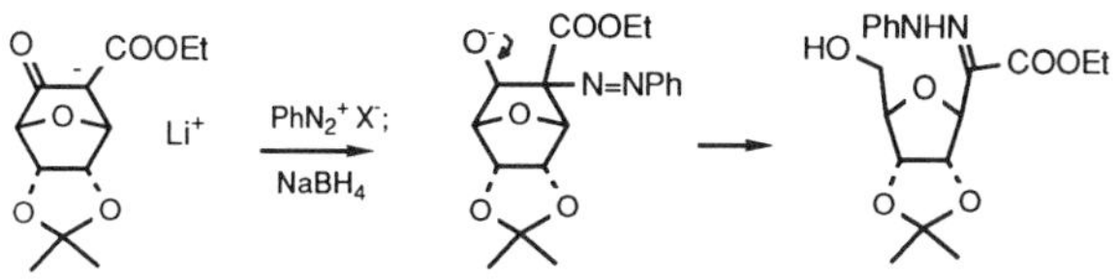

Nitrosation of ketones at a disubstituted α-carbon atom usually is followed by fragmentation. As tautomerization of such nitroso ketones to the oximino ketones is prohibited, the lowest energy reaction pathway available to them is fragmentation.

2-Hydroxyisocarbostyrils have been prepared from 2-substituted 1-indanones in one step [Moriconi, 1966]. Heterolysis of the α-nitroso ketones afforded ketoximes which recombined with the acylium ion in an intramolecular reaction.

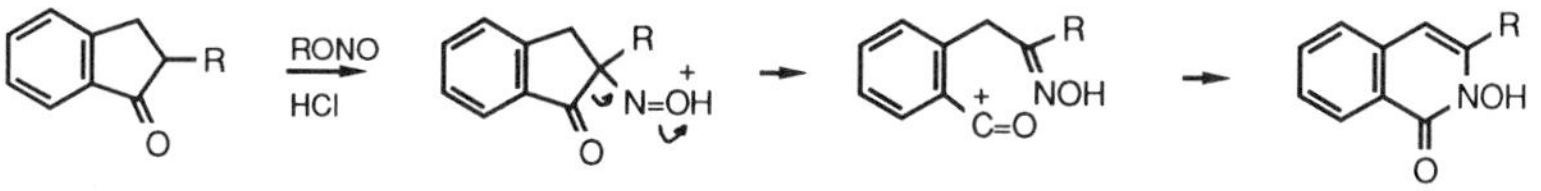

The value of this nitrosation–fragmentation as a synthetic method may be illustrated by its application to the classic synthesis of quinine [Woodward, 1945].

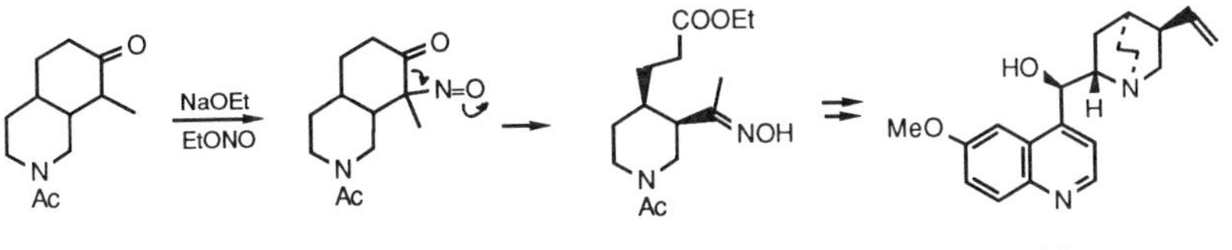

quinine

An approach to (+)-*trans*-chrysanthemic acid from (−)-carvone [Ho, 1982a] also employed this process to cleave the cyclohexanone moiety after the three-membered ring was formed.

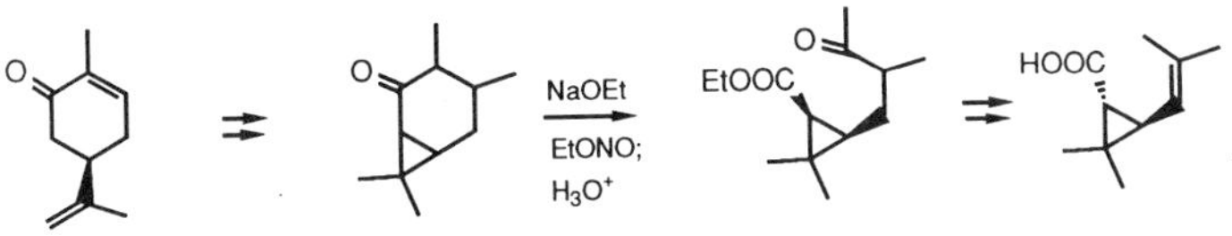

As a degradation device its use in clarifying the structural feature in the vicinity of the exocyclic methylene group of aromadendrene was critical [Birch, 1953].Thus, reaction of apoaromadendrone with amyl nitrite resulted in a cyclopentanone oxime.

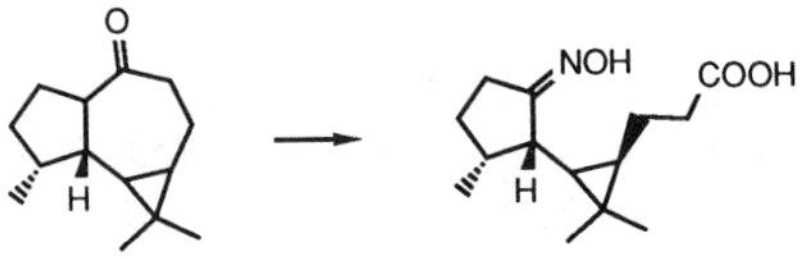

Although it is generally true that fragmentation of nitroso ketones requires nontautomerizable feature, under certain experimental conditions tautomerizable intermediates could be persuaded to undergo CC bond fission [Rogic, 1975].

The respective formation of $\gamma$- and $\delta$-lactones from the epoxidation reactions of isomeric silylallenes [Bertrand, 1980] can be envisaged as stepwise isomerization of the allene oxides to the cyclopropanones, and intramolecular attack of the carbonyl by the hydroxyl group. The direction of the ring opening is controlled by the silyl group. A typical **a1a** situation prevails.

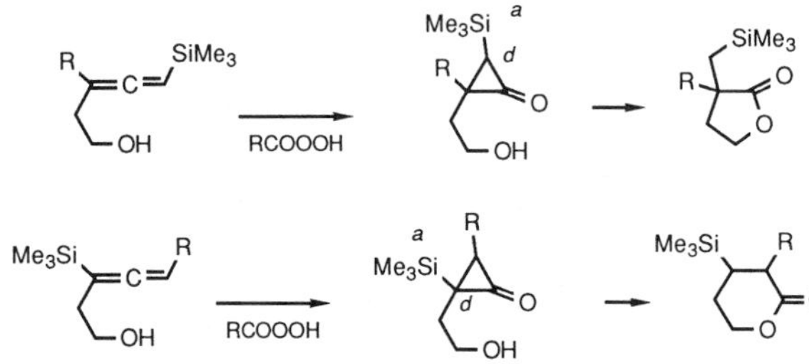

$\alpha,\alpha$-Dihalocycloalkanones of five-membered and smaller rings may be fragmented on exposure to a strong base. Strictly speaking, the capability of chlorine atoms to promote such fragmentation is inconsistent with the usual donor behaviour of these halogen atoms; the factors of inductive effect and ring strain may be of a dominant influence.

The facile fragmentation of the cyclobutanone derivatives has great importance in synthesis. Because the availability of $\alpha,\alpha$-dichlorocyclobutanones from [2 + 2]cycloaddition of dichloroketene with olefins, combination of the two reactions enables significant abbreviation of synthetic operations and expanded usage of olefins in synthesis. In this respect a synthesis of multifidene [Boland, 1978] from cyclopentadiene may be mentioned.

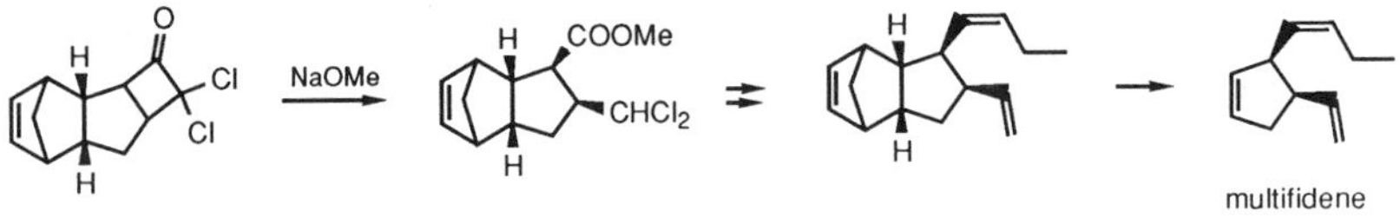

multifidene

A spiroannulated dibromocyclobutanone of a totally different pedigree has been transformed into grandisol [Trost, 1975b] via fragmentation.

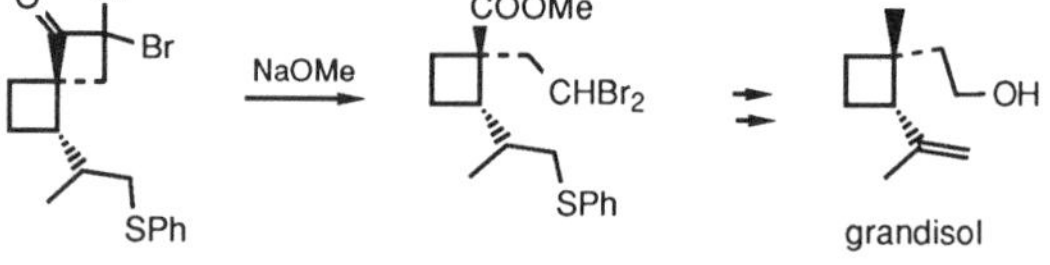

grandisol

Contrapolarization resulting from oxidation eases the fragmentability of *vic*-dioxygenated substances including 4-hydroxy-isopulegone [Torii, 1982]. Thus electrochemical oxidation of this compound provided a useful precursor of rose oxide.

Protodeacylation of an optically active α-silylalkyl phenyl ketone by amide bases in hot benzene [Paquette, 1989] has been interpreted in terms of frontal protonation of the incipient anion species with the benzamide coproduct before the departure of the latter.

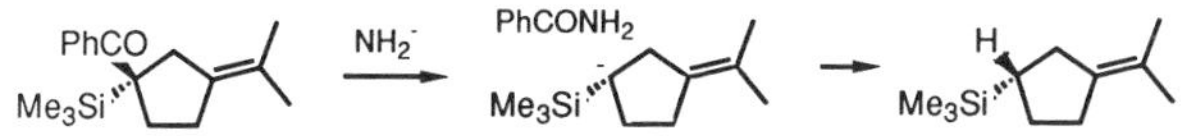

## 3.6. CLEAVAGE OF MISCELLANEOUS a1a SYSTEMS

A galaxy of α-functionalized carbanions have been generated by the reaction of *gem*-bifunctional methanes with organometallic reagents. The metallophilic group of these bifunctional methanes must be an acceptor.

In view of the trends that carbanions which are stabilized by an α-acceptor and destabilized by an α-donor, the former class of anions is particularly readily accessible by the transmetallation technique. Substrates such as $(RSe)_2CH_2$ [Seebach, 1969, 1974; Dumont, 1974], $(PhTe)_2CH_2$ [Seebach, 1975], $(Ph_2As)_2CH_2$ [Kaufmann, 1978] and $(Ph_2Sb)_2CH_2$ [Kaufmann, 1978] are known to undergo transmetallation with organolithium reagents. In the cases of unsymmetrical systems, the transmetallation is often selective, as shown by the preferential attack on selenium of $PhSCH_2SePh$ [Anciaux, 1975] and $Me_3SiCH(R)SeMe$ [Dumont, 1976], and on tin of $ArSCH_2SnPh_3$ [Kricheldorf, 1977; Kuwajima, 1982] and $Me_3SiCH_2SnBu_3$ [Seitz, 1981].

The use of 5-trimethylsilylcyclopentadiene in Diels–Alder reactions has several advantages. The silyl group facilitates the cycloaddition, also it can be replaced by other electrophiles via a rearrangement [Fleming, 1978] in which the crucial event is the generation of a fragmentation-prone silyl carbocation. An excellent synthesis of verrucarol [Roush, 1983], has incorporated such a reaction sequence.

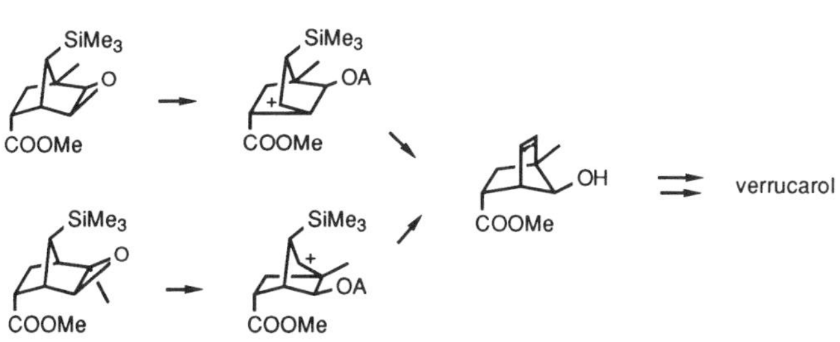

The acetolysis of 4,4-dimethylsilacyclohexyl tosylate has its major course in the generation of an aliphatic compound [Washburne, 1971]. Ionization of the tosyloxy group is attended by 1,2-hydride shift. The ensuing carbocation undergoes fragmentation.

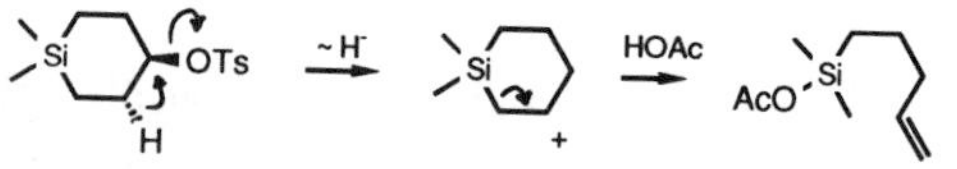

Although iminium ions have much positive charge residing at the carbon site, there is definite modification of the nitrogen atom regarding its donor character. Actually, by conventional formula an iminium species is indicated as if all the positive charge is carried by the nitrogen atom. We must consider this somewhat confusing situation in terms of the affected neighbouring atoms. As the trigonal carbon atom is concerned, the nitrogen is a donor (—C$^+$—N—), but the other carbon substituents must sense certain acceptor character of the same nitrogen (—C=N$^+$—). This latter scenario of identifying the nitrogen as acceptor and therefore the existence of an **a1a** system is essential to generation of ylides from iminium salts (including pyridinium and other heterocyclic systems).

Besides deprotonation, another method for iminium ylide formation is by desilylation of the *N*-(trimethylsilyl)methyl derivatives with fluoride ion. This alternative procedure eschews introduction of strong bases which may be detrimental to other functional groups of a complex substrate. Syntheses of retronecine [Vedejs, 1980] and eserethole [R. Smith, 1983] have been published which are based on the novel **a1a** fragmentation.

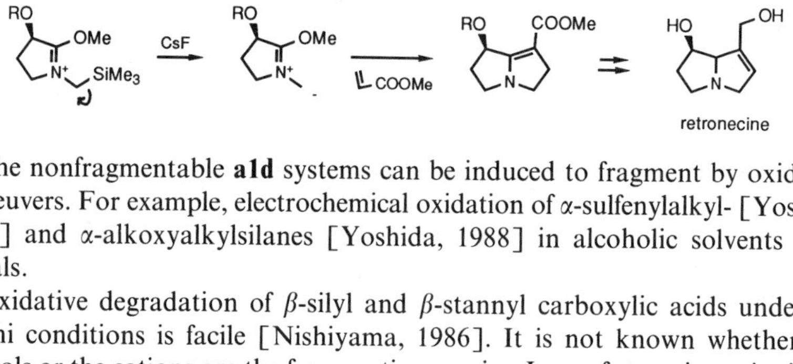

retronecine

The nonfragmentable **a1d** systems can be induced to fragment by oxidative maneuvers. For example, electrochemical oxidation of α-sulfenylalkyl- [Yoshida, 1987] and α-alkoxyalkylsilanes [Yoshida, 1988] in alcoholic solvents gives acetals.

Oxidative degradation of β-silyl and β-stannyl carboxylic acids under the Kochi conditions is facile [Nishiyama, 1986]. It is not known whether free radicals or the cations are the fragmenting species. Loss of stereointegrity in the products is observed.

# 4

# a2d FRAGMENTATIONS

There are two ways to initiate fragmentation of the **a2d** system: by attack from an external donor on the acceptor site, or employment of a powerful acceptor to help ionize the donor group. Consequently, the fragmentation is rather facile when the donor-substituted atom is a tertiary carbon.

Decarboxylation of 7-cycloheptatrienecarboxylic acid [Rüchardt, 1966] and cyclopropylacetic acid [Hart, 1962] via acid-catalyzed decomposition of their peroxy esters fall into this category. The cycloheptatriene and cyclopropylmethyl moieties are excellent acceptors as formation of tropenium ion and cyclopropylcarbinyl cation are very favorable.

## 4.1. Het—C—C—O AND RELATED SYSTEMS

An application of this fragmentation is embodied in a method for carboxyl and amino protection by the 2-silylethyl and 2-silylethoxycarbonyl groups. This method has been incorporated into a synthesis of (−)-curvularin [Gerlach, 1977]. Other applications are too numerous to be listed here.

$$RCOOCH_2CH_2SiMe_3 + F^- \rightarrow RCOO^- + CH_2{=}CH_2 + Me_3SiF$$

$$RNHCOOCH_2CH_2SiMe_3 + F^- \rightarrow RNH^- + CO_2 + CH_2{=}CH_2 + Me_3SiF$$

An unexpected reward is an efficient macrolactonization during demasking of the 2-(trimethylsilyl)ethyl protecting group [Vedejs, 1984].

The same group serves as an excellent device for anomeric protection of sugars [Jansson, 1988]. These glycosides can be transformed into 1-*O*-acylsugars by treatment with boron trifluoride etherate and carboxylic anhydrides.

When the silicon of the silylethyl group is replaced by even stronger acceptors such as Hg, Ge, Sn, or Pb at C-2, the resulting compounds are very labile, and fragmentation occurs in the presence of water [Whitmore, 1948; R.B. Miller, 1974].

The regiochemistry of the Beckmann fragmentation can be controlled by placing a silyl group at the $\beta$-position of an oxime. Trimethylsilyl triflate is an excellent catalyst for such transformation [Nishiyama, 1983]. The reaction may proceed via a normal rearrangement but the incipient iminocarbenium ion suffers loss of the silyl group. Accordingly, the position of the double bond of the product(s) is predetermined. This silicon-directed fragmentation has been used in a synthesis of a potato tuber worm moth sex pheromone [Nishiyama, 1988]. It should be emphasized that analogous substrates lacking the silyl group undergo normal Beckmann rearrangement. (The **a2d** pathway formulated by the authors cannot be differentiated from a concerted **a4d** fragmentation.)

AcO–N=, SiMe3 —Me3SiOTf→ Me3SiO+, O–N, SiMe3 → +N, SiMe3 → CN

The tin analogs fragment in a similar fashion [Bakale, 1990].

The high affinity of silicon to fluoride ion coupled with a powerful electrofugal group such as the triflyl group makes possible a generation of benzyne from 2-(trimethylsilyl)phenyl triflate at room temperature [Kobayashi, 1983].

$$\text{SiMe}_3,\ \text{OTf} + \text{Bu}_4\text{N}^+\ \text{F}^- \rightarrow \text{(benzyne)} + \text{Bu}_4\text{N}^+\ \text{TfO}^- + \text{Me}_3\text{SiF}$$

The 2-haloethyl groups are useful for protection of active hydrogens. The halogen atom which can assume an acceptor role in the presence of strong donors may be abstracted. The detached carbon atom with a net electric charge is prone to expel the donor substituent at a neighbouring atom. In reality the scission of the X—C and C—O bonds must be more or less concerted.

The best known protecting group of this series is the 2,2,2-trichloroethyl group. This group was developed during a synthesis of cephalosporine-C [Woodward, 1966]. Its great merit is the fragmentability under conditions (e.g. Zn, $H_2O$—HOAc) that would not disturb most other sensitive functionalities.

O, O, CCl3, O, N, OAc, HN, H, O, CHCOOCH2CCl3, NHCOOCH2CCl3 —Zn, aq. HOAc→ O, OH, O, N, OAc, HN, H, O, CHCOO⁻, NH3⁺

A brilliant scheme for synthesis of the aspidosperma alkaloids [Gallagher, 1982, 1983] involves in situ generation of the quinodimethane system which is trapped intramolecularly by an alkene linkage (Diels–Alder reaction). The quinodimethane formation has been effected by reaction of appropriate imines derived from 2-methylindole-3-aldehydes with trichloroethyl chloroformate. The carbamate group of the tetracyclic intermediates can be removed by fragmentation.

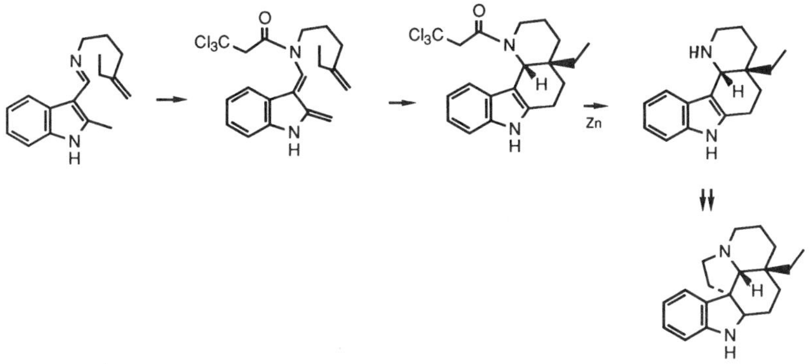

Reductive fragmentation of the X—C—C—O system in which X = Br,I is more facile, the presence of only one halogen atom being sufficient. This behavior is in agreement with the increasing acceptor character of the heavier atoms and it enables double protection of certain unsaturated alcohols.

In a magnificent synthesis of picrotoxinin [Corey, 1979] the angular hydroxyl group emerged from a bridged ketone quite early. Simultaneous protection of which and the isopropylidene double bond was achieved in the form of a $\beta$-bromo ether. In the last step of the synthesis, treatment with zinc regenerated the two functionalities.

It should be emphasized that the choice of this particular protection is ideal because derivatization of tertiary alcohols (intermolecularly) is notoriously difficult and often beset by side reactions such as dehydration; their unmasking can also be complicated.

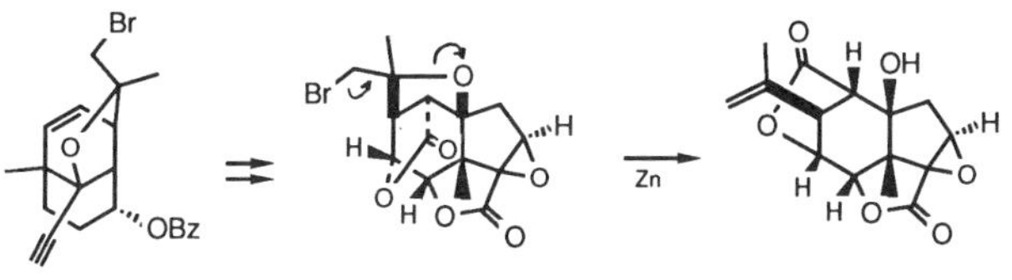

picrotoxinin

Internal protection of a double bond and a primary hydroxyl group by the same method was also essential in the syntheses of anguidine [Roush, 1987] and verrucarol [Schlessinger, 1982]. Deprotection was performed with zinc–silver couple in ethanol and sodium in ethylamine, respectively. (Note that many reductions, particularly those using alkali metals/solvated electrons, involve multiple one-electron transfers. Their inclusion in this book usually denotes that the fragmentation step occurs after the formation of an anionic species. However, the possibility of bond scission immediately succeeding the free radical formation cannot be eliminated.)

Phorbol was synthesized [Wender, 1989] by a route that involved incorporation of the tertiary hydroxyl group at C-9 in the form of a cyclic ether. Release of this function was achieved by a fragmentation protocol after an iodine atom was introduced.

The conversion of gibberellic acid into methyl esters of 12-hydroxy gibberellins [Chu, 1988] via remote oxidation was mediated by a bromohydrin derived from the exocyclic methylene group. The double bond was regenerated by zinc reduction.

Perhaps the most eminent classic example of the Br—C—C—O fragmentation is the unfolding of an octalone intermediate en route to reserpine [Woodward, 1958]. In this case the bromo ether subunit was introduced separately and therefore the overall transformation is not merely a protection–deprotection routine.

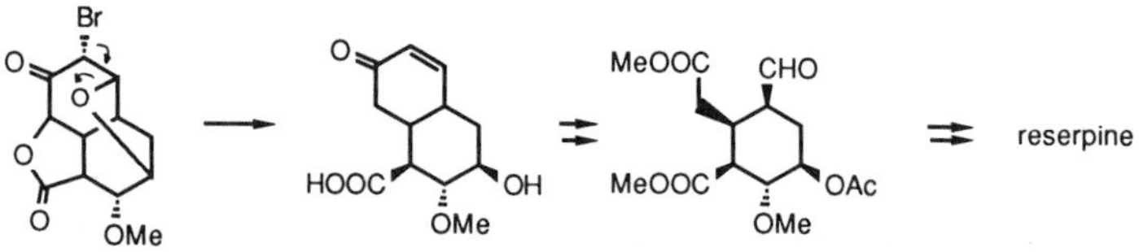

Variations on the same theme of reductive fragmentation include an integral part of a carbonyl protection scheme involving the formation of 5,5-dibromo-1,3-dioxanes [Corey, 1976a] and 4-bromomethyl-1,3-dioxolanes [Corey, 1973].

Relevant to the above reactions is the Boord olefin synthesis [Dystra, 1930] from aldehydes which consists of enol etherification, bromination, Grignard reaction, and zinc dust-promoted fragmentation.

## 4.2. O=C—C—C—X SYSTEMS

Carbonyl compounds with an nucleofugal substituent at the β-carbon avail themselves to at least three reaction courses that involve the carbonyl group. Elimination is blocked when the α-carbon is fully substituted or the segment is placed in a special environment such that the elimination process engenders a highly strained unsaturated system. The two remaining pathways are fragmentation and internal displacement after attack of a nucleophile on the carbonyl function.

There is an interesting divergence of the major pathways depending on the nature of the nucleofugal group. For example, in the aliphatic aldehyde series those substrates with a β-halogen tend to undergo fragmentation whereas the arenesulfonyloxy analogs give mainly the lactol ethers [Nerdel, 1965; Janculey, 1967].

A synthetically useful preparation of citronellic acid is by treatment of pulegone hydrochloride with alkali [Plesek, 1956].

The leaving group dependence is best explained by invoking the Hard and Soft Acids and Bases (HSAB) Principle [Ho, 1977]. The $S_N2$ displacement is particularly favored by the sulfonates because they are harder and the attacking alkoxide ions are hard bases (symbiosis).

The complex interplay of various reaction parameters governing the outcome may be appreciated by examining the ring size dependence of fragmentation versus internal alkylation of monocyclic α-alkyl-α-tosyloxymethylcycloalkanones [Heinz, 1990]. It appears that only in the cyclopentanone series does fragmentation predominate.

Several fragmentation protocols to convert the cyclopentadiene-phenylsulfenylmethylketene adducts into *vic*-disubstituted cyclopentenes have been developed [Ohshiro, 1982]. Thus, treatment with dimethyloxosulfonium methylide and reaction of the corresponding sulfoxides with methoxide ion are two variations.

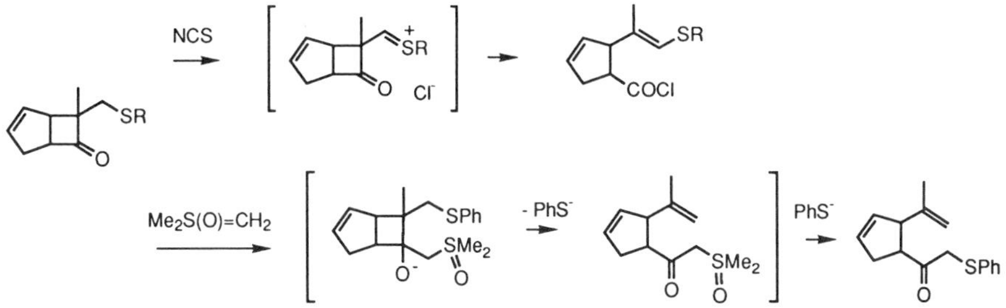

Condensation of succinoin derivatives with ketals furnishes versatile intermediates for 1,3-cyclopentanediones and γ-keto esters [Shimada, 1984]. The access to the latter class of substances means achieving an overall reductive succinylation of carbonyl compounds. A synthesis of lanceol serves to demonstrate this chain elaboration.

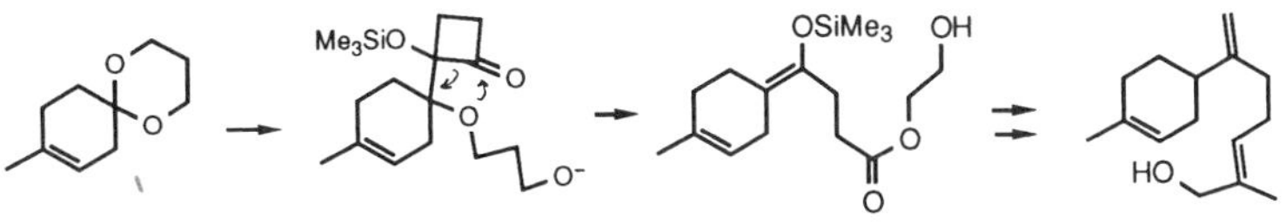

The stereochemical requirements for solvolytic fragmentation of β-tosyloxy carbonyl compounds has been investigated with cyclohexanecarboxaldehydes which are also substituted with a *t*-butyl anchor [Nerdel, 1969a]. The results indicate a favourable fragmentation occurs when the formyl and the tosyloxy groups are *trans*. The *cis*-isomer affords many products.

A surprising observation is the high yield of the cyclohexene obtained from the *trans* compound in which the 5-*t*-butyl group is *cis* to the formyl substituent. The conclusion is that this compound reacts in the boat conformation.

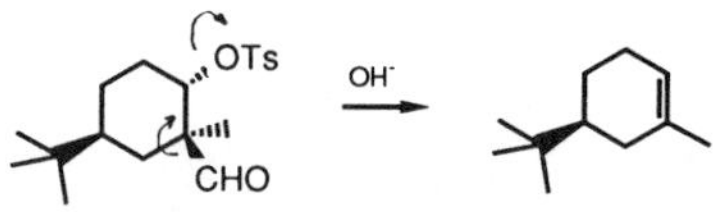

The main products from alkaline treatment of 2-alkyl-2-heteromethylcyclohexanones are 6-alkyl-6-heptenecarboxylic acids [Nerdel, 1967], although bicycloheptanones are also formed. Apparently these reaction conditions favor the fragmentation pathways.

In an approach to the pyrethric acids [S. Julia, 1966] 6,6-dimethylbicyclo[3.1.0]hexan-2-one was induced to fragment after α-acylation and hydroxymethylation.

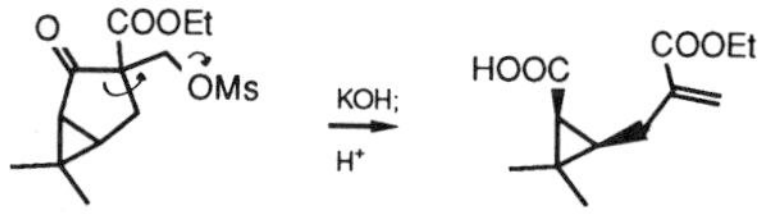

The related *cis*-chrysanthemic acid has been acquired by alkaline treatment of a keto mesylate with a bicyclo[3.1.0]hexane skeleton [Krief, 1988].

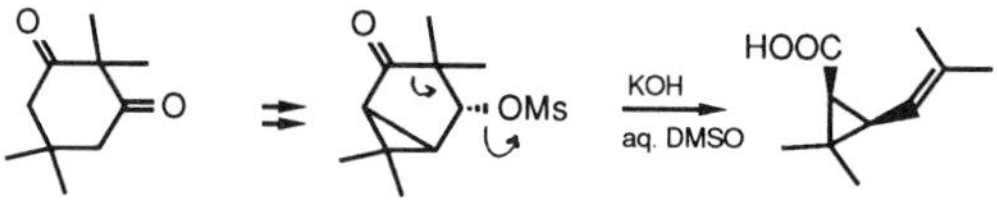

Several types of products were generated when 2,2,6,6-tetrakis(tosyloxymethyl)cyclohexanone was reacted with alcoholic sodium hydroxide [Nerdel, 1970]. Bridged ketone intermediates in which a tosyloxymethyl group was still present at the bridgehead would undergo fragmentation.

Degradation of a spirocyclic aldol tosylate [Carruthers, 1977] led to a cyclopentenebutanoic ester.

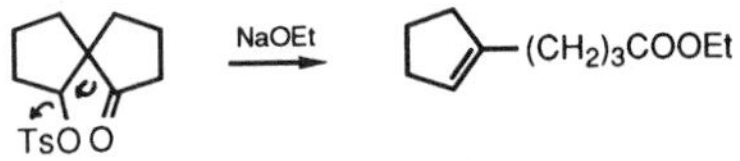

The fragmentation pathway is also preferred by a bridged dibromoketone [Goldstein, 1982] on its exposure to cyanide ion because of the inherent strain.

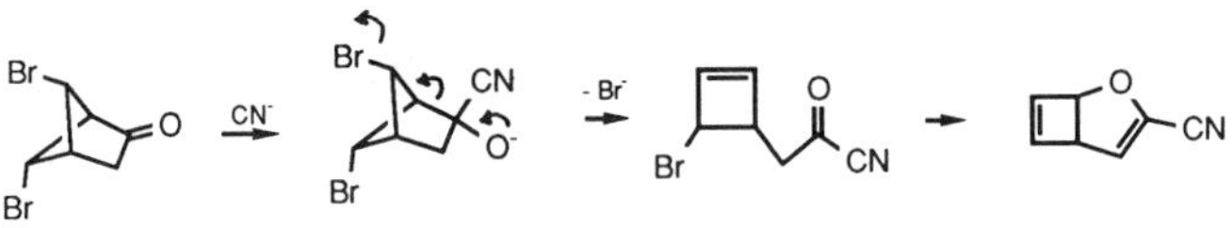

A bridgehead halogen as leaving group in a **d3d** fragmentation has been reported [House, 1979].

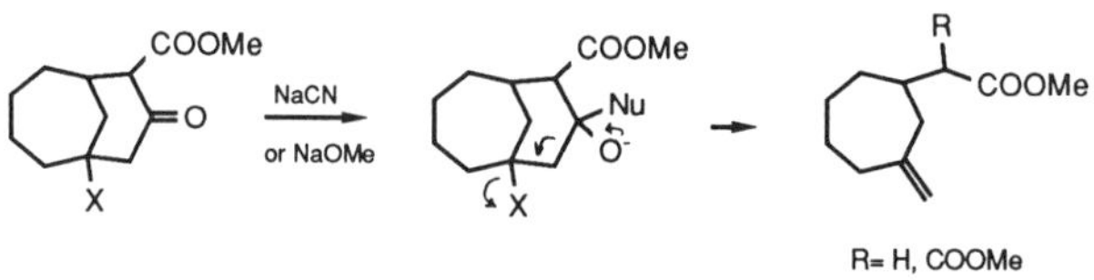

In a synthetis of laurenene [Wender, 1988] based on *meta*-photocycloaddition of aromatic compounds a benzosuberane derivative was needed. From the functional group distribution pattern it seemed expedient to employ a fragmentation route to acquire the intermediate.

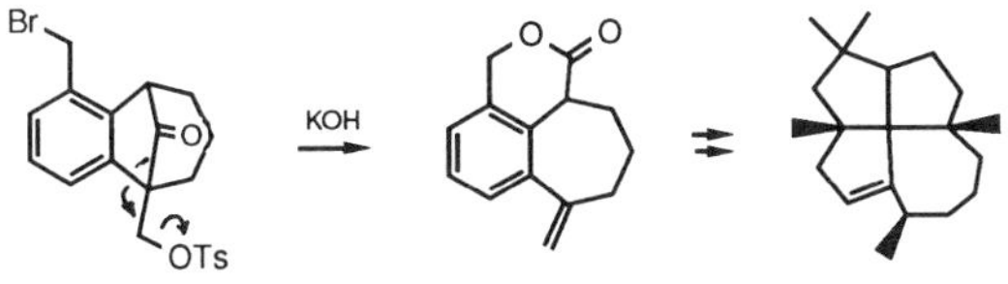

laurenene

The ring expansion exemplified by the above work is very general. Thus it has been demonstrated a formation of 5-methylenecyclononane-carboxylic acid derivatives [Marschall, 1974] and the critical unfoldment of a functionalized 8:5-fused ring system in a synthesis of ceroplastol-I [Boeckman, 1989]. In the latter reaction the ester group generated was to participate in a Dieckmann condensation to complete the tricyclic array of the diterpene molecule.

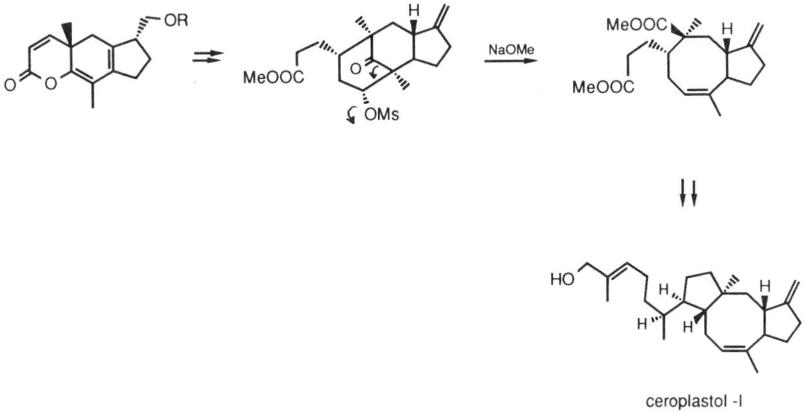

ceroplastol -I

Epimeric bicyclic $\beta$-keto tosylates behaved differently toward base [H.L. Brown, 1979], and the differences are due to stereoelectronic effects. The axial tosylate underwent retro-Claisen fission, a synperiplanar decarboxylation, and detosyloxylation.

5-Cyclodecenyl-1-methanol can be prepared in the (*E*)- or (*Z*)-form using the concerted fragmentation process on the 11-oxobicyclo[4.4.1]undecan-2-yl mesylates [Garst, 1982].

Epimeric keto tosylates of the bicyclo[3.2.1]octane series fragment in different manners [G.L. Buchanan, 1966b]. The equatorial tosylate afforded a *gem*-diester, whereas the axial tosylate is stereoelectronically unfit to fragment concertedly and it underwent retro-Claisen reaction. Finally an E1cB reaction completed the process.

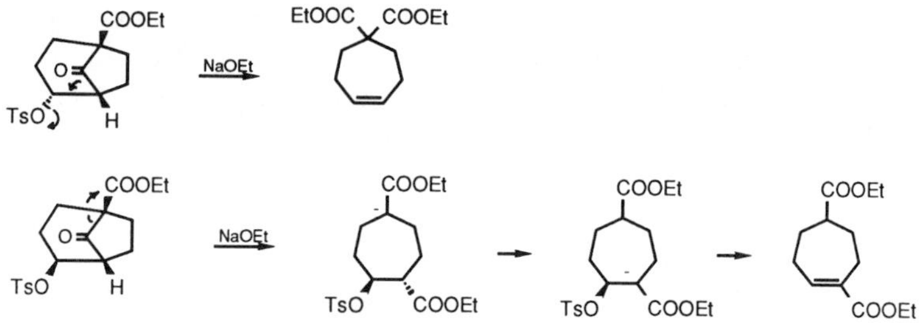

The presence of a bridgehead ester group is crucial to the fragmentation of the axial tosylate. In a homologous system without the ester group, only one of the two epimeric keto tosylates was found to fragment; the other epimer, presumably the axial tosylate, suffered elimination only [Martin, 1964].

The major product of intramolecular Friedel–Crafts acylation of 3-cyclohexene-1-acetyl chloride was shown to be *anti*-6-chlorobicyclo[2.2.2]octan-2-one [Monti, 1975] by chemical transformations including complex hydride reduction to give 2-(3-cyclohexenyl)ethanol. The isomeric *endo*-2-chlorobicyclo-[3.2.1]octan-8-one furnished 4-cycloheptenecarboxylic acid [Fickes, 1973] on alkaline treatment.

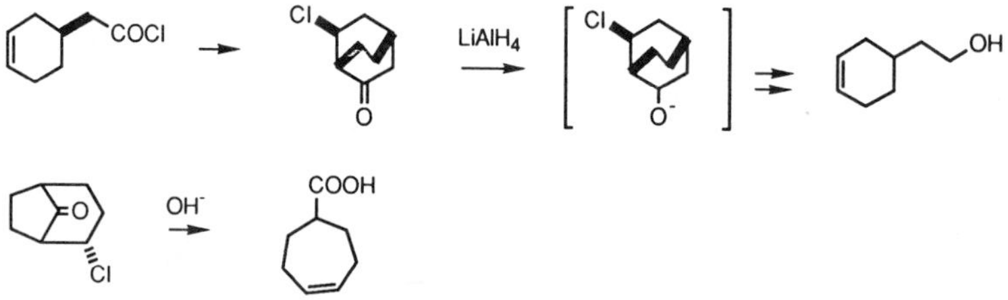

It is remarkable that bridgehead tosylate of a bicyclo[2.2.2]octane framework did not show any reluctance as leaving group in the fragmentation reaction [W. Kraus, 1970], contrary to its behavior in substitution and elimination reactions. The antiperiplanar alignment of the C-2/C-3 bond and the C-4/OTs bond is perfect for fragmentation.

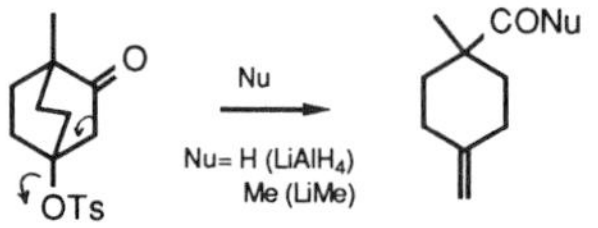

From (−)-*β*-pinene a synthesis of (+)-hinesol was accomplished [Chass, 1978] which was based on a skeletal rearrangement and ring exchange. Cleavage of the existing bridged system established an isopropylidene pendant useful for introduction of the tertiary hydroxyl.

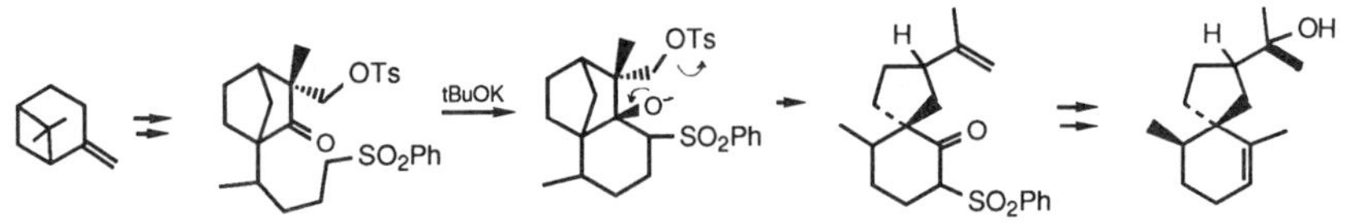

hinesol

A convenient method for synthesis of bicyclo[3.3.1]nonene derivatives is via fragmentation of 4(*eq*)-mesyloxyadamantan-2-one [Sasaki, 1971]. This facile fragmentation makes the preparation of 4(*ax*)-deuterioadamantan-2-one less straightforward [Numan, 1978], as deuteride displacement of the equatorial mesylate is impossible. Note that lithium aluminum deuteride reduction of the 4-*axial* mesyloxy ketone proceeded without skeletal change.

Bridged ring systems offer excellent sterocontrol on their peripheries during reactions. This benefit can be transferred to simpler ring systems when means of degradation such as fragmentation are provided. Among the many examples are syntheses of *threo*-juvabione [Larsen, 1979] and α-*trans*-bergamotene [Larsen, 1977]. The latter work featured a transformation of a bicyclo[3.2.1]-octane to a bicyclo[3.1.1]-heptane via an intramolecular alkylation and fragmentation reaction sequence.

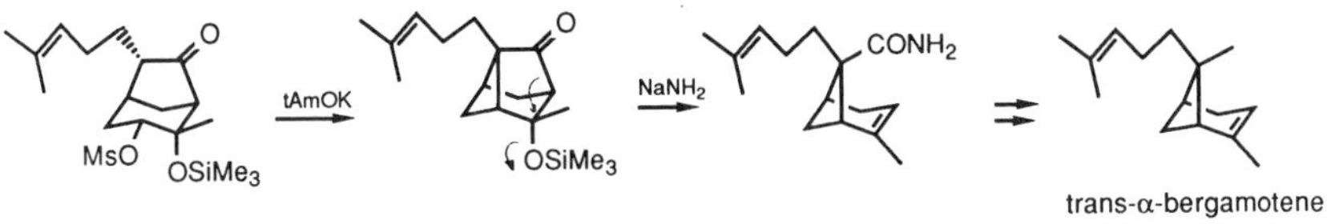

*cis*-Hydrazulene derivatives may be assembled by Stork's method [Hendrickson, 1971], that is the fragmentation of a cyclopentano-bicyclo[3.2.1]octanone.

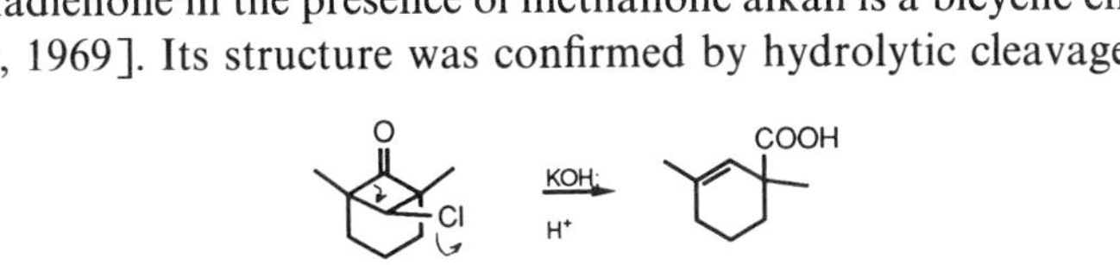

One of the hydrogenation products of 6-dichloromethyl-2,6-dimethyl-2,4-cyclohexadienone in the presence of methanolic alkali is a bicyclic chloroketone [Brieger, 1969]. Its structure was confirmed by hydrolytic cleavage.

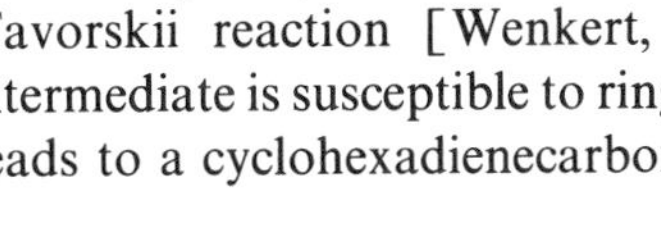

6-Dichloromethyl-3,6-dimethyl-2-cyclohexenone eliminates hydrogen chloride by way of a homo-Favorskii reaction [Wenkert, 1971]. The bridged β-chlorocyclobutanone intermediate is susceptible to ring fission in two directions. One of the pathways leads to a cyclohexadienecarboxylic acid.

Interestingly, an acyclic ester has been obtained [Wenkert, 1970] from treatment of 2-dichloromethyl-2-methylcyclohexanone with potassium *t*-butoxide. It is the (*E*)-isomer.

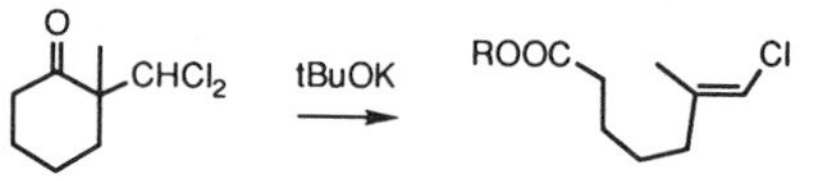

The treatment of 5-bromocamphor with 10% acetic acid led to a condensed $\gamma$-lactone [Hirata, 1971], as a result of a fragmentation-induced Nametkin rearrangement.

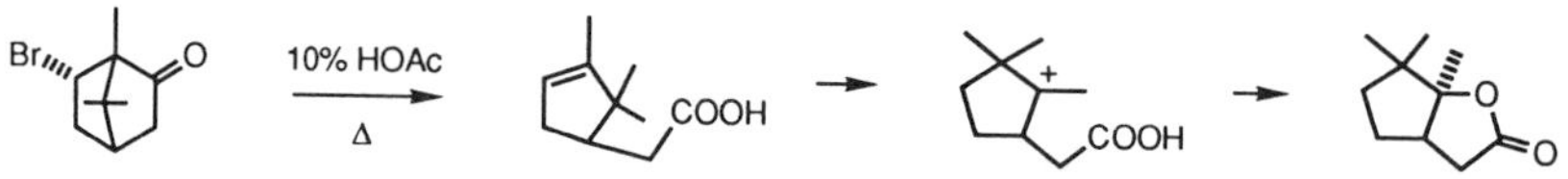

Under formylation conditions fragmentation of a bridged ketol was reported [Saha, 1985]. Apparently the initial reaction product is capable of transferring the *C*-formyl group to the angular hydroxyl, rendering the latter nucleofugal. Stereoelectronically this system is unsuited for fragmentation.

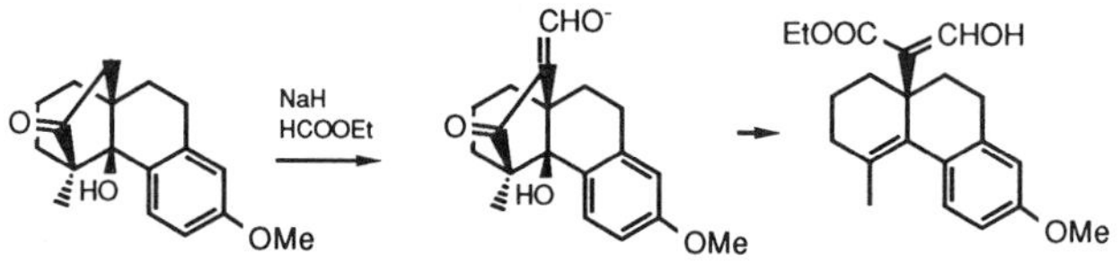

A fragmentative step is the key to a synthesis of $\alpha,\beta$-unsaturated carbonyl compounds [Ksander, 1977]. The $\alpha$-alkylidenation involves activation by oxalylation and treatment with an aldehyde. The $\alpha$-oxo-$\beta$-acyl-$\gamma$-butyrolactone intermediate possesses an **a2d** circuit for fragmentation (which can be classified under the **a1a** system).

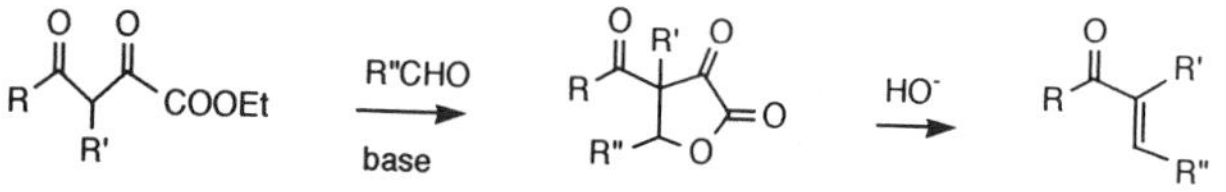

In a synthesis of trihydroxydecipiadiene [Greenlee, 1981] the geminal sidechains were created from a secoalkylation process on a cyclobutanone. Thus, treatment of the $\alpha$-dimethylaminomethylene derivative with acidic methanol effected an addition–fragmentation sequence. Both reactions were aided by angular strain of the system. Ordinarily, enaminoketones (vinylogous amides) do not add alcohols.

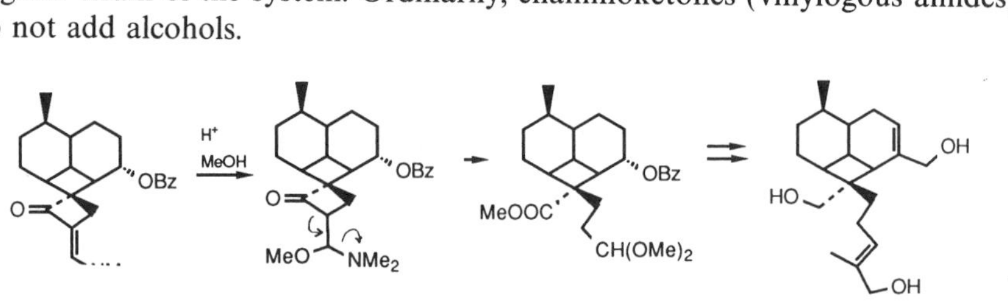

trihydroxydecipiadiene

Fragmentation of $\gamma$-oxosulfonium methylides can lead to two types of products [Watanabe, 1992] as shown in the following equation. Ring size effect of this fragmentation has been examined.

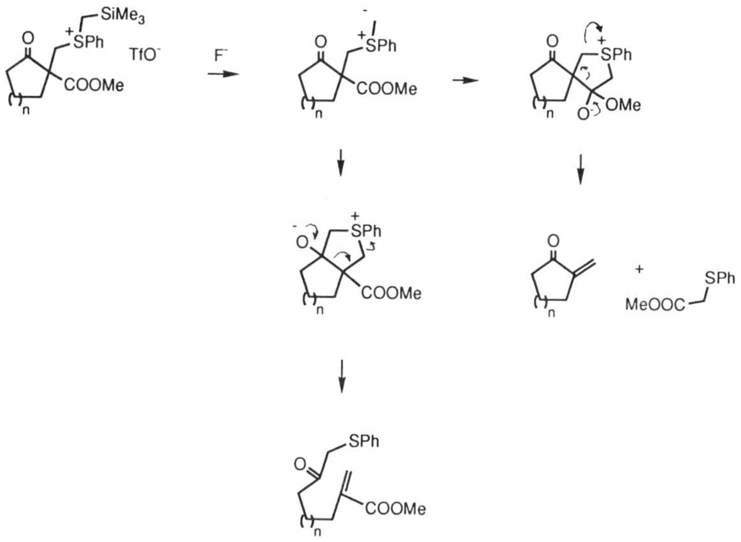

A convenient synthesis of indigo [Baeyer, 1882] involves condensation of 2-nitrobenzaldehyde with acetone in the presence of aqueous alkali. An intramolecular redox transformation between the nitro group and the benzylic alcohol of the aldol occurs and the resulting nitroso diketone is poised to undergo an aldol-type reaction. It is thought that the *N*-hydroxyindoxyl and its anhydro form combine, and the adduct reorganizes by $C \rightarrow O$ transacetylation to give an *N*-acetoxy-2-acetylindoxyl derivative which is prone to lose the element of acetic anhydride on exposure to hydroxide ion.

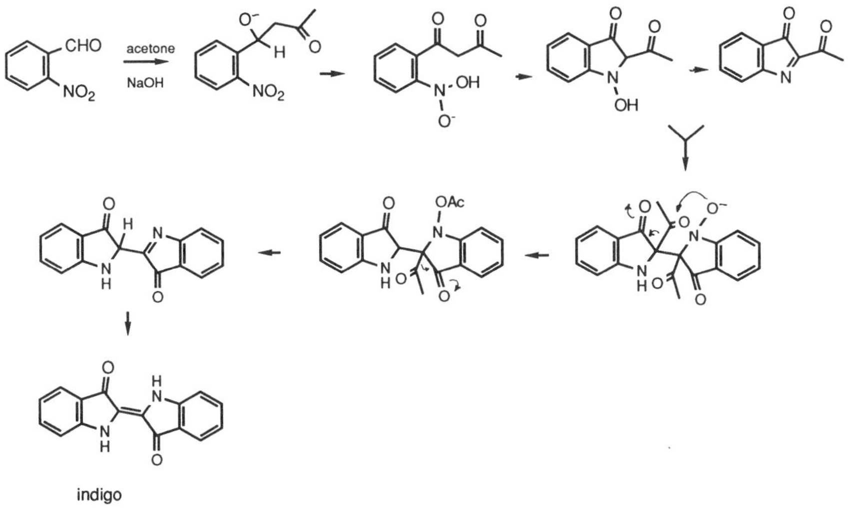

indigo

Acetylenes have been generated from $\alpha,\beta$-unsaturated $\beta$-halocarbonyl compounds [Bodendorf, 1965]. The deacylchlorination by base also happens in the second stage of the Favorskii rearrangement of $\alpha,\alpha'$-dibromoketones which gives rise to $\alpha,\beta$-unsaturated carboxylic acid derivatives [Wagner, 1950], the halogen atom of the $\alpha$-bromocyclopropanone intermediates is also at a $\beta$-position. $\alpha,\beta$-Dibromoketones yield the unconjugated products [Stork, 1960]. The position of the double bond in the primary products is subject to isomerization, as shown in the cases of steroidal 17,21-dihalo-20-ones and 16,17-dibromo-20-ones [Romo, 1957], both leading to $\Delta^{17(20)}$-21-carboxylic acids.

Br O Br NaOMe Br COOMe

Br O Br NaOMe Br COOMe

2,2,6-Trimethyl-2,3-dihydro-$\gamma$-pyrone is a valuable synthon for terpenes such as *ar*-turmerone. This pyrone can now be prepared from 2-chloro-3,3-dimethylcyclopropanone generated in situ from a Favorskii-type reaction [Sakai, 1985].

O Cl X + O COOtBu base O O⁻ COOtBu Cl O O COOtBu O

O COOtBu O O

*ar*-turmerone

Perhaps the best known example of Favorskii rearrangement is the transformation of pulegone dibromide into pulegenic acid [Wallach, 1918]. Carvone tribromide reacts with primary amines to give iminolactones [Wolinsky, 1968].

O Br Br HO⁻ O Br COOH

pulegenic acid

The Favorskii rearrangement of pulegone oxides has been studied [Cavill, 1967; Reusch, 1967]. The retention of a hydroxyl group in substantial amounts of the product mixtures must be due to equilibration of the cyclopropanone intermediates with their conjugate bases; departure of an oxide dianion in these conjugate bases is not feasible under the reaction conditions. The direction of ring opening is, however, still under the influence of this donor function, as the incipient tertiary carbanions are being generated.

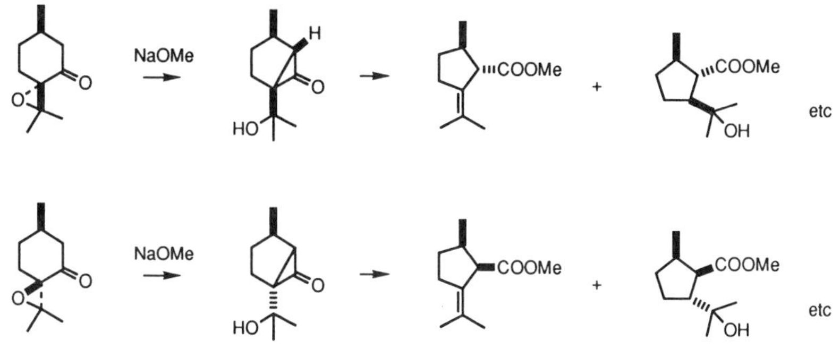

NaOMe → H HO → COOMe + COOMe OH etc

NaOMe → HO → COOMe + COOMe OH etc

5β,6β-Epoxyandrostane-4,17-dione but not the 5α,6α-epoxide was found to undergo Favorskii rearrangement [Hanson, 1978]. Reaction of the latter isomer would involve an intermediate with a boat-like A-ring.

The Grignard reaction of chalcone epoxide [Kohler, 1931] is abnormal. For example, one of the products from the reaction with phenylmagnesium bromide is trityl alcohol, presumably arising from in situ cleavage of the initial adduct and further reaction of the resulting benzophenone.

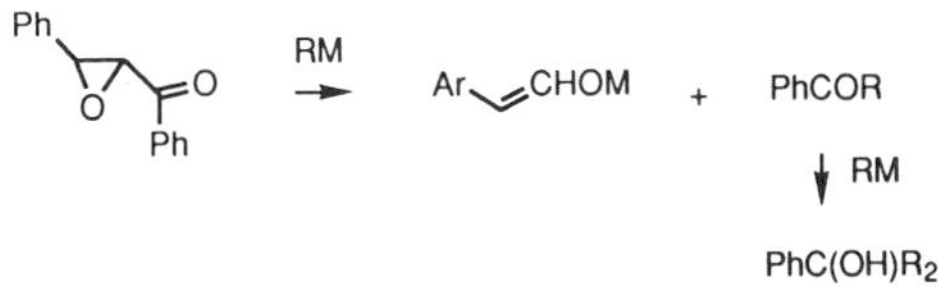

Ph O Ph →(RM) Ar CHOM + PhCOR ↓ RM $PhC(OH)R_2$

A convenient method for synthesis of fluorescamine [Weigele, 1976], which finds extensive use in fluorometric quantitation of primary amines, is via epoxidation of 2-benzylidene-1,3-indandione, hydrolytic cleavage of the epoxide, and formylation.

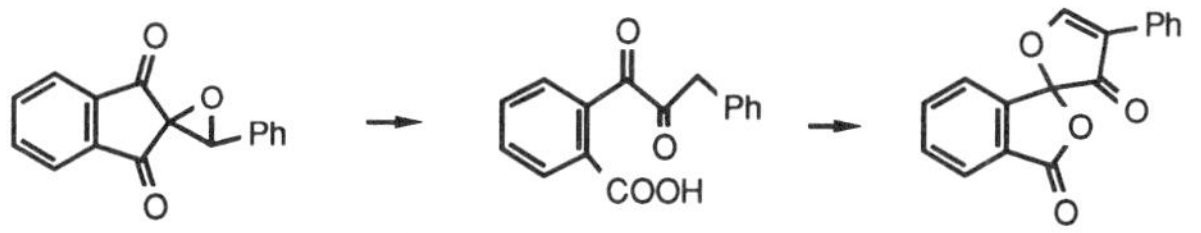

O O Ph O → O Ph O COOH → O Ph O O O

Chemical studies in the oleanic acid series have uncovered a unique fragmentation of an epoxy dione [Ruzicka, 1943]. Cleavage of the E-ring of siaresinolic acid [Barton, 1952] was also effected by alkali treatment of an epoxy dione.

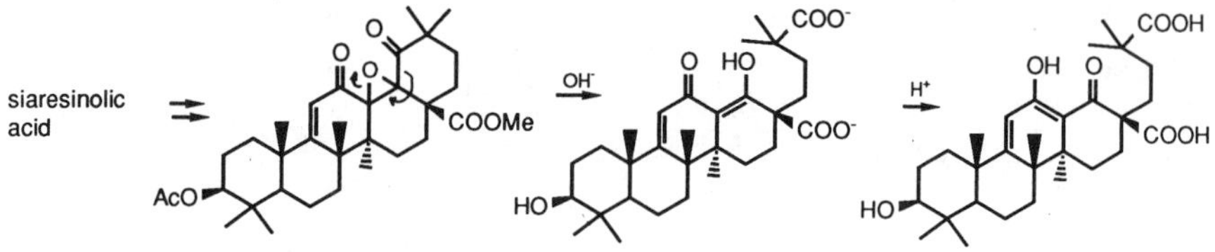

siaresinolic acid

The ready transformation of a ketone to a cyclobutanone in two steps, that is reaction with a cyclopropylidene sulfurane and Lewis acid-catalyzed rearrangement, is extendable to $\alpha,\beta$-epoxy ketones [Trost, 1978a]. The products, having a donor substituent $\beta$ to the strained carbonyl group, are very susceptible to attack by nucleophiles. Thus, $\gamma,\delta$-unsaturated $\varepsilon$-hydroxy esters can be acquired by reaction of such cyclobutanones with alkoxide ions. A synthesis of the dimethyl ester of a Monarch butterfly pheromone highlights the utility of this method.

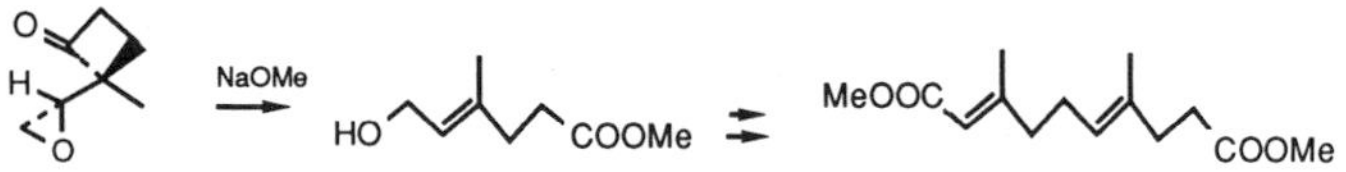

A *cis*-hydrindane $\beta,\gamma$-epoxy ketone fragmented on reaction with boron ttifluoride etherate [Halsall, 1978]. The seco acylium ion recyclized and the diquinane intermediate underwent a retro-Prins reaction. Interestingly, the monocyclic acylium ion derived from the *trans*-hydrindane did not recyclize, the product being a diester.

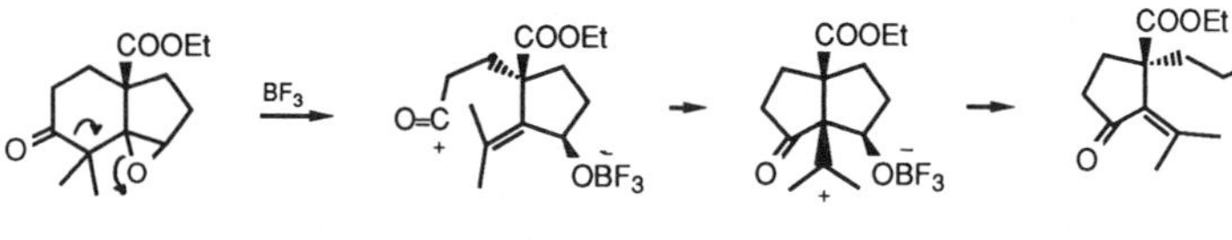

Asymmetric synthesis of many interesting compounds has been facilitated by the ready availability of both (+)- and (−)-camphor and the possibility of functionalizing C-9 and C-10 of the monoterpene molecule without loss of enantiometric integrity. For example, a synthesis of the California red scale pheromone from (+)-camphor has been accomplished [Hutchinson, 1985]. Two fragmentations were involved: degradation of 9,10-dibromocamphor to a methylenecyclopentane, and the deannulation of an $\alpha$-bromomethylcyclopentanone at a more advanced stage.

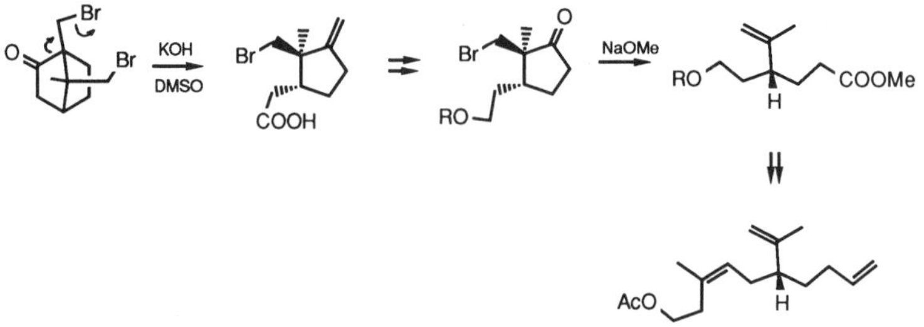

Other applications of the methylenecyclopentane intermediate include an assembly of *ent*-estrone methyl ether [Hutchinson, 1985] and a synthesis of (+)-ophiobolin-C [Rowley, 1989].

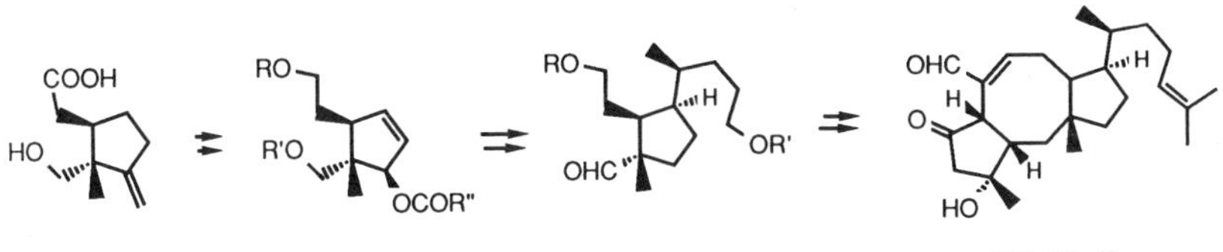

ophiobolin- C

10-Camphorsulfonic acid formally contains an **a2a** system. However, the contrapolarizability of the sulfone group can modify its reactivity to that of an **a2d** system. Alkali fusion of the camphorsulfonic acid leads to the monocyclic α-campholenic acid. The (−)-isomer of the latter compound has found service in syntheses of (−)-khusimone [H. Liu, 1979] and (−)-quadrone [H. Liu, 1988].

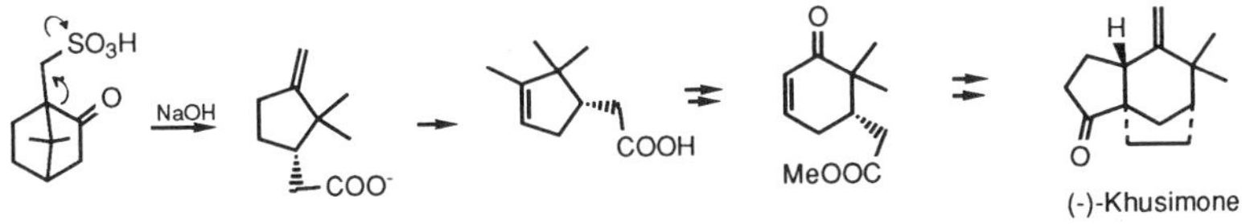

(-)-Khusimone

Fragmentation becomes a viable process even in simple aliphatic systems in which elimination pathways are blocked or strongly disfavored [Nerdel, 1960, 1965].

Oxidation of 3-pyrazolidinones with mercuric oxide leads to olefins [Kent, 1968]. The pyrazolenone intermediates are susceptible to fragmentation in the presence of a nucleophile (e.g. water).

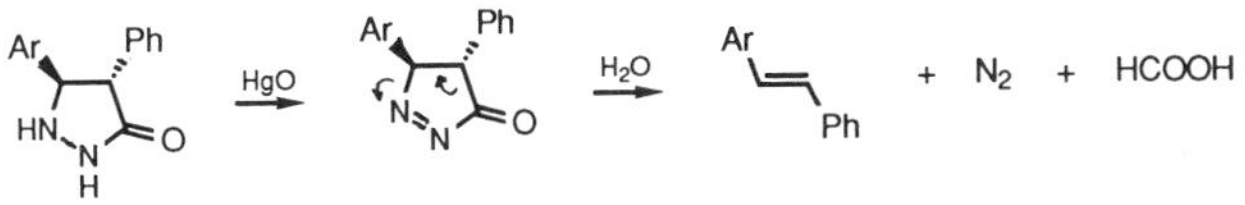

A synthetic precursor of *exo*-brevicomin is 6-nonyn-2-one. This compound can be prepared from dihydroresorcinol via *C*-ethylation, chlorination, and fragmentation of the chloro enone [Coke, 1977]. Effecting the last reaction involved treatment of the chloroenone with methyllithium and pyrolysis of the adduct.

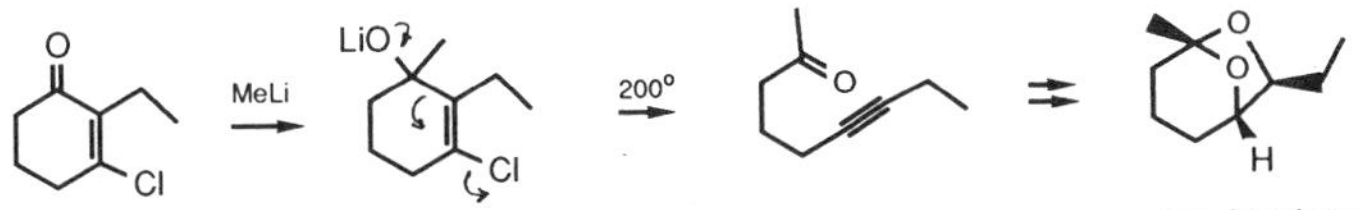

*exo*-brevicomin

The interconversion of trevoagenin-A and -B [Francisco, 1983], which are 21-epimers, in hot hydrochloric acid presumably proceeded with participation of the angular hydroxymethyl substituent, establishing an **a2d** array for fragmentation.

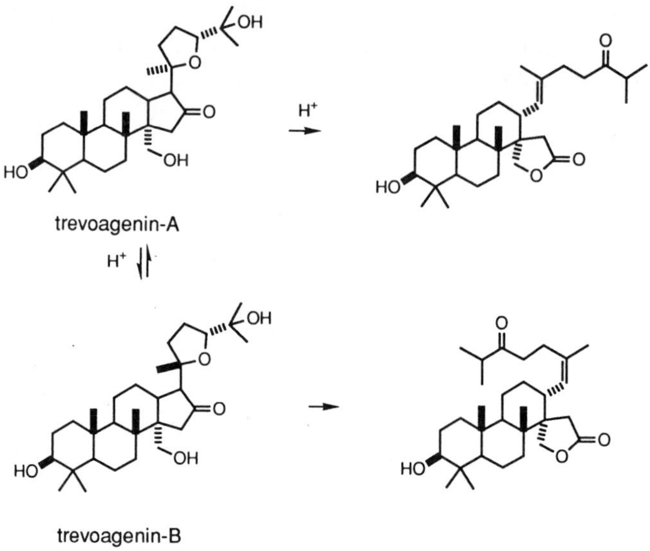

trevoagenin-A

trevoagenin-B

## 4.3. O=C—C—O—O SYSTEMS

The α-peroxy carbonyl system is highly prone to fragmentation which results in O—O bond cleavage. Thus, treatment of hanalpinol-9-one with boron trifluoride etherate led to alpinolide [Itokawa, 1984]. The isomerization probably involved heterolysis of the peroxy linkage under fragmentative participation of the carbonyl group, lactonization, and epimerization.

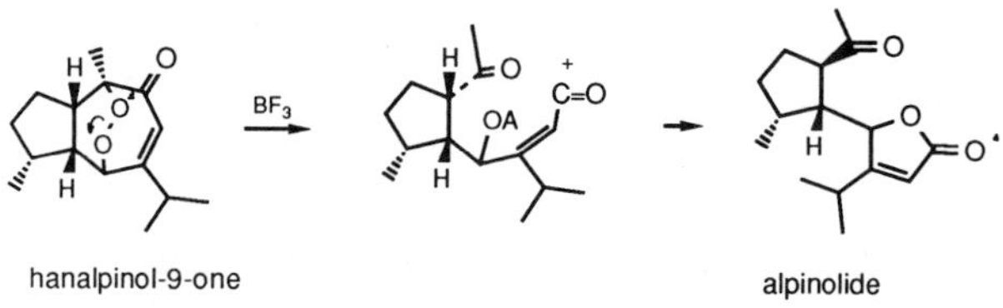

hanalpinol-9-one

alpinolide

Substantial amounts of methyl 2-benzoylphenylacetate were produced in the Favorskii rearrangement of diphenyl-α-halopropanones with sodium methoxide in methanol [Bordwell, 1970]. Cyclization of the oxyallyl zwitterion intermediate and tautomerization would furnish the indanone. Autoxidation and an **a2d** fragmentation followed.

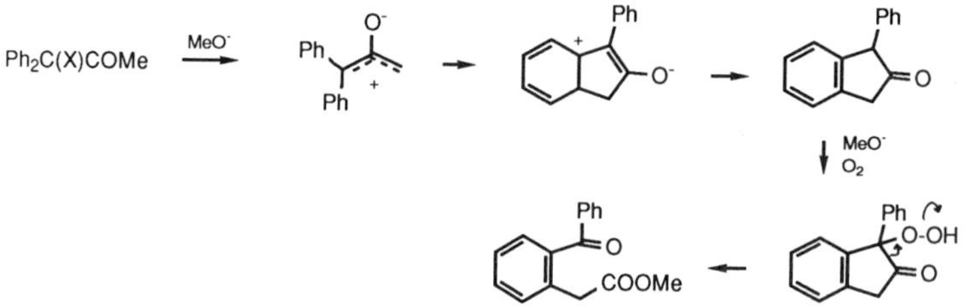

$Ph_2C(X)COMe$

Certain dihydroquinazolinones are obtainable from indole-1,2-carboximides via ozonation [Ishizumi, 1973]. Hydrolytic decomposition of the ozonides generates the 2-ureidophenones which undergo cyclization readily.

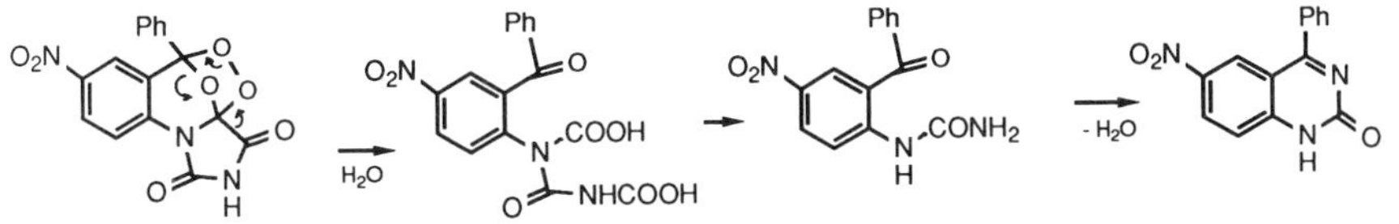

Benzylic autoxidation is extremely facile in 2-(2-pyridyl)-cyclopentanone [Houminer, 1985]. The hydroperoxide disintegrates in the presence of silica gel by two pathways, leading to a keto acid and an α-hydroxy ketone. Formation of the acid is a fragmentation process.

*N*-Benzoyl-3-piperidinone undergoes degradation on treatment with sodium methoxide in the presence of air [Schirmann, 1983]. The captodative carbon atom is an excellent site for free radical reactions and autoxidation opens a way for removal of that carbon.

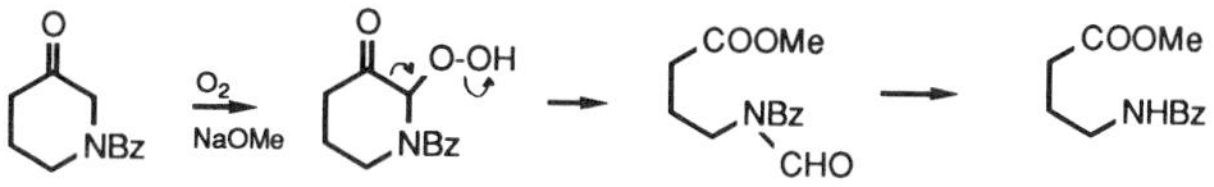

Degradation of 3-alkyl-1,2-cyclohexanediones by singlet oxygen [Utaka, 1985] involves the formation of bicyclic peroxy adducts. Decomposition of these adducts may be initiated by the existing hydroxyl group or by addition of solvent on the carbonyl. This reaction has been used in a synthesis of jasmone lactone [Utaka, 1986a].

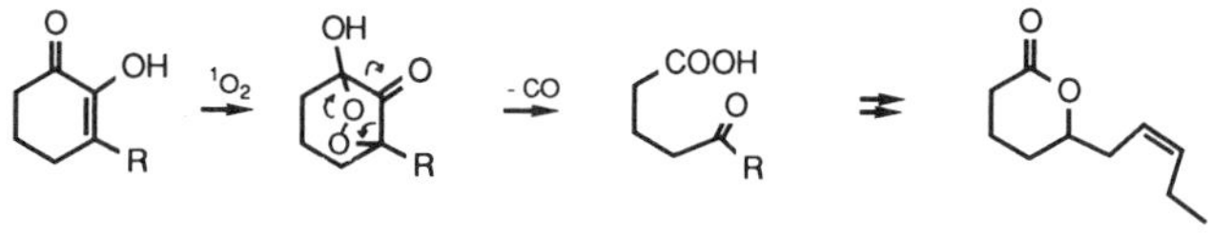

jasmone lactone

Interestingly, reaction of the cyclopentanediones led mainly to glutaric acid monoesters [Utaka, 1986b] because of steric effects. It seems that the carbonyl group in such adducts is more exposed.

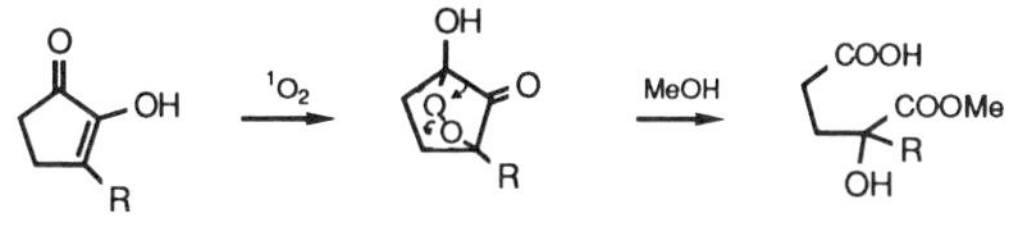

## 4.4. O=C—C—O—X AND O=C—C—N—X SYSTEMS

A novel access to 2-benzoylquinoline and 1-benzoylisoquinoline [Sliwa, 1987] involves fragmentation of the *N*-phenacyloxy iminium salts by treatment with a base. Intramolecular attack of the ylide on the carbonyl initiates the extrusion of acetone.

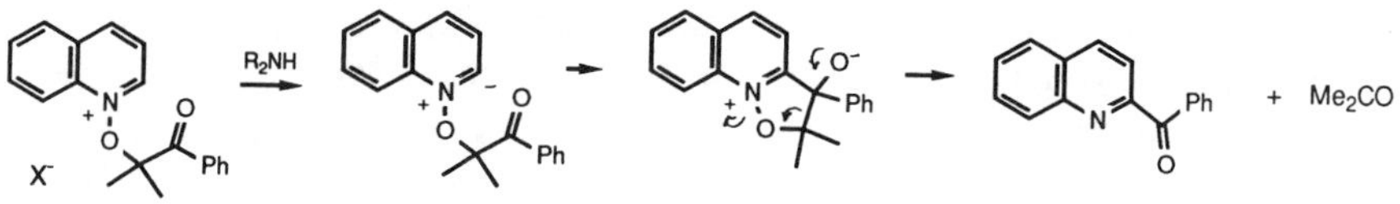

The Hofmann degradation of α-keto amides [Arcus, 1954] differs from the more conventional cases as fragmentation is possible in the presence of an acceptor three bonds away from the nucleofugal halide of the first intermediate.

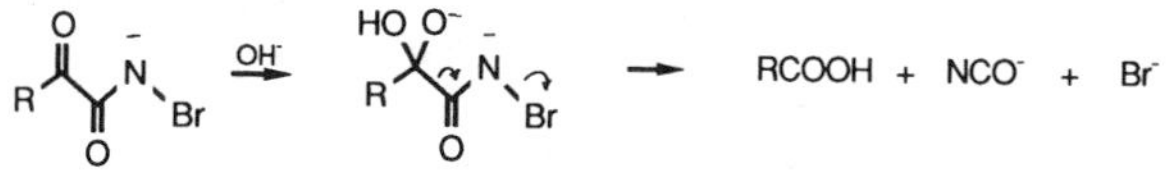

The oxidative fragmentation of quinuclidin-3-one with sodium hypochlorite [Dennis, 1967] started from *N*-chlorination which transformed the nitrogen atom into an acceptor site.

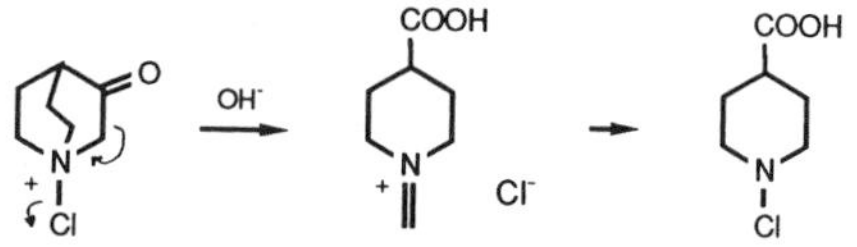

The reaction of 2-naphthol-1-sulfonic acid with diazonium salts in basic solutions to give phthalazine derivatives was described more than sixty years ago [Rowe, 1926; Jaecklin, 1971]. It is analogous to the Japp–Klingemann reaction.

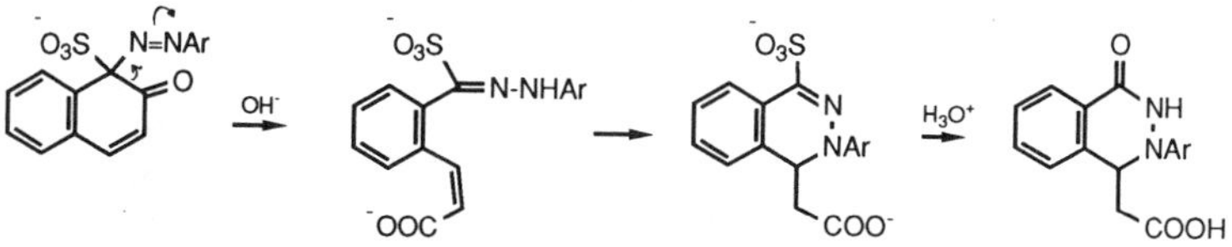

The Beckmann fragmentation is greatly facilitated by the presence of an α-acyl substituent. A classical example is the degradation of isonitrosostrychnine which gave a seco nitrile on reaction with thionyl chloride [H. Wieland, 1933]. This seco nitrile is a precursor of the well-known Wieland–Gumlich aldehyde.

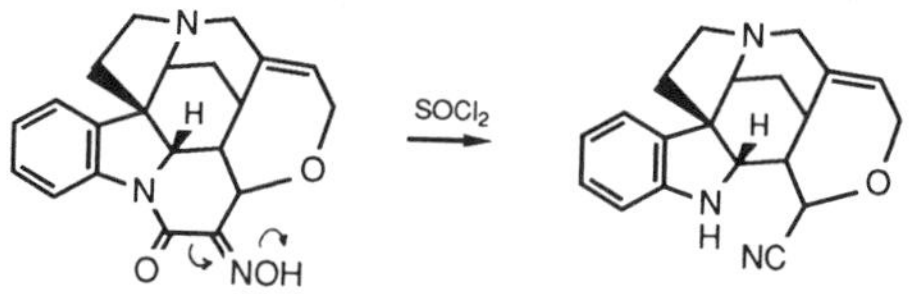

During structural investigations of gibberellic acid, the points of attachment of a $CH_2CO$ bridge to a hydrofluorene nucleus were determined by converting gibberic acid into a spirocyclic diketone and finally a substituted benzoic acid [Cross, 1958]. Gibberic acid is an acid hydrolysis product of gibberellic acid.

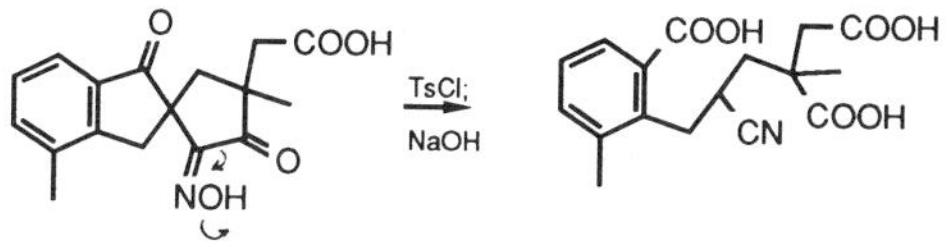

Various 2,4-bisoximino-3-keto steroids afford 2,3-seco-A-nor steroid 2,3-dinitriles on treatment with acetic anhydride in pyridine [Cava, 1966]. A double Beckmann fragmentation took place, although the order of the cleavage cannot be determined.

Symmetrical monoximes of 1,2,3-triketones fragment on reaction with cyanide ion [Crabtree, 1967].

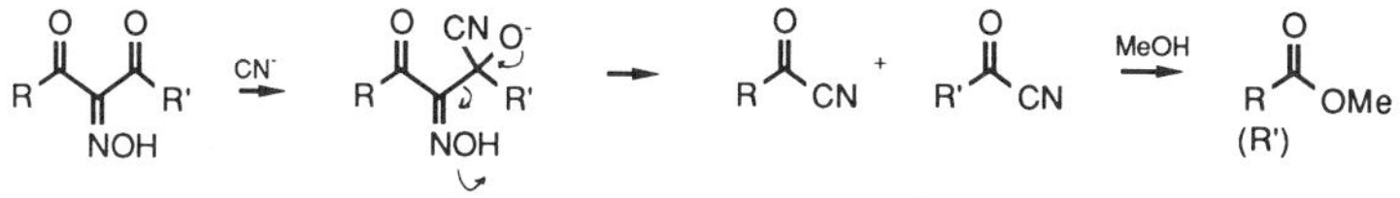

The thermal decomposition product of 4-methoxy-1,2-benzoquinone 1-(*O*-phenylsulfonyl)oxime in aqueous pyridine is $\gamma$-cyanomethyltetronic acid methyl ether [Fleming, 1972]. A Beckmann fragmentation was followed by an intramolecular Michael addition.

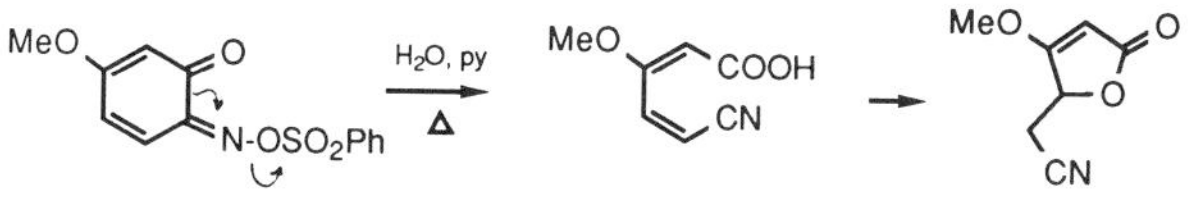

3-Acylisoxazoles are susceptible to degradation on treatment with a base [Borsche, 1912]. The fragmentation generates $\alpha$-cyanoketones and carboxylic acids or derivatives.

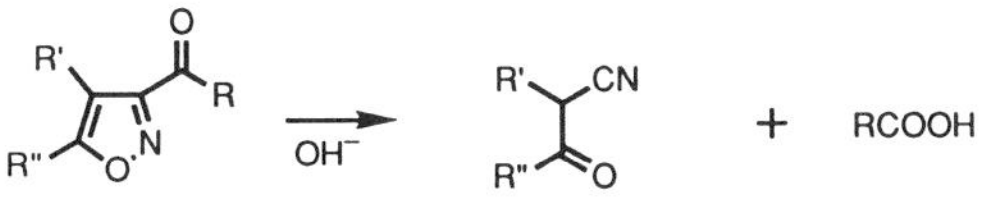

The cleavage of a ketone or aldehyde containing an $\alpha$-methylene group by nitrosylpentaamineruthenium(II) ion in the presence of alkali [Schug, 1979] apparently involved nitrosation and isomerization to a Ru-complexed monoxime of the $\alpha$-dicarbonyl compound. Intramolecular reaction provided an oxazetene which decomposed into a carboxylic acid and a nitrile.

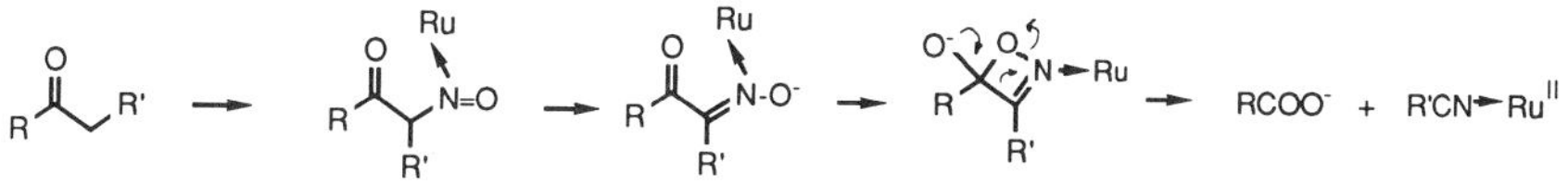

In the late stages of vitamin-$B_{12}$ biosynthesis deacetylation of corrin-8 following the ring contraction is logical [Scott, 1989]. The C—C bond cleavage is mediated by valence change at cobalt in the MeCO—C—N—Co(II) segment.

The Japp–Klingemann reaction has been discussed under the **a1a** fragmentations, although it might be classified as an **a2d** system. We should mention also its extensive use in synthesis of carbazole alkaloids such as murrayanine [Chakraborty, 1968].

Generally the cleavage of the central single bond of a conjugate system is energetically demanding. However, when the substrate is energized and the cleavage is accompanied by release of energy, such reaction would proceed smoothly. The zwittazido cleavage [H.W. Moore, 1979] is prototypical of this family of reactions, as the extrusion of dinitrogen from α-azido enones leading to an acylium ion and an α-cyano carbanion is a very favorable process. There is enormous synthetic potential in zwittazido cleavage.

One of the earlier examples of this cleavage is the conversion of 2-azidotropone to 2-cyanophenol [Hobson, 1967]. Many other ring contractions based on the same principle have been reported subsequently. These include the synthesis of 2-cyano-2-methyl-4-cyclopentene-1,3-diones [Weyler, 1973], the sulfonylimino derivatives and the tetrahydroquinoline-2,4-dione ring system from the azidoazepinediones [H.W. Moore, 1979], a pyrazolidinedione [Sasaki, 1973], an azacyclopentadiene *N*-oxide [Abramovitch, 1973], and 5α-cyano-4-keto A-norsteroids [E.M. Smith, 1974]. Thermolysis of 3-azido-4-phenylcyclobutenedione affords phenylcyanoketene [DeSelms, 1969; Schmidt, 1969].

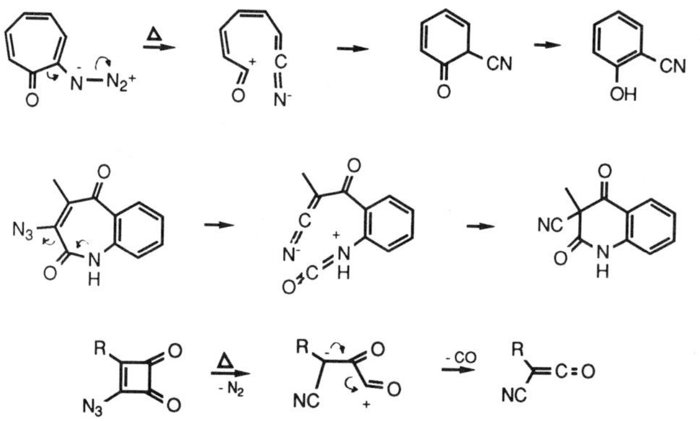

The cyanoketene synthesis is most efficiently achieved by heating 2,5-diazido-1,4-benzoquinones [Weyler, 1975]. The availability of such reactive ketenes has broadened the vista of cycloaddition approaches to synthesis of numerous ring systems [Dorsey, 1986; K. Chow, 1987].

## 4.5. N=C—C—C—X SYSTEMS

The preparation of unsaturated aldehydes by condensation of acetaldimine with ketones [Wittig, 1964] is complicated by side-reactions during the hydrolysis–dehydration step. Hydration of the imines furnishes the fragmentation-prone 1,3-diol systems.

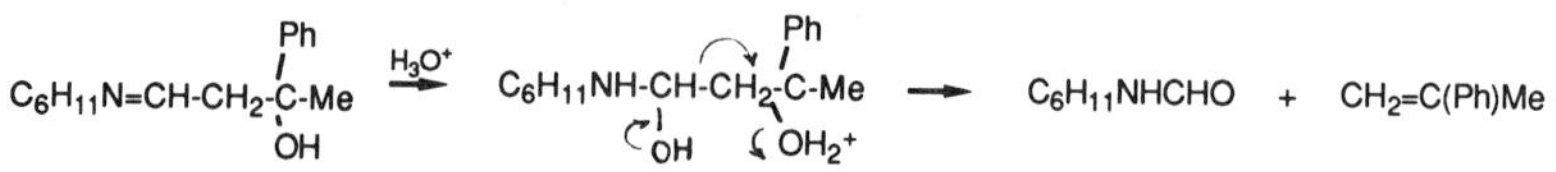

$\beta,\beta$-Disubstituted indolenines in which the substituent contains a nucleofugal group may fragment to give anilides [M.F. Bartlett, 1958].

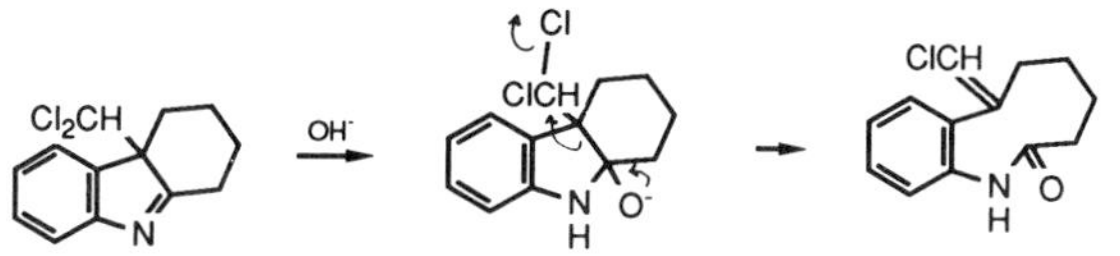

*N*-[(Dichloromethyl)alkylidene]anilines undergo Favorskii-type reactions [DeKimpe, 1974]. Unsaturated iminoesters are readily obtained.

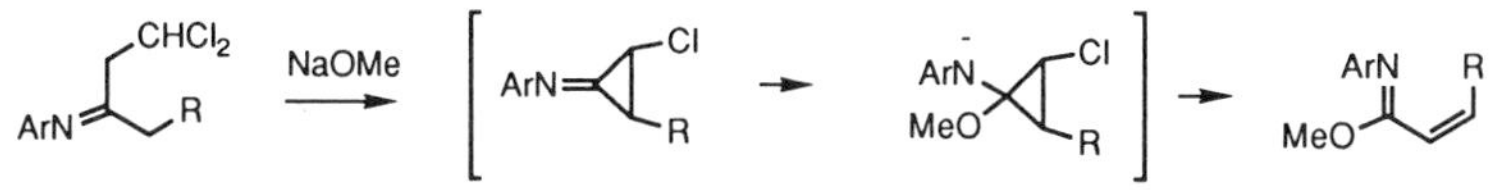

The reverse reaction of olefin nitrosochlorination can be effected by exposure of the adducts to alkoxides [Pritzkow, 1966]. As a synthetic method this is of little significance.

Pyridinium salts in which the *N*-substituent also carries a nucleofuge at the α-carbon are liable to undergo ring fission on reaction with hydroxide ion [Kuhn, 1968]. Fragmentation is directed inwards the imininium system.

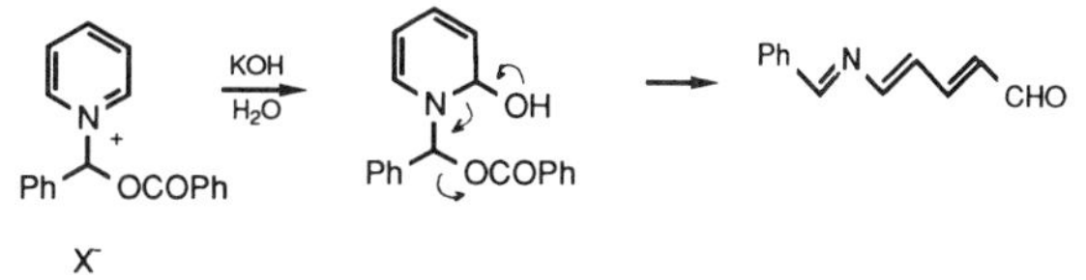

*N,N'*-Benzaldipyridinium salts decompose on heating with water. Since these salts are subject to deuteration via the ylides, a convenient preparative method for 1-deuteriobenzaldehyde (and other araldehydes) presents itself [Olofson, 1967]. The decomposition is initiated by hydration of one of the pyridinium moiety, the other heterocycles acts as nucleofugal group.

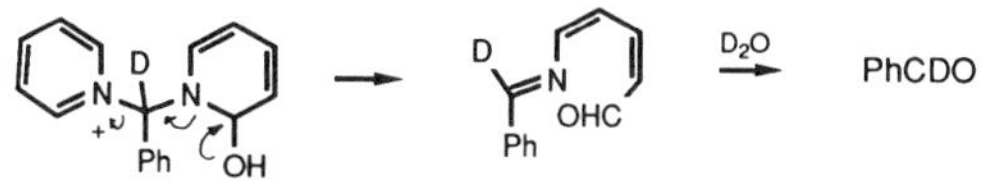

## 4.6. OTHER a2d SYSTEMS

The fragmentable systems described in this section contain acceptor centers which may either be a carbon or an heteroatom. The quaternary carbon of a trityl group is such an acceptor, and an example of the A—C—C—X fragmentation involving it is the formation of a 2-cyanomethyltetrahydrofuran from a 5-alkenyl-3-tritylisoxazoline [Kurth, 1989].

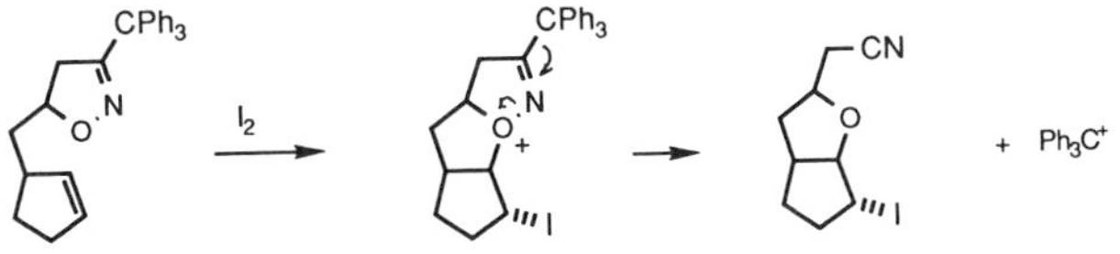

An unusual fragmentation process is that which generates a carbamoyl anion during cleavage of an α-acyloxycarboxamide [Kozikowski, 1990].

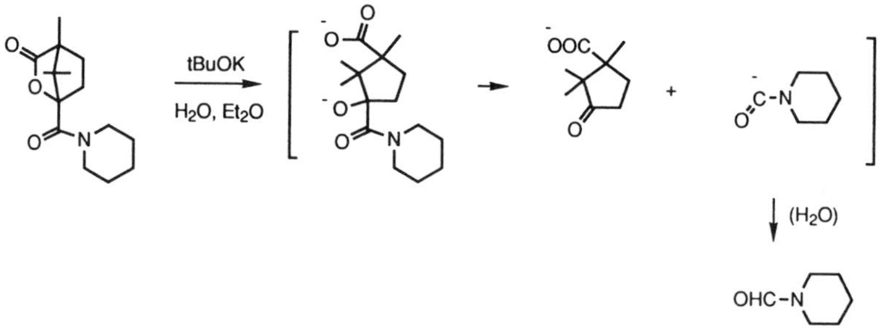

The borane intermediate derived from 2,3-dicarbomethoxy-2,3-diazabicyclo-[2.2.1]hept-5-ene can be transformed into either the secondary alcohol or a cyclopentene [Allred, 1966] by selecting the proper nucleophile in the reaction media. Hydroxide ion strongly favors the formation of the fragmented product.

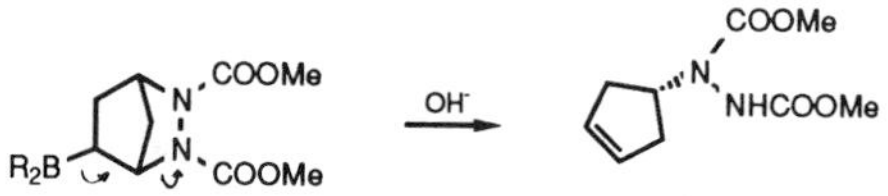

Quantitative generation of 1,2-dihydronaphthalen-1-ol from the hydroboration of 1,4-epoxy-1,4-dihydronaphthalene with 9-borabicyclo[3.3.1]nonane [H.C. Brown, 1985] is due to a facile $C \rightarrow O$ transfer of the boryl residue in the organoborane product.

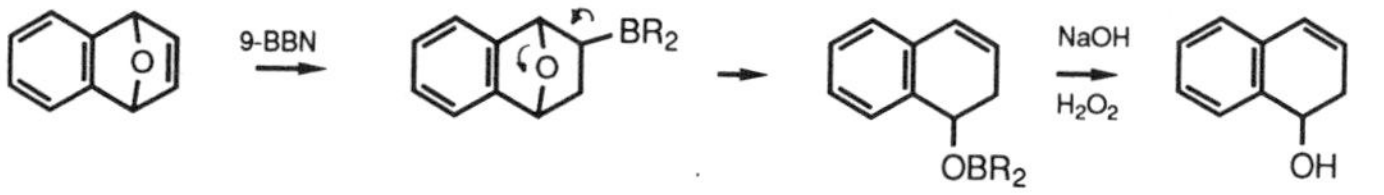

The decomposition of 2-chloroethylarsenic dichloride in the presence of sodium hydroxide [Sommer, 1970] apparently proceeds by hydroxide ion attack at the arsenic atom to initiate a fragmentation.

The fluoride ion-promoted elimination of β-silylalkyl halides is a mild method for synthesis of strained (transient) olefins [Chan, 1977]. A highlight of its application is the generation of 1,2,3-cycloheptatriene [Zoch, 1983].

In the fragmentation of 2-(trimethylsilyl)ethylsulfonamides [Weinreb, 1986] the sulfonyl group acts as a donor and the silicon atom as an acceptor. There is reduction at sulfur when the sulfonyl function is extruded as sulfur dioxide.

The convenient synthesis of *erythro*-1,3-diols from *Z*-allyl alcohols which are also silylated at the other allylic position provides an access to 1, *endo*-3-

dimethyl-2,9-dioxabicyclo[3.3.1]nonane [Mohr, 1987]. The latter compound is produced by Norway spruce upon attack by the ambrosia beetle.

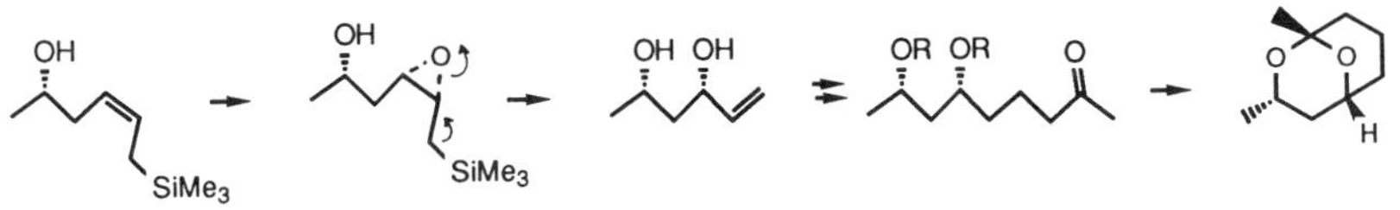

*N*,*N*-Dimethyl-2*Z*,5*E*-heptadienylamine is available from a 2-(1-silylethyl)-pyridine via reduction and fluoride ion-promoted fragmentation [Bac, 1982]. This amine is a useful precursor of a lepidoptera pheromone.

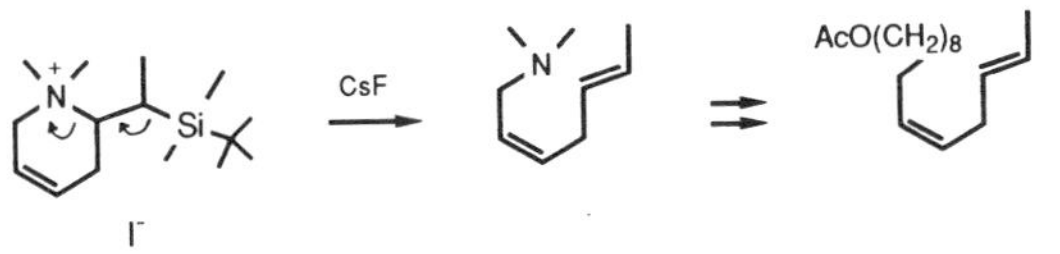

1,3-Desilylhydroxylation of 3-silyl-2-phenylsulfenylalkanols proceeds under acid catalysis [Brownbridge, 1976]. The sulfur substituent is shifted to the originally hydroxylated carbon atom and a double bond is created in its wake. Formation of the episulfonium intermediates initiates fragmentation.

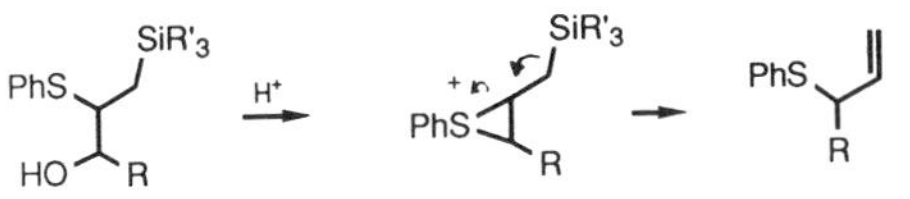

Malaldehydic acid has been obtained from 2-trimethylsilylfuran by reaction with singlet oxygen [W. Adam, 1981]. The endoperoxy adduct decomposes by elimination with the silyl group as an acceptor.

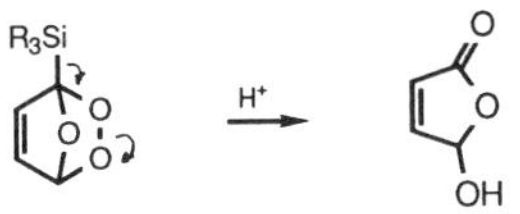

A peptide synthesis is based on reaction of amine with *N*-tosyl-5-oxazolidones [Dane, 1959].

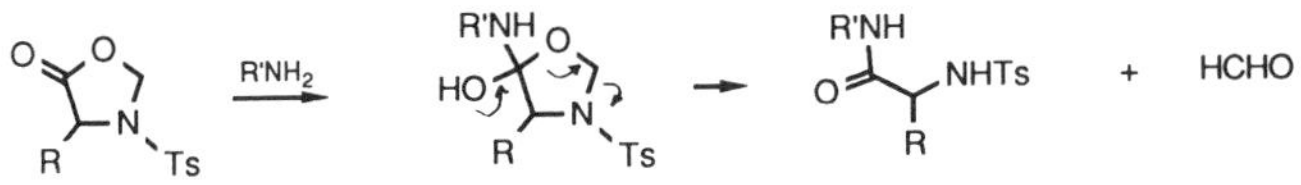

Reaction of berberine with acetic anhydride gave substituted 1-(α-naphthyl)-3,4-dihydroisoquinolines [Shamma, 1977]. Introduction of a ketene element to the meso bridge of the alkaloid via an iminium–enamine shuffle was followed by a retro-Michael reaction. This process seems to be quite general for isoquinolinium salts.

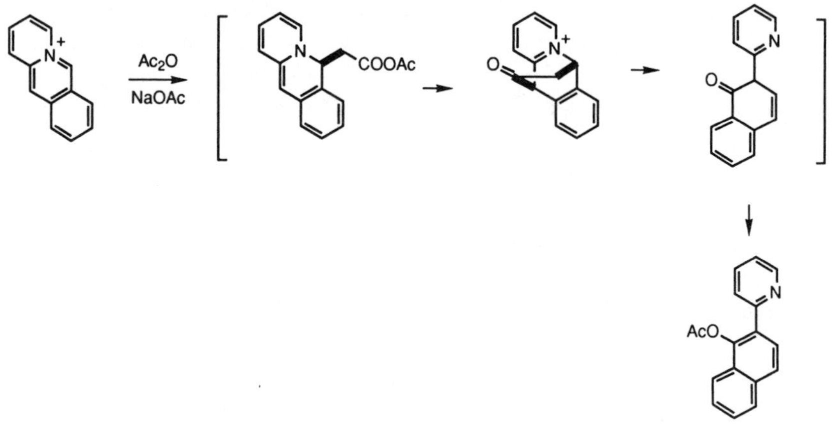

N+ — $Ac_2O$ / NaOAc → [ N, COOAc → O, N+ → O, N ] → AcO, N

The reductive elimination of α-nitroketone dithioketals by exposure to a combination of a thiol and a Lewis acid such as $AlCl_3$ shows a HSAB-specific process [Node, 1982]. Attack on a sulfur atom of the thioketal function by the thiol reagent is in consort with the departure of the nitro group which is coordinated by the Lewis acid.

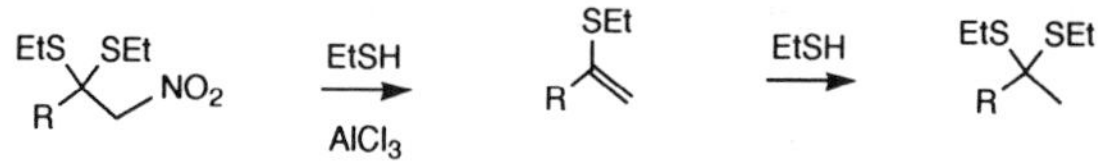

EtS SEt, R, $NO_2$ — EtSH / $AlCl_3$ → SEt, R — EtSH → EtS SEt, R

# 5

# FRAGMENTATION OF HIGHER a|n|x SYSTEMS

## 5.1. a3a SYSTEMS

Many strained polycycles undergo isomerization on treatment with heavy metal ions [Paquette, 1973]. For example, secocubane furnished 1,5-cyclooctadiene, among other products, on exposure to silver tetrafluoroborate [Wrister, 1970]. Similarly, diademane was converted into triquinacene [deMeijere, 1974]. Here a strained cyclobutane C—C bond undergoes heterolysis on coordination with the metal ion and thereby an **a3a** system is created. Subsequent demetalation results in two double bonds.

$AgBF_4$ Ag

The vinylogous Wolff rearrangement of $\beta',\gamma'$unsaturated $\alpha$-diazoketones [Smith, 1976] may actually involve cyclization of copper carbenoids and subsequent fragmentation [Alonso, 1987]. As shown in the example, the scission of the C—Cu bond is coupled to the neutralization of the carbenium ion.

$N_2$ O $CuSO_4$ Cu O + O Cu MeOH COOMe C O

## 5.2. a3a=d SYSTEMS

δ-Diketones and ketoesters containing at least one α- or γ-hydrogen atom belong to the **a3a=d** family. Unless a compound of this class suffers from steric inhibition of deprotonation its fragmentation can generally be induced by a base. Thus, an interesting synthesis of a bicyclo[3.3.1]nonenedicarboxylic acid [Clarke, 1987] that served as a chlorismate mutase transition state inhibitor was via a Diels–Alder reaction and treatment of the tricyclo keto diester with an alkoxide base. The smoothness of the ring cleavage must be due to strain relief and excellent orbital alignment.

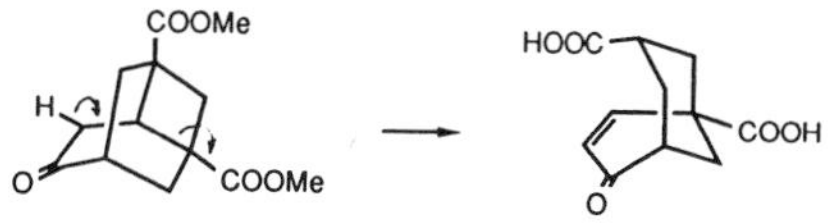

In the classic synthesis of longifolene [Corey, 1964b] some tactical problems were encountered whhich were related to the Michael cyclization leading to a bridged tricycle. The low yield of the desired diketone was due to the reversibility of the reaction and the presence of a new circuit for retro-Michael degradation. By-products with the bicyclo[2.2.1]heptane skeleton have been identified.

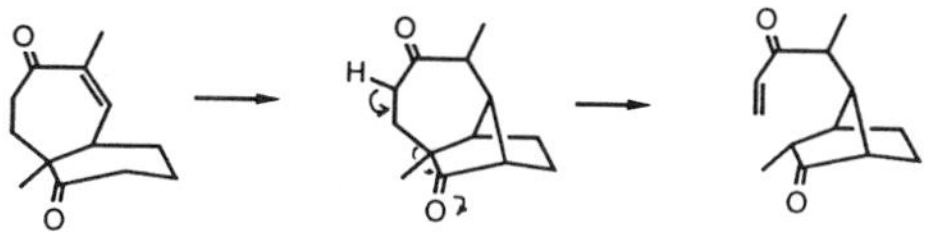

A half-cage δ-diketone relieved ring strain by undergoing ring cleavage [Marchand, 1990] on acid treatment. The γ-diketone possessing the same skeleton remained unchanged under the same conditions.

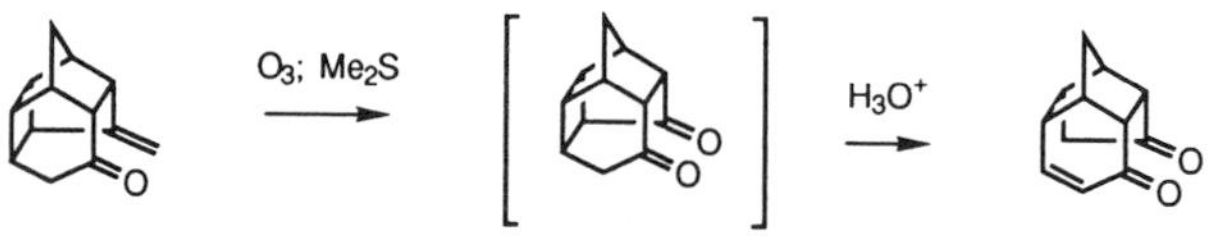

The Diels–Alder adduct of 2,6-dimethyl-1,4-benzoquinone with a 4-methyl-2,4-pentadienyl ether underwent dehydrogenation in situ (by way of the quinone) and this product was transformed into a hydroxynaphthaldehyde on acid hydrolysis [Kienzle, 1985].

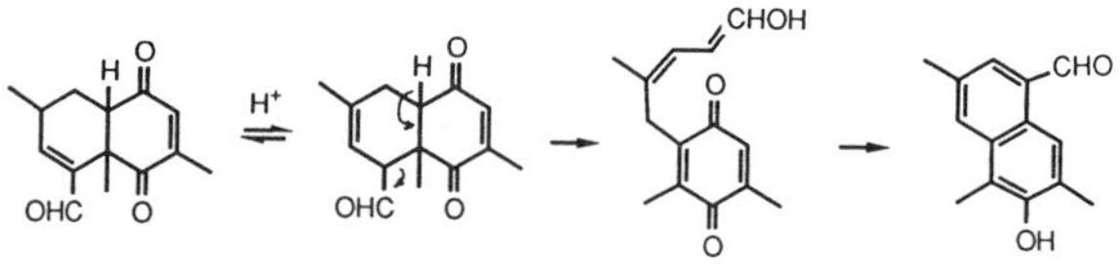

An alkylation route to the triquinacene skeleton failed. Instead, an abnormal Favorskii rearrangement occurred by which a conjugate enedione system emerged.

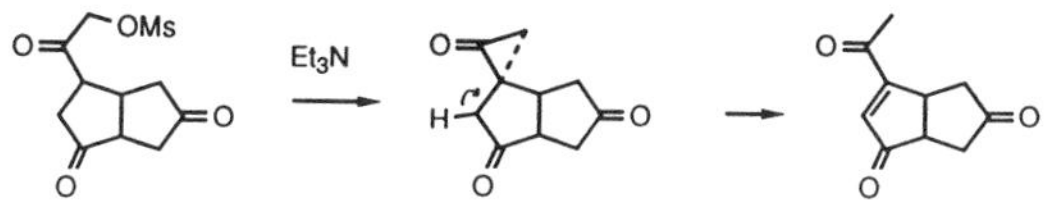

Epoxidation of (15*R*)-prostaglandin-A esters in the presence of larger amounts of a base resulted in the cleavage of the five-membered ring [Schneider, 1978]. The seco $\beta$-ketoaldehyde underwent deformylation in situ.

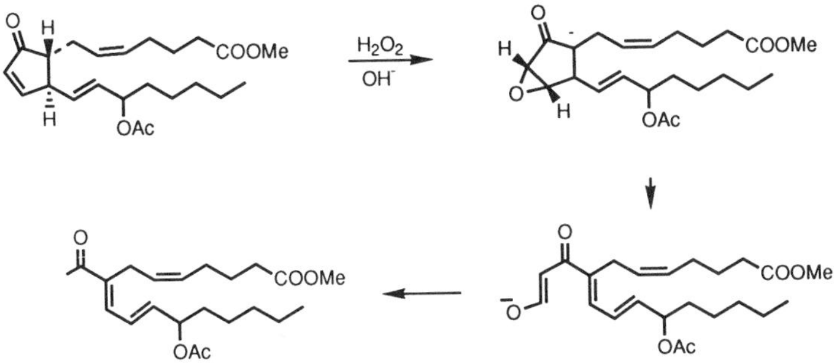

Two carbon frameworks related to sesquiterpenes have been unraveled from a common precursor [Deslongchamps, 1977]. Kinetically controlled conditions led to a ketonorcedrene and prolonged base treatment gave a norpatchoulane derivative.

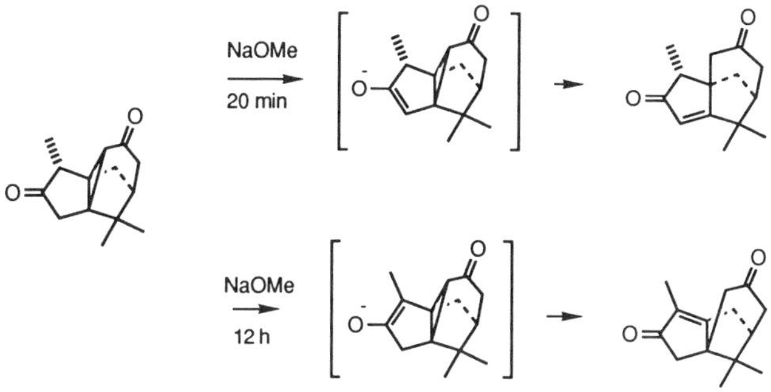

An excellent building block for brefeldin-A synthesis [Corey, 1976b] is a symmetrical oxobicyclo[3.1.0]hexanedicarboxylic ester whose retro-Michael ring opening provided an enone. A similar step was involved in the preparation of a cyclohexenone in a synthesis of trichodermin [Colvin, 1973].

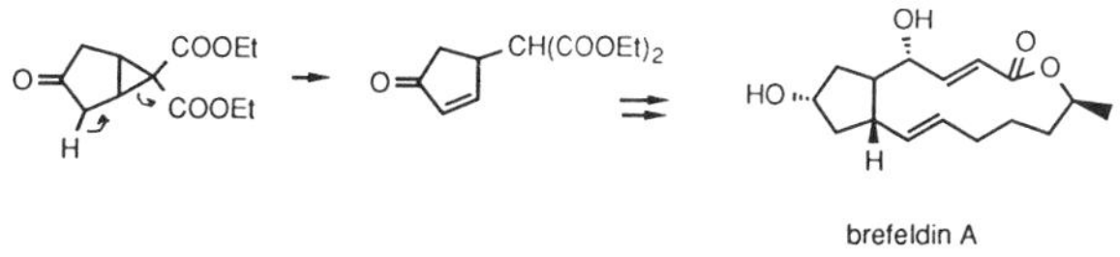

brefeldin A

Reversible formation of a spirocyclic nitro diketone and an anthraquinone was observed during synthesis of daunomycinone [Ashcroft, 1984]. As expected, the nitro group functioned in exactly the same manner as a carbonyl.

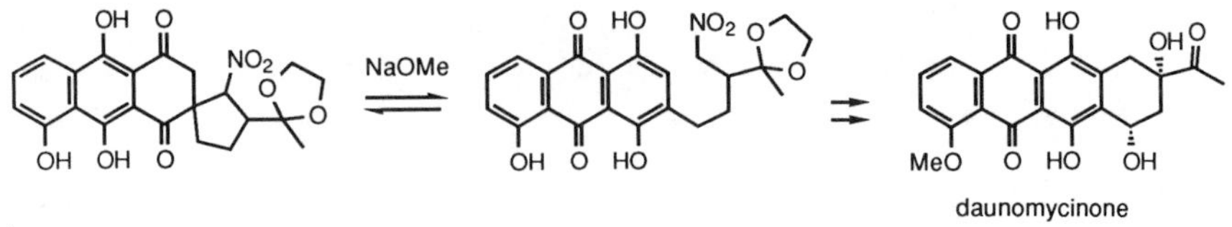

daunomycinone

The major biosynthetic processes for the construction of polyketide natural products are the aldol and Claisen condensations. Such compounds when retaining oxygen functionalities can often be degraded by acids or bases. Thus, alkali treatment of deoxypurpurogenone [Roberts, 1971] gave rise to formic acid and numerous pigments including 1,4-dihydroxy-2-methylanthraquinone. The complex network of the natural compound was disentangled by two retro-Claisen reactions which was followed by aromatization.

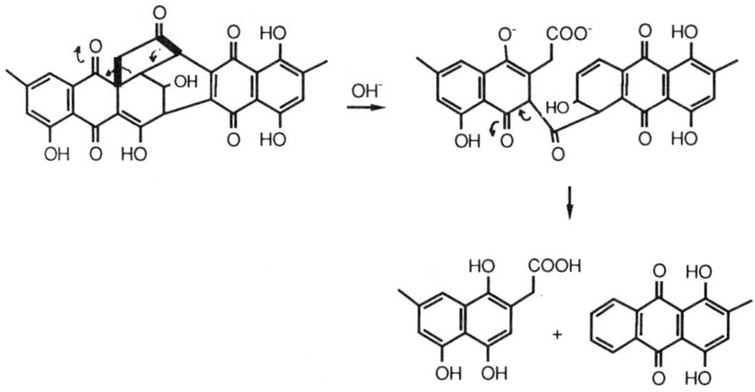

Vigorous saponification of gambogic acid [Ahmad, 1966] yielded many small molecules. The more interesting pieces arose from the bridged ring portion. Thus, retro-Claisen degradation enabled opening of the 4-chromanone as the retro-Michael process was no longer forbidden (Bredt's Rule). The phloroglucinol system (in the keto form) could act as an electron sink and deacylation occurred under the drastic conditions.

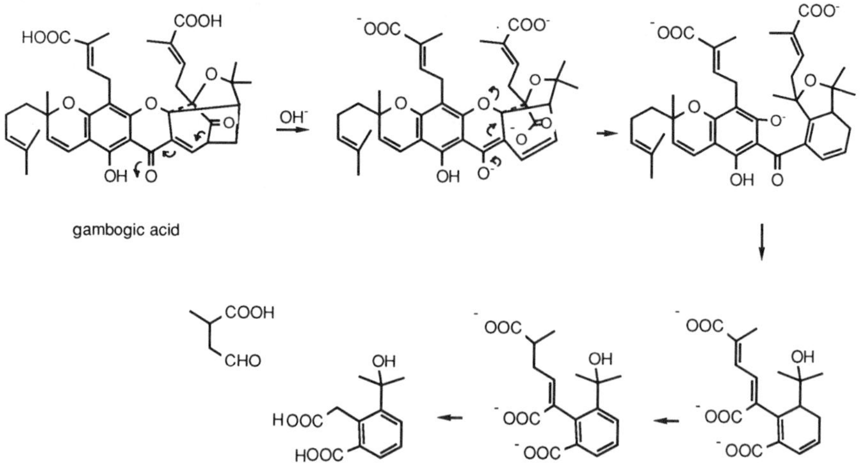

gambogic acid

The relationship of the ketone group of limonin with the α,β-epoxy-δ-lactone subunit was established [Arigoni, 1960] by a fragmentation reaction on the corresponding α,β-unsaturated lactone.

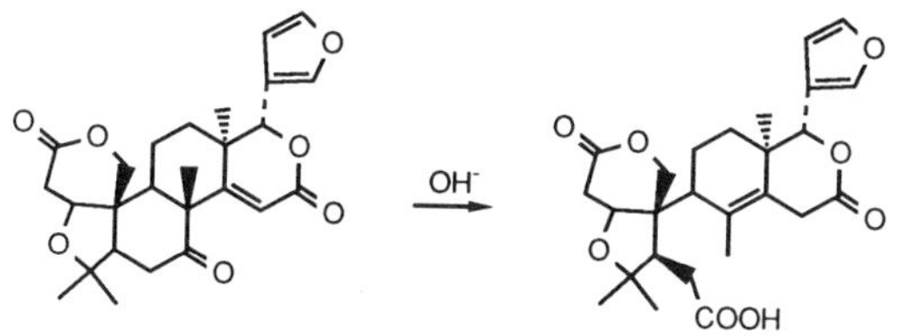

The AB-ring functionalities of withaferin-A were deduced by oxidation of the steroid to an enedione [Lavie, 1966] and acid treatment that led to loss of formaldehyde and water.

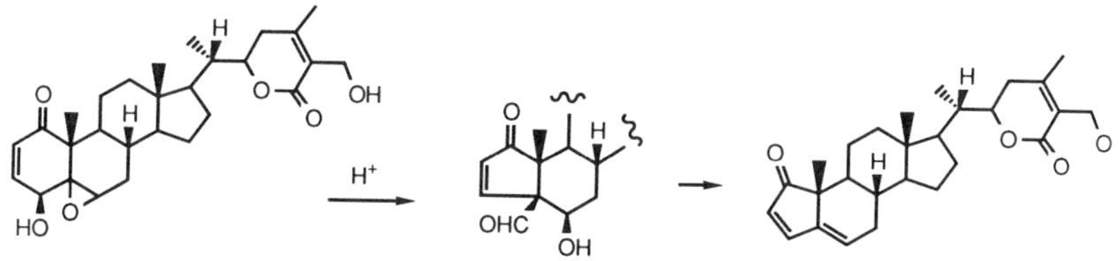

An interesting strategy for synthesis of chlorophyll-*a* [Woodward, 1960] pertained to closure of the macrocycle which must provide a γ-substituent to participate in the formation of the cyclopentanone moiety. Because of interfering reactivity, functional groups with one or two carbon atoms were far less suitable to be installed in that position. On the other hand, a succinoyl group attached to the D-ring fulfilled all the requirements and it could serve as a pivot in the redox sequence to establish the chlorin nucleus. The two terminal carbon atoms of the succinoyl group were destined to leave in the form of an oxalate residue.

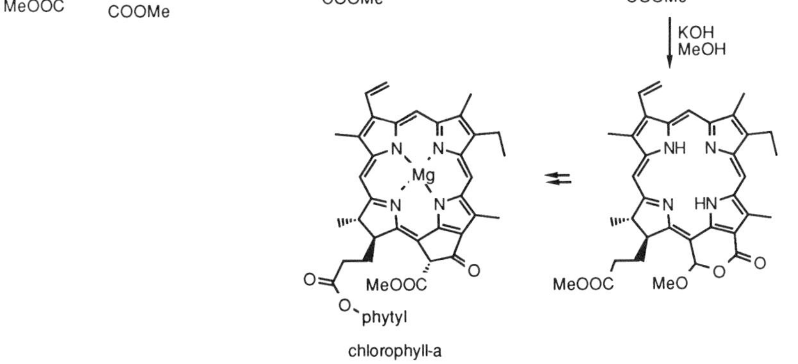

chlorophyll-a

Humulone was converted into humulinic acid [Howard, 1957] via an acyloin rearrangement and retro-Claisen reaction.

humulone

humulinic acid

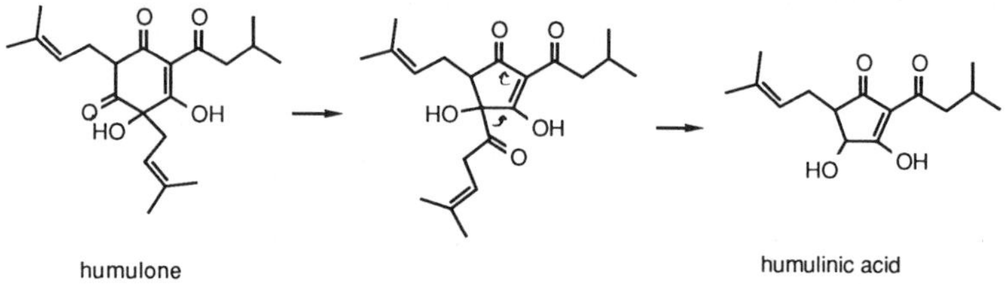

The isomerization of usnic acid to usnolic acid [Dean, 1957] by sulfuric acid involved ring opening.

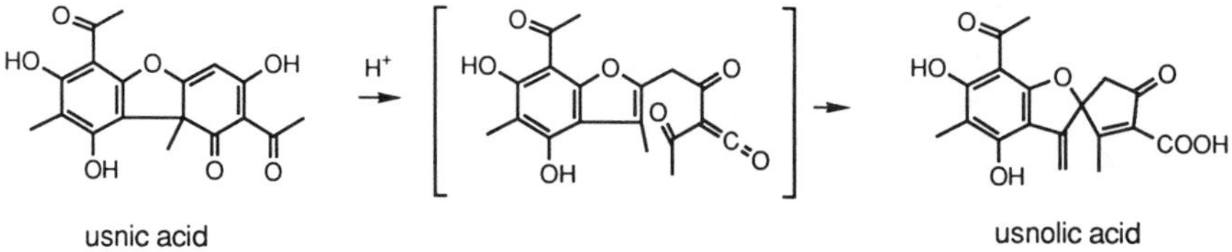

usnic acid

usnolic acid

Guaiol has been synthesized [G.L. Buchanan, 1973] from a hydrazulenone derivative which was bridged by a carbonyl group. Dismantling of this bridge was aided by the cyclopentenone subunit. In other words, this reaction and those described in the previous paragraphs are vinylogous retro-Claisen reactions.

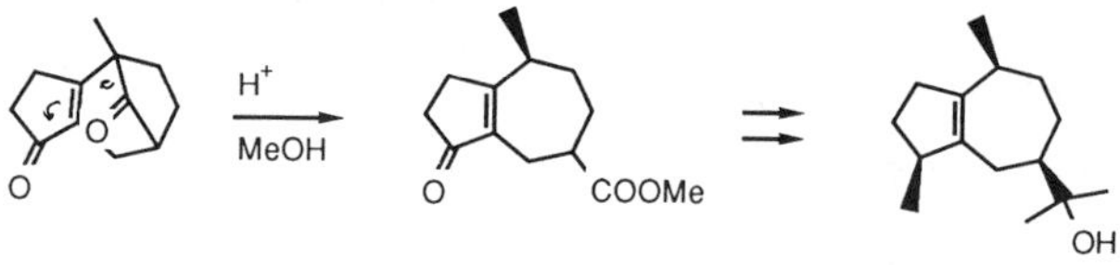

guaiol

Conversion of $\alpha,\beta$-epoxy ketones to $\beta$-hydroxy ketones may be effected by reaction with a sodium phenylseleno(triethoxy)borate complex [Miyashita, 1988] which is generated from diphenyl diselenide and sodium borohydride. However, uncomplexed sodium phenylselenide in ethanol causes deoxygenation. When this deoxygenation was attempted on a decalindione epoxide in which the two ketone group is 1,5-related, an in situ fragmentation occurred.

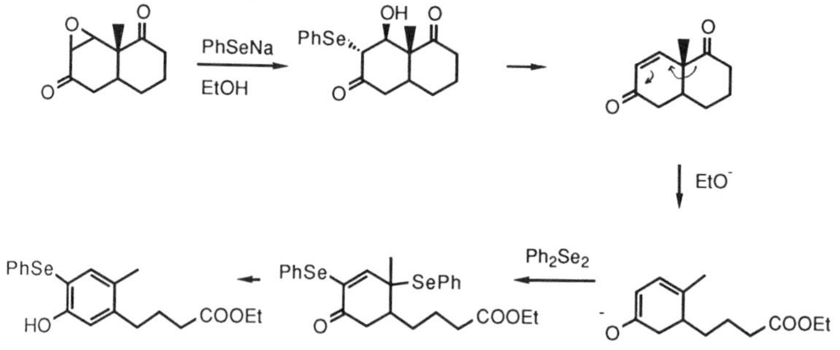

The dienone-phenol rearrangement of (2*H*,8*aH*)-3,4-dihydro-3,3,8a-trimethylnaphthalene-1,6-dione [Waring, 1985] may very likely involve a fragmentation step. This fragmentation was clearly shown in the reaction of an analogous ketol benzoate [Banerjee, 1985].

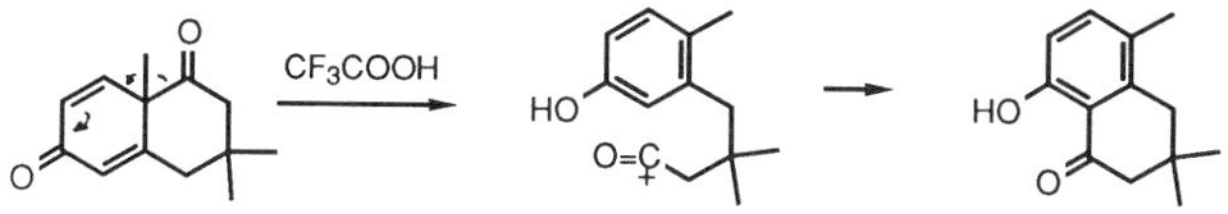

A transannular cyclization occurred when trimethylbrazilone was exposed to acid [Robinson, 1958]. Fragmentation of the intermediate was inevitable as it was essential for rearomatization of the resorcinyl moiety.

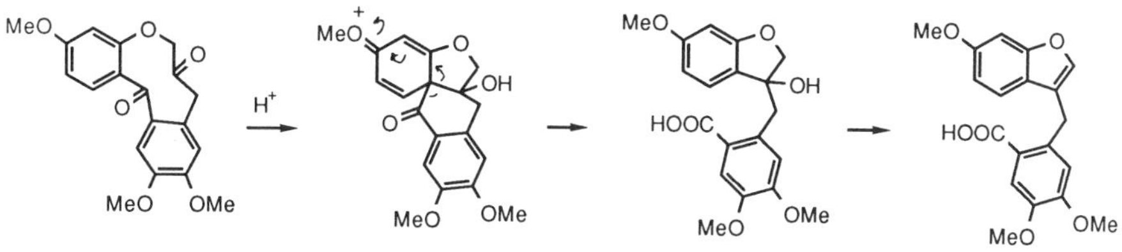

Reaction of the isophorone–diketene head-to-head photoadduct with aniline resulted in a bicyclo[3.3.1]nonane derivative [T. Kato, 1978b]. Opening of the $\beta$-lactone ring triggered fragmentation of the fused cyclobutane. Intramolecular aldolization and enamine formation completed the transformation.

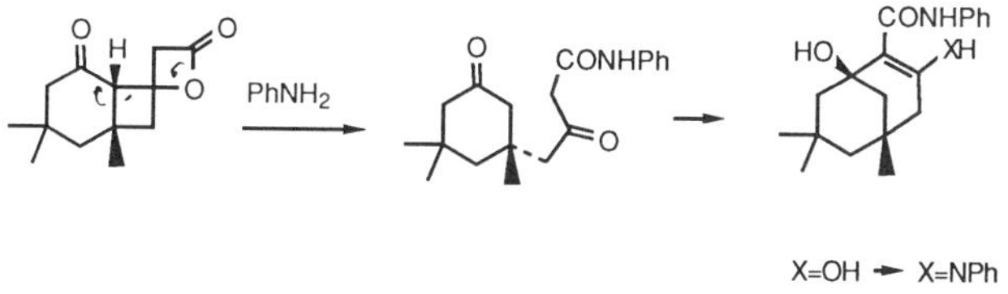

Oxidative degradation of the purple pigment obtained from methyl 2-(2-oxindolin-3-yl)glyoxylate and acetic anhydride led to a spirocyclic lactone [Ballantine, 1977]. The lactone underwent a deep-seated rearrangement on reaction with sodium methhoxide, resulting in a 2,4-dihydroxyquinoline derivative.

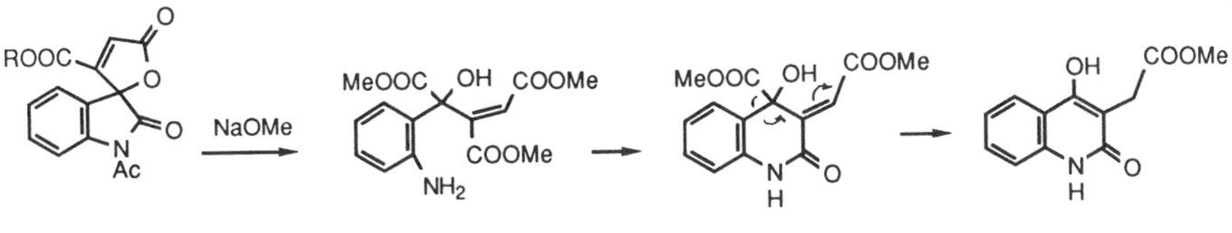

*N*,*N*-Dimethyl furano[4,5-*b*]quinoxaline-3-carboxamide is sensitive to various acidic and basic reagents [Kurosawa, 1981]. Even aqueous alcohols are capable of degrading the furan moiety of the molecule.

During condensation of the same furanoquinoxaline with cyanoacetic ester there was a destruction of the furan ring while spiroannulation of a cyclobutanone at the quinoxaline nucleus took place [Kurosawa, 1984]. The reaction is very unusual as the proposed intermediate could have cyclized to give a spirocyclic system containing two six-membered rings.

The retro-Michael reaction in the last stage of the reaction is a **3a=d** fragmentation.

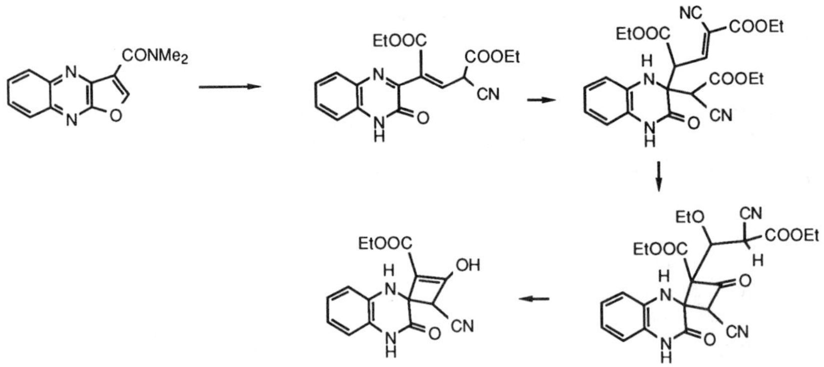

Pyrolysis of hypaphorine gave rise to many products, and among these is a dimeric substance with a 1-(indol-3-yl)-1,2,3,4-tetrahydrocarbazole-4-carboxylic acid structure [Raverty, 1977].

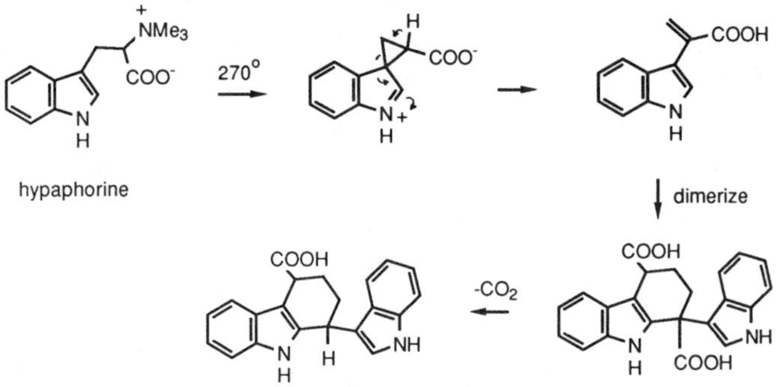

The photocycloadducts of allylsilanes with naphthoquinones are cleaved readily in the presence of Lewis acid catalysts [Ochiai, 1981]. Thus, an indirect allylation method has evolved. The adduct from a symmetrical disilyl isobutene underwent fragmentation and cyclization. The silyl group is the acceptor end of the **a3a=d** system.

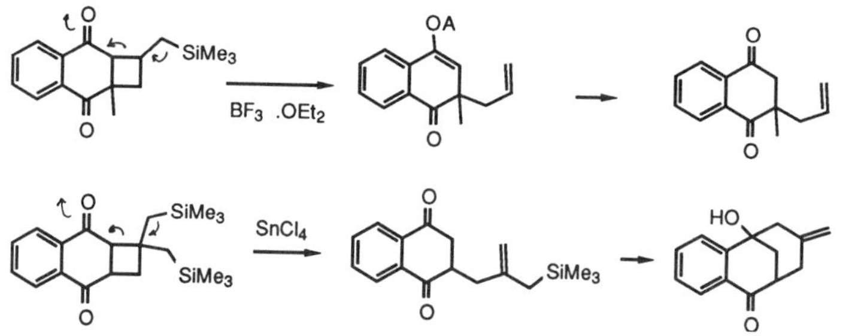

A unique feature of an aphidicolin synthesis [Ireland, 1984] based on spiroannulation via a Claisen rearrangement is the selective cleavage of a pentacarbocyclic intermediate. The fragmentation was directed by the silicon and the carbonyl group. The intricate ring system was formed by intramolecular ketene–alkene cycloaddition.

SiMe3 SiO2 H H H H HO HO OH HO HO

aphidicolin

A total synthesis of brefeldin-A [Corey, 1976b] started from a bicyclo[3.1.0]-hexene *gem*-diester. Hydroboration gave the desired hydroxy diester but a side-product is the cyclopent-2-en-1-ylmalonic ester. The latter compound arose from the regioisomeric borane which is fragmentable because it embodies an **a3a=d** array.

COOEt COOEt BH3; COOEt COOEt + COOEt COOEt HOOH, NaOH COOEt COOEt HO COOEt COOEt brefeldin A

A route to 7,12-secoishwaran-12-ol was based on an intramolecular 1,3-dipolar cycloaddition to form the bridged ring system [Funk, 1983]. The required allyl substituent at the angular position was introduced by an allene–enone photocycloaddition which was followed by conversion of the exocyclic methylene group into a primary bromide and reductive cleavage. As mentioned in Chapter 1, redox processes always involve contrapolarization and consequently the halogen atom in a $\gamma$-haloketone becomes an acceptor toward external electron sources. The **a3a=d** system is one that fragments readily.

O Br O Li NH3 O HO

Fragmentation of $\gamma$-haloketones under reductive conditions is quite general, and it is particularly facile when the functional groups are incorporated in a strained environment, as the above example shows. In a synthesis of α-acoradiene [Oppolzer, 1983] an allylic chloride was employed as an intramolecular photoaddend. The tricyclic adduct turned out to be an ideal precursor as expected.

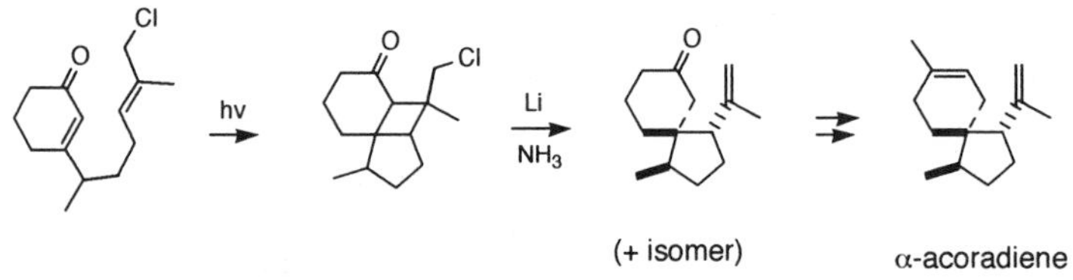

(+ isomer)

α-acoradiene

Certain 1,4-diketones undergo reductive C—C bond cleavage. For example, the head-to-head photo adduct of a 2-substituted 2-cyclopentenone with 1-chloro-1-octen-3-one proved to be a useful intermediate for 11-deoxyprostaglandin-$E_1$ [Bagli, 1972], owing to its susceptibility to reductive fragmentation.

π-Bromocamphor is subject to fragmentation to give dihydrocarvone [Hamon, 1975b]. Similar reduction was also conducted during a synthesis of iceane [Hamon, 1975a], and an application to natural product synthesis concerns with (−)-furodysinin [Richou, 1989].

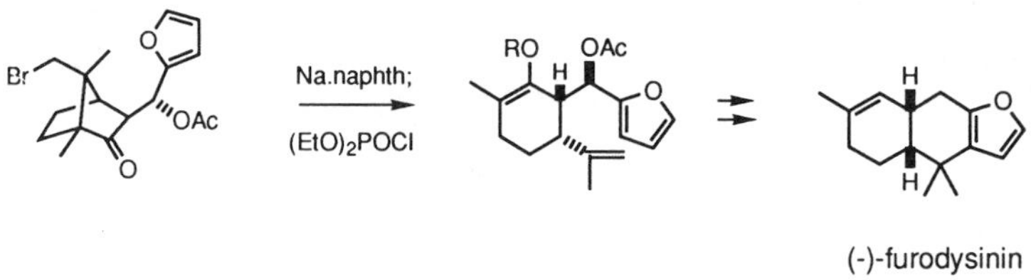

(-)-furodysinin

Sulfenyl groups are also capable of contrapolarization. Thus, reductive fragmentation of the [2 + 2]cycloadduct of 1,1-dimethoxyethene with a γ-phenylsulfenylated bridgehead enone has been demonstrated [G. Kraus, 1987b]. The bicyclo[5.3.1]undecene nucleus generated is a structural subunit present in the taxane diterpenes.

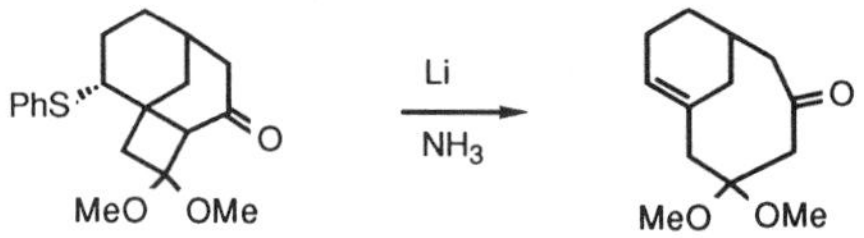

Several fascinating polycyclic substances have been obtained by intramolecular photocycloaddition of two enone moieties followed by reduction. Since the central C—C bond of such diketone photoadducts is incorporated in a cyclobutane, the reduction requires mild conditions (e.g. Zn, HOAc). Peristylane [Eaton, 1977], $C_{16}$-hexaquinacene [Paquette, 1978] and certain caged compounds [Mehta, 1985] represent only a few examples that have been efficiently synthesized using such a protocol.

COOMe
COOMe
Zn
HOAc
O
O
O
O
$C_{16}$-hexaquinacene

Certain 1,4-cyclohexanediones also undergo C—C bond cleavage during Clemmensen reduction [J.G.S.C. Buchanan, 1967; 1979]. This behavior is rare among unstrained diketones.

A critical requirement for synthesis of dodecahedrane is the skeletal C—C bond formation from the *endo*-face of a precursor in each step. The escalating contrathermodynamics in placing functional groups inside the developing molecular cavity cannot be compromised in order to achieve the goal. This consideration was the guidance of the elegant work [Ternansky, 1982] that employed a domino Diels–Alder reaction to establish four of the cyclopentane rings present in the target. The advantage of this approach also lies in the presence of a locking C—C bond which served to maintain a desired topology of advanced intermediates. Since each terminus of this extra bond is substituted with an ester group, a means of its cleavage is available. The projected reductive cleavage was accomplished during the formation of the seventh ring. This step must have involved *exo*-protonation of at least one the cleaved ester groups and further reductive coupling of the *endo*-oriented ester with the chlorinated carbon atom culminated in the very difficult task.

E
E
Cl
E
Cl
E
PhO
COOMe
E = COOMe
dodecahedrane

1,3-Cycloheptanedicarboxylic acid derivatives are needed as precursors of bridgehead, bridgehead-disubstituted bicyclo[4.1.1]octanes. An expedient synthesis of such a diester started from dimethyl 3,6-dihydrophthalate [Warner, 1981], the intracyclic bond of the norcarene intermediate was severed reductively.

COOMe
COOMe
Li
$NH_3$
COOMe
COOMe

A reductive C—C bond cleavage of $\gamma$-keto acid and derivatives was a crucial step in a photocycloaddition approach to the ophiobolane diterpenes [Coates, 1985].

R
COOH
H
O
Li
$NH_3$
COOH
R
H
O

In the following reactions activated methine or methylene hydrogens were identified as the acceptors. A lactol derivative of (−)-2$\beta$-hydroxy-9-oxoverrucosane was found to undergo aromatization on alkaline treatment [Takaoka, 1979]. 2-Chloroquinoline was converted to 3-cyanoquinaldine [Moon, 1983] on exposure to potassioacetonitrile in liquid ammonia. The heterocycle experienced opening and reclosure in the process. The loss of the ethylene moiety from the tryptamine portion of a pentacyclic tetrahydro-$\beta$-carbolinium salt posed an intriguing problem concerning its mode of genesis [Incze, 1985]. Fragmentation of an indolenine and disengagement of the *N*-vinyl substituent in the resulting zwitterion seem to be involved.

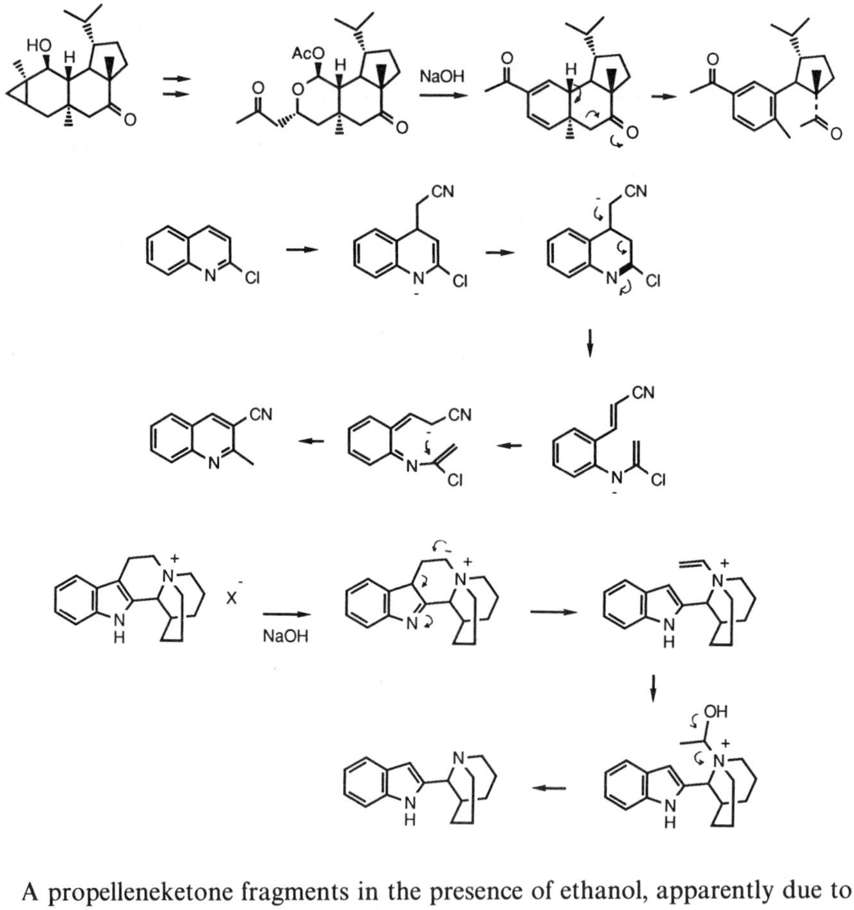

A propelleneketone fragments in the presence of ethanol, apparently due to an intramolecular attack at the $\gamma,\delta$-double bond by an acetal derived from another ketone group [Farina, 1990].

Retro-Diels–Alder reactions of furan and fulvene adducts are very smooth, mainly because the odd-membered bridge is a donor and conjoint with the acceptor of the functionality originated from the dienophile. The acceleration of such reaction by a trimethylsilyl group at C-7 of bicyclo[2.2.1]heptene derivatives has been witnessed [Magnus, 1987a]. It is interesting to compare this mode of cycloreversion with the cheletropic process when the bridging atom is an acceptor (e.g. C=O, $SO_2$).

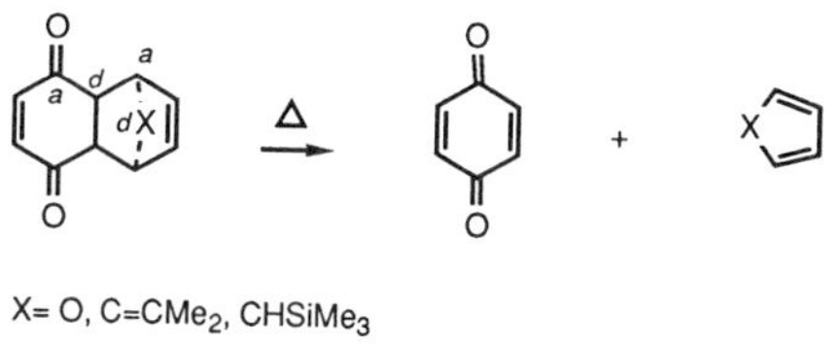

X= O, C=$CMe_2$, $CHSiMe_3$

### 5.3. a4a=d SYSTEMS

The highly oxygenated CD-ring system of aconitine was transformed into an aromatic nucleus when oxonitine (an amide derivative) was submitted to hydrolysis, periodate cleavage, and base treatment [Wiesner, 1963]. The last reaction accomplished aldolization, dehydration, retro-Claisen fission and autoxidation.

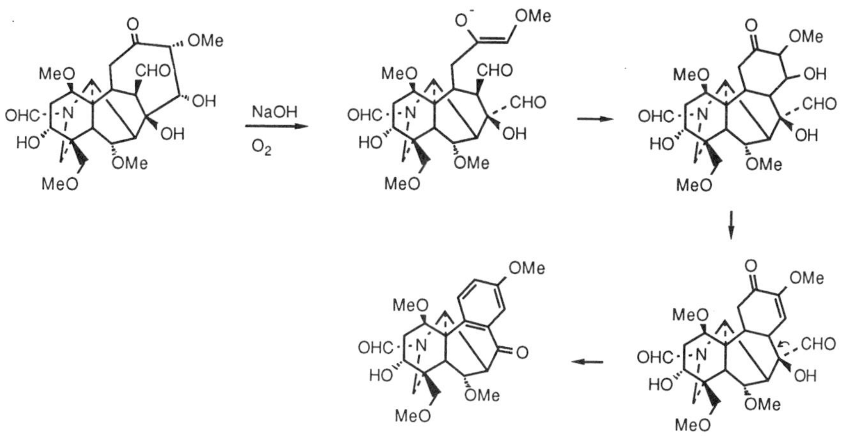

NaOH, $O_2$

One of the major products from Zn–HOAc reduction of akuammicine in the presence of copper(II) sulfate is deformylgeissoschizine [Hinshaw, 1971]. Picraline-type intermediates are susceptible to fragment the carbocycle in order to regain the aromatic indole moiety.

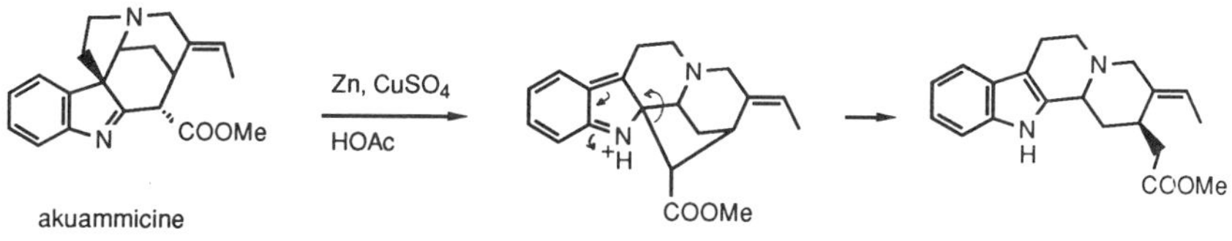

Zn, $CuSO_4$, HOAc

akuammicine

Reductive cleavage of the C-9/C-10 bond of $\Delta^{1,4}$-3,11-diketo steroids [Ananthanarayan, 1983] represents an expedient route to vitamin-D intermediates.

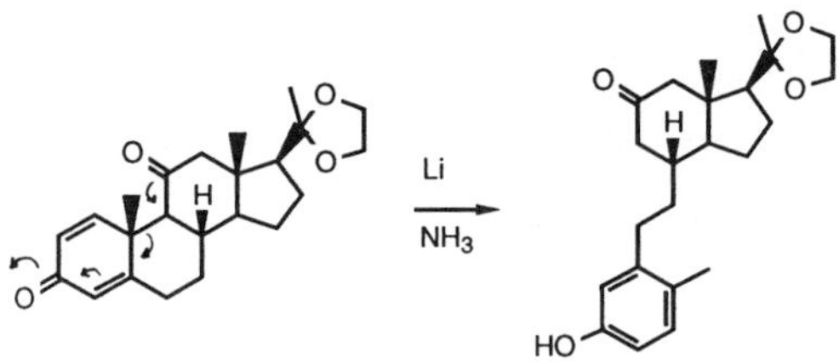

Li

$NH_3$

## 5.4. a4d SYSTEMS

We consider fragmentation involving deprotonation of carbonyl compound and related substances in this section. An unusual reaction was discovered when *cis*-1,6α-epoxy-3α-methyloctahydropentalen-2-one was treated with dilute methanolic sodium carbonate [Tsuzuki, 1977]. An acyclic keto ester was produced which has been converted into geraniol.

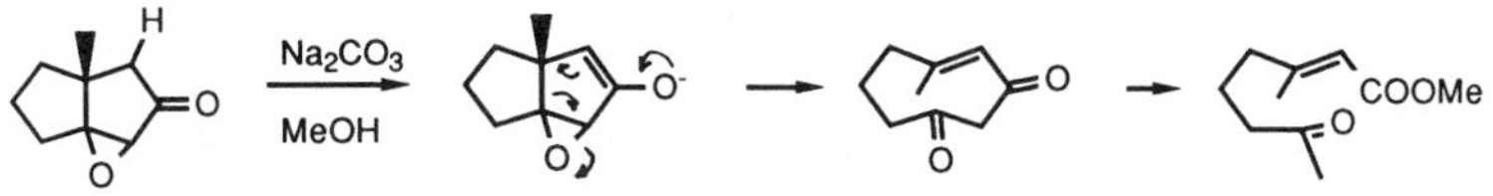

$Na_2CO_3$

MeOH

Dimerization of 2-alkyl-1,4-naphthoquinones with *N*-methylcyclohexylamine in the presence of air [Baxter, 1972] resulted in pentacyclic epoxytetralones. These compounds have been converted into diquinones on exposure to alkali.

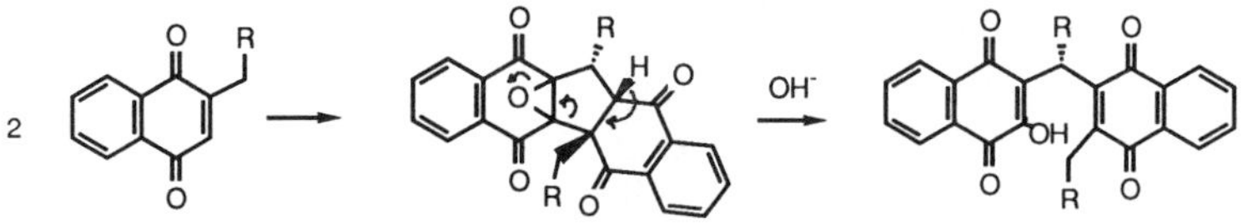

$OH^-$

The formation of prostaglandin-$A_1$ methyl ester from a bicyclic glycol on mesylation [Schneider, 1969] probably involved elimination of an epoxide intermediate.

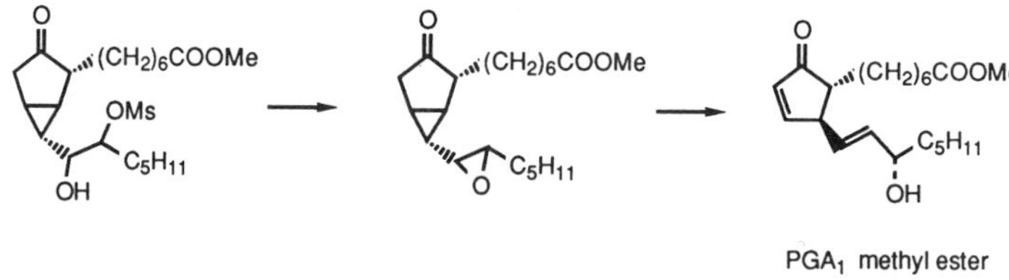

$PGA_1$ methyl ester

The unraveling of the spirocyclic system of agarospirol [Mongrain, 1970] was by acid treatment of a tricarbocyclic ketal diol. Thus, ionization of the cyclopropanecarbinol initiated a fragmentation.

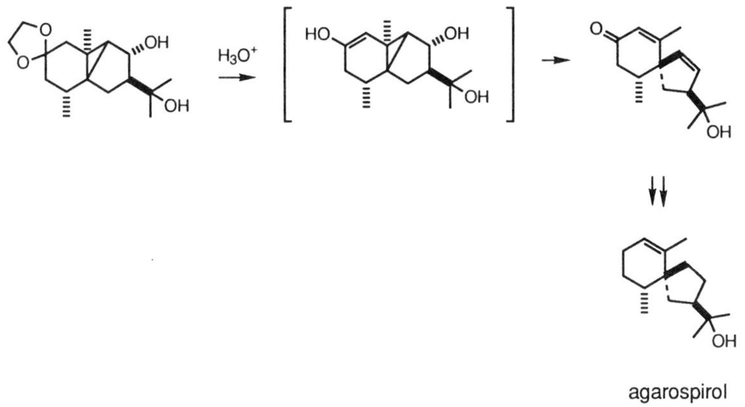

$H_3O^+$

agarospirol

Ketones with a himachalane skeleton have been acquired from ω-bromolongifolene on treatment with base [Mehta, 1974]. The fragmentation is a strain-relieving process. A similar process which competed with the methylation of π-bromocamphor might be responsible for the low yield of the desired product [R.V. Stevens, 1985].

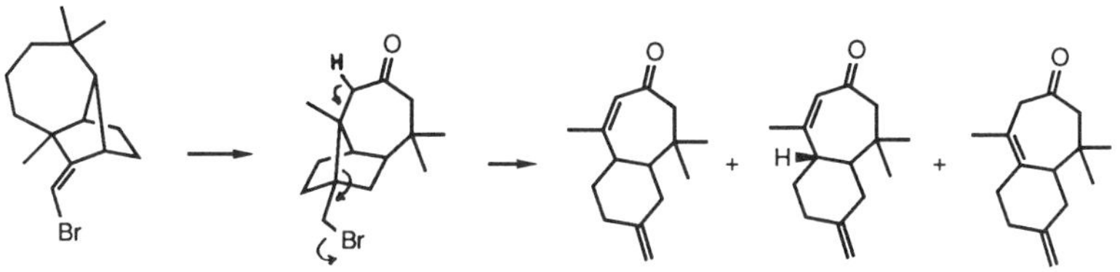

The conversion of a bridged dichloromethylbicyclo[3.3.1]nonanone to a phenol [Wenkert, 1969] was initiated by fragmentative elimination of hydrogen chloride.

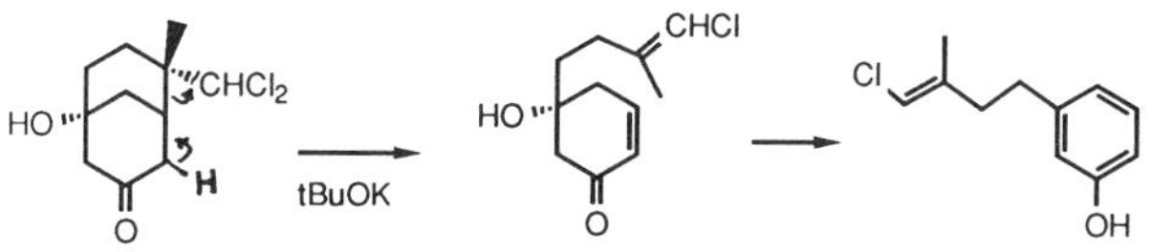

tBuOK

The diverse behavior of the *cis*-decalone tosylate derived from the Wieland–Miescher ketone [Heathcock, 1966] and its 7,7-dimethyl analog [Mukharji, 1969] seems to be due to conformational effects. The former compound underwent intramolecular alkylation predominantly, whereas the fragmentation of the latter substance was claimed.

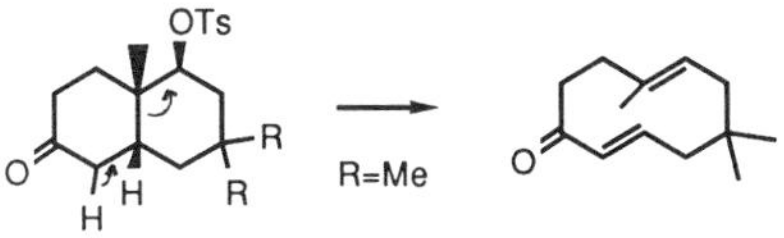

R=Me

$\delta$-Elimination processes have been designed in ring severance maneuver [Marshall, 1971b; J.M. Brown, 1977], such as in the creation of a germacrene precursor.

TsO
iPr$_2$NLi
COOMe
COOMe

An example of elimination involving sulfonyl activation and nitrogen (as sulfonamide) leaving group is a ring contraction of a cycloadduct of 8-azaheptafulvene with a sulfene [Truce, 1977]. The reaction proceeded via a norcaradiene tautomer.

RN-SO$_2$ R' nBuLi RN-SO$_2$ R' R' SO$_2$NHR

The most intriguing chemistry of santonin concerns with the formation of santonic acid and the latter compound's transformations. The formation of $\gamma$-metasantonin is a cycloreversion [Woodward, 1963b]. The net result of the base and acid treatments of santonin is the double bond migration to a remote site.

OH$^-$ H$^+$ COOH OH

$\gamma$-metasantonin

A formal retro-Fries rearrangement was suggested for the transformation of 4-benzoyl-4-methyl-2,5-cyclohexadienone to *p*-cresyl benzoate [L.B. Jackson, 1985] on treatment with a secondary amine. Conformationally the carbinolamine adduct is well-juxtaposed for transannular $C \rightarrow O$ benzoyl migration.

COPh R$_2$NH R$_2$N H Ph -R$_2$NH PhCOO

3-(Tosylazo)-2-cyclohexenones tend to lose dinitrogen and the tosyl group on reaction with alkoxide ions [Friary, 1986]. This reaction is somewhat akin to the Eschenmoser–Tanabe fragmentation (*vide infra*).

Ph NaOMe COOMe Ph N=N-Ts

A *trans*-decalindione monoxime tosylate in which the two $sp^2$-hybridized carbon atoms are separated by the two angular atoms underwent a Beckmann-type fragmentation in a concerted manner [Grob, 1968].

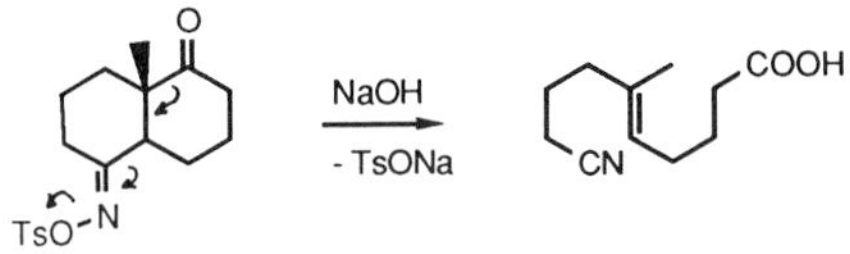

The major difficulties in the construction of mesocyclic dienes are regio-selectivity and stereoselectivity. Fragmentative approaches often provide excellent if not unique solution to such problems, as demonstrated in a synthesis of hedycaryol [Wharton, 1972]. The **a4d** fragmentation of octalinyl sulfonates is also adaptable to the creation of 1,6-cyclodecadienes [Marshall, 1974a] and the acquisition of a compound of the latter type paved the way to complete a synthesis of globulol.

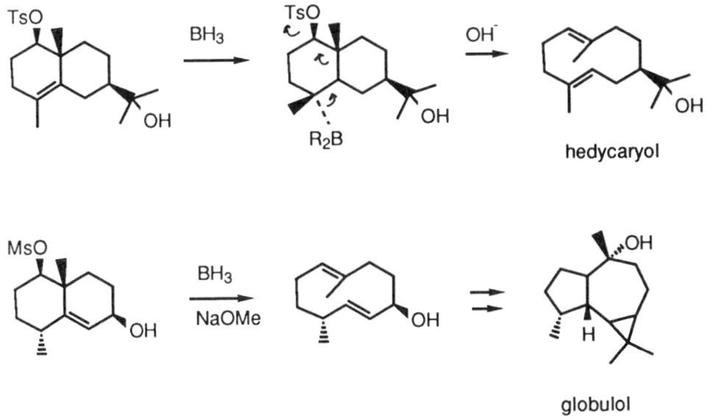

A stereochemical correlation of fragmentation versus cyclization (1,3--elimination) in a benzohexalin mesylate system has been reported [Takeda, 1973].

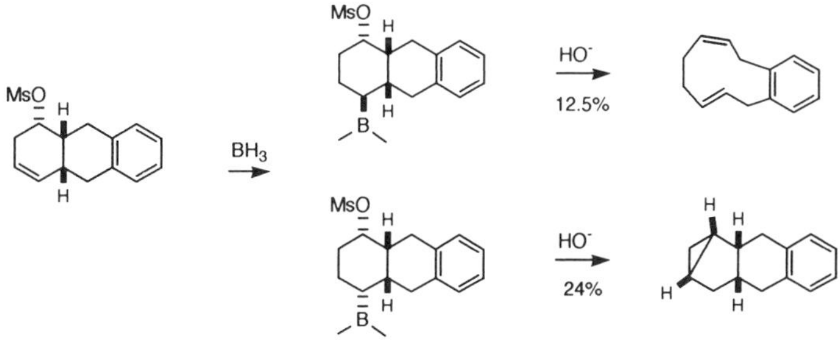

A tricyclic intermediate which has been converted into (+)-3-deoxyaphidicolin [Koyama, 1985] is obtainable from (−)-abietic acid through manipulation of the C-ring to a cyclopentenone. The photocycloaddition of the enone with allene is followed by reduction, mesylation, and hydroboration. Alkaline decomposition of the borane led to cyclobutane cleavage with establishment of the allyl chain at C-9.

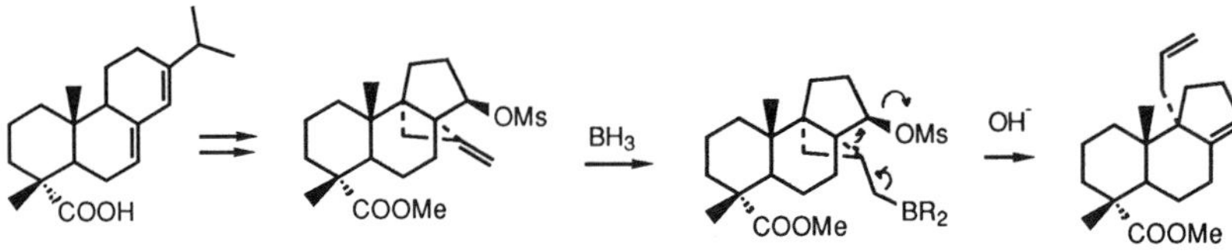

Ethyl 2-iodobenzeneazocarboxylate gave a certain amount of benzyne [Hoffmann, 1965] on reaction with ethoxide ion. The attack occurred at the carbonyl group and fragmentation followed. However, it is not possible to distinguish between a concerted **a4d** and a stepwise **d3d** mechanism.

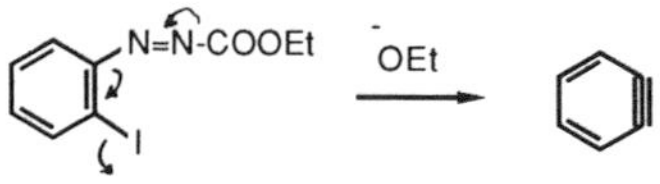

The stereoselective formation of tetrahydronaphthalenes from *o*-quinodimethane intermediates generated by the benzo-Peterson reaction found an application in a synthesis of sikkimotoxin [Takano, 1987]. An intramolecular version culminated in the access of 11α-hydroxyestrone methyl ether [Djuric, 1980].

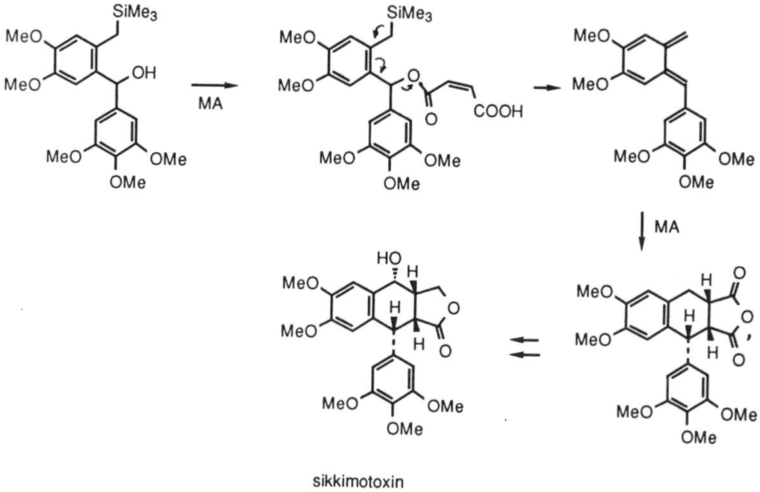

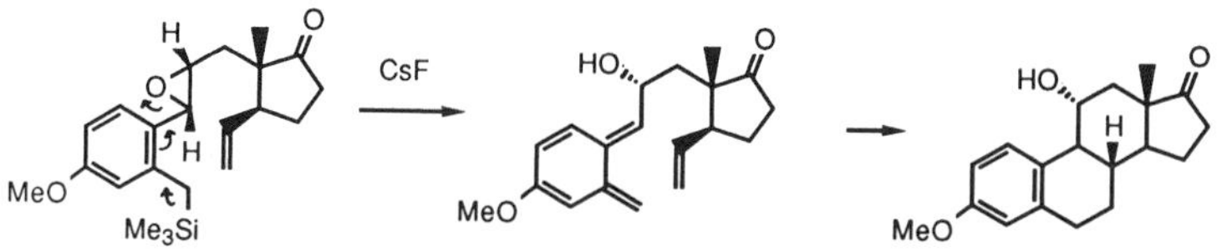

sikkimotoxin

Further variation of this **a4d** system provided a route to the azaquinodimethanes and a synthesis of gephyrotoxin [Ito, 1983] fully exploited the fragmentation–Diels–Alder reaction tandem.

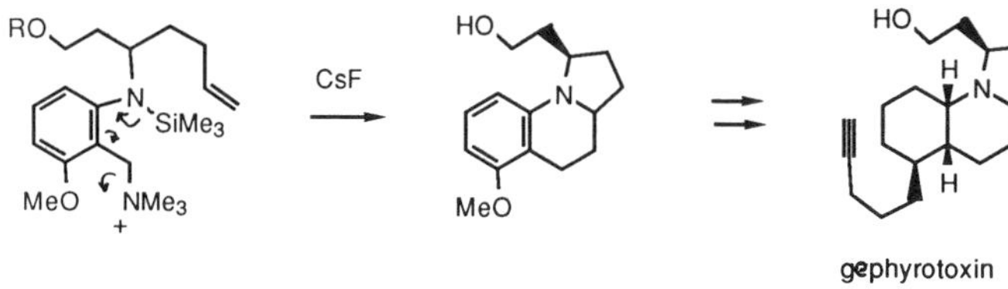

gephyrotoxin

2-Trimethylsilyltetrahydrothiophenes were fragmented on exposure to fluoride ion [Aono, 1986]. The vinylic sulfide may be trapped by alkylating agents such as iodomethane.

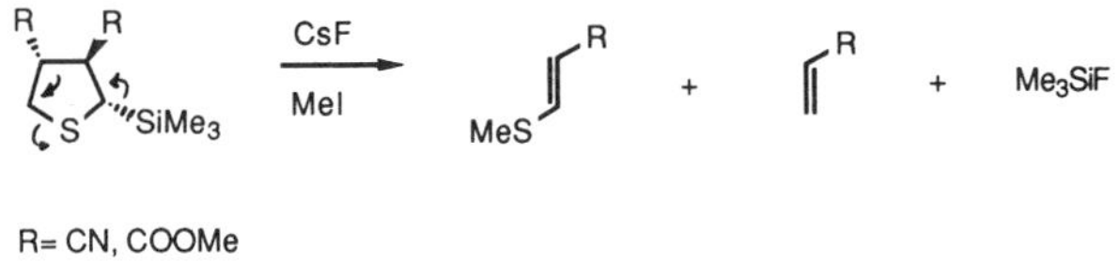

R= CN, COOMe

A carboxyl group protected as the 4-trimethylsilyl-2-buten-1-yl ester may be converted to the trimethylsilyl ester by treatment with palladium(O) catalyst [Mastalerz, 1984]. The second step of this transformation can be virewed as an **a4d** fragmentation, regeneration of the Pd(O) species is triggered by attack on silicon with the carboxylate ion.

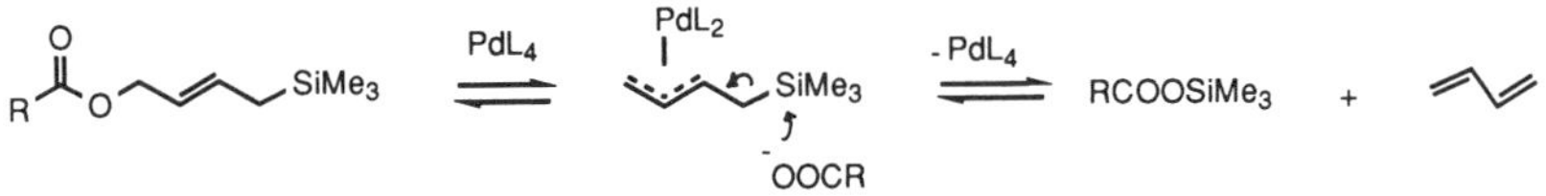

(2-Silylmethyl)cyclopropylcarbinols are prone to fragment upon exposure to an acid, giving 1,4-dienes [Wilson, 1988]. Thus, a desilylative fragmentation follows the generation of a carbenium ion.

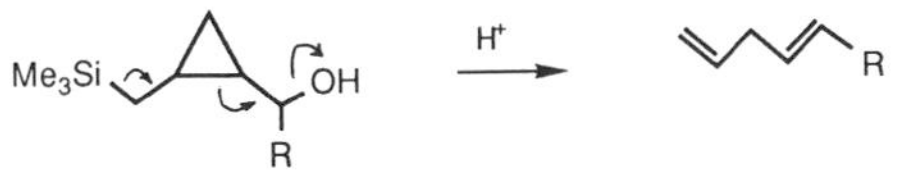

There is definite acceleration in the solvolysis of 2-adamantyl brosylates when C-5 is substituted with a trimethylstannyl group [Adcock, 1990]. The major product of this reaction arises from fragmentation.

The relative rates of solvolysis (in 97% trifluoroethanol and 3% water) for the *trans*-isomer is 7000, and for the *cis*-isomer, 98 times faster than the unsubstituted adamantyl brosylate. Interestingly, no fragmentation is detectable for the corresponding silyl compounds.

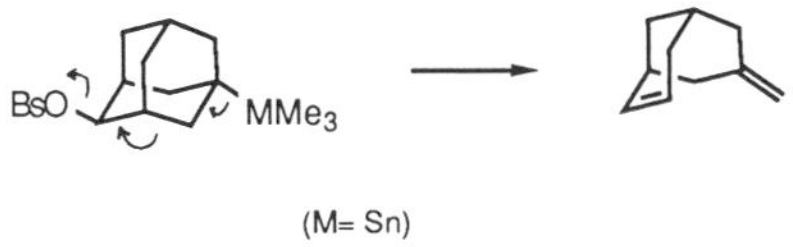

(M= Sn)

The Pd(II)-catalyzed Cope rearrangement [Overman, 1982] at room temperature could proceed via palladiocyclohexyl intermediates. Although the overall reaction is not a fragmentation process, the regeneration of the dienes from the palladiocyclohexanes may be regarded as such.

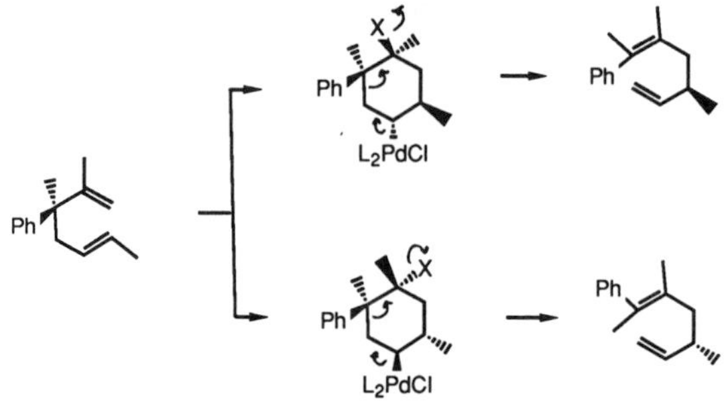

Numerous 1,4-dihalo compounds are susceptible to fragmentative dehalogenation by active metals. The two halogen atoms play different roles, one of which undergoing contrapolarization during reduction. Among the theoretically significant dienes that have been created by this dehalogenation are the head-to-head [4 + 4] dimer of cyclopentadiene [Doecke, 1978], precursors of trishomocubane [Underwood, 1970] and homopentaprismane [Marchand, 1976], the hydrocarbon hypostrophene [McKennis, 1971; Paquette, 1974], and bicyclo-[4.2.2]deca-1,5-diene [Wiseman, 1978]. The trihalo tetracyclic substrates for the preparation of the last compounds are available from dihalocarbene reaction with 7-halonorbornadienes.

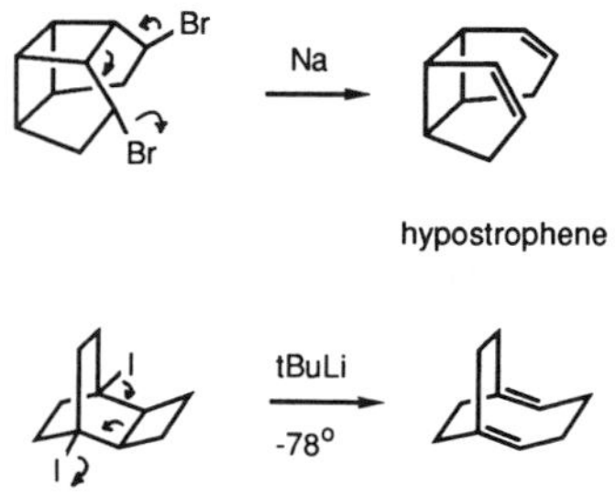

hypostrophene

A route to tetramethyl homonorcaradiene starting from tetramethyl-1,4-benzoquinone [Birladeanu, 1970] also depended on reductive dechlorination. 1,5-Dimethylene-*cis*-decalin has been generated from a tricyclic dibromide [Brinker, 1985].

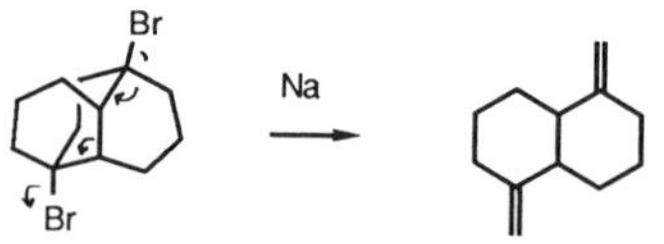

Dechlorination accompanied the unraveling of the iceane skeleton [Spurr, 1983] and the formation of a chlorosemibullvalene [Kwantes, 1978] have also been observed.

4-Haloalkyl sulfonates behave in an analogous manner as the corresponding dihalides toward reducing agents. In these cases, the halogen atom is the contrapolarizing site. Thus fragmentation of 1-bromoadamantan-4-yl mesylate with chromium(II) ion occurred when the mesyloxy group is equatorial [Babler, 1976].

The ready access of bicyclo[3.3.0]octane-3,7-diones by the Weiss condensation has made the synthesis of semibullvalene tetraesters very simple [L.S. Miller, 1981].

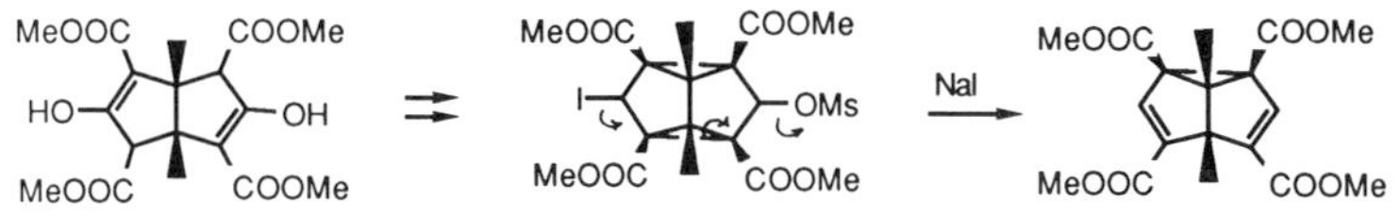

$\gamma$-Hydroxyalkylstannanes contain an **a3d** array and are therefore not fragmentable as such. However, alkenyl carbonyl compounds may be prepared from these substrates by an oxidative fragmentation pathway. For example, lead tetraacetate contrapolarizes the hydroxyl group and the evolved **a4d** system would decompose immediately. In terms of synthetic value, this fragmentation has been used in a synthetic of brefeldin-A [Nakatani, 1985], *exo*-brevicomin [O'Shea, 1989], macrocyclic enones [Posner, 1986], and unsaturated macrolides [Posner, 1987]. Iodine(III) reagents appear to function as well as lead tetraacetate [Ochiai, 1987].

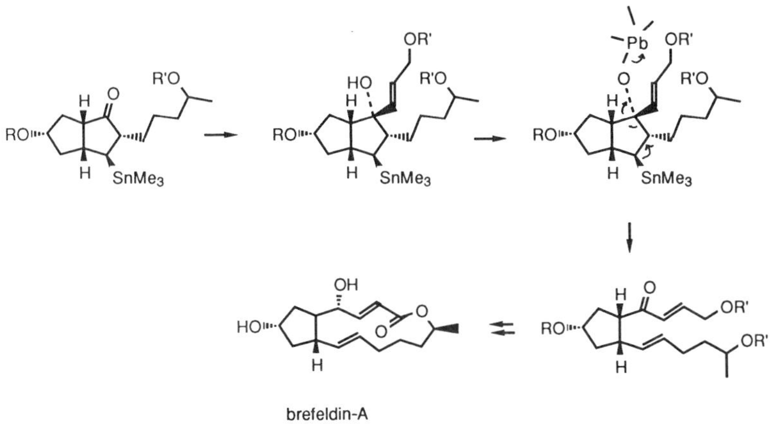

brefeldin-A

($5\alpha,7\alpha,11\beta$)-3,6-Dioxogermacr-1-en-1,3,7-olide which is useful for synthesis of germacrane and guaiane sesquiterpenes, may be obtained by photooxidation of a $5\alpha$-hydroxy-2-oxoeudesmanolide [Harapanhalli, 1988].

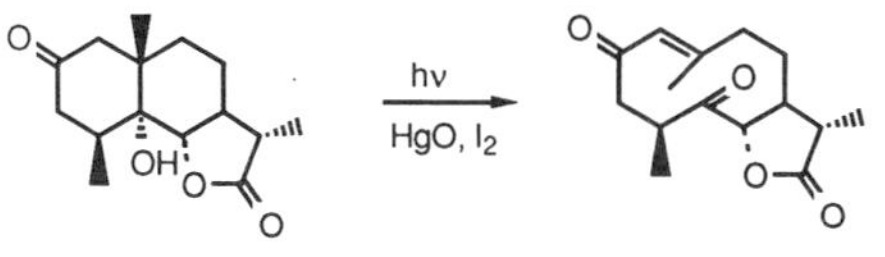

A skeletal amendment was required during a synthesis of ibogamine [Büchi, 1965]. The azabicyclo[3.2.1]octene framework was opened by zinc reduction, allowing reestablishment of an azabicyclo[2.2.2]octane subunit. Thus the fragmentation–recyclization of a **d3a=d** system resulted in a conjoint **d2a=d** array.

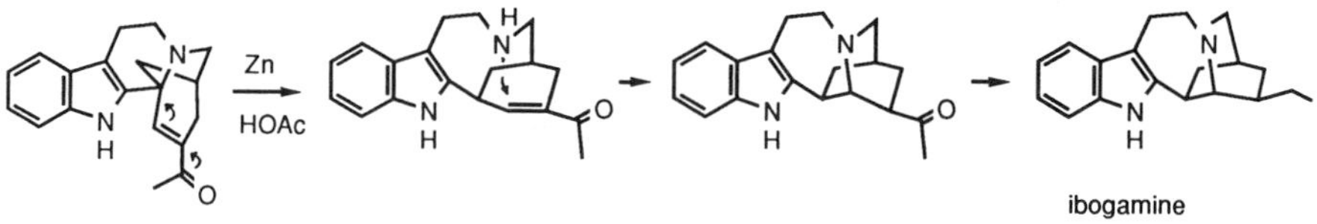

Zn
HOAc

ibogamine

## 5.5. a5a=d SYSTEMS

The bridged ring system of mexicanolide was unhitched on treatment with potassium hydroxide in methanol [J.D. Connolly, 1968]. A similar cleavage of augustinolide was also observed [Lavie, 1970] although a double bond migration must precede the main event.

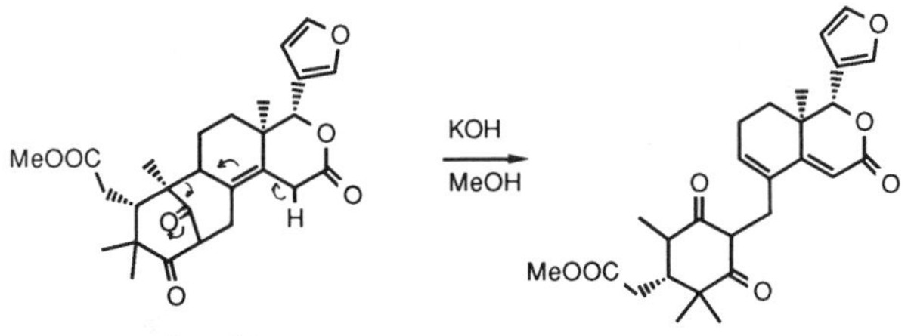

KOH
MeOH

mexicanolide

In an attempted synthesis of chelidonine [Cushman, 1980] from a diazoketone, an apparent migration of an aromatic substituent took place. This was the result of a combined spiroannulation and fragmentation.

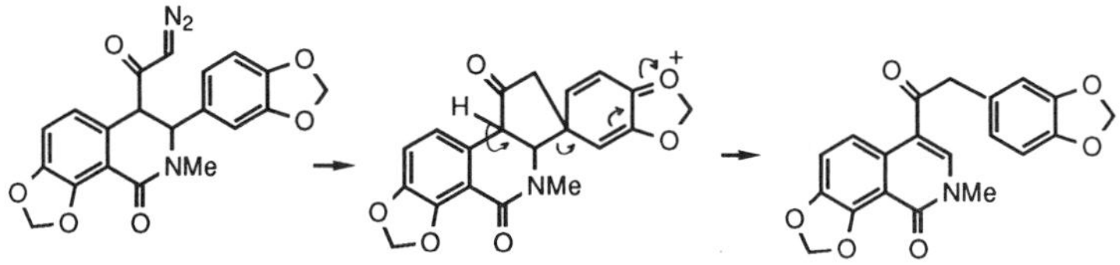

An important degradation sequence in the structural determination of culmorin [Barton, 1968] was the ozonation of a bicyclic enone ester and saponification of the ozonide to give tetrahydroeucarvone. The decomposition generated initially a mixed anhydride of acetic acid and tetrahydroeucarvone-α-carboxylic acid.

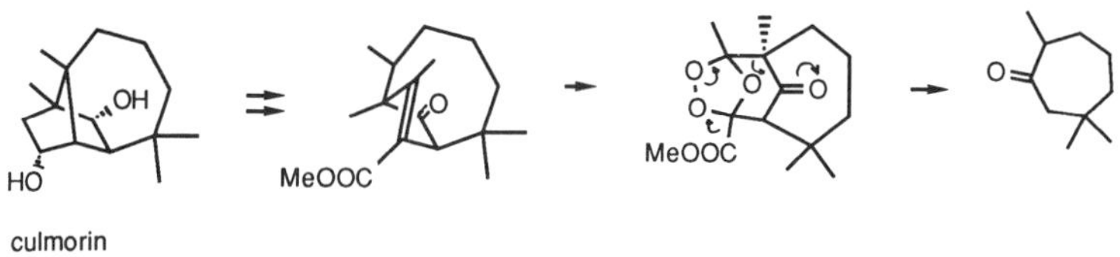

culmorin

## 5.6. a6d AND a8d SYSTEMS

The complex natural product isoasatone is considered as biosynthesized from two $C_6$–$C_3$ subunits, probably 2,6-dimethoxy-4-allylphenol. It is therefore very interesting to note that reduction of tetrahydroisoasatone [Yamamura, 1973] with zinc in acidic media led to 2,6-dimethoxy-4-propylphenol in good yield.

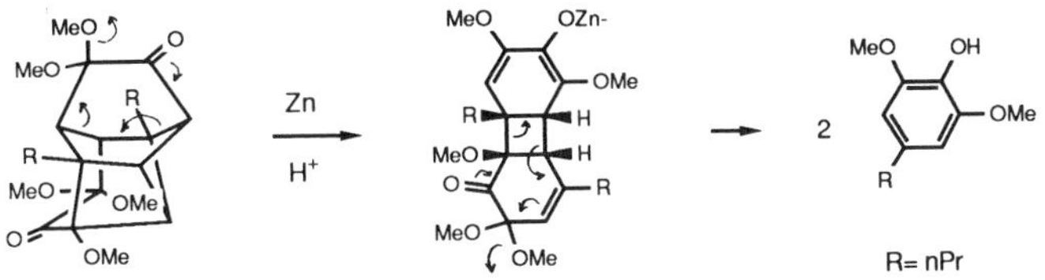

R= nPr

isoasatone

# 6

# d1d AND d2a FRAGMENTATIONS

## 6.1. d1d SYSTEMS

Generally, fragmentation of the **d1d** systems such as acetals, thioacetals, and aminals seems trivial, and their cleavage will not be discussed. Setting up a fragmentable **d1d** array from other systems is often the most interesting operation.

The thermal decomposition of diazo(2-furyl/2-thienyl)alkanes affords carbenes which spontaneously undergo heterocyclic opening [Hoffman, 1978]. The carbene intermediates are in resonance with zwitterionic species; fragmentation of the latter is probably the only way to stabilize such molecules.

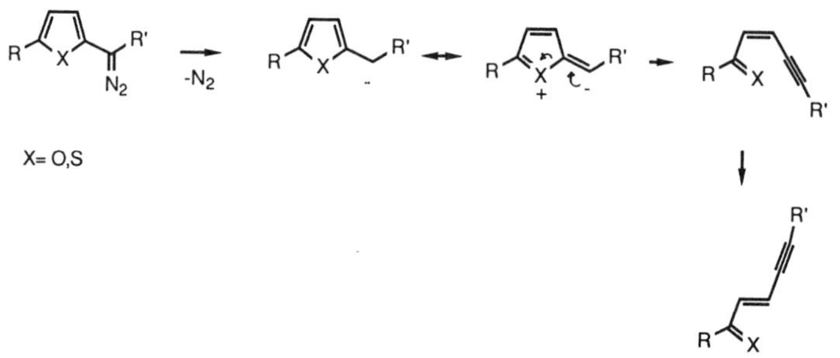

-N$_2$

X= O,S

α-Amino acids contain a disjoint array of polar functions. However, transformation of the carboxyl group into an acyl halide [Wasserman, 1984], mixed anhydride [R.T. Dean, 1978; Feldman, 1987; Bates, 1979], or acyluronium species [van Tamelen, 1970] enables the **d1d** fragmentation via decarbonylation to occur.

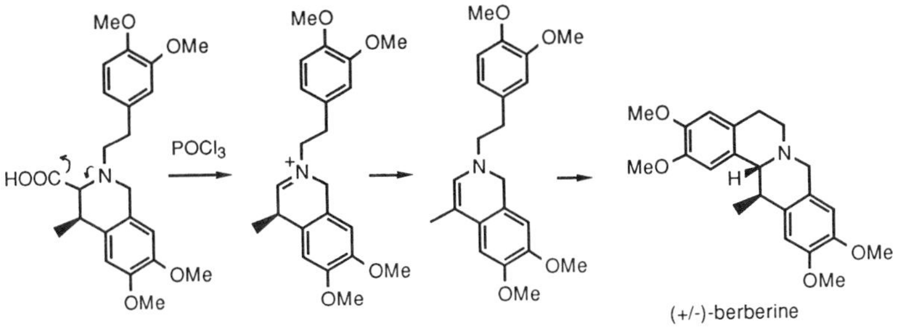

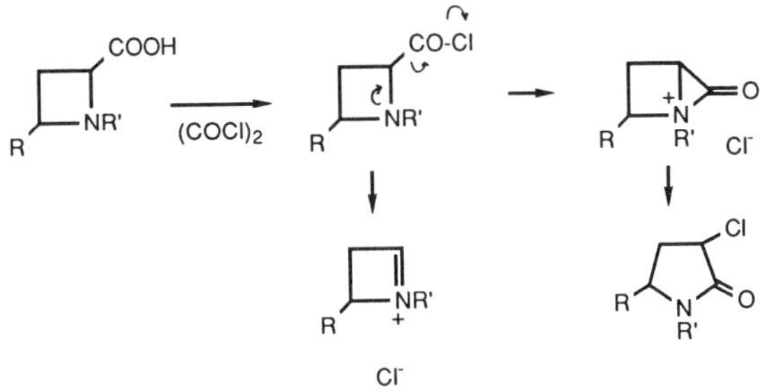

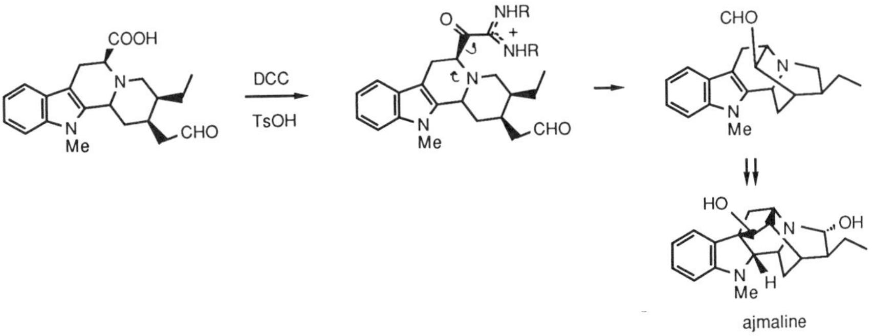

(+/-)-berberine

ajmaline

Decarbonylation from ketonic intermediates is implicated in a synthesis of corydaline [Saá, 1986] and the decomposition of a hydroperoxide [Hayakawa, 1985].

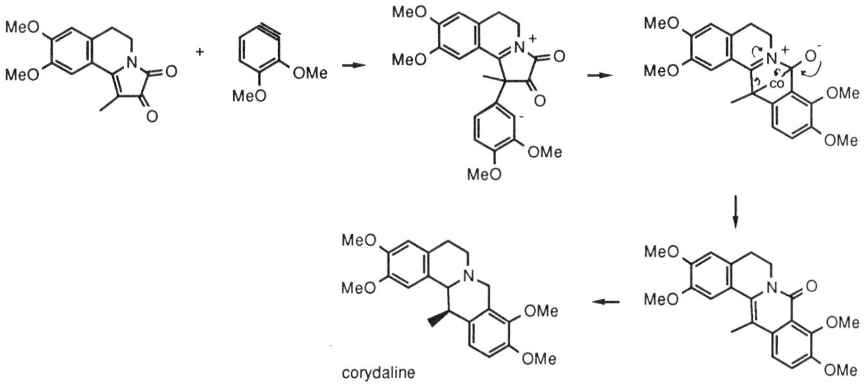

corydaline

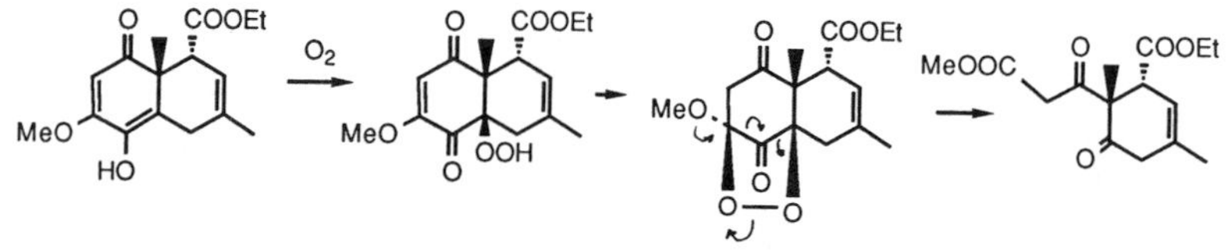

Rather unexpectedly, decarboxylation instead of a Friedel–Crafts acylation occurred when an α-oxy-γ-arylbutyryl chloride was reacted with silver triflate [Genot, 1987].

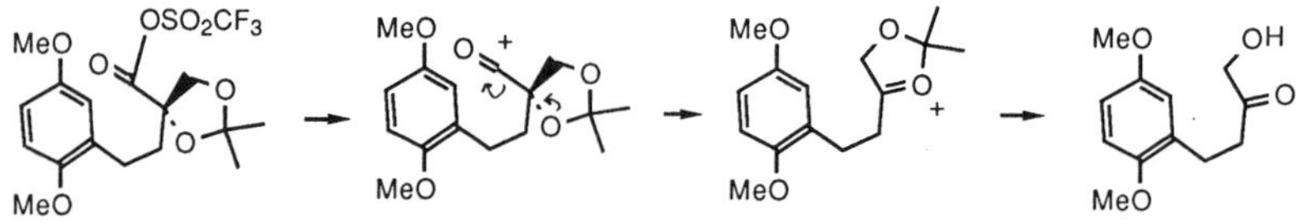

The attack by nucleophiles after selective activation of (methylthio)methyl esters via *S*-methylation [Ho, 1973a,b] can be viewed as proceeding by reaction at the carbonyl group of the sulfonium salts which provokes the fragmentation.

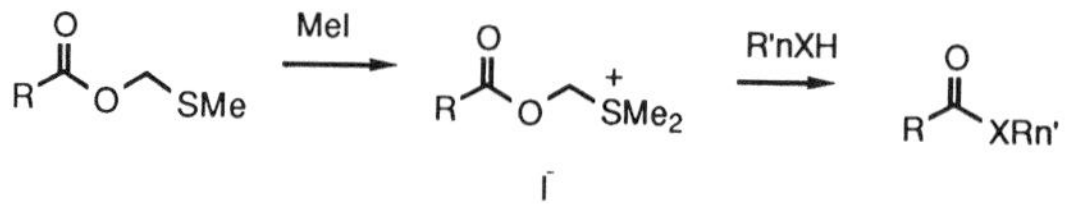

Related to this reaction are intramolecular transacylation reactions which have been used to regulate ring size and functionality. These include translactonization, $S \rightarrow O$ lactone exchange, and translactamization ("zip" reaction).

The classical method for primary amine preparation uses hexamethylenetetramine and alkyl halides [Delepine, 1895, 1897]. Fragmentation of the quaternary ammonium salts during the hydrolytic work up is a crucial feature.

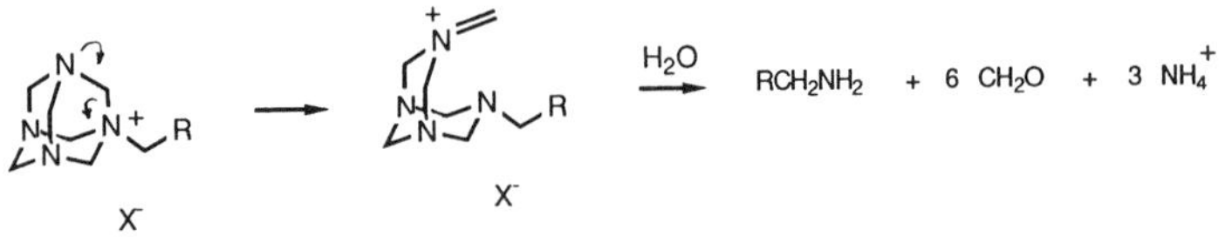

Racemization of Troger's base in acidic media is the result of a reversible fragmentation-cyclization process [Prelog, 1944].

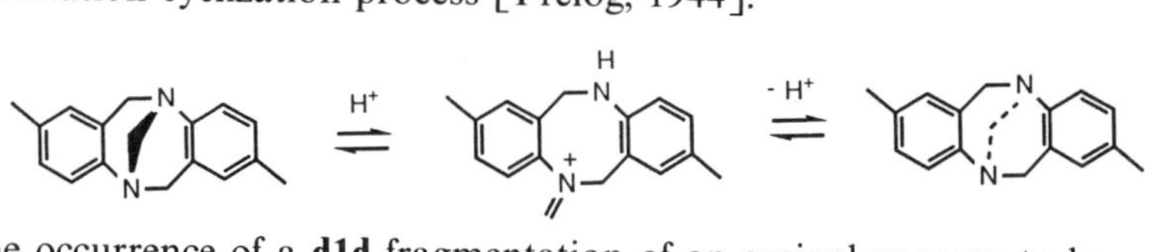

The occurrence of a **d1d** fragmentation of an aminal appears to be critical to the rectification of the tetracyclic structure of a predecessor of dasycarpidone [N.D.V. Wilson, 1968].

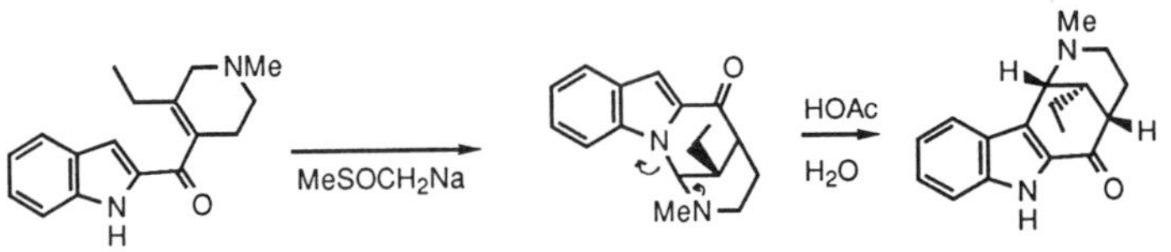

A similar type of transformation is involved in the unfoldment of the protoberberine skeleton on acid treatment of the photochemical products of *N*-(tetrahydroisoquinolin-2-yl)methylphthalimides [L.R.B. Bryant, 1984].

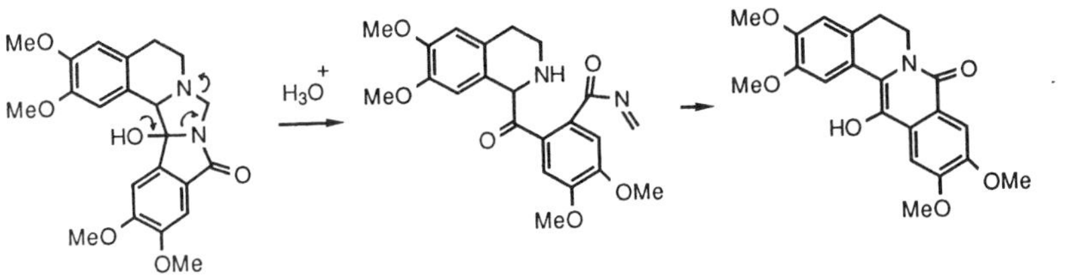

Of great synthetic import is the generation of iminium species by the fragmentation of α-cyanoamines. For example, it provides the trigger to the tandem cationic aza-Cope rearrangement/Mannich cyclization which has been extensively exploited [Overman, 1992].

## 6.2. d2a SYSTEMS

There is no great differences between **d2a** and **a2d** fragmentations. The separate classification is solely based on the ffact that an acceptor group is preexistent in the **a2d** system whereas a **d2a** fragmentation is concerned with an in situ generation of the acceptor site.

Many 1,3-diols are susceptible to fragmentation on acid treatment. For example, the decomposition of tetramethyl-2,4-pentanediol into acetone and 2,3-dimethyl-2-butene was reported [Slawjanow, 1907] many years ago.

In unsymmetrical 1,3-diols, fragmentation is determined by the relative stability of the two possible incipient carbocations [English, 1952].

An exceptionally facile fragmentation of the 1,3-diol system in which the nucleofugal group is benzylic has been discovered [M. Kato, 1989]. This observation is particularly noteworthy as the fragmentation of a related substance in which the benzylic alcohol was replaced with a primary tosylate did not occur (in the presence of a base). Apparently, ionization of the benzylic alcohol plays a crucial role in the reaction.

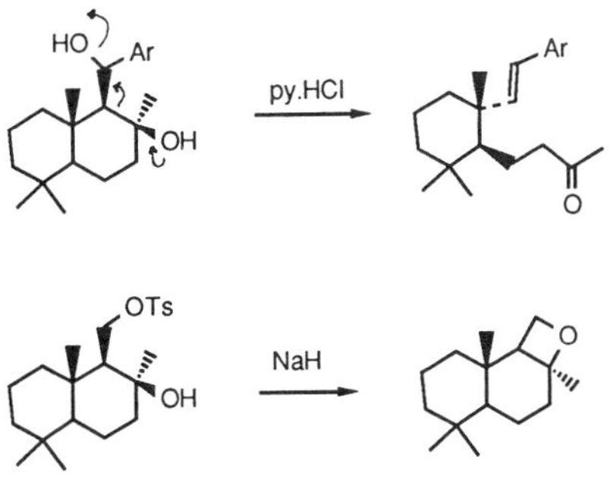

The conversion of *trans*-1,3-dihydroxy-2,2,4,4-tetramethylcyclobutane to an aliphatic aldehyde [Hasek, 1961] is to be contrasted to the stability of the *cis*-diol. The relative ease of the fragmentation has been attributed to backside solvation of the incipient cyclobutyl cation.

OH H$^+$ HO HO + CHO

*cis*-1,2-Diphenyl-2,3-epoxy-1-cyclobutanol gave 2,3-diphenylfuran on acid treatment [Padwa, 1967]. This transformation removes the enormous strain of the ring system while producing an aromatic substance.

1,3-Amino alcohols can be induced to undergo fragmentation by nitrosation [English, 1956]. The amino group is rendered nucleofugal in the process. It must be said also that other reaction pathways may compete with the fragmentation.

$$R_2C(OH)\text{-}CH_2\text{-}CR_2(NH_2) \xrightarrow{HNO_2} R_2C(HO)\text{-}CH_2\text{-}CR_2(N_2^+) \longrightarrow R_2C{=}O + CH_2{=}CR_2 + N_2$$

The instability of $\alpha,\alpha,\alpha',\alpha'$-tetraphenylspiro[2.6]nona-4,6,8-triene-1,2-dimethanol toward acids [Toda, 1972] is due to stabilization in forming a tropenium salt. However, this ion avails itself a pathway of eliminating benzophenone. The result is the formation of 8-(2,2-diphenylvinyl) heptafulvene.

OH Ph Ph Ph HO Ph H$^+$ (SiO$_2$) + Ph Ph Ph HO Ph Ph Ph

The reaction of 3,3-dimethoxycyclopropene with tetracyclone gives a 1:1 adduct which has been formulated as a spirocyclic ortho ester [Albert, 1977]. A zwitterionic intermediate is expected to fragment (**d2a** system) and internal ortho ester formation follows.

Ph$_4$ =O + OMe OMe Ph$_4$ O$^-$ + OMe OMe Ph$_4$ O$^-$ + OMe OMe Ph$_4$ O OMe OMe

Although *anti*-7-hydroxy-2,4,6,6-tetramethyl-2-norbornene gains much anchimeric assistence in ionization of the hydroxyl group from the double bond, the molecule is also quite strained and it can be converted into a fragmentable species upon protonation of the same double bond. The equilibrium may then be displaced toward formation of the monocyclic product [V.K. Jones, 1970].

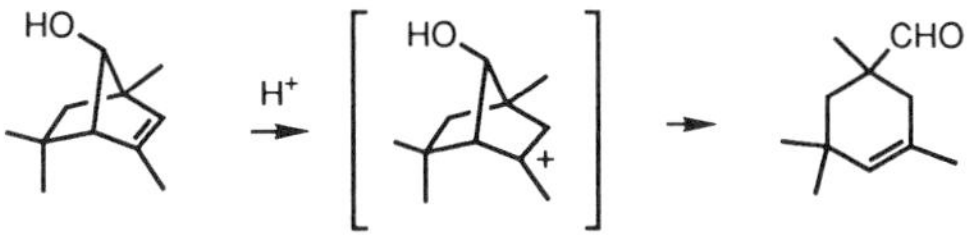

The 1,5-dimethyl-6-methylidene-*exo*-8-alkyl/aryl-tricyclo[3.2.1.$0^{2,7}$]-oct-3-en-*endo*-8-ols are prone to aromatization on exposure to acid [Mukherjee-Müller, 1977]. Protonation of the exocyclic double bond readily generates cationic species which spontaneously collapse to the norcaradiene ketones and, thence, substituted benzyl ketones.

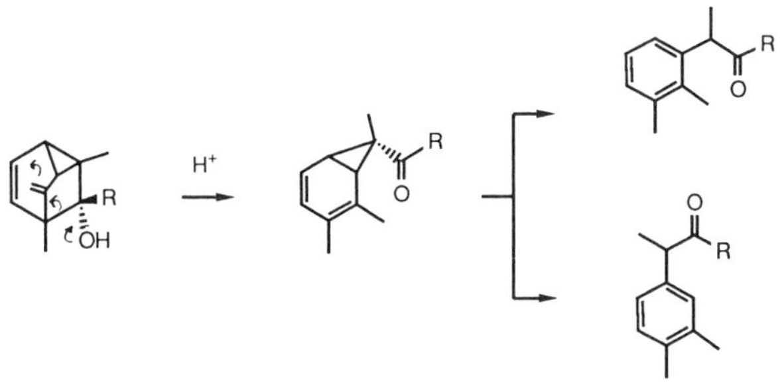

Hetisine gave two acetals with a dioxaadamantanoid skeleton on reaction with acid [Pelletier, 1981, 1983]. The homoallylic hydroxyl groups are conducive to fragmentations, and because of the proximity of these functionalities, acetalization and internal hydration are expected.

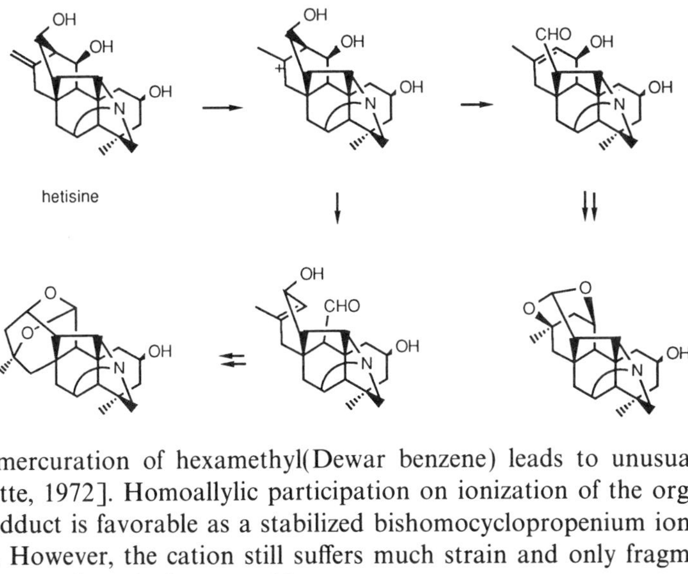

hetisine

Oxymercuration of hexamethyl(Dewar benzene) leads to unusual results [Paquette, 1972]. Homoallylic participation on ionization of the organomercurial adduct is favorable as a stabilized bishomocyclopropenium ion may be formed. However, the cation still suffers much strain and only fragmentation to a ketone solves the latter problem.

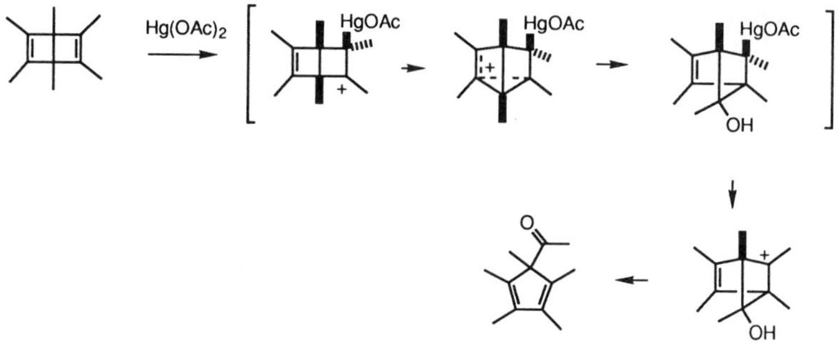

The above reaction was rationalized on the basis of *exo*-attack by the reagent. A rearranged diol has also been obtained from hydroxylation from the *endo* face [Krow, 1972]. The diol has been converted into the same ketone on melting.

2-Cyclohexenols may undergo photoinduced fragmentation [Marshall, 1971a], A carbocation generated $\gamma$ to the hydroxyl group provokes C—C bond cleavage. Depending on the medium, various derivatives of the unsaturated aldehyde may be obtained.

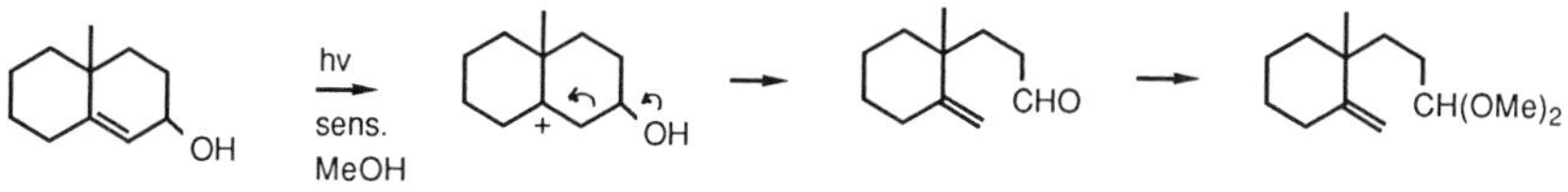

During ethyleneketalization of 6-methylbicyclo[2.2.1]hept-5-en-2-one [Lal, 1989] fragmentation is a competing process. Protonation of the double bond to give a fragmentable **d2a** system is the cause of this untoward side-reaction.

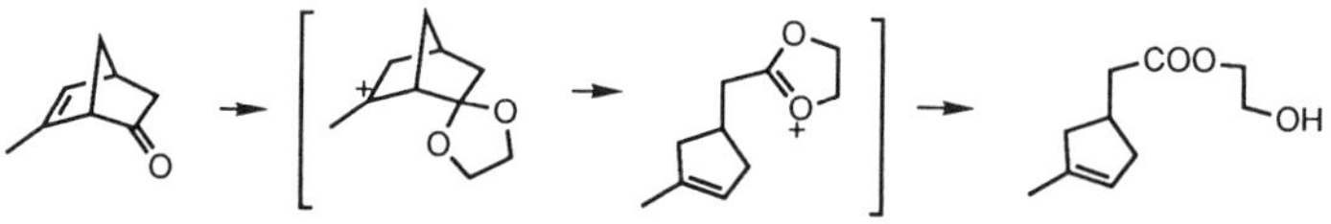

The solvolysis of 10,10-dibromo[4.3.1]propellane [Reese, 1972] and 11,11-dihalo[4.4.1]propellane [Warner, 1977] gives, in each case, a monocyclic ketone besides other products. Along the reaction pathway is a cyclopropyl $\beta$-hydroxy carbenium ion.

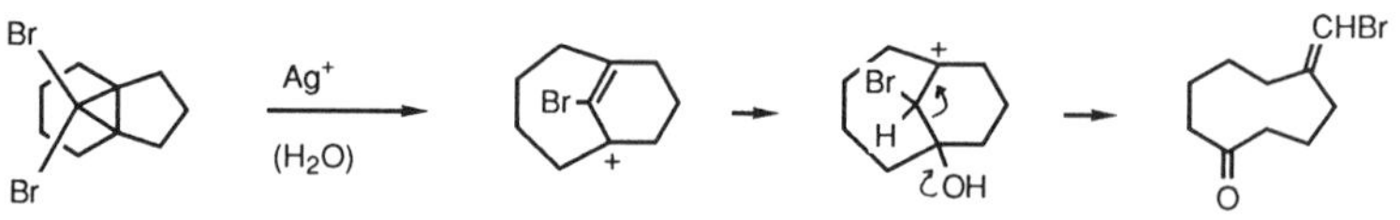

Two different $\beta$-hydroxy carbenium ions mediated the solvolysis of a bicyclic hydroxy tosylate [Marshall, 1967].

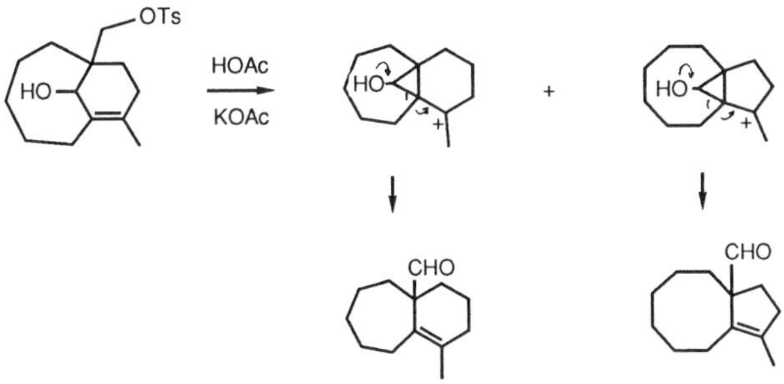

A novel cyclization with rearrangement of a secoibogamine derivative under solvolytic conditions has been observed [Szantay, 1985]. The initially formed pentacyclic base is liable to fragmentation and recyclization occurs at an allylic position to afford a less strained product.

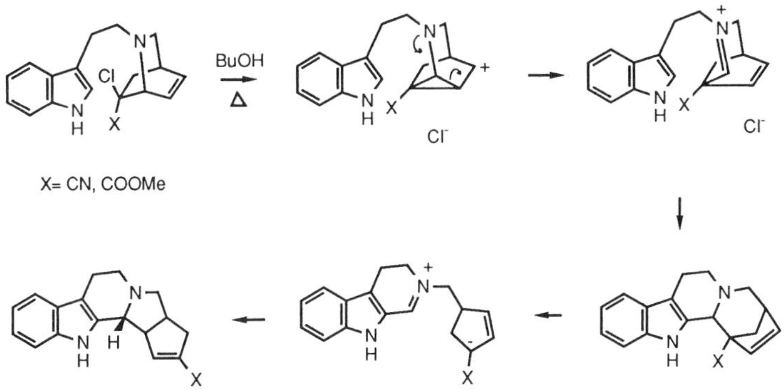

X= CN, COOMe

Chlorination of *exo*-5-hydroxyisolongifolene leads to a bicyclic chloro aldehyde [Yadav, 1982]. The reaction can be viewed as involving the generation of a tertiary carbocation and, since this cationic center is $\gamma$ to the hydroxyl group, fragmentation is inevitable. A variant of this fragmentation is observable from an epoxy alcohol.

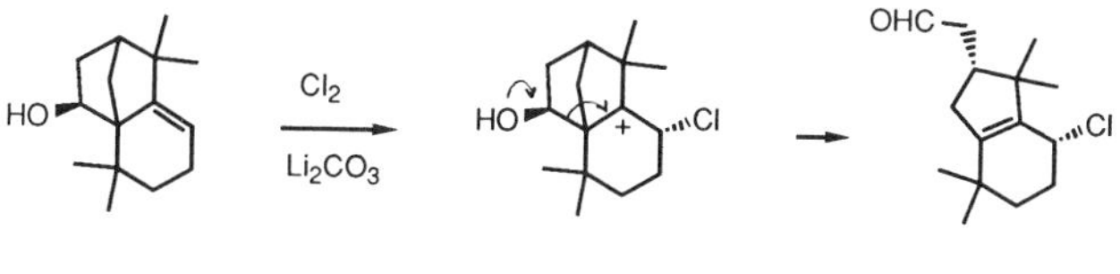

An excellent route to pleuromutilin [Gibbons, 1982] involved construction of a tetracyclic skeleton and controlled fragmentation of which to create a bridged cyclooctenone. A $\beta$-ketol was set up, but retroaldol fission failed. The problem was solved by brominating the double bond at the $\gamma,\delta$ position of the hydroxyl group.

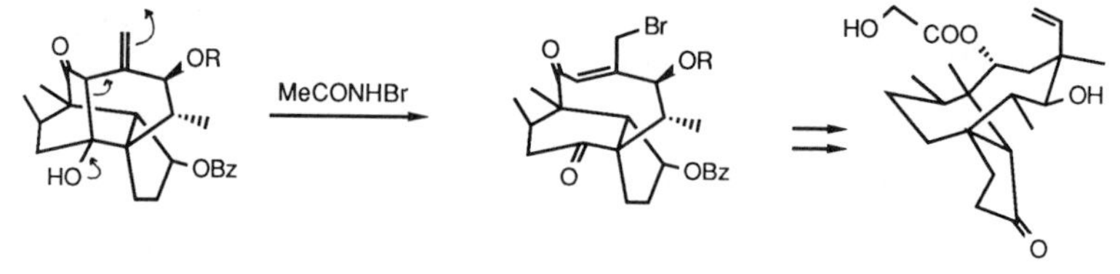

MeCONHBr

pleuromutilin

Bromine initiated fission of a tetracyclic photocycloadduct of a methoxy-α-curcumene [Wender, 1982]. The bromo ketone possesses the carbon framework of cedrene.

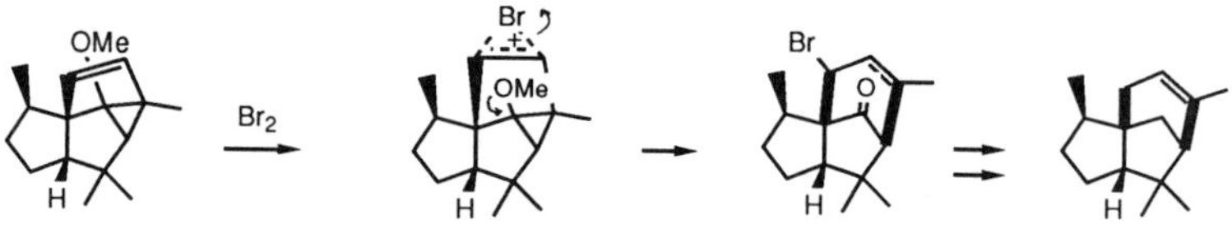

$Br_2$

cedrene

Treatment of grayanotoxin-II with lead(II) acetate in methanol at room temperature furnishes leucothol-D [Kaiya, 1979]. Palladation of the double bond of the toxin creates an electron deficient center at the γ position to the hydroxyl group, thus enabling a fragmentation to occur. The allylpalladium species being electrophilic, serves well as an alkylating agent for the ketone, therefore recyclization is facile.

Interestingly, thallium(III) acetate causes fragmentation of grayanotoxin-II to give the tricyclic grayanol-B [Kaiya, 1981].

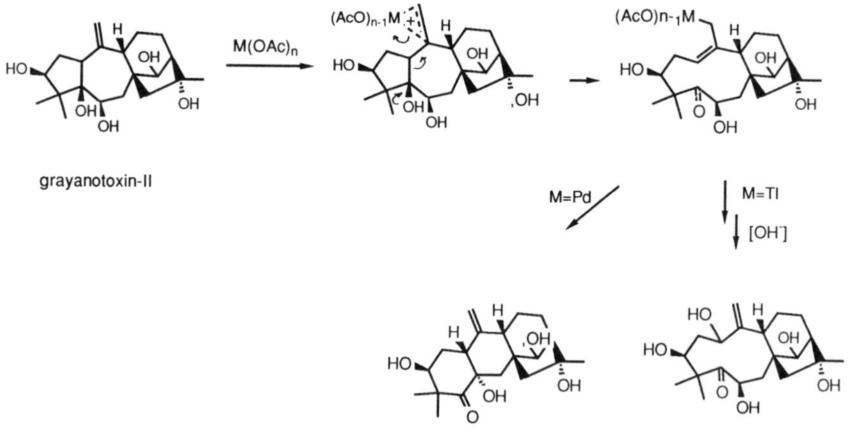

$M(OAc)_n$

grayanotoxin-II

M=Pd

M=Tl

$[OH^-]$

grayanol B

Camphenilone is convertible to 4′-methylacetophenone by sulfuric acid [Noyce, 1950]. In situ generation and fragmentation of a **d2a** system is indicated.

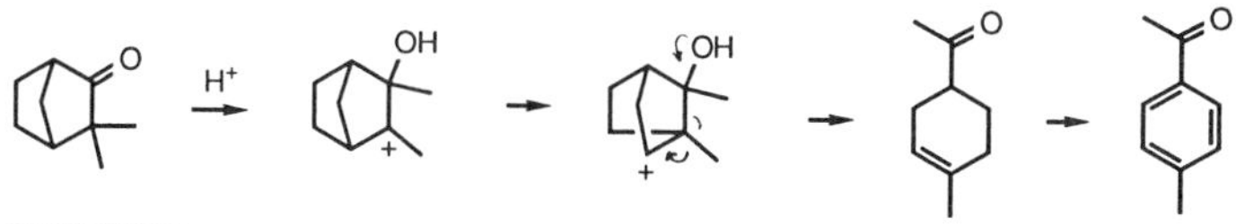

camphenilone

One of the solvolytic products of *endo*-3,3-dimethoxybicyclo[2.2.1]heptan-2-yl triflate is methyl 3-cyclohexene-1-carboxylate [Creary, 1977]. The simplest explanation of its genesis entails a Wagner–Meerwein rearrangement and methoxy-directed fragmentation.

The pinacol obtained by reduction of santonic acid undergoes rearrangement on acid treatment [Hortmann, 1970]. The ring contraction actually is the result of a **d2a** fragmentation.

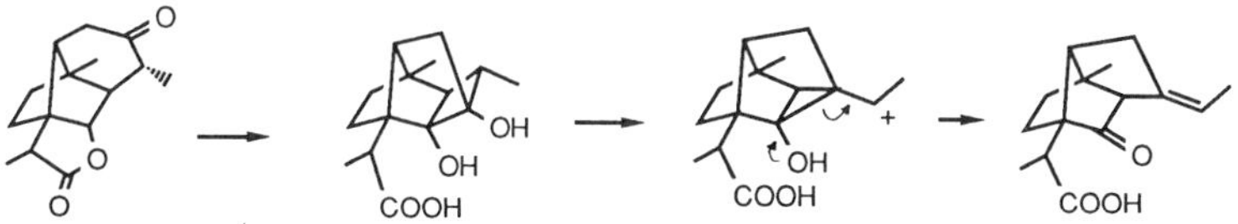

A deep-seated rearrangement occurred when epoxyisogermacrone was exposed to boron trifluoride etherate [Tsankova, 1982]. A mechanism accounting for the formation of a cyclohexene requires attack of the epoxy oxygen atom on the complexed carbonyl group which in turn provokes a transannular cyclization. A tertiary carbenium ion is now generated at a $\gamma$ position to the oxygen atoms and fragmentation ensues.

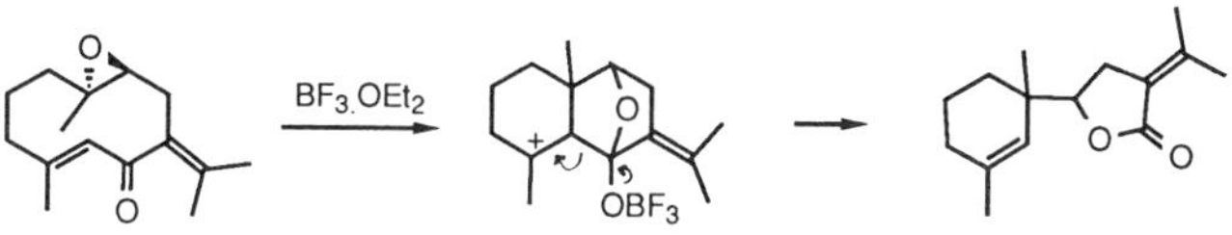

The formation of 6,7-dimethyl-6-octenal by acetolysis of 1-*t*-butylcyclohexene oxide [Corona, 1984] is due to a 1,2-methyl shift to generate a fragmentable **d2a** system.

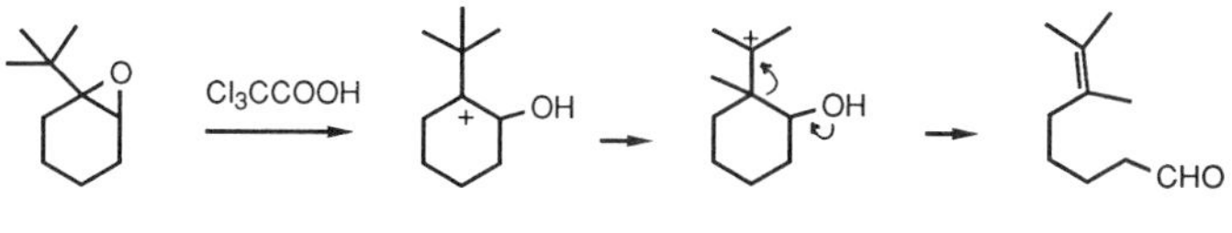

Lewis acid-catalyzed rearrangement of epoxides leads to carbonyl compounds by a 1,2-shift mechanism. When the migrating group is not originally attached to an epoxy terminus there ensues a $\beta$-oxycarbocation. Naturally, $C_\alpha$–$C_\beta$ bond cleavage follows [Hikino, 1969].

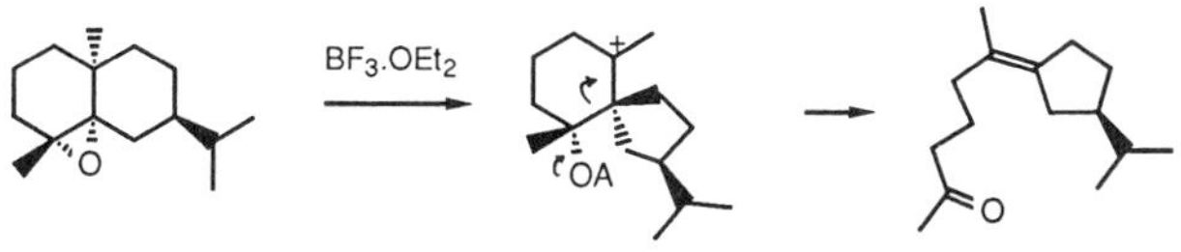

2α,3α-Epoxypinane readily undergoes ring opening to give campholenic aldehyde [Farges, 1962; Hartshorn, 1964]. The ring opening step is also accompanied by a skeletal rearrangement which results in a **d2a** system.

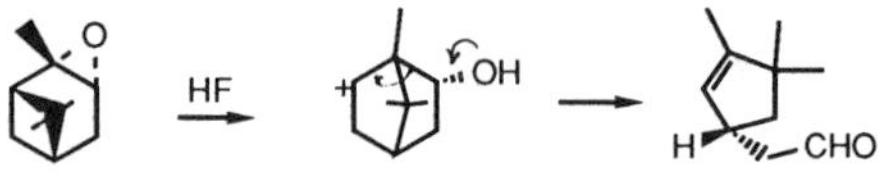

HF

OH

H CHO

The epoxide of methyl (−)-kaur-9(11)-en-19-oate furnishes a tricyclic aldehyde on treatment with a Lewis acid [Nakano, 1985]. Epoxide opening induces a Wagner–Meerwein rearrangement to generate a β-oxy carbenium ion which collapses by C—C bond fission.

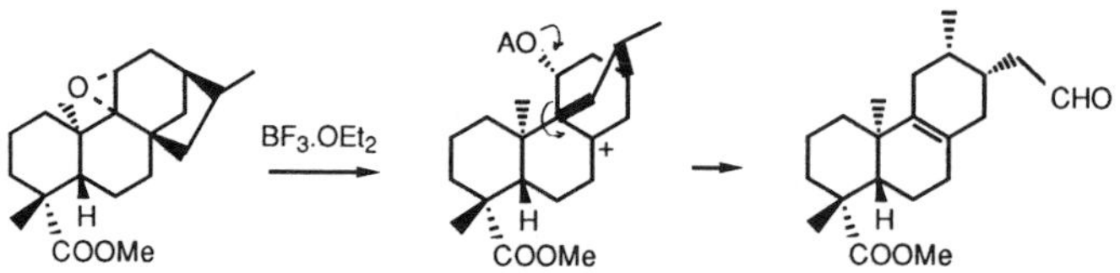

O

H COOMe

$BF_3.OEt_2$

AO

H COOMe

CHO

H COOMe

The photo isomer of methyl levopimarate forms an epoxide; the latter compound is liable to rearrange and fragment on contact with boron trifluoride etherate [Herz, 1975].

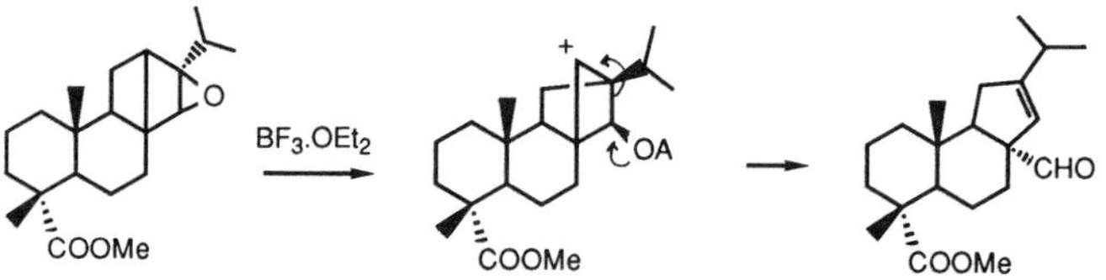

O

COOMe

$BF_3.OEt_2$

OA

COOMe

CHO

COOMe

Treatment of 24,28-epoxystigmast-5-en-3β-yl acetate with boron trifluoride led to a diene with loss of acetaldehyde, in addition to two carbonyl compounds [Ohtaka, 1973]. The substitution pattern around the epoxide ring seems to be critical for the occurrence of the fragmentation pathway.

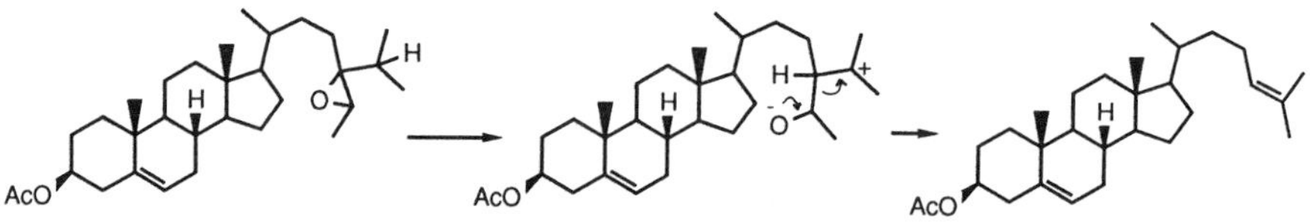

H O H

AcO

H H O

AcO

H

AcO

One of the products of the $BF_3$ reaction with lipiferolide is xanthanolide [Wilton, 1983]. Transannular cyclization to the hydrazulene skeleton initiates a 1,2-hydride shift which allows for fragmentation to take place.

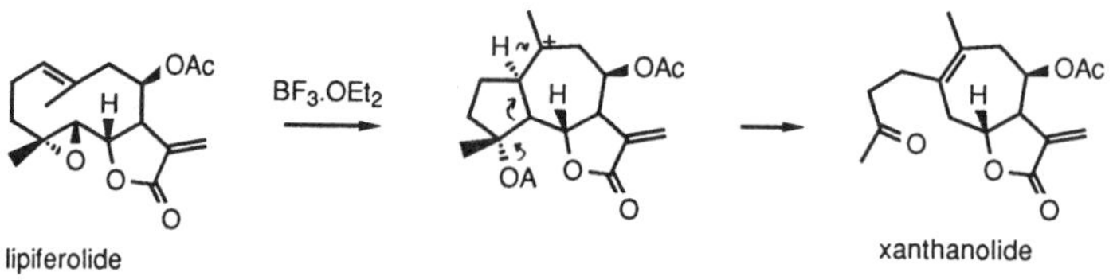

OAc H O O O

lipiferolide

$BF_3.OEt_2$

H OAc H OA O O

OAc H O O O

xanthanolide

Cyperene oxide gives three rearrangement products on reaction with stannic chloride [Bang, 1973]. A bicyclic ketone results from a sequence of alkyl and hydrogen migrations which ultimately places a cationic site $\gamma$ to the oxygen atom. Patchoulene oxide also furnishes a ketone [Bang, 1972].

Dieldrin is isomerized in acetic anhydride in the presence of sulfuric acid [ApSimon, 1974] to a half-cage carbocation because the proximity of two bridges enables intramolecular hydride transfer so readily as to preclude fragmentation prior to the C—C bond formation.

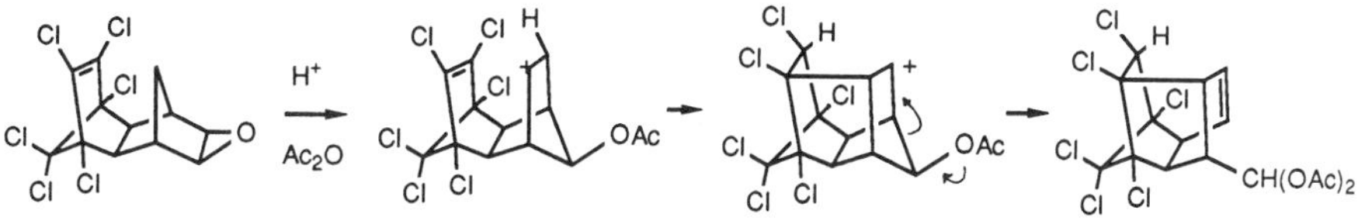

Peracid oxidation of norbornadiene leads to bicyclo[3.1.0]hex-2-ene-*endo*-6-carboxaldehyde rather than the expected epoxide [Meinwald, 1963]. The latter compound is actually the primary reaction product but it is extremely prone to ring opening and fragmentation.

Thermolysis of the formal [4 + 2]cycloadduct of dimethyl phthalate with acetylene oxide gives a cycloheptatrienecarboxaldehyde [Lewars, 1977]. Fragmentation occurs after epoxide opening which is attended by a C—C bond migration.

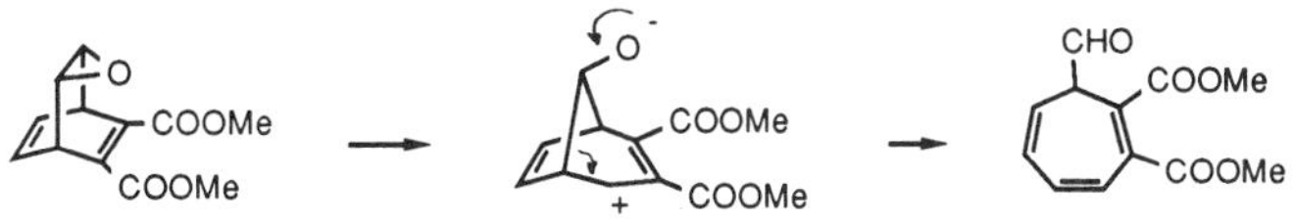

An epoxide obtained from Corey reaction of the Wieland–Miescher ketone does not undergo the expected isomerization on exposure to boron trifluoride etherate. Instead, a hydrazulenone arising from a rearrangement–fragmentation pathway results [Geetha, 1979].

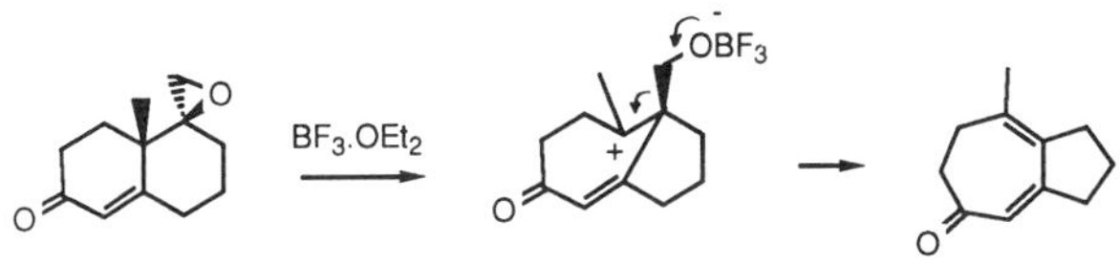

12-Methyl-$\Delta^{11}$-togogenin acetate epoxide affords C-nor-D-homoalcohols, among other products, on treatment with to boron trifluoride [Coxon, 1969]. Noteworthy is that both $\alpha$- and $\beta$-alcohols are formed. The $\beta$-isomer is derivable from a Prins-type reaction through the cleavage–recyclization sequence.

The epoxide of 12-methylenetigogenin rearranges to a mixture of epimeric aldehydes and a C-nor-D-homo olefin [Coxon, 1967]. The last compound arises from migration of the intracyclic bond and collapse of the ensuing cation by loss of formaldehyde.

*endo*-2-Hydroxy-*exo*-2-phenyl-5,6-epoxybicyclo[2.2.1]heptane has been transformed into a cyclopentene [Waddell, 1987]. Apparently an aluminum isopropoxide catalyzed oxirane–oxetane exchange was followed by heterolysis to form a benzylic cation which could undergo fragmentation.

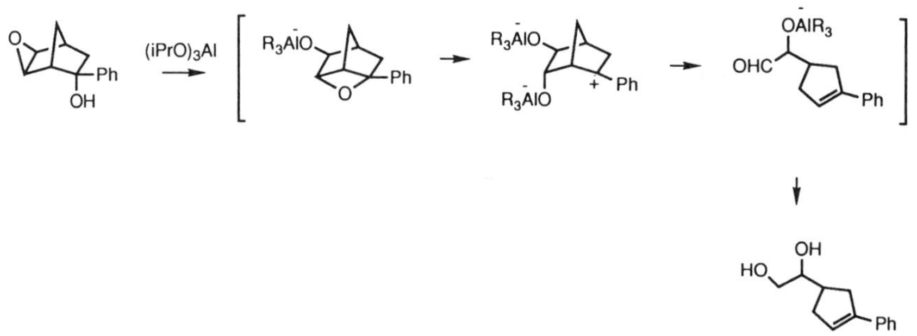

(iPrO)$_3$Al

R$_3$AlO

Ph

OAlR$_3$

OHC

HO

OH

A synthesis of ryanodol [Deslongchamps, 1984] features conversion of a $\beta$-mesyloxy lactol to an unsaturated lactone. Although it has been formulated as a direct fragmentation, the bond alignment of the substrate suggests a more reasonable mechanism involving participation of the ortho ester subunit.

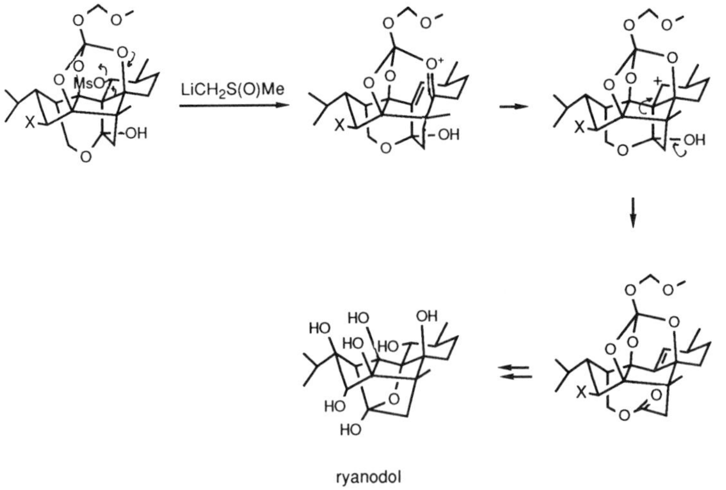

LiCH$_2$S(O)Me

ryanodol

Ultraviolet irradiation effects an alkenyl migration of vomifoliol acetate [Katsumura, 1982]. A zwitterionic intermediate may be formed after electron demotion. This zwitterion can either collapse directly to give the 1,4-diketone product, or after isomerization, to a cyclopropanol.

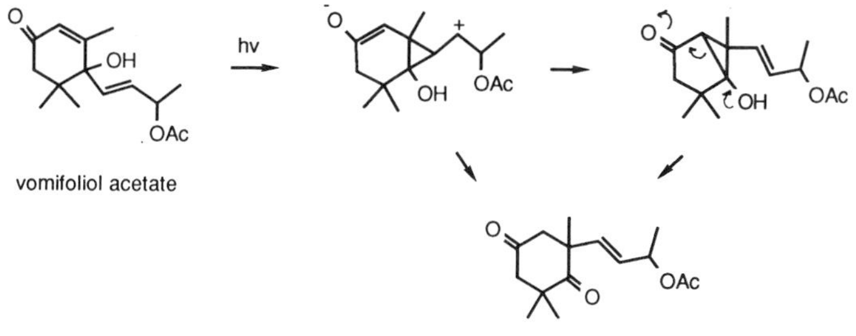

vomifoliol acetate

Electrolytic generation of (*E*)-5-cyclodecenone [Wharton, 1963] from a hydroxy decalin carboxylic acid proceeds in excellent yields. A incipient carbenium ion triggers ring opening. The stereoselectivity of the reaction is remarkable.

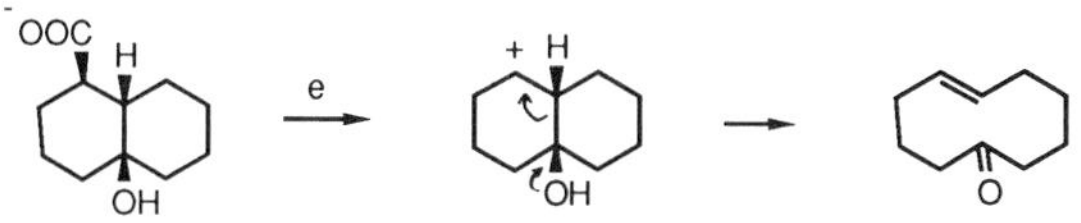

The reaction of thebaine with phenylmagnesium bromide leads to the tricyclic phenyldihydrothebaine [Bentley, 1952]. Opening of the dihydrobenzofuran ring is accompanied by an alkyl migration. The latter process creates a fragmentable situation.

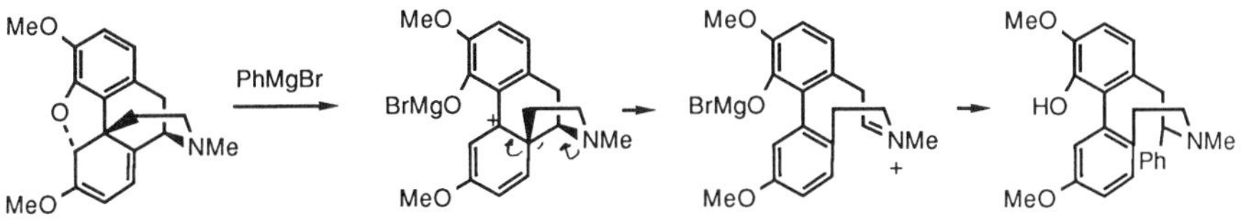

As a model reaction for the biosynthesis of the erythrina alkaloids a hydroxy-*N*-mesylnorsalutaridine analog was treated with a Lewis acid [B. Franck, 1971]. The treatment accomplished a C—C bond migration and fragmentation to give a dibenzazonine.

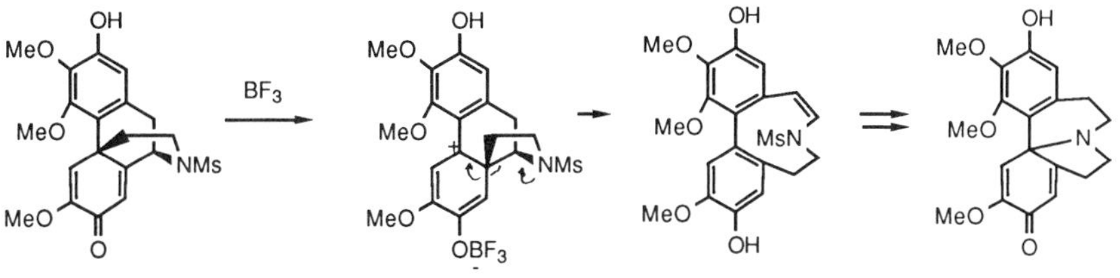

Amurine rearranges under acidic conditions to a phenanthrene [Döpke, 1968]. Protonation of the cross-conjugated dienone system induces a 1,2-alkyl shift which places a cationic center at the $\gamma$-carbon of the basic nitrogen atom, enabling fragmentation to proceed.

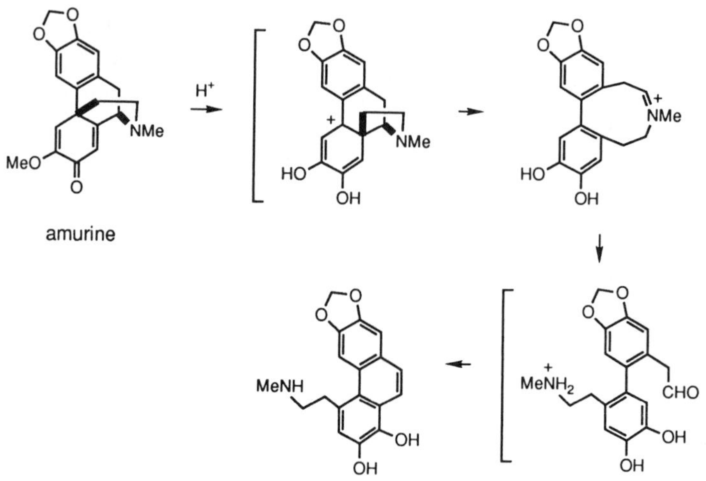

amurine

The fragmentation pathway sought by 4-iodomethyl- and 4-tosyloxymethylquinuclidines in alkaline media [Grob, 1962, 1970] is interesting in view of their substitution reactions occurring in acids. The formation of a disjoint *N,C*-dication in the latter situation is unfavorable.

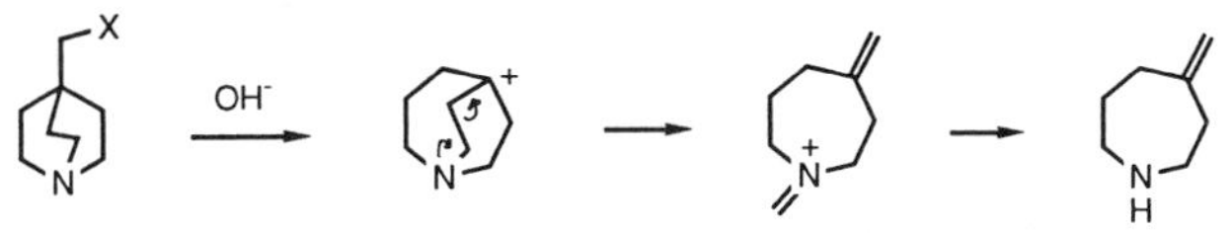

Photolysis in aqueous medium of a chloroacetamide in which the amine is part of a piperidine ring peri-annulated to an anthracene skeleton gives an imide [Bremner, 1989]. The nitrogen atom of the product is at a bridgehead position. It is reasonable to assume that the photochemical reaction involves intramolecular alkylation, hydration, and fragmentation.

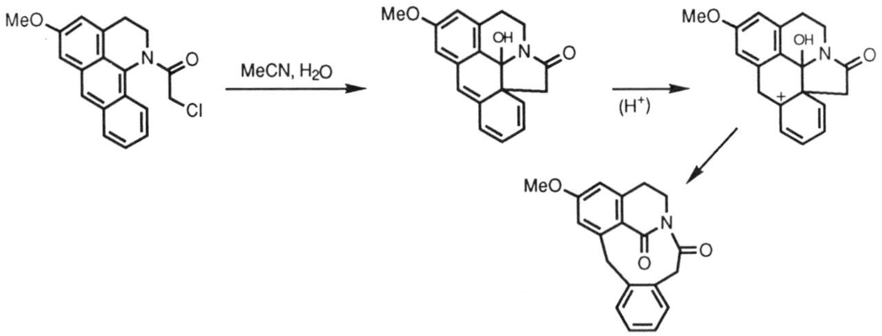

The fragmentation is possible only after rearrangement of the incipient carbenium ion. It should be noted that a competing deprotonation cannot occur because of strain (Bredt's Rule).

Chrysanthenol affords a monocyclic aldehyde on refluxing with methanolic potassium hydroxide [Hurst, 1960]. This transformation may be considered as a **d2a** type fragmentation although it could be a concerted process.

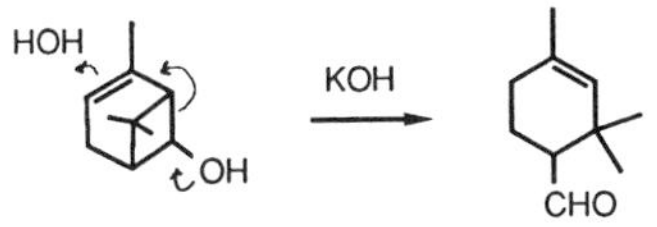

One of the photochemical products of 2-(2-hydroxy-4-penten-2-yl)-1,4-benzoquinone is 3-(2-oxopropyl)-2,3-dihydro-5-hydroxybenzofuran [Bruce, 1971]. Apparently isomerization of the Paterno–Büchi adduct to a cyclobutanol precedes the fragmentation.

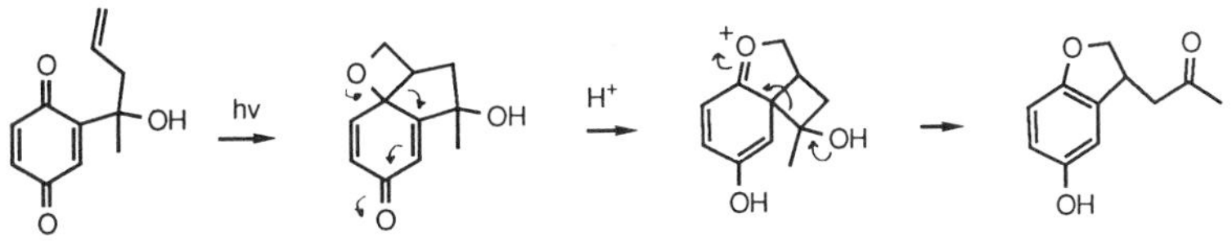

A convenient method for the preparation of 1-deuteriobenzaldehyde is by treatment of benzil with cyanide ion in heavy water [Burgstahler, 1972]. Rearrangement of the acyl cyanohydrin adduct leads to a **d2a** system.

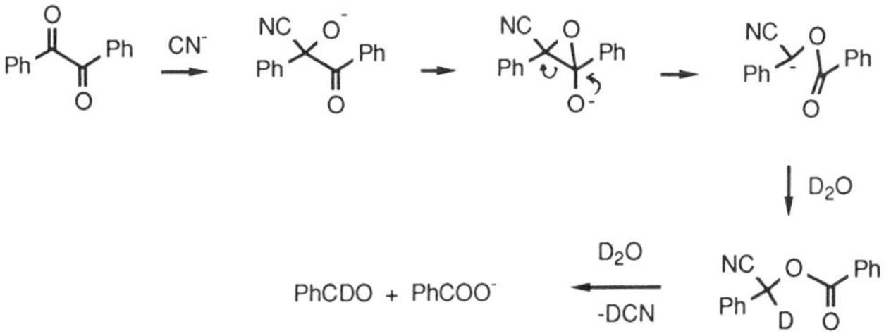

α-Bromocamphoric anhydride provides some laurolenic acid on heating with water or aqueous sodium carbonate [Fittig, 1885, Aschan, 1894]. Methyl migration is likely to accompany the departure of the bromide ion in order to avoid generation of an α-acylcarbenium ion. The rearranged cation is ready to fragment.

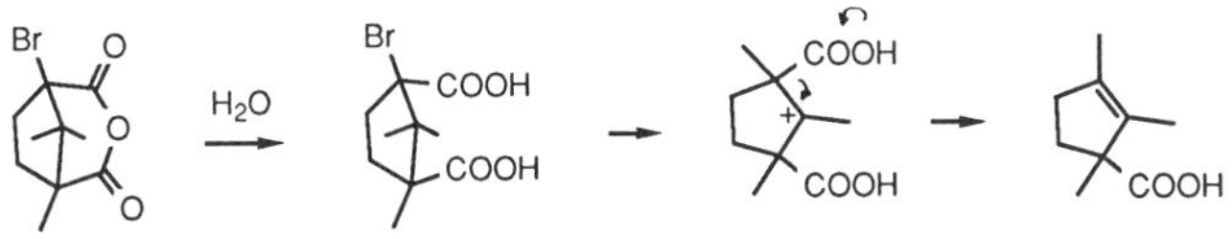

A similar sequence of events occurs in the cedrene series [Stork, 1953]. Thus, treatment of bromonorcedrenedicarboxylic acid with base led to a monocarboxylic acid and the clarification of this transformation contributed to the structural assignment of cedrene.

The photooxygenation of thujopsenol to form mayurone [Takashita, 1969] may have some biosynthetic implications. Loss of formaldehyde from the hydroperoxy compound is easy.

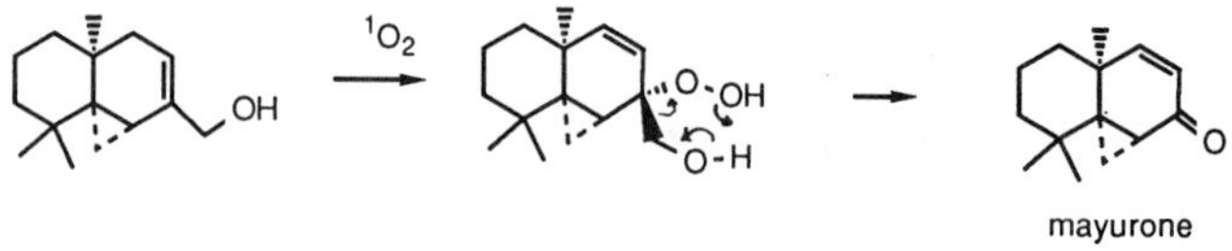

mayurone

The instability of actinidol is due to air oxidation [Sakan, 1967]. Oxidative cleavage of the ethanol chain results in actinidiolide.

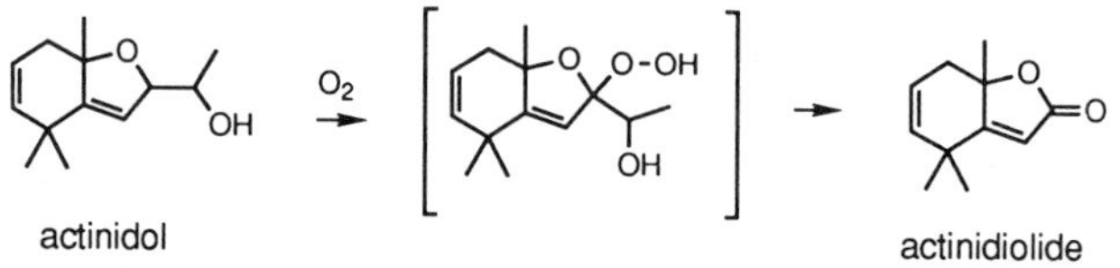

actinidol actinidiolide

A rearranged tetracyclic imine was obtained by pyrolysis of a dibenzocyclooctenyl azide [Jung, 1981]. It seems possible that a zwitterion was formed and it equilibrated with the cyclopropylcarbinyl cation, the latter species is an apparent precursor of the observed product.

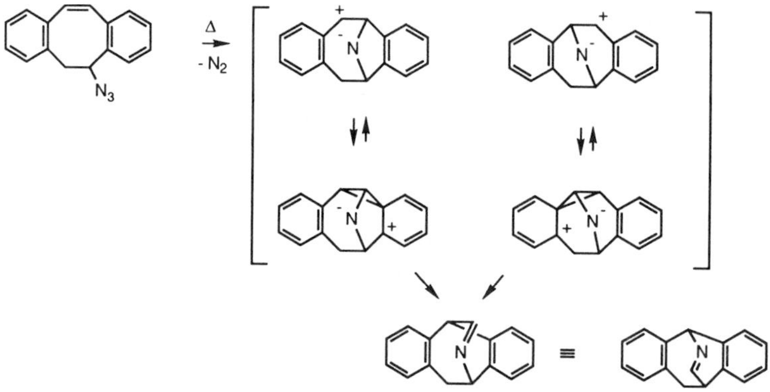

Mannich reaction of 1-hydroxybenzyltetrahydroisoquinolines in refluxing propanol also provides cleavage products [Kametani, 1977] in addition to the protoberberines. The cleavage is a **d2a** fragmentation.

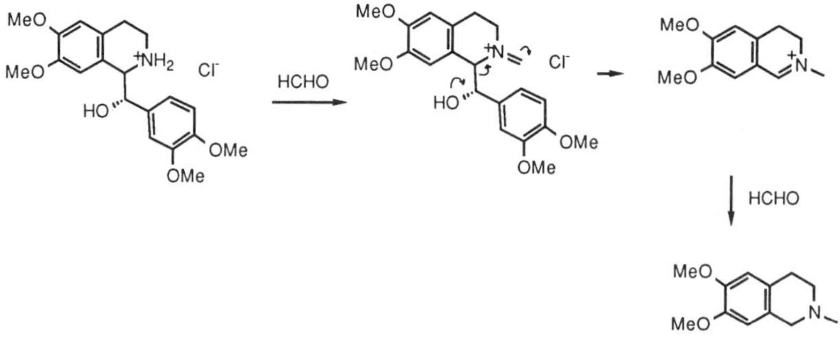

Sulfur atom-assisted ring expansion of 1-vinylcyclobutanols [Cohen, 1985] constitutes a new synthetic method for cyclohexanone formation. The sulfenyl group stabilizes the incipient carbanions generated from fragmentation of the cyclobutoxides.

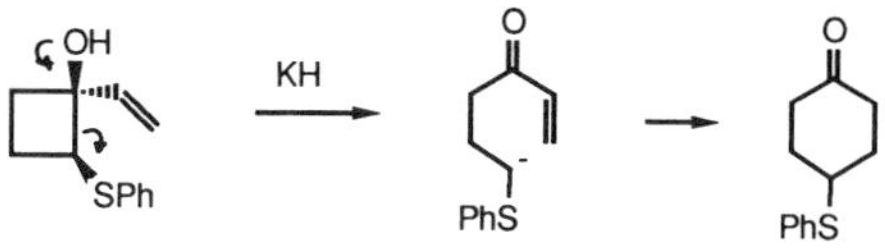

OH KH PhS PhS SPh

A synthesis of the vetispirane sesquiterpenes (agarospirol and hinesol) [Trost, 1975c] serves to demonstrate the usefulness of a "secoalkylation" protocol by which a carbonyl group is reductively replaced by two functionalized carbon chains. Thus, reaction of a carbonyl compound with a cyclopropylsulfonium ylide and rearrangement of the oxaspirane product leads to a cyclobutanone. For synthesis of the vetispiranes α,α-bissulfenylation and subsequent reaction with a methylmetallic reagent transform the cyclobutanone into a fragmentable species.

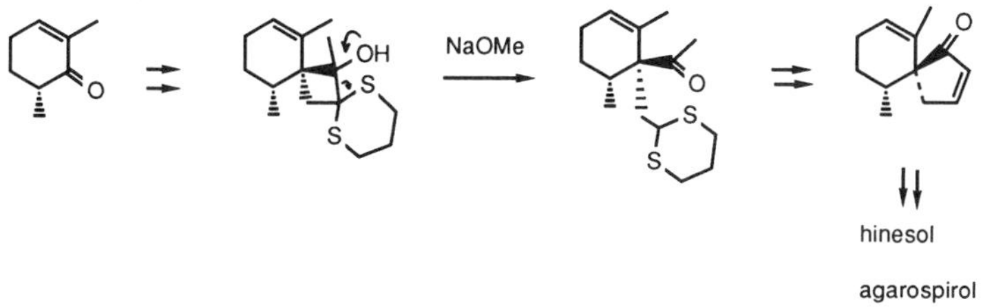

NaOMe

hinesol

agarospirol

The ability of the sulfinate anion to cleave epoxides [Kocienski, 1983] is due to the contrapolarizability of the sulfonyl group ($d \rightarrow a$).

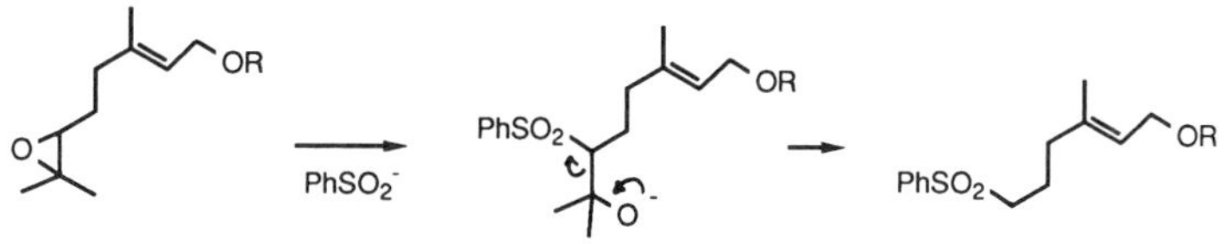

$PhSO_2^-$

Indoles undergo autoxidation readily. This transformation has been gainfully employed in a biomimetic synthesis of camptothecin [Boch, 1972]. Decomposition of the dioxetane intermediates to the keto amides may be viewed as a **d2a** fragmentation.

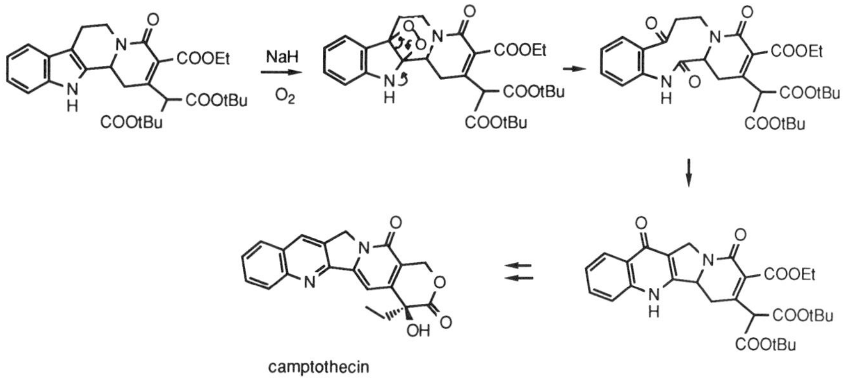

NaH $O_2$

camptothecin

Pyrolysis of the lithium salt of 8,9-dehydro-2-adamantanone tosylhydrazone furnishes *endo*-3-ethynylbicyclo[3.2.1]oct-6-ene [Murray, 1977]. Cleavage of the three-membered ring in the carbene intermediate sets up a fragmentation.

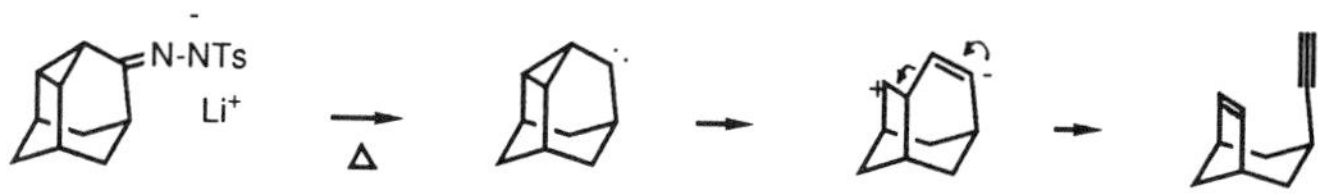

Reaction of proline with methyl propiolate in the presence of formaldehyde gives rise to an azacyclooctadiene [Ardill, 1987]. The first intermediate is an iminium ylide derived from methylenative decarboxylation. The iminium carboxylate from proline contains a **d2a** dircuit.

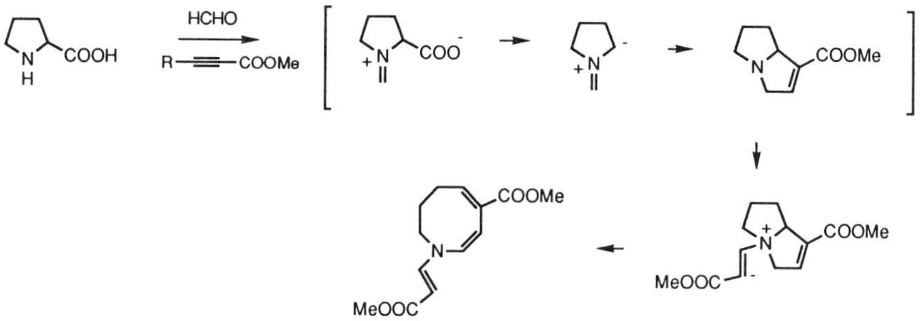

The formation of a tropolone ring during synthesis of colchicine [J. Schreiber, 1961] by osmium tetroxide oxidation of a cycloheptatrienecarboxylic acid involves decarboxylation. The proximal ethereal oxygen atom of the cyclic osmate ester acts as an acceptor. The osmium atom is further reduced.

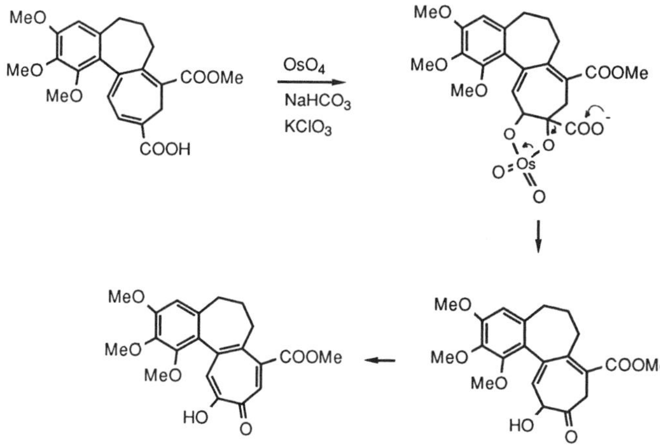

# 7

# d2a=d FRAGMENTATIONS

The separation of a donor and a doubly bonded acceptor site by two atoms gives rise to a situation conducive to fragmentation. Thus the fragmentability of $\beta$-oxy- and $\beta$-amino-carbonyl compounds and their frequent encounter by organic chemists mean that a wealth of fragmentation reactions would be found in the literature. Additionally, many other bifunctional substances belonging to the same **d2a=d** category contribute to the richness and variegation of these fragmentations.

## 7.1. DECARBOXYLATION OF $\beta$-KETOACIDS AND ANALOGS

The well-known instability of $\beta$-ketoacids indicates the energy gain by extrusion of carbon dioxide. The decarboxylation usually proceeds via a cyclic transition state from which the enol form of the ketone is produced. This requirement explains the resistance to decarboxylation by certain members in which the functional array is placed in a bridged ring system.

The monodecarboxylation of bicyclo[2.2.2]octane-2,5-dione-1,4-dicarboxylic acid is intriguing, because Bredt's Rule is apparently violated. However, the same rule is obeyed after the first molecule of carbon dioxide is removed! A rationalization [G.L. Buchanan, 1975] could be put forward when it was observed that the (+)-acid decarboxylated with racemization. Thus a ring opening (from a dimer adduct) precedes the loss of carbon dioxide from the ensuing cyclohexanecarboxylic acid and recyclization.

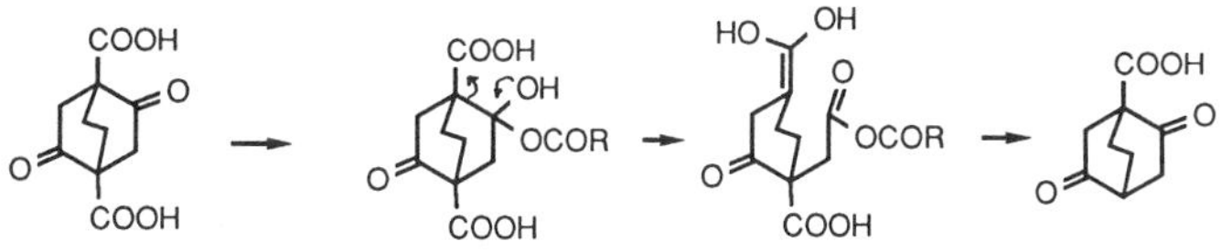

The generation of Hagemann's ester from acetoacetic ester and formaldehyde in the presence of an amine catalyst is interesting as only one of the ester groups of the intermediate disappears. The logical explanation entails the formation of a bicyclic lactone that undergoes $\beta$-elimination and then decarboxylation [McAndrew, 1979].

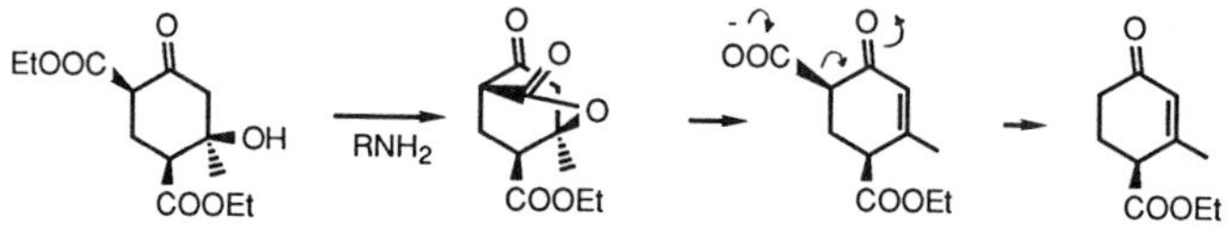

The intramolecular Friedel–Crafts acylation of a thienloxy *m*-phthalic acid is accompanied by the loss of a carboxyl group [Watthey, 1982]. The intermediate is a $\beta$-keto acid whose decarboxylation is favored by the prospect of rearomatization.

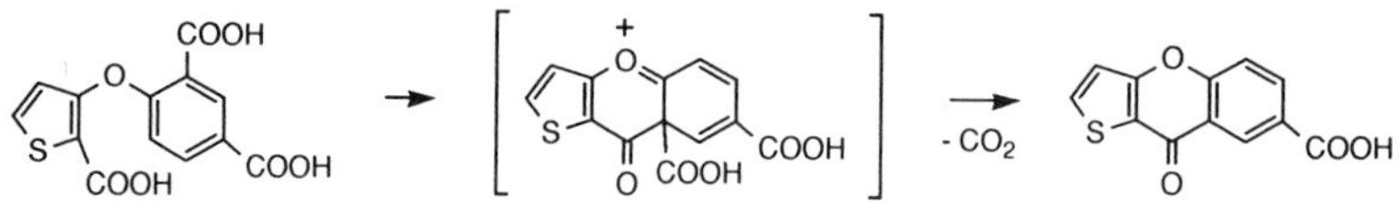

A synthesis of yohimbone [Swan, 1950] started from a condensation of tryptamine with *m*-methoxyphenylpyruvic acid. The Pictet–Spengler reaction product contains a carboxyl group which decarboxylates readily. Removal of an extraneous carboxyl group from a $\beta$-imino ester in a separate step was required in a synthesis of pumiliotoxin-C [Habermehl, 1976].

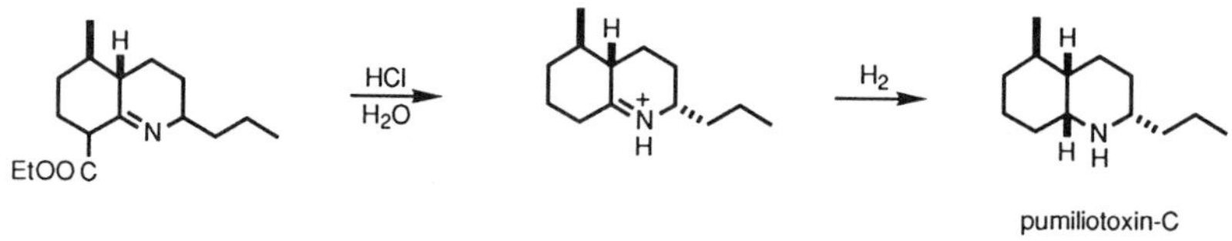

pumiliotoxin-C

On treatment with aqueous alkali a 1,2-dihydroquinolin-4-one underwent a Dieckmann cyclization, saponification, and decarboxylation [Nicholson, 1989]. The tetracyclic product deannulated to generate a phenolic ring.

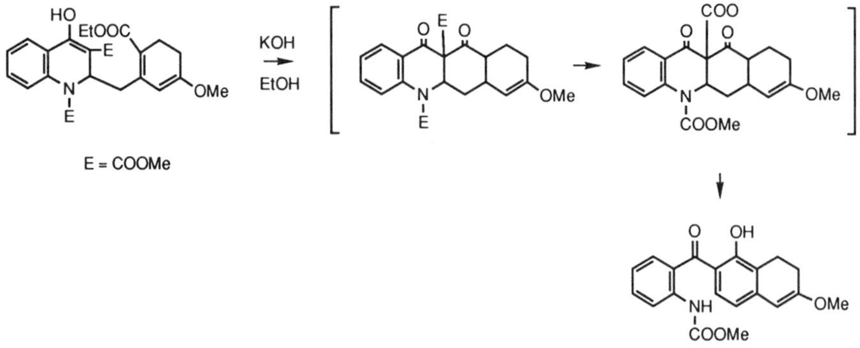

E = COOMe

During a synthetic study of kopsine [Magnus, 1989a] a hexacyclic chloro acid was exposed to silver acetate. Among the three different reaction pathways following the C—Cl bond heterolysis is *N*-assisted 1,2—CC bond migration. The resulting α-carboxy iminium ion rapidly lost carbon dioxide.

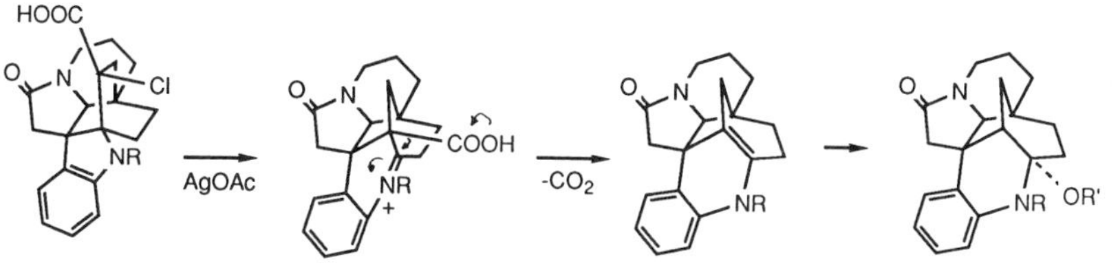

Oxidation of α-hydroxylaminocarboxylic acids with bromine leads to oximes [Pritzkow, 1967]. Decarboxylation occurs at the level of the α-nitrosocarboxylic acids.

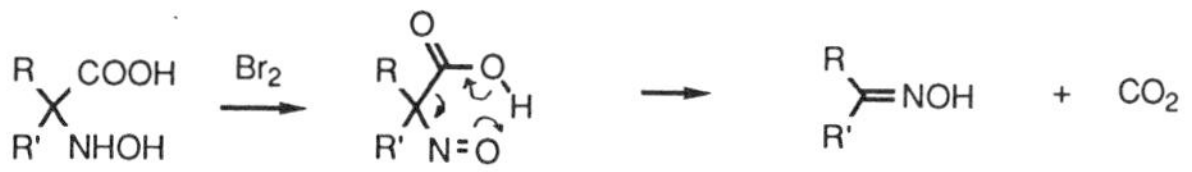

## 7.2. RETRO-ALDOL FISSIONS

The most widely recognized **d2a=d** system is an aldol. Retro-aldol cleavage is a well-known reaction which is usually carried out in neutral or basic conditions, when the competing dehydration is disfavored or prohibited by steric and/or electronic reasons. On many occasions, protected aldols in the form of ethers, esters also undergo this process of C—C bond cleavage.

Thermal decomposition of β-hydroxy ketones is unimolecular and involves a cyclic transition state [B.L. Yates, 1969]. The antibiotic X-537A is conveniently split into two portions by a thermal or based-catalyzed retroaldolization [Westley, 1970]. This process facilitated structural elucidation of the natural product.

A retro-aldol reaction initiates the conversion of aquayamycin to an anthracycline when treated with barium hydroxide [Sezaki, 1970].

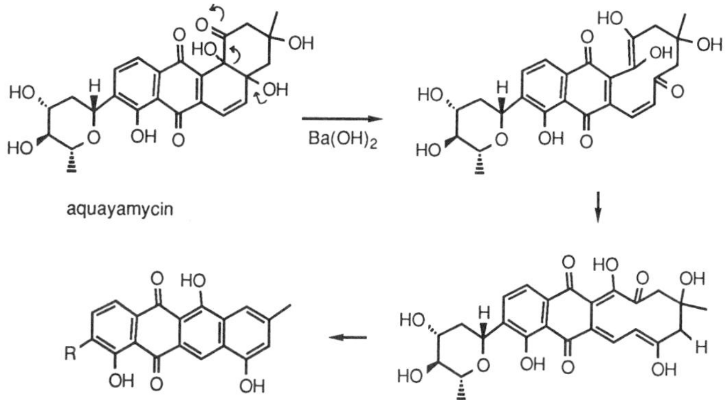

Besides hydrolysis of the primary amide, the antibiotic vancomycin undergoes a conformational change on heating at 80°C at pH 4.2 for 3 days. The change is a rotation of ca. 180°C of the aromatic ring III, likely occasioned by a retro-aldol/aldol reaction tandem which involves the α-hydroxy-*m*-chlorotyrosyl residue [Williamson, 1981].

In the alkaline hydrolysis of heteratisine [Aneja, 1964] retro-aldol reaction unravels a cycloheptanone and the bridged lactone via a decarboxylation process.

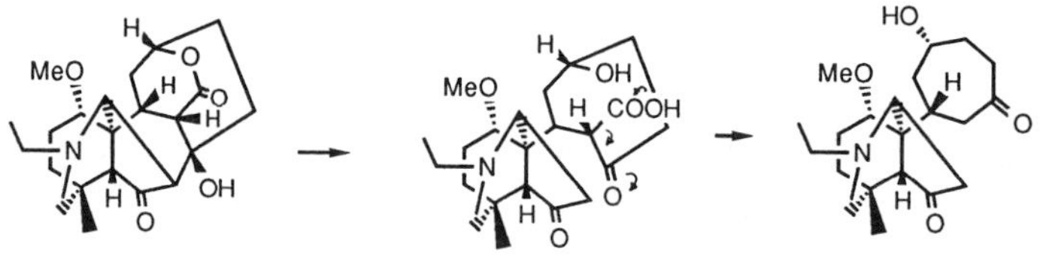

heteratisine

The stereochemistry of the ring fusion in ishwarone, especially the relationship between the cyclopropane and the tetralone system, was clarified by degradation via isoishwarone and a tricyclic ketol [Fuhrer, 1970]. Retro-aldol cleavage of the latter compound afforded a ketoaldehyde which on Wolff–Kishner reduction was converted into (+)-nootkatane.

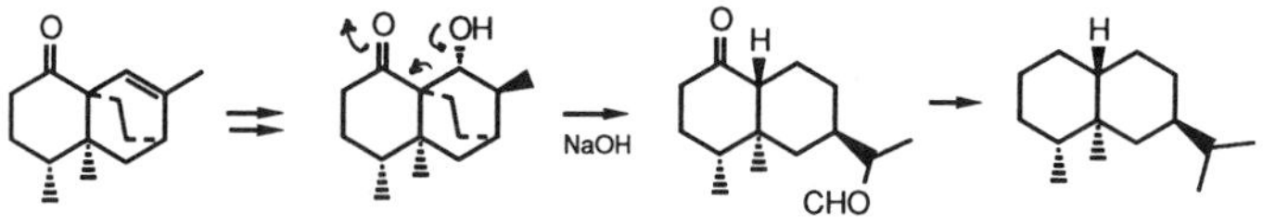

In synthesis the retro-aldol process sometimes helps and sometimes impedes attain the goal. For example, a synthesis of (−)-phytuberin [Findley, 1980] involved α-hydroxymethylation of dihydrocarvone. After separation of the 2:3 mixture of diastereomers, the major, undesired isomer was recycled by thermal equilibration (retro-aldol/aldol reaction tandem) to give a 5:4 mixture enriched in the useful product.

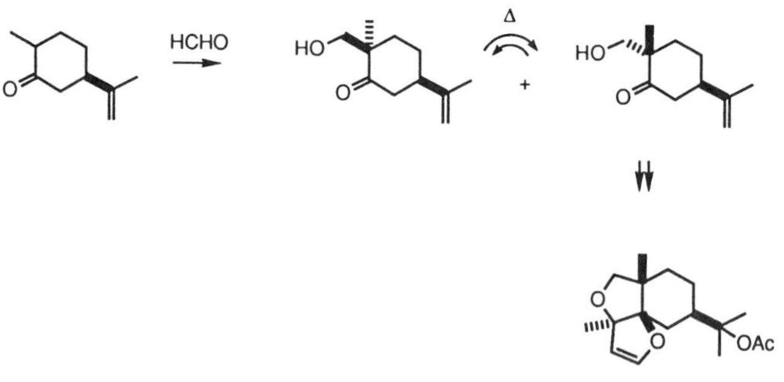

(-)-phytuberin

On the other hand, a disastrous retro-aldol fission intervened during the Wittig reaction in a route to (+)-hanegokedial [Taylor, 1983].

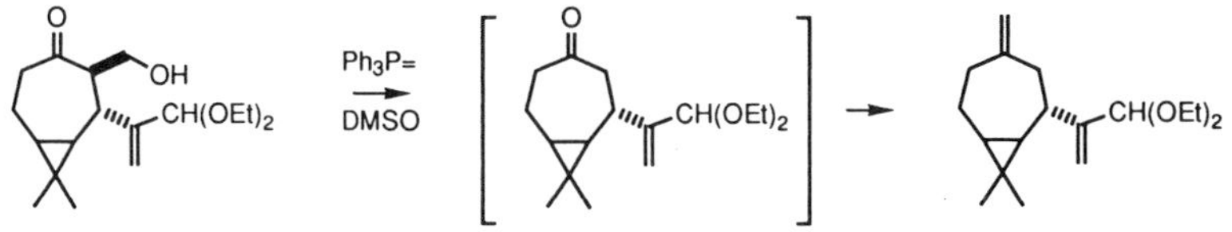

Saponification of the steroidal compound BR-1 pentaacetate leads to a complete loss of the A-ring [Miyahara, 1980]. Two retro-aldol fission processes, made possible by the presence of hydroxyl groups at C-1 and C-4, are responsible for the extensive degradation.

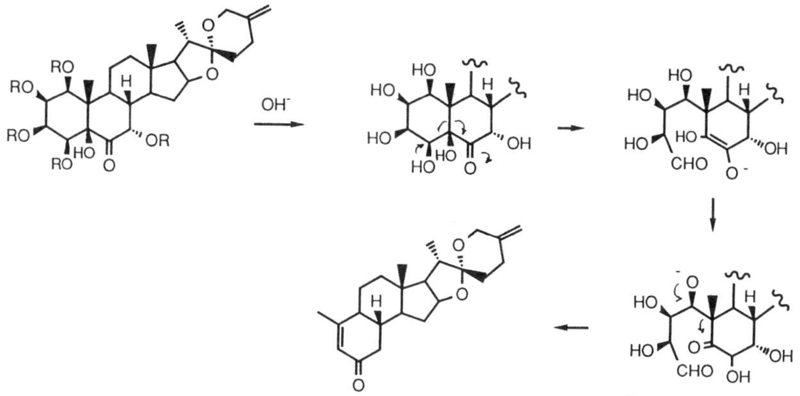

The diketone obtained from the Westphalen diol diacetate by ozonation suffers extensive degradation by bases [Morzycki, 1989]. The disengagement of the A-ring is resultant of a retro-aldol reaction, but the mechanism might not be a simple one.

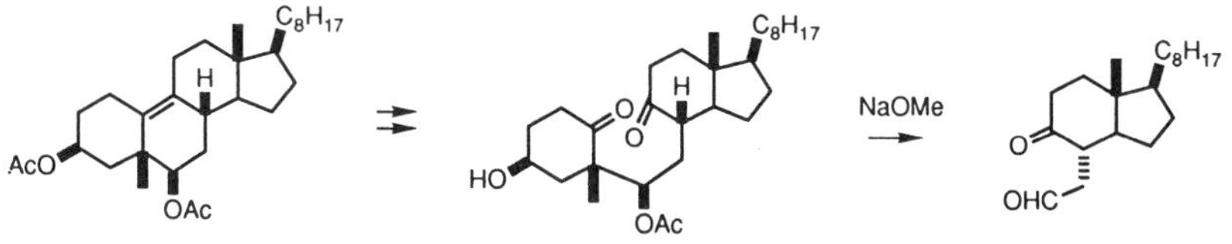

16-Acetoxy-17α-hydroxy-20-keto steroids expand the D-ring in the presence of base [Wendler, 1960]. This transformation involves a retro-aldol/aldol reaction sequence.

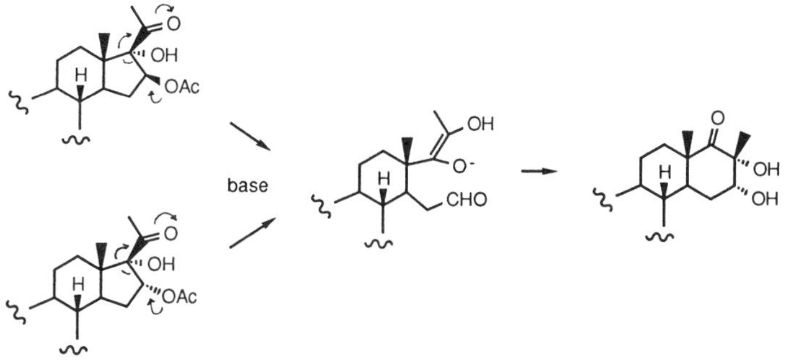

Certain corticosteroids have been degraded by reaction with formaldehyde in the presence of sodium bicarbonate [Noguchi, 1963]. An aldol condensation enables a tautomeric shift of the carbonyl group to C-21, and thence a retro-aldol fission.

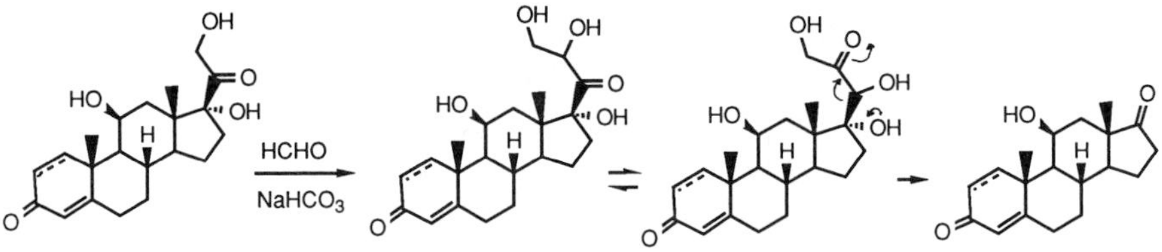

HCHO
NaHCO3

Saponification of trichokaurin gives mainly a tricarboxylic aldehyde [Fujita, 1969b] as the result of a retro-aldol cleavage. After equilibration a new lactol is formed.

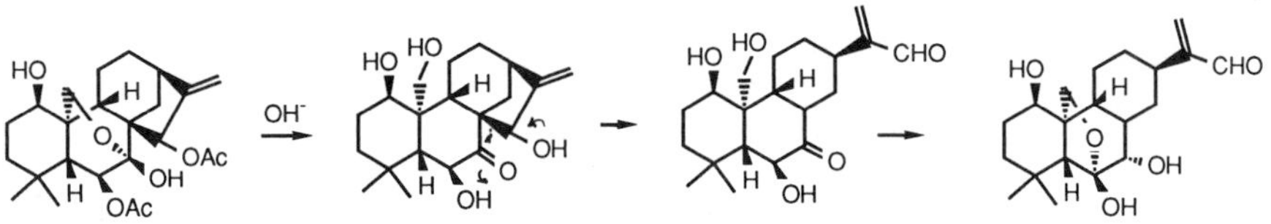

Epimerization of the endo hydroxyl group of the bridged ring system of various enmein derivatives can be effected under weakly basic conditions [Fujita, 1969a]. It is a thermodynamically controlled retro-aldol/aldol reaction sequence whereby the eclipsing interaction between the hydroxyl and the secondary methyl group is removed.

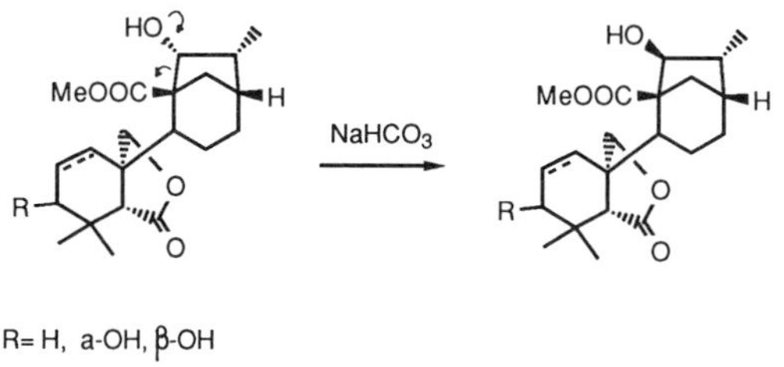

NaHCO3

R= H, a-OH, β-OH

Bicyclo[3.1.1]heptan-2-ol-6-one affords 3-(2-oxocyclobutyl)propanal on heating [P. Yates, 1966]. This transformation relieves a portion of the molecular strain.

In the last stages of a synthesis of the dehydrofuropelargones [Büchi, 1968] the acetylcyclopentene moiety was constructed by base treatment of a ketol. A retro-aldol/aldol tandem accomplished this goal.

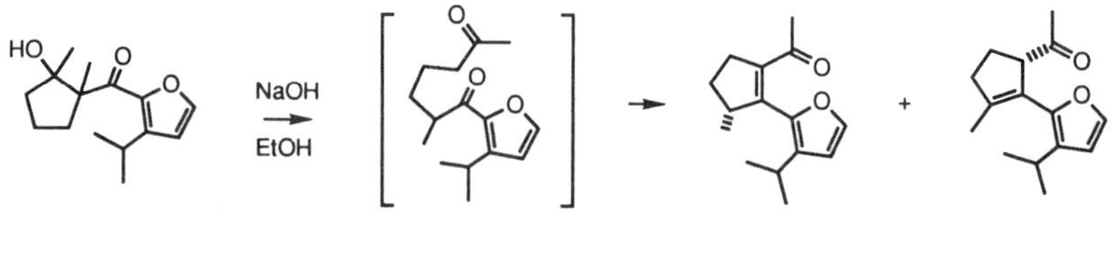

NaOH
EtOH

The facile substitution of the *exo*-4-acetoxy group of a peristylane-2,6-dione by various nucleophiles with retention of configuration has been interpreted [Eaton, 1979] as involving a retro-aldol and Michael reactions.

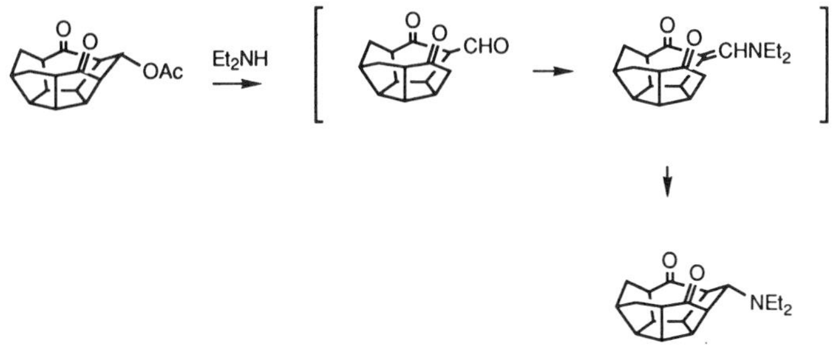

In the preparation of a key hydrazulenone intermediate for synthesis of helenalin [M. Roberts, 1979], and confertin and damsin [Quallich, 1979], a retro-aldolization was involved and it was followed by the Horner–Emmons olefination.

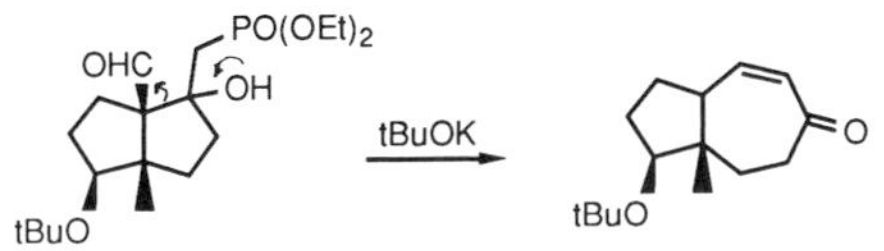

Consisting in the major pathway for the reversible isomerization of 2,3-dimethyl-3-alkylthio-2,3-dihydro-1,4-napthoquinones is also a retro-aldol/aldol reaction sequence [Silverman, 1981].

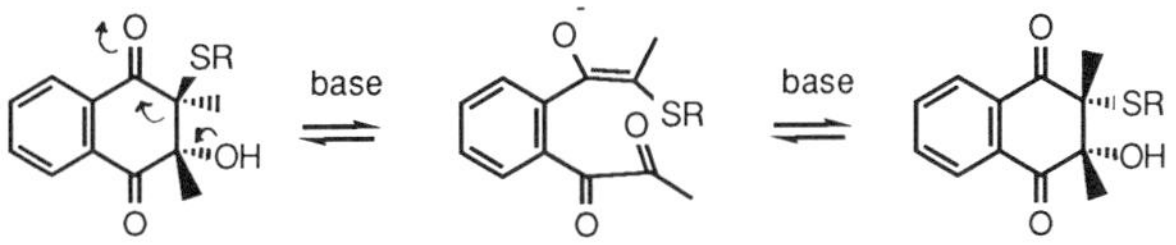

Ring-chain tautomerism has been revealed when $\beta$-hydroxy-$\alpha$-phenylsulfenyl ketones were exposed to base [Caine, 1983]. The equilibrium is strongly influenced by the substitution patterns of the substrates. Opening of a ketol to a diketone is evidently more favorable than the opening which results in a ketoaldehyde. However, the equilibrium can be shifted by irreversible trapping of the aldehyde, for example by a Wittig reagent, as shown in a synthetic route to ( − )-selinene starting from ( + )-carvone [Caine, 1987b].

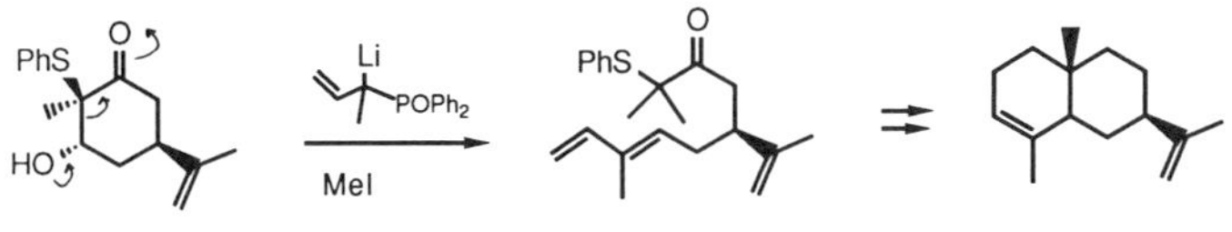

a-selinene

A model study for the assembly of the 2-methylcyclonon-2-ene-1,6-dione system was attempted with a retro-aldol ring fission in mind [Caine, 1987a].

The fission indeed occurred, but an aldol condensation in an alternative manner could not be avoided.

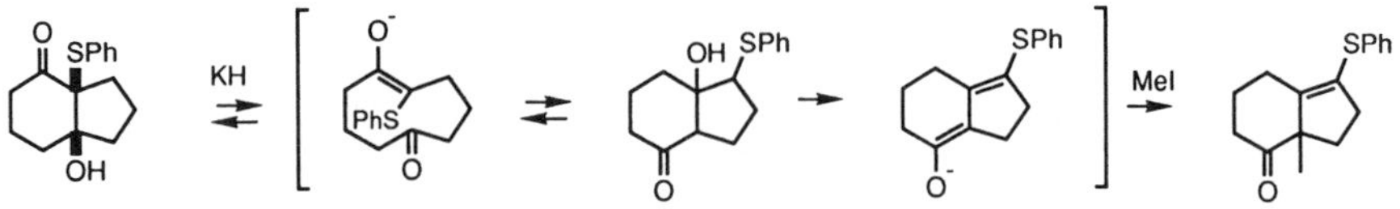

*O*-Methylation of both epimeric 4-hydroxy-2-adamantanones results in the same product [Duddeck, 1979]. Apparently methylation of the equatorial alcohol group is rapid, and the less reactive axial alcohol is equilibrated with the equatorial alcohol via a retro-aldol/aldol reaction pathway.

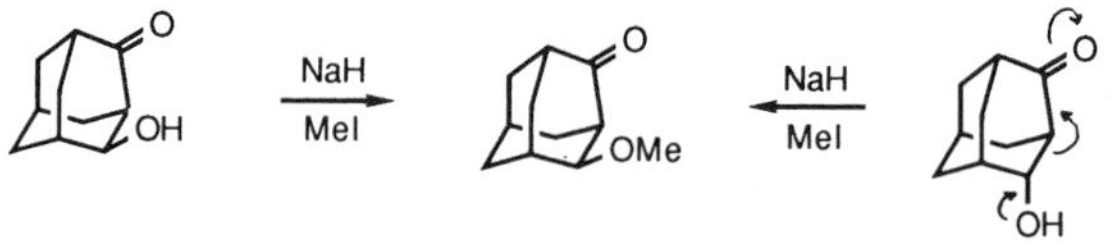

Skeletal reorganization was observed when the following cage ketol was treated with base [Mehta, 1981]. The fragmentation also caused the establishment of a conjugate ketone system, allowing an internal Michael addition to take place.

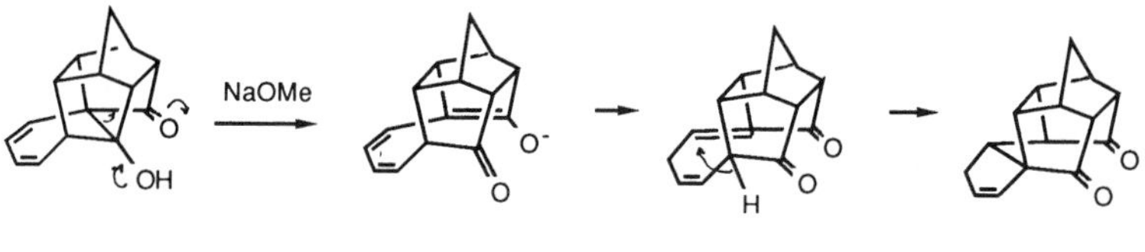

In a velbanamine synthesis the assembly of an iboga skeleton was deemed expeditious [Büchi, 1970]. Provision of a $\beta$-oxy ketone subunit permitted the required C—C bond cleavage at a later stage.

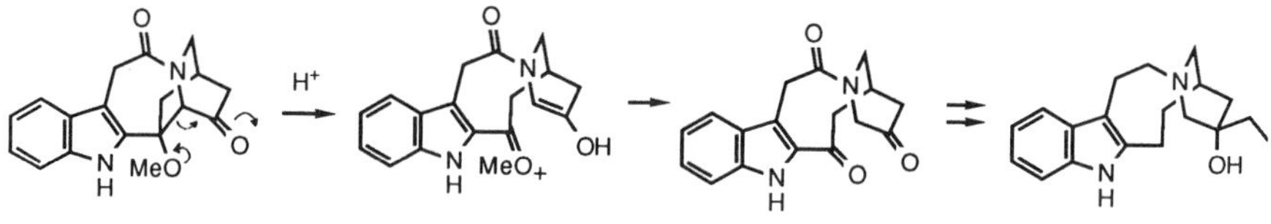

The push–pull mechanism for the consecutive fragmentation of donor-substituted cage diketones was established [Okamoto, 1984, 1985] by kinetic studies. The cycloreversion is a very fast process in the presence of a Lewis acid.

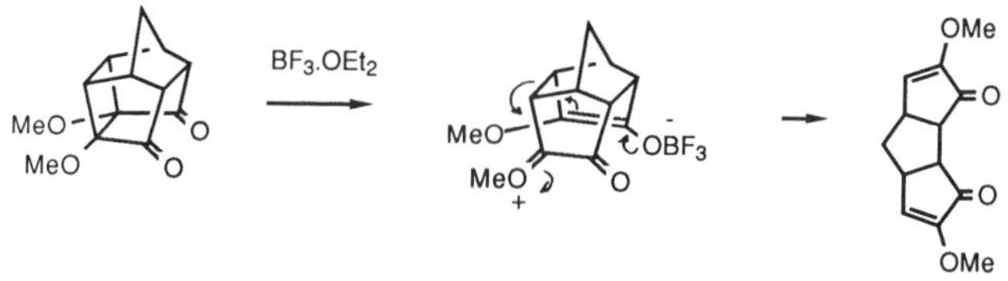

Diels–Alder adducts of thebaine and derivatives with enones are prone to fragment on contact with acid [Bentley, 1958]. Further transformations may follow.

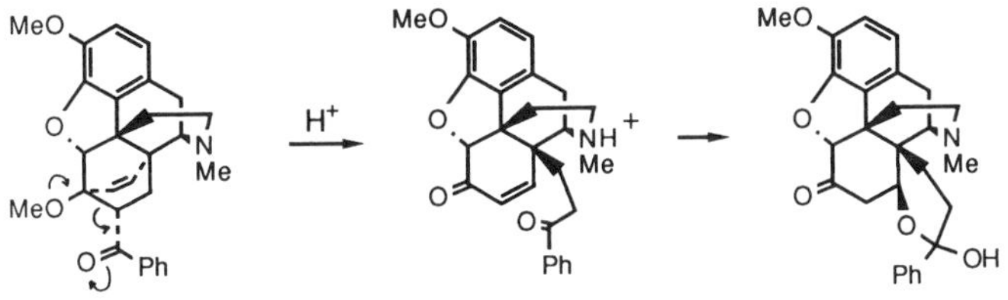

The Diels–Alder approach to 4-substituted cyclohexenones has been exploited in numerous syntheses, for example in an approach to *threo*-juvabione [A.J. Birch, 1969].

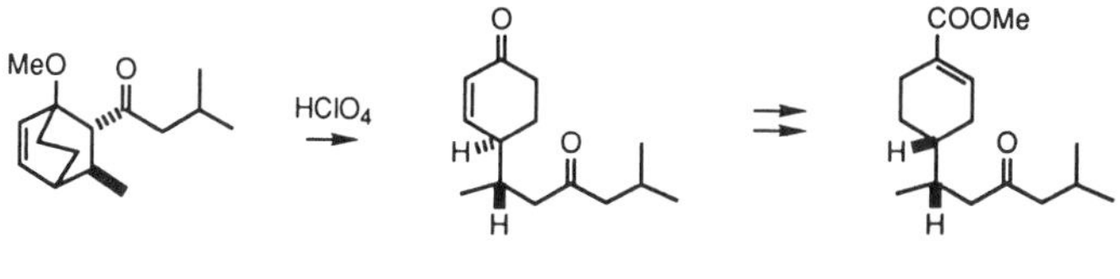

*threo*-juvabione

A new route to certain Amaryllidaceae alkaloids is based on the Diels–Alder reaction-fragmentaton protocol [Bird, 1989]. Aromatization provides a driving force for the retro-aldol-type fission.

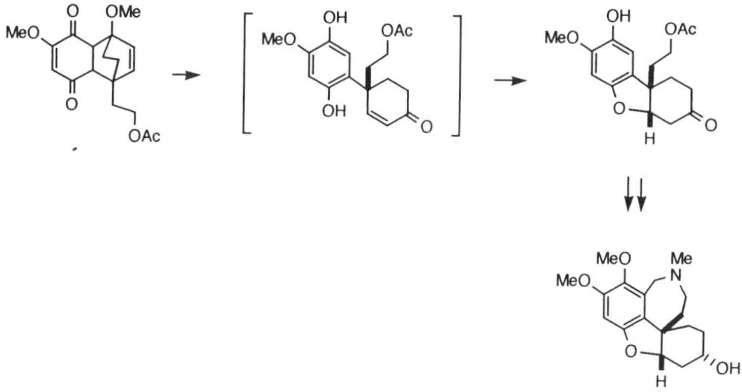

The aldol subunit is latent in benzobarrelene bridgehead ethers, and treatment of such a compound with acid promoted a retro-Friedel–Crafts reaction to give a biphenyl [Heaney, 1978].

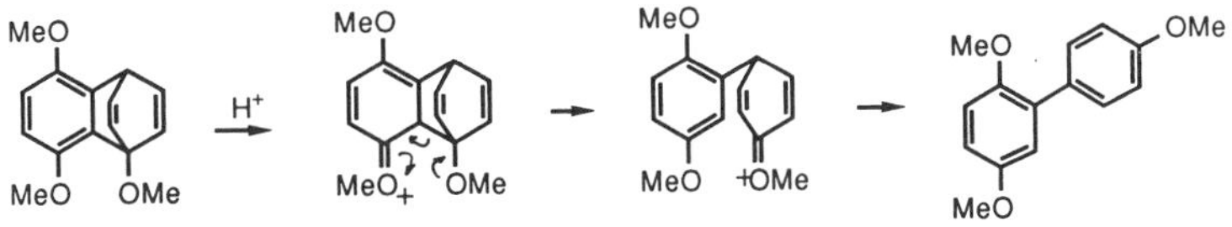

### 7.2.1. Fission of Nascent Aldols

Aldols in latent forms or emerged during reactions are liable to fragment. A well-known reaction of rotenone is the formation of rotenol and derritol [LaForge, 1929, 1930] on treatment with zinc and alkali. Loss of the two-carbon unit from the benzopyran ring is consequential of a retro-Michael reaction, reductive C—O bond cleavage, hydration of the enone, and retro-aldol cleavage.

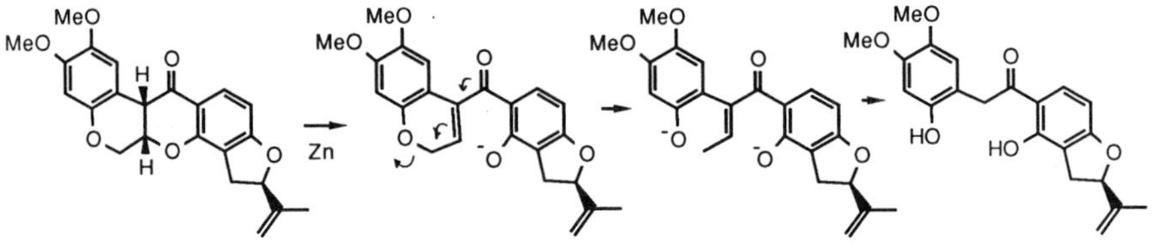

rotenone

derritol

The alkaline degradation of α-toxicarol leads to apotoxicarol and acetone [Heyes, 1935]. A retro-aldol reaction is also a key step.

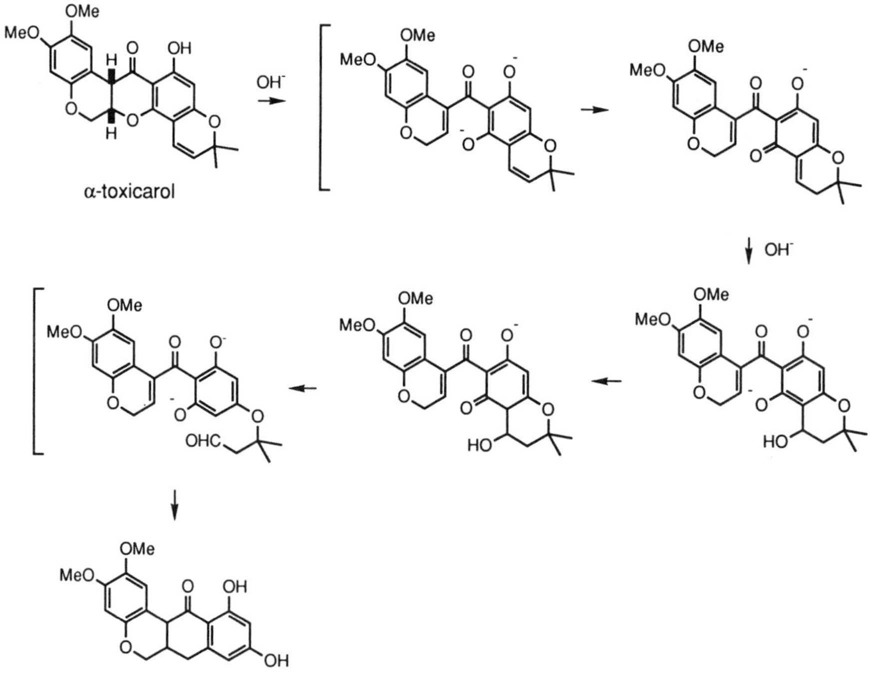

α-toxicarol

apotoxicarol

Destruction of the cyclohexenone system of the flavothebaone trimethyl ether methine base [Bentley, 1957] on reaction with alcoholic potassium hydroxide originates from hydration and retro-aldol fission.

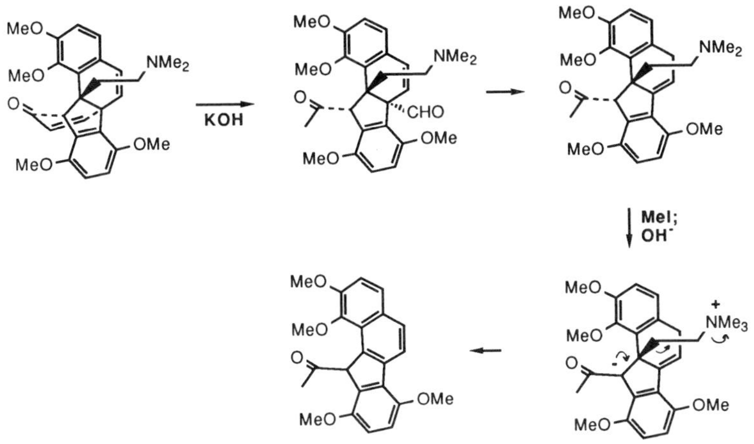

The enone transposition of 1-methyl-17$\beta$-acetoxyandrost-1-en-3-one with base [Bohlmann, 1964] is the result of a retro-aldol/aldol process. The driving force is perhaps the steric interaction of the vinylic methyl group with the C-11 methylene in the $\Delta^1$-enone.

Cyclization of 1-(2-oxoalkyl)-1-cyclopropanols on treatment with sodium hydride proceeds via an aldol condensation with the homoenolate anions which are derived from ring fission. However, this reaction cannot be extended to the preparation of angularly alkylated hydrazulenones in which the alkyl group is larger than methyl [Reydellet, 1989]. Such ketol anion intermediates appear to be extremely prone to fragmentation.

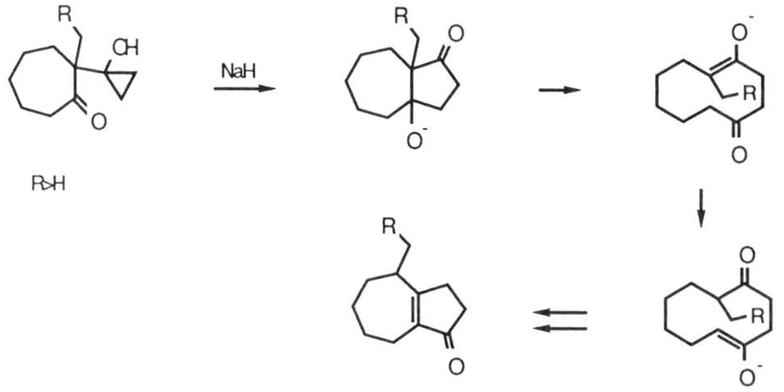

Oudenone contains an enolized triacylmethane function. Its hydration at high temperatures is followed by a fragmentation, producing 1,3-cyclopentanedione and $\gamma$-propyl-$\gamma$-butyrolactone [Ohno, 1971].

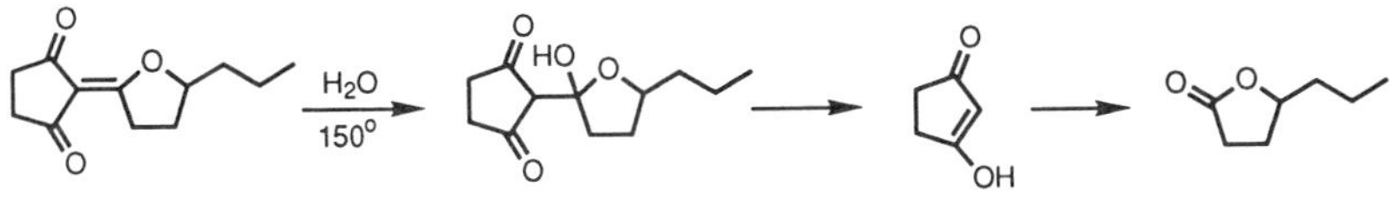

Ring cleavage of 2,2-disubstituted 1,3-cyclohexanediones has also been observed under the ketalization conditions [Jayaraman, 1989].

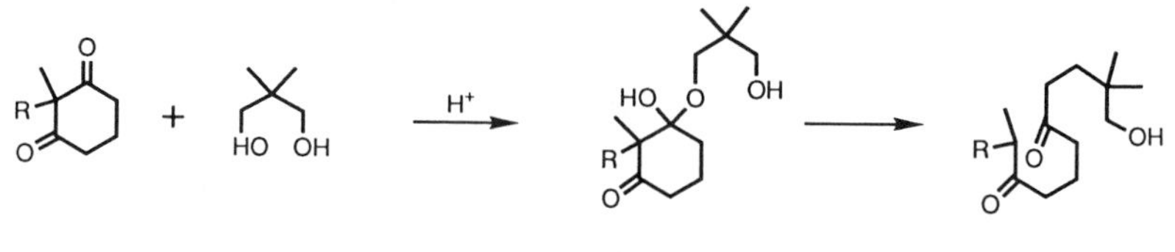

Vigorous saponification of forskolin yields a rearranged product [Saksena, 1985b]. The inversion of the C-1 configuration indicates a hidden retro-aldol/aldol reaction sequence.

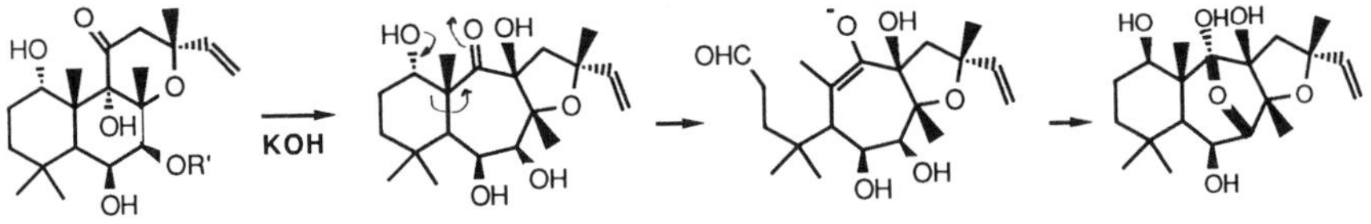

The Jones oxidation of a forskolin derivative [Khandelwal, 1986] gave both the expected diketone and one with a rearranged skeleton. Regarding the genesis of the latter compound, a retro-aldol/aldol route has been postulated, although a simple pinacol rearrangement before or after the oxidation step can also account for the result.

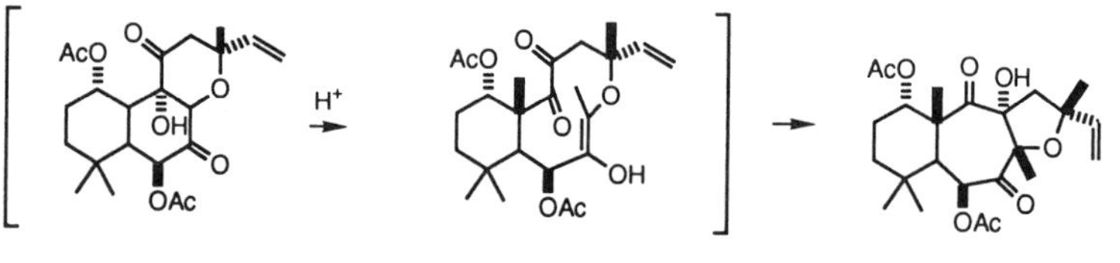

Tenulin undergoes rearrangement to afford desacetylneotenulin [Herz, 1962] in the presence of a mild base. Deacetylation is apparently the step preceding the retro-aldol fragmentation, double bond migration and aldolization at the $\gamma$ site of the new enone system.

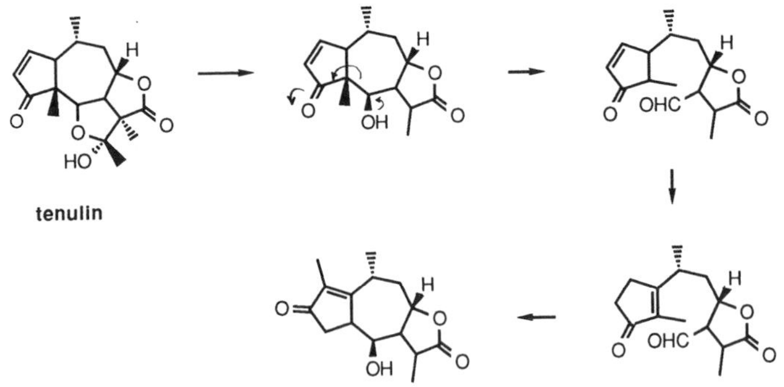

tenulin

Tetrodotoxin is a fragile molecule. It has been converted into a 2-aminobenzopyrimidine in alkaline solution [Goto, 1965]. The loss of a glyoxalic acid chain occurs after consecutive dehydration of the 1,2,3-triol subunit to generate an enone. The carbonyl group, being adjacent to the angular carbon, is capable of promoting the fragmentation.

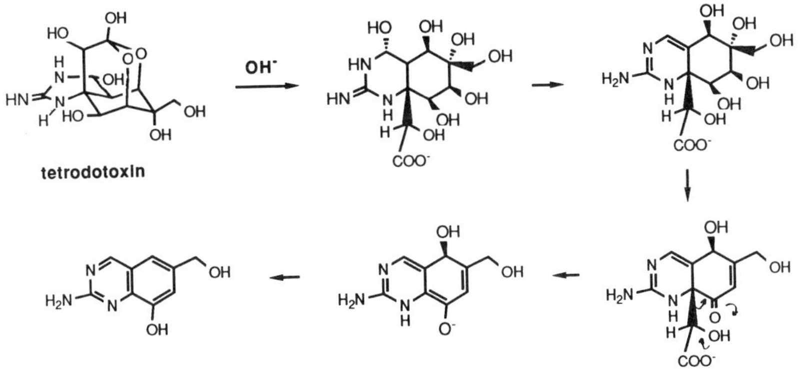

tetrodotoxin

Alkaline degradation of 4-deoxynivalenol results in three products [Young, 1986]. All of them are derivable from loss of formaldehyde via a retro-aldol process after isomerization of the hydroxy cyclohexenone system.

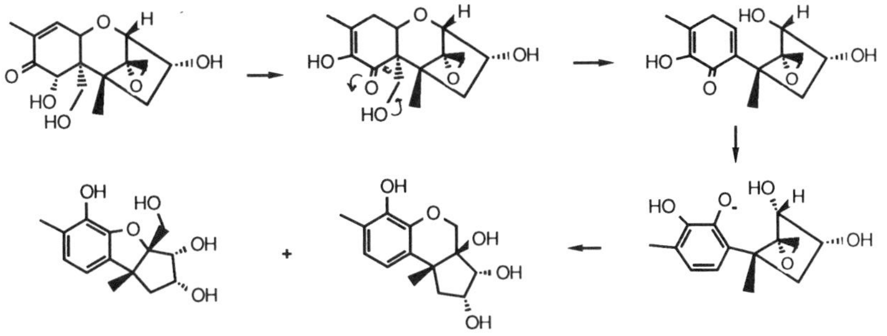

In alkaline medium the dihydroxydiketone monoaroate derived from illudin-S furnishes a red pigment [Nakanishi, 1965]. The structure of the pigment was shown to contain an extended orthoquinonemethide chromophore which arose from isomerization of the cyclopropylcarbinol and loss of the hydroxymethyl group after saponification of the ester.

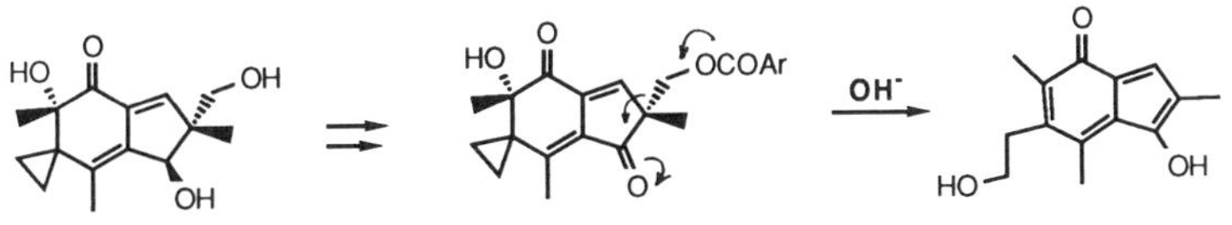

illudin S

An intriguing degradation of dihydroxymutilin ensues on treatment with alkali [Arigoni, 1962]. A transannular hydride transfer between the carbinol (C-11) and the ketone (C-3) triggers the release of glycolic aldehyde.

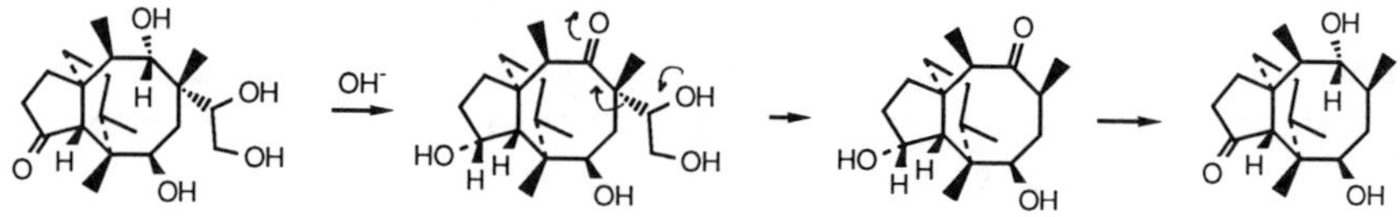

Fragmentation of the hydroxy ketal system shown now necessitates a 1,3-hydride shift [Botta, 1985] which has been authenticated by deuterium labeling studies.

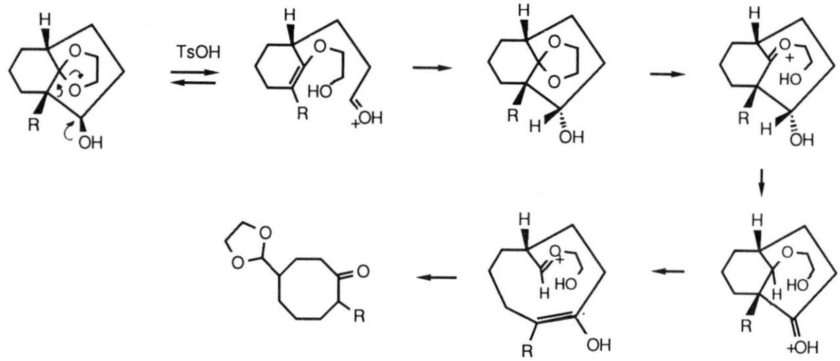

A synthetic study on the estrogen mirestrol [M. Miyano, 1972] focuses on the von Pechmann condensation to assemble the ABC-ring system. A desirable ketoester should be derivable from carbethoxylation of 2,2-dimethyl-1,3-cyclohexanedione. However, the reaction product of the diketone with diethyl carbonate is a dimer of the diketone. Under the reaction conditions an aldolization prevails and it is followed by a retro-aldol fission.

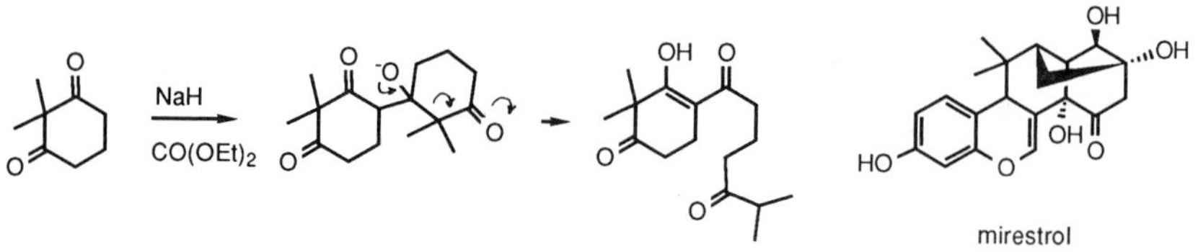

mirestrol

The ring cleavage of 1,3-cyclohexanediones in the aldol step of Robinson annulation on these substrates has been shown to involve oxatwistane and $\delta$-lactone intermediates [Muskopf, 1985].

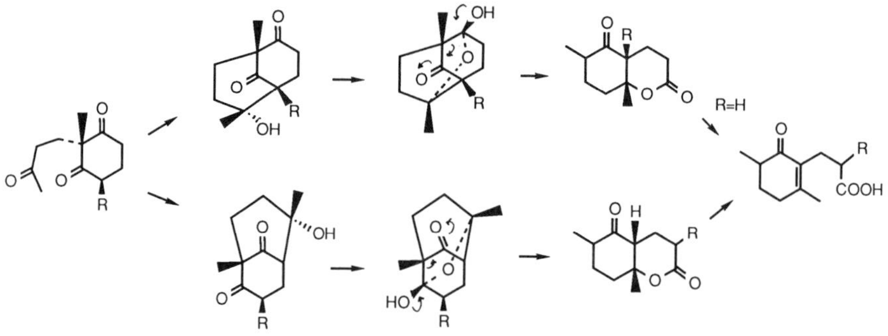

Rearranged *N*-acyl-1*H*-pyrroles arise from condensation of cyclic 2-(acylmethyl)-1,3-diones with ammonia [Sammes, 1984]. Fragmentation of the $\beta$-ketol intermediates and reorganization of the monocylic keto enamides represent the preferred pathway. The thermodynamically controlled process shunts the formation of the 3*H*-pyrrole system.

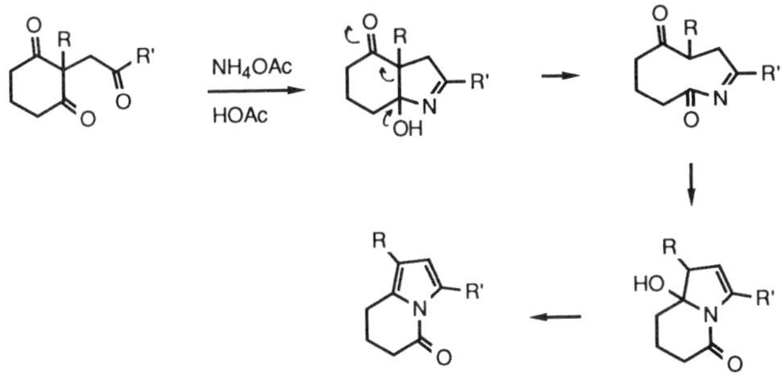

Base degradation of pseudopelletierine methiodide furnishes acetophenone and 1-acetylcyclohexene [Meinwald, 1956]. A possible mechanism embodies double elimination, hydration, retro-aldol/aldol reaction sequence, and disproportionation.

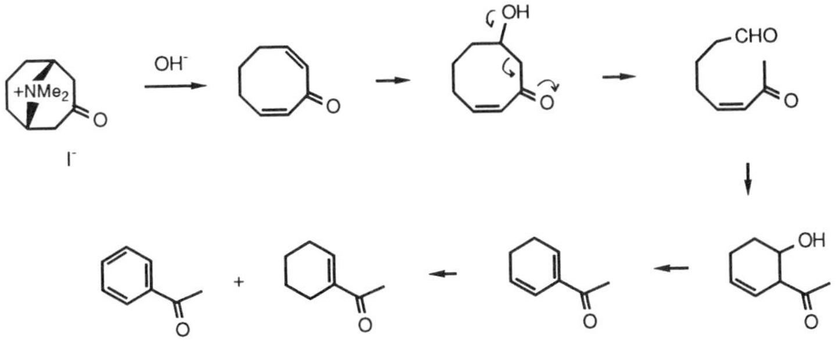

Photolysis of *N*-chloroacetyl-*O*-methyl-L-tyrosine leads to a pyrroloazepinonecarboxylic acid [Yonemitsu, 1967]. As the initial product cannot rearomatize (Bredt's Rule) without introducing excessive strain to the system, the hydration–retro-aldol fission process becomes the favorable conduit.

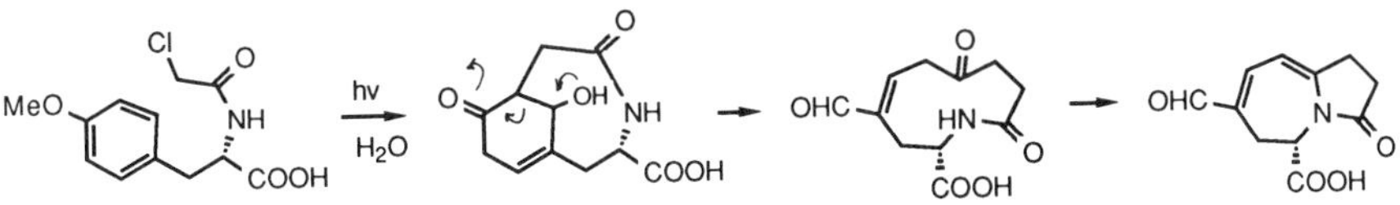

The production of (+)-*cis*-homocaronic acid from ozonation of (+)-4$\alpha$-acetoxymethyl-3-carene and decomposition of the ozonide with alkaline hydrogen peroxide [W. Cocker, 1974] involves a series of reactions by the iniitally formed keto acid. Subsequent elimination of acetic acid, epoxidation, retroaldolization and cleavage of the $\alpha$-diketone can be expected.

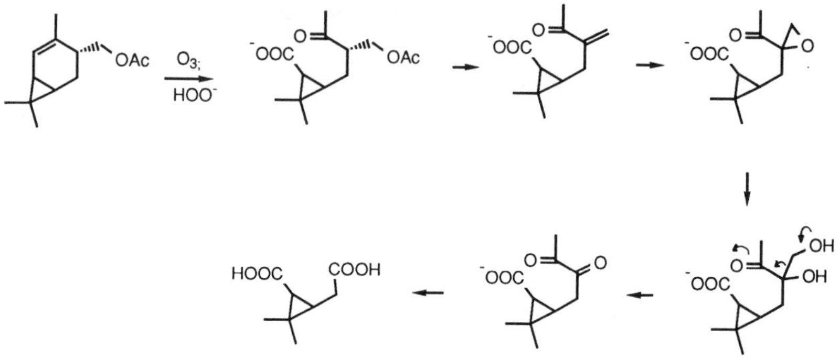

Application of the secoalkylation method to $\alpha,\beta$-epoxy ketones with proper modification has resulted in the formation of cyclohexenones [Trost, 1972]. It should be noted that such cyclohexenones are isomeric with those elaborated by the Robinson annulation on the parent enones.

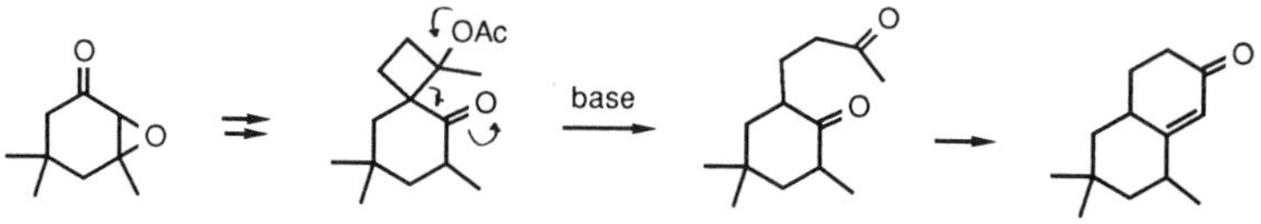

Indirect alkylation of isophorone at C-4 has been achieved via the Diels–Alder reaction of its enol acetate and cleavage of the adducts by the **d2a=d** fragmentation [Leffingwell, 1970].

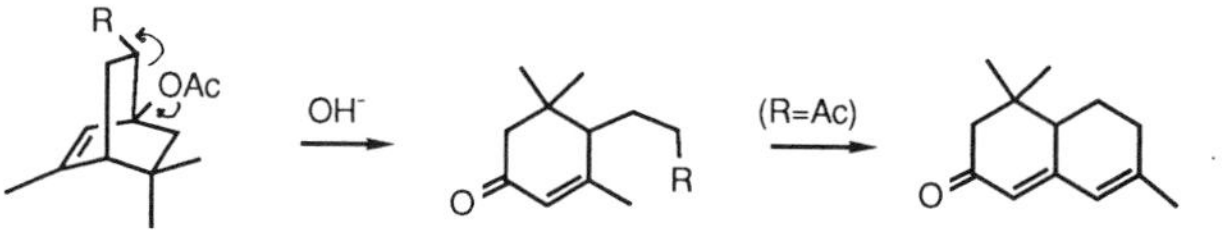

The Oppenauer oxidation of fucoxanthin yields paracentrone [Hora, 1970], albeit in low yield. The epoxy diketone intermediate can undergo $\beta$-elimination and a retro-aldol fission, resulting in the loss of the trimethylcyclohexane moiety.

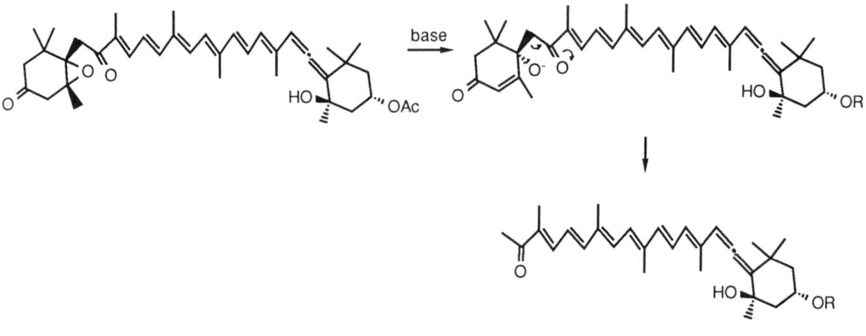

paracentrone (R=H)

The functional groups of kessyl glycol were interrelated [de Mayo, 1957] by the extrusion of acetone from the derived monoketone. With the establishment of the gross structure the thermal decomposition of a seco ketoacid is easily explained.

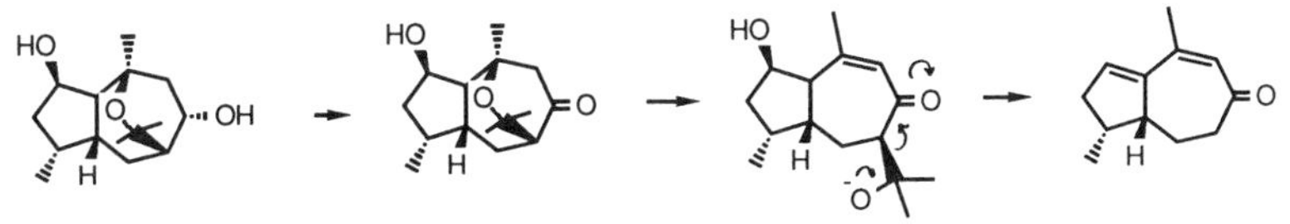

The structure of the major cycloadduct of dimethylketene and methylcyclopentadiene was shown [Huber, 1970] by its oxidative degradation to a seco ketoacid which split into 3-oxo-4-methylcyclopentanecarboxylic acid (*cis* + *trans*) and acetone.

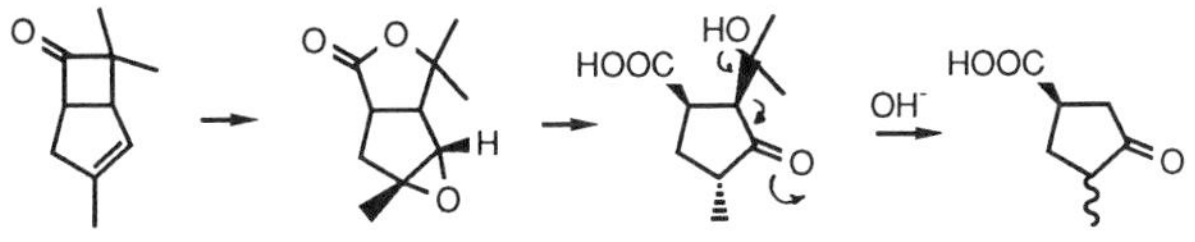

In a photochemical route to reserpine [Pearlman, 1979] the skeletal carbon atoms for ring-D were derived partly from angelicalactone and a dihydrobenzoic acid, with the latter supplying the ring-E template. The synthesis requires cleavage of the four-membered ring of the photoadduct and removal of an extraneous $C_2$ unit. The disjoint relationship of the functional handles is not suitable for fragmentation therefore an oxidative maneuver (Baeyer–Villiger reaction) was instituted.

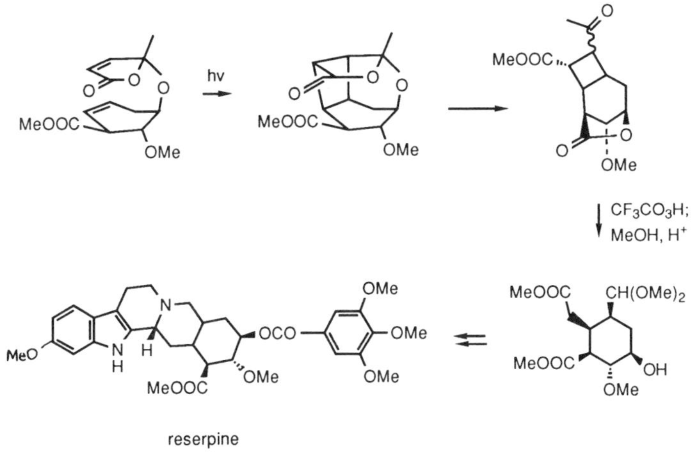

reserpine

An AB-seco epoxide derivative of totarol affords a naphthol on reaction with acid [Cambie, 1969]. This interesting process consists of a transannular cyclization and retroaldolization.

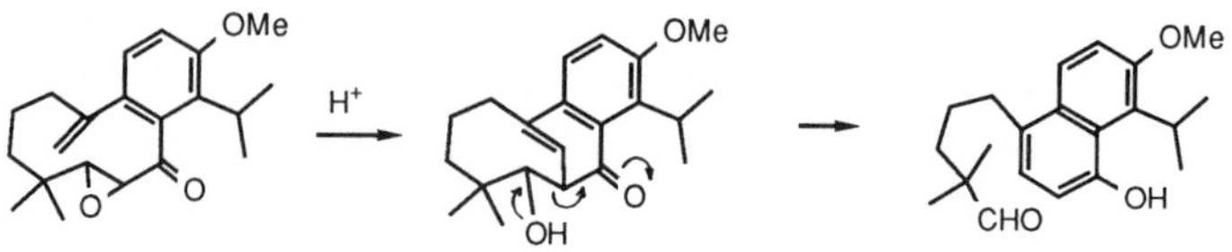

The "criss-cross" annulation via $\beta$-carbinolamine formation and reorganization is a useful strategy to gain access to certain nitrogen heterocycles. For example, by manipulation of a 1,3-cyclopentanedione it has been possible to synthesize the Nuphar indolizidine [Ohnuma, 1983b] which is a component of the dried scent gland of the Canadian beaver. A similar approach led to a mitomycin model system [Oda, 1984; Ohnuma, 1979] and dihydrodesoxyotonecine [Ohnuma, 1983a].

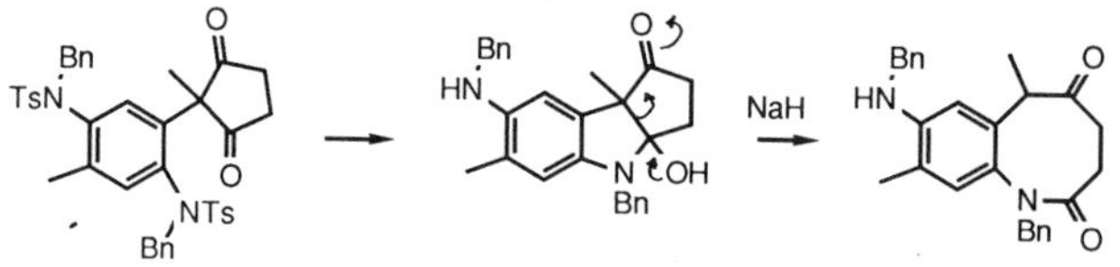

The combination of strain and the availability of a fragmentation pathway contributes to the lability of the following cage ketone. On standing at room temperature in toluene it dismutates into a linear tricycle. A twofold fragmentation mechanism accounts for this result [Klunder, 1984].

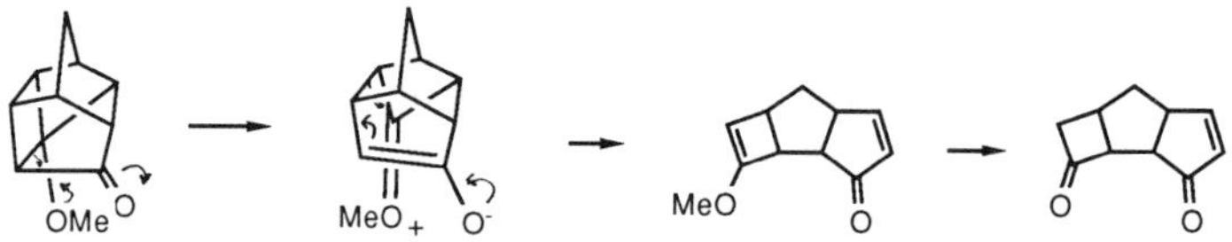

The facile conversion of a tertiary ether to the alcohol of an iboga skeleton [Büchi, 1970] is expected on the basis of a fragmentation pathway because it is $\beta$ to a ketone group.

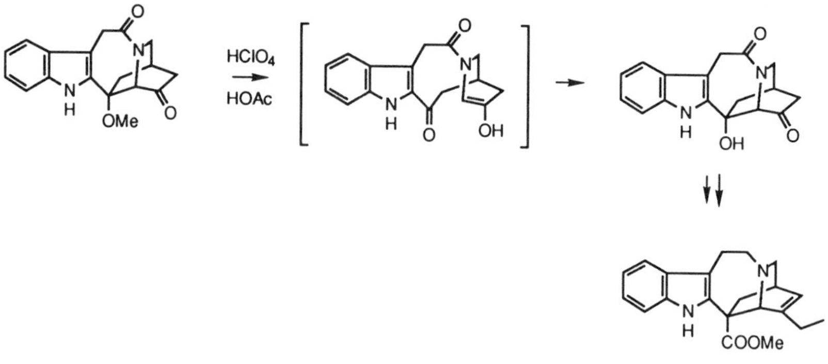

catharanthine

Exposure of the Diels–Alder adduct of 1-methoxy-4-methyl-1,3-cyclohexadiene and methyl vinyl ketone to acid affords 10-methyl-3,7-decalindione and two bridged ketols [A.J. Birch, 1966]. The intermediate for all these products is a cyclohexenone generated by fragmentation.

MeO O $H^+$ O O O O + OH O

A ketone derived from dihydroincensole oxide suffers cleavage of the macrocycle in the presence of an acid [Nicoletti, 1968]. Participation of the tetrahydrofuran oxygen in a retro-aldol scission is conceivable.

O O O HO O O HO O O

$\beta$-Cyano alcohols share many reactivity patterns with $\beta$-ketols. Fragmentation of this system takes place when dehydration is forbidden or energetically less favorable, such as its incorporation into a small ring. An (−)-acorenone synthesis [S.W. Baldwin, 1982c] took full advantage of this possibility.

O O CN OH CN CHO O

(-)-acorenone

A BC-ring synthon for the more highly oxygenated quassinoids such as bruceantin has been prepared [Hamilton, 1986] from an alkoxybenzoic acid via reductive alkylation and further transformations. A bridged ketol intermediate had to be converted to the condensed ring system by mild base treatment.

$Me_3SiO$ OMe COOMe OH OH O OMe COOMe $K_2CO_3$ OMe COOMe O

A temperature-dependent Robinson annulation sequence at high pressure has been discovered [Dauben, 1983]. The bridged ketol undergoes retro-aldol cleavage and linear condensation at higher temperatures.

COOMe O O 15 kbar MeCN, $Et_3N$ 35-40° COOMe O H OH 15 kbar MeCN, $Et_3N$ 65-70° COOMe O

Acetolysis of the endo-tosylate of 1,5-dimethyl-9-oxobicyclo[3.3.1]nonan-2-ol affords two products with rearranged skeletons, in addition to simpler compounds [J. Martin, 1972]. The mechanism suggests a 1,2-acyl shift upon ionization which precedes another rearrangement or hydration–retro-aldol cleavage.

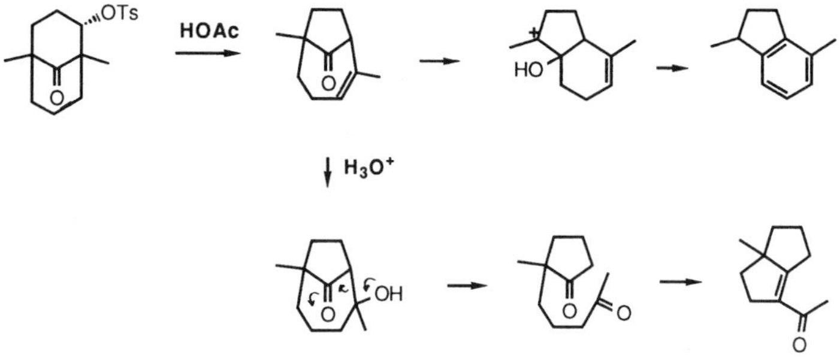

The attempted preparation of a methyl ketone from camphene-1-carbonyl chloride [Sobti, 1970] yielded a monocyclic diketone on acid hydrolysis of the crude product. It is envisaged that the molecule can sense a way of strain relief on hydration of the double bond because the $\beta$-ketol system is well disposed to fragmentation.

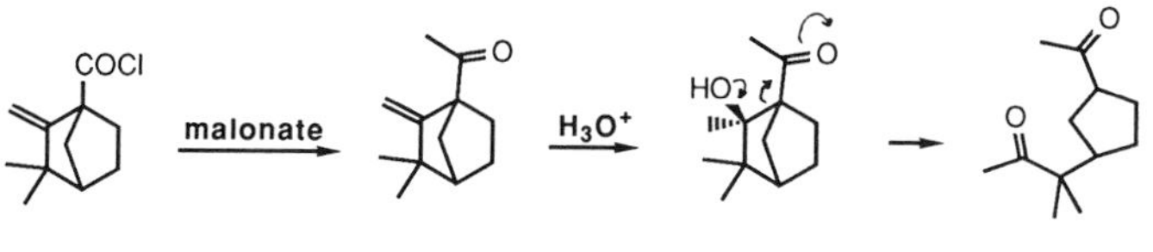

Pyroheteratisine undergoes decarboxylative degradation on boiling with alkali (pH 9) [O.E. Edwards, 1964]. The bridgehead double bond is prone to hydration and the molecule is relieved of much strain by a retro-aldol ring fission.

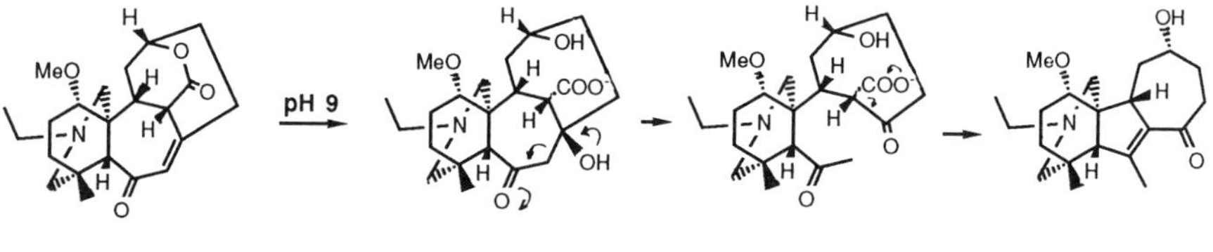

**pyroheteratisine**

Mercuric oxidation of tetracycline results in a B-seco quinone [Schwarz, 1967b] due to a retro-aldol reaction. The generation of allotetracycline by heating tetracycline in dioxane in the presence of triethylamine [Schwarz, 1967a] can be similarly rationalized.

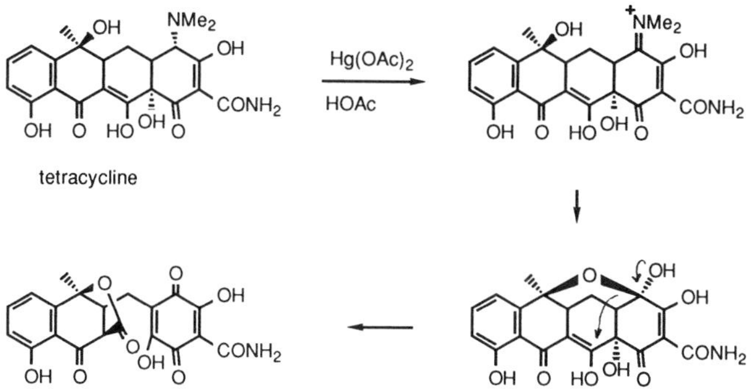

tetracycline

Reaction of a chalcone epoxide with sodium hydroxide in aqueous acetone does not furnish the dihydroflavonol. Instead, benzalacetone, 7-bromo-4,6-dimethoxyaurone, and 7-bromo-2-(2-hydroxyisopropyl)-4,6-dimethoxycoumaranone have been obtained via a 5-*exo-tet* mode cyclization, dehydration or retro-aldol cleavage. The coumaranone and benzaldehyde each condenses with acetone [Donnelly, 1979b].

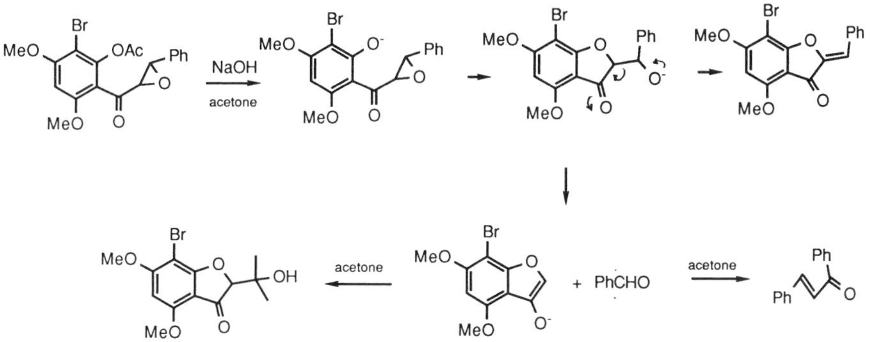

Alkaline degradation of verrucosidin [Ganguli, 1984] is initiated by pyrone ring opening. A retro-aldol cleavage follows the internal alkylation with the epoxide.

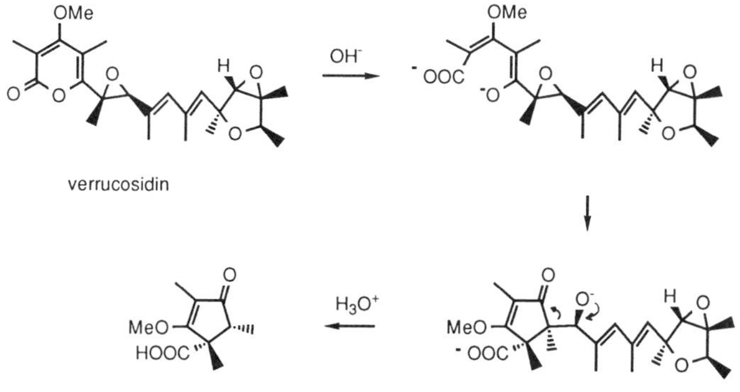

verrucosidin

One of the three products obtained from the action of alkali on polyporic acid is α-benzylcinnamic acid [Kögl, 1928]. The transformation apparently involves benzilic acid rearrangement and retro-aldol ring fission, a second benzilic acid rearrangement, and extrusion of oxalic acid.

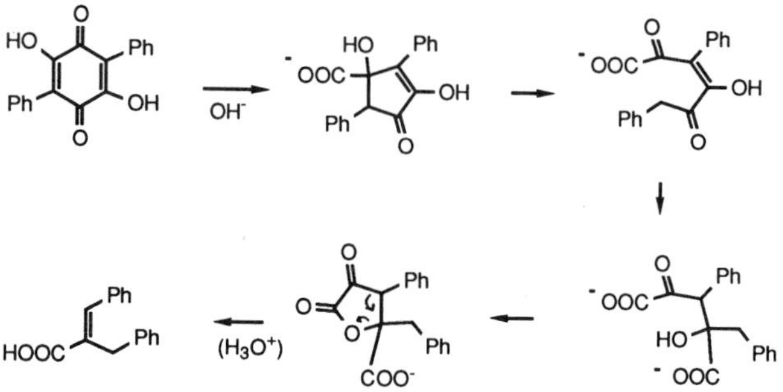

The base-catalyzed rearrangement of pseudoanisation to a γ-lactone [Kouno, 1984] features configuration inversion of the *vic*-glycol system. A possible mechanism entails pinacol rearrangement, retro-aldol fragmentation, equilibration of the monocyclic cyclopentanolone, reclosure of the six-membered ring and reversed steps.

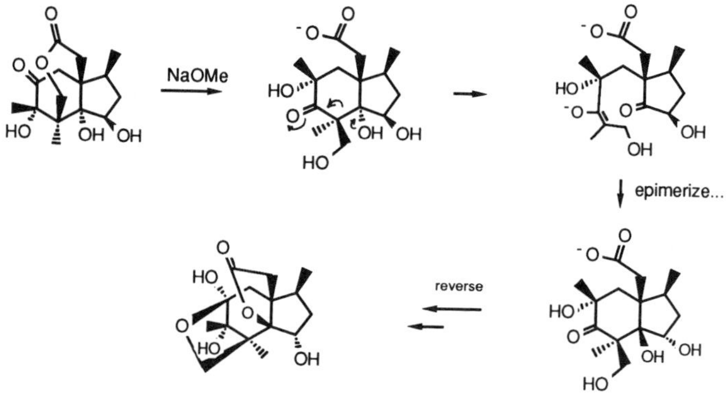

Condensation of 3,4-dihydroisoquinolines with unsymmetrical, nonenolizable 1,3-diketones takes two different courses [von Strandtmann, 1968]. The carbinolamine intermediates fragment to give amide/lactam products.

Of significance to prostaglandin biosynthesis is the disproportionation of an endoperoxide precursor. A model study [Zagorski, 1980] shows that the β-ketol anion arising from deprotonation of at the bridgehead may fragment or protonate, depending on reaction conditions.

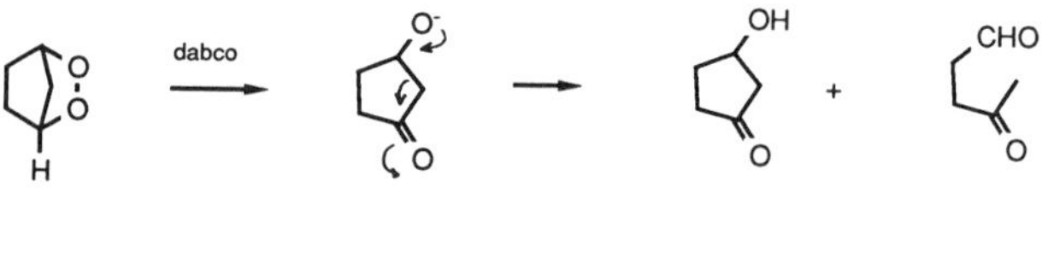

In the course of a synthetic study on anguidine [Ziegler, 1990], photooxygenation of a cyclopentadiene was complicated by the presence of a free alcohol in the six-membered ring. Apparently the internal base effected the decomposition of the endoperoxide.

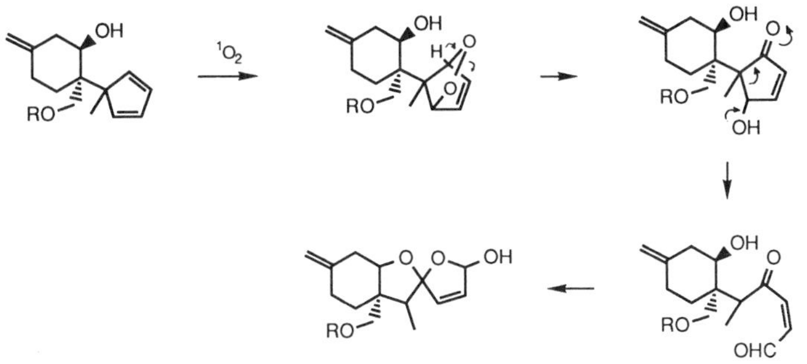

The methoxycarbonyl shift in the monoenol acetate of the Diels–Alder adduct of butadiene and 2-carbomethoxy-1,4-benzoquinone may be considered to be proceeding via an ene reaction, followed by fragmentation and acetylation [Akpuaka, 1982].

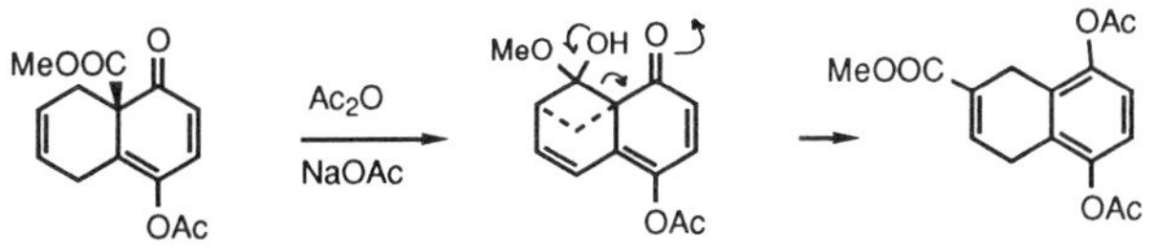

### 7.2.2. Cleavage of Three-Membered Rings

The propensity of cyclopropanols to undergo ring opening is largely due to ring strain and energy gain in the generation of carbonyl products. The presence of an appropriate exocyclic substituent such as C=O further facilitates the C—C bond fission. Needless to say, the direction of the cleavage would be governed by the acceptor group.

Aromatization with acetyl migration in a triketone as indicated in the following equation is easily rationalized in terms of an intramolecular alkylation and retro-aldol fission [F.B.H. Ahmad, 1981]. Mechanistically this reaction is analogous to the conversion of 2-(2-oxoalkyl)-indan-1,3-diones into 2-acyl-1,4-dihydroxynaphthalenes [Koelsch, 1940, 1943; Radulescu, 1927].

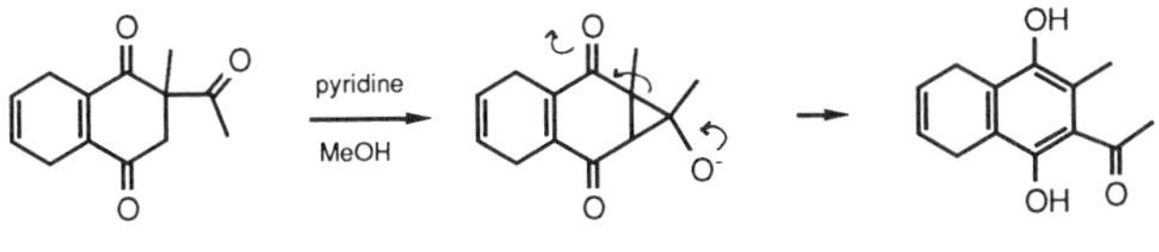

The formation of *m*-hydroxybenzaldehydes [Sampath, 1983; Föhlisch, 1985; Mann, 1985] from 6,7-disubstituted 8-oxabicyclo[3.2.1]octan-3-ones plausibly proceeds via acylcyclopropoxide intermediates.

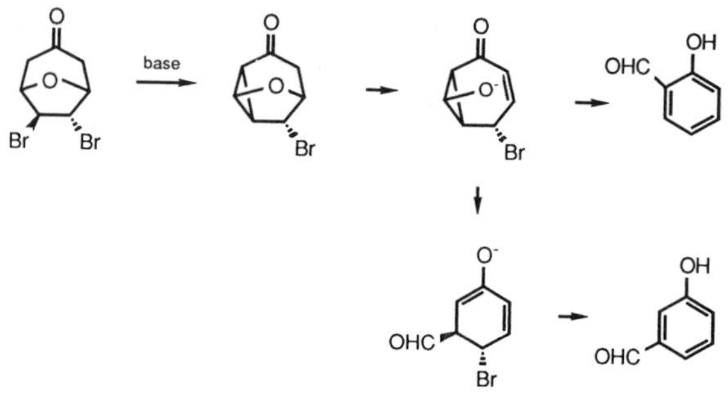

Conversion of 1,3,3-triphenyl-1,4-pentanedione to the 1,4,4-triphenyl isomer [P. Yates, 1971] involves formation and fission of cyclopropanol intermediates.

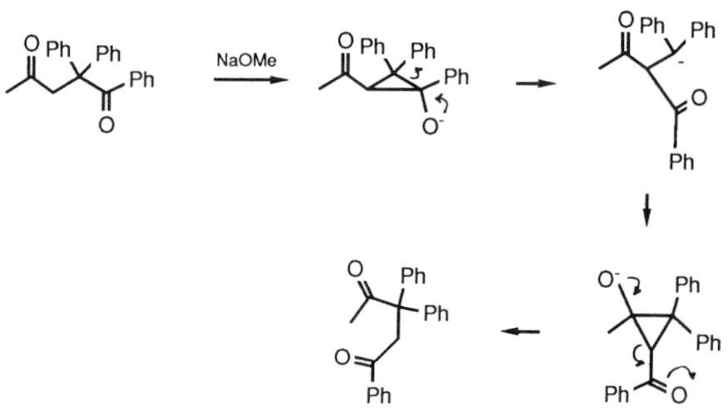

The formation of parasantonide from acid treatment and pyrolysis of santonic acid has been explained [Woodward, 1963b]. According to this mechanism fragmentation of an acylcylopropanol is the last step of the transformation.

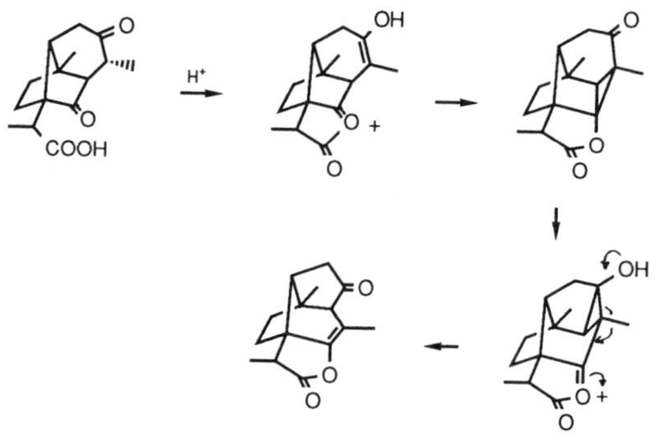

Decomposition of diazoketones in the presence of enol silyl ethers generates siloxycyclopropyl ketones. With subsequent silatropic retro-aldol rearrangement this reaction constitutes an excellent method for the preparation of monoprotected 1,4-diketones [Coates, 1975].

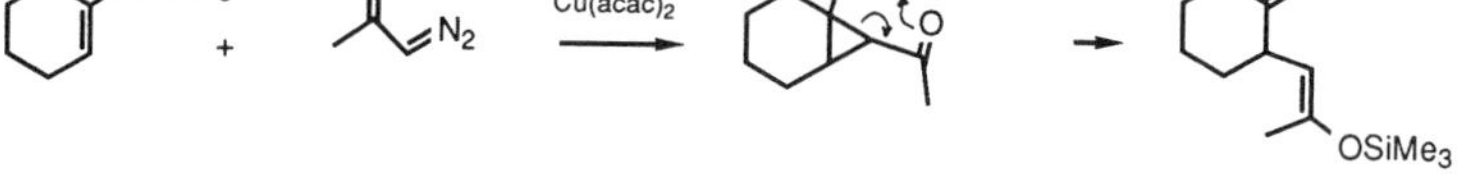

Photochemical reactions of cross-conjugated dienones often proceed via zwitterionic intermeidates. For example, a bicyclic oxyallyl species arising from 4-hydroxy-2,4,6-tri-*t*-butyl-2,5-cyclohexadienone proves capable of dissipating its charges by rearrangement to a hydroxy enone [Matsuura, 1968]. Relief of molecular strain can then be achieved by fragmentation.

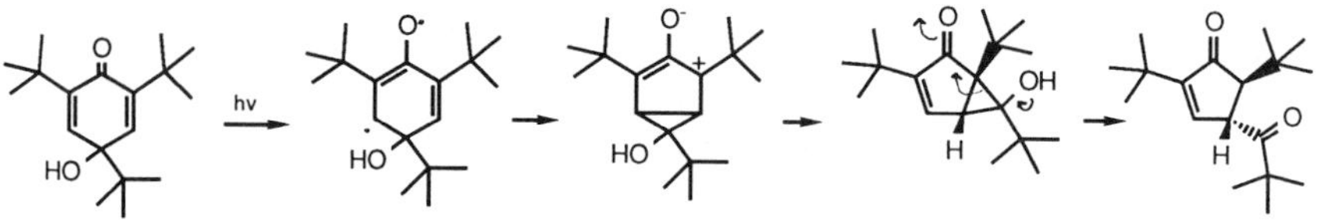

As shown by the following reactions an ethereal function in the ring is sufficient to effect facile cleavage of acylcylopropanes. Since the alkoxycyclopropyl ketones and esters are readily available from vinyl ethers and diazo ketones/esters, 1,4-dicarbonyl compounds are conveniently obtained. Thus, precursors of α-cuparenone, acorone, β-vetivone [Wenkert, 1978a], hydrourushiol [Wenkert, 1983], and prostaglandin-$E_1$ [Wenkert, 1985] have been assembled by this method.

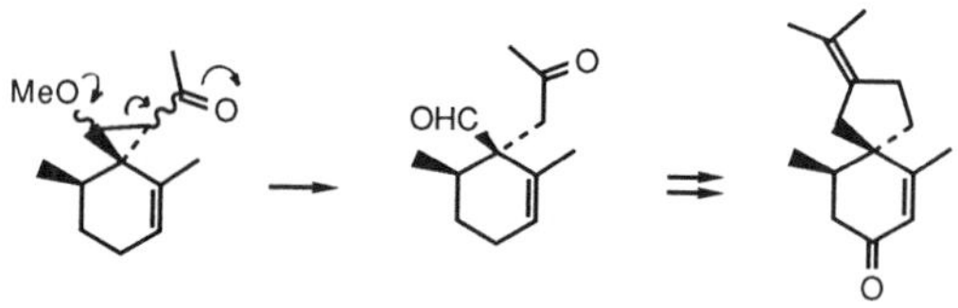

β- vetivone

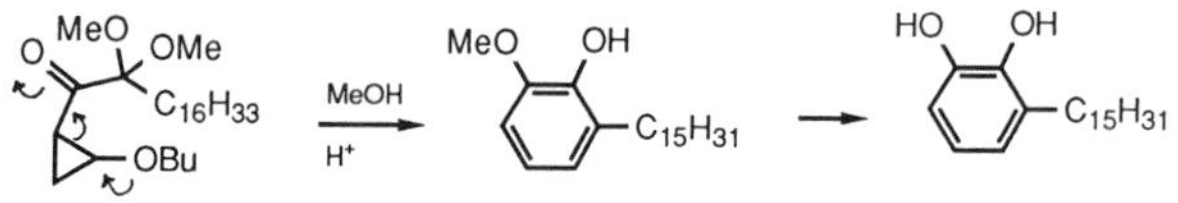

hydourushiol

An intramolecular version of this reaction sequence accomplishes a cycloalkylation of furans [Padwa, 1986].

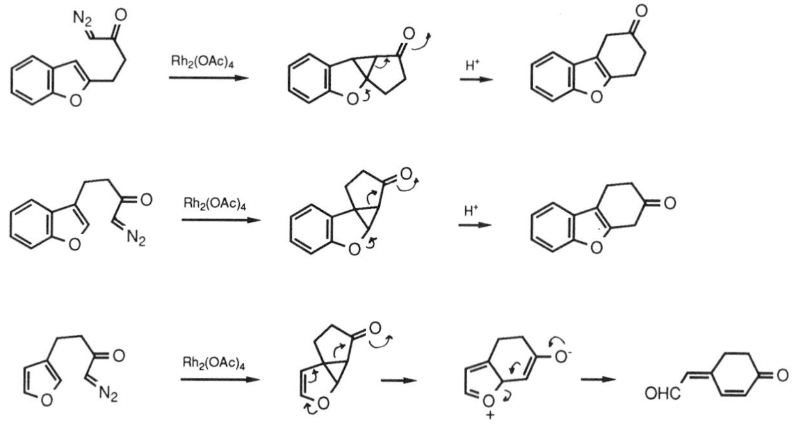

A spirocyclic ortho ester emerged when 3-methylidenebenzo-1,5-dioxepine and diazoacetone were mixed with rhodium acetate [Dowd, 1985b]. It is interesting to note that the corresponding primary adduct from methyl diazoacetate does not undergo fragmentation.

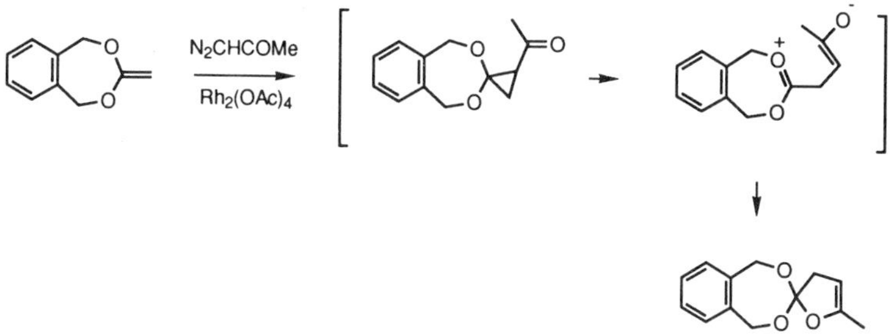

The proposed mechanism for the conversion of 2,2-dichlorocyclopropyl ketones to the $\gamma$-keto esters consists of dehydrochlorination and methanol addition followed by ring fission [Banwell, 1983]. An $\alpha$-substituent prohibits the reaction.

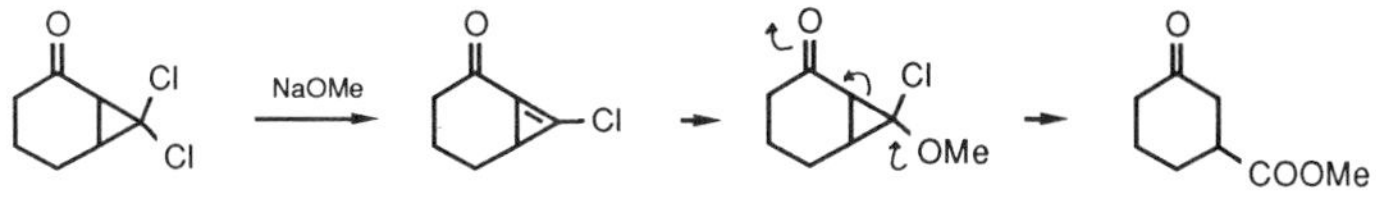

The major thermolysis products of azido cephems in the presence of methanol are ring expanded lactim ethers [Spry, 1984]. Azirenes are probable intermediates of these reactions.

The Thiele acetylation of *p*-tropoquinone [Hirama, 1977] furnishes a benzaldehyde derivative. Ring contraction via a bicyclo[4.1.0]heptenedione is conceivable.

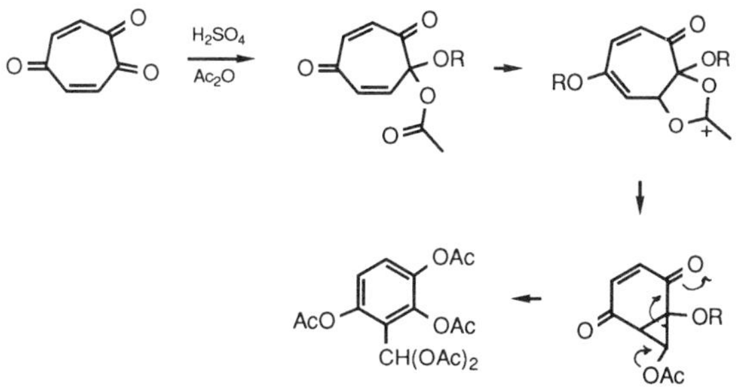

The photoinduced transformation of 1-methoxybicyclo[2.2.2]oct-5-en-2-ones to diketodiquinanes [Demuth, 1985] involves tricyclic intermediates in which the two oxygen functions are separated by two CC bonds. Spontaneous fragmentation of such intermediates is due to strain of such systems. Construction of a precursor of pentalenolactone-G [Demuth, 1984] took full advantage of the reaction sequence.

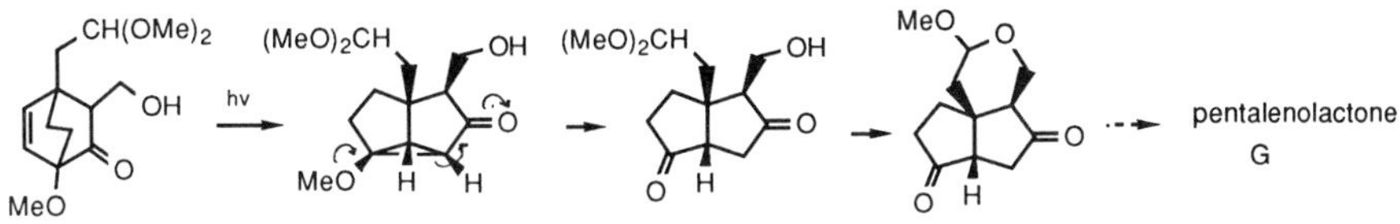

In vitro realization of a skeletal rearrangement which was proposed to be of biosynthetic significance for averufin, that is the creation of the benzofuran subunit [Ahmed, 1983], gives credence to the possible intervention of a spiroannulated oxycyclopropyl group.

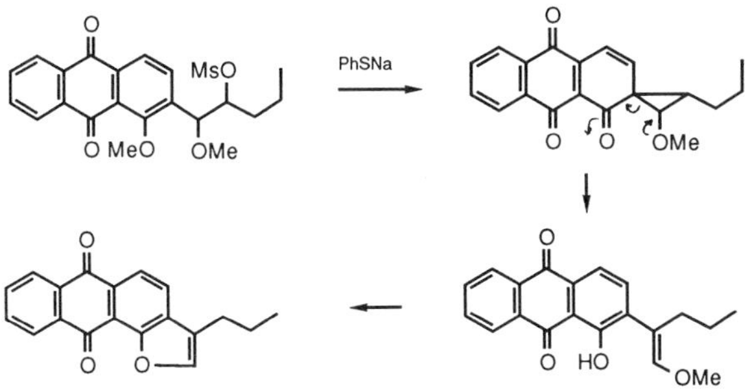

Flavothebaone is a product from alkaline treatment of thebainequinol [Barneis, 1968]. It is believed that an intramolecular alkylation of the phenoxide ion initiates the transformation. A similar reaction occurs when the Diels–Alder adduct of thebaine with 2-thiolane-4-one-1,1-dioxide is exposed to a base [Tolstikov, 1986].

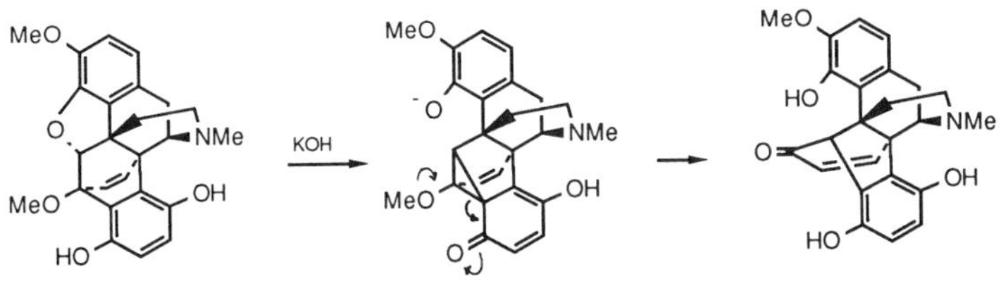

2-Ethoxycyclopropanecarboxaldehyde which has been prepared by oxidation of the corresponding alcohol or reduction of the ester, fragments spontaneously to afford 2-ethoxy-2,3-dihydrofuran [Menicagli, 1985]. The cyclobutane homolog is more stable as its isomerization at room temperature takes much longer time (about one month) [Menicagli, 1983].

One of the most remarkable biochemical processes is the interconversion of malonyl-CoA and succinyl-CoA mediated by a vitamin-$B_{12}$ coenzyme. This rearrangement proceeds via an alkylcobalamine intermediate and a cyclopropylcarboxylic ester. A model of this reaction has been validated in vitro [Lowe, 1970].

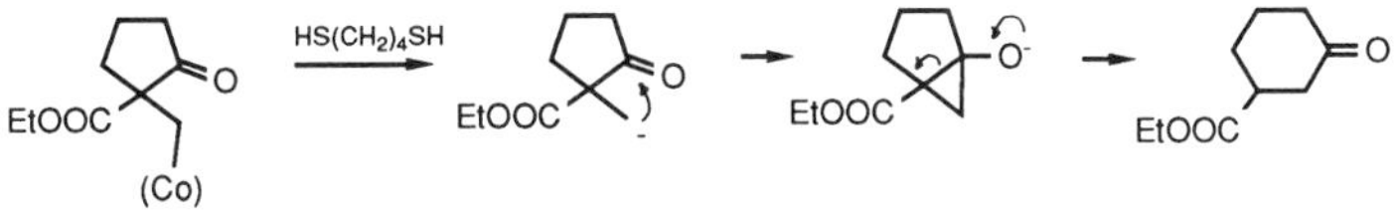

The highly strained 1,5-dicarbomethoxy-2,4-dimethyl-3-oxaquadricyclane fragments in the presence of trichloroacetic acid to give a tricyclic enol ether [Hogeveen, 1978]. The preservation of the remaining cyclopropane is probably due to a delicate balance of strain and electronic factors.

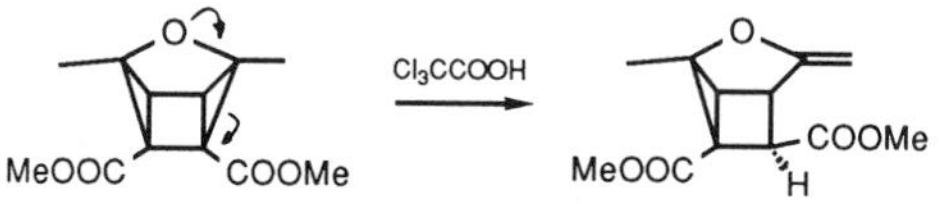

$\gamma$-Keto esters are most readily accessible by way of carbenoid addition which leads to 2-oxycyclopropylcarboxylates. Among the numerous applications we may mention an angular alkylation [House, 1968] and the synthesis of an eburnamonine intermeidate [Wenkert, 1978b]. The latter report also describes the effectiveness of NH donor in place of oxygen (cf. indirect acetonylation of 1-carbethoxy-1,2,3,4-tetrahydropyridine at C-3 [Inokuchi, 1987]).

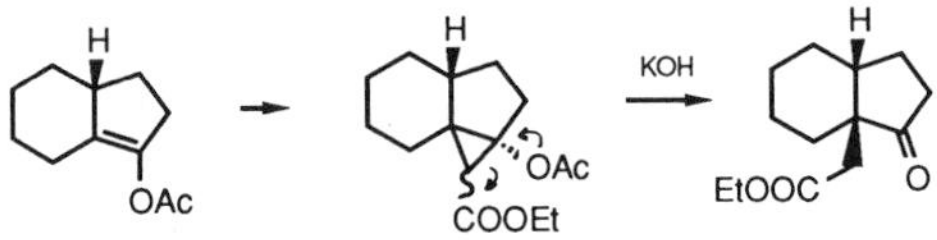

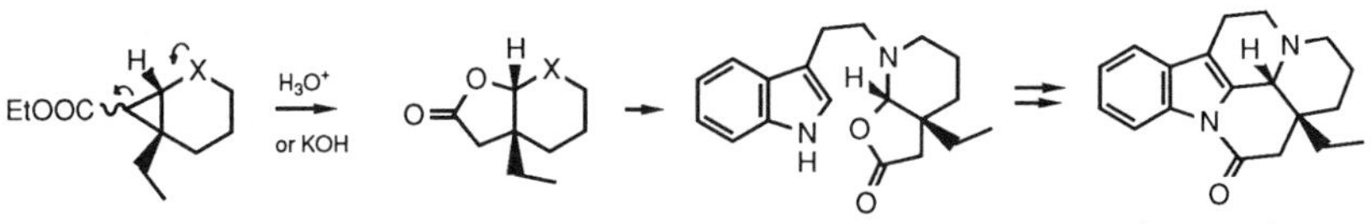

X=O, NCOOMe

eburnamonine

Extremely mild conditions are required for the ring fission of siloxycyclopropylcarboxylic esters. Iodotrimethylsilane is generally an effective catalyst [Reissig, 1985], while fluoride ion desilylation provides a general route to keto esters which may be elaborated further, as demonstrated by a synthesis of the prostaglandins [Marino, 1984].

Trapping of the keto ester anions is possible and the strategy serves to simplify synthetic operatons. An approach to pentalenolactone-E methyl ester [Marino, 1987] and an annulation that has potential for synthesis of compactin [Marino, 1988] are exemplary.

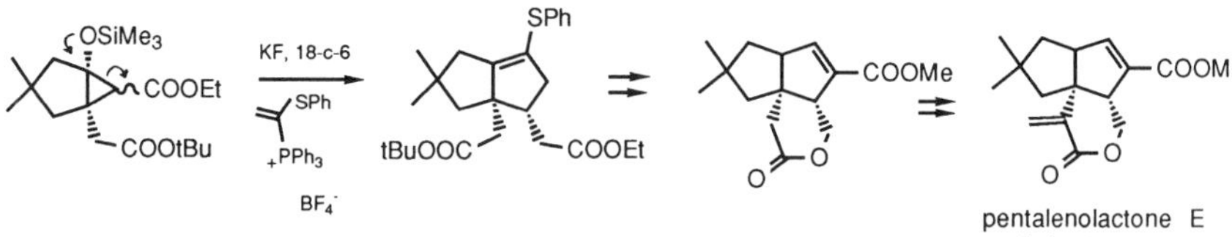

pentalenolactone E methyl ester

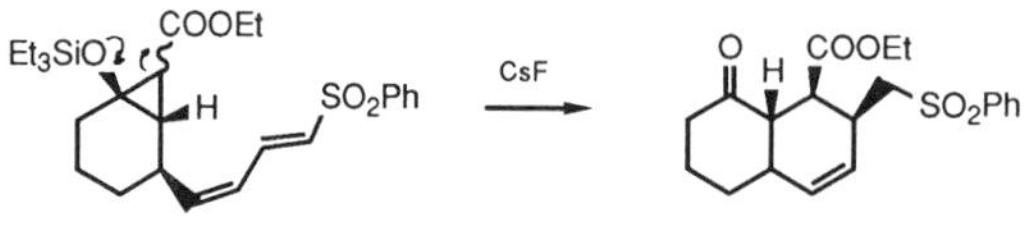

A one-pot preparation of $\gamma$-lactones [Grimm, 1985] from siloxycyclopropylcarboxylic esters has been reported. The process is a combination of fragmentative ring opening, reduction, and lactonization.

OSiMe3 COOMe KBH4 TsOH O O

Methylenecyclopropanecarboxylic acids decompose thermally in the presence of water to give ketonic products [Goldschmidt, 1983]. Hydration of the double bond is followed by fragmentaton and decarboxylation.

HOOC COOH $H_2O$ HOOC COOH OH COOH COOH O $-CO_2$ COOH O

Reaction of a 2-isopropylidenecyclopropanecarboxylic ester with peracid yields a $\delta$-lactone [Crandal, 1978]. A great relief of strain attends both the opening of the epoxide with the help of the ester group and the cyclopropane ring.

COOEt O COOEt O⁻ EtO O + O O O

Other donor-substituted cyclopropyl carbonyl and iminocarbonyl compounds exhibit similar reactivities of ring scission. Thus, 4-oxoketene diphenylthioacetals are readily formed by exposing *gem*-bis(phenylthio)cyclopropyl ketones to an acid [Kulinkovich, 1982]. Reactions or equilibria involving carbanions as donors are the following [Kierstead, 1952; Ullman, 1959; Danishefsky, 1975]. For cyclobutyl analogs, see: Ingold, 1922; Eschenmoser, 1951.

$(EtOOC)_2CH^-$ + COOEt COOEt → COOEt EtOOC COOEt COOEt $H^+$ COOEt EtOOC COOEt COOEt

COOMe MeOOC COOEt COOEt base ⇌ COOMe MeOOC COOEt COOEt

Diacylcarbinols are subject to facile rearrangement thermally or by treatment with mild base [Blatt, 1936; Karrer, 1950; House, 1958; Rubin, 1979b]. It is mechanistically akin to the reaction of 2,3-epoxyisoflavones with primary alkylamines [Yokoe, 1985].

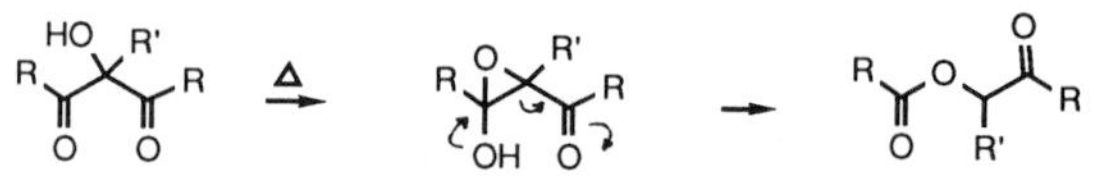

Epiflavipucine but not flavipucine is isomerizable to a pyridone on exposure to boron trifluoride [Girotra, 1979]. Possibly the coordination of the two ketone groups by the Lewis acid is essential for the fragmentation to take place.

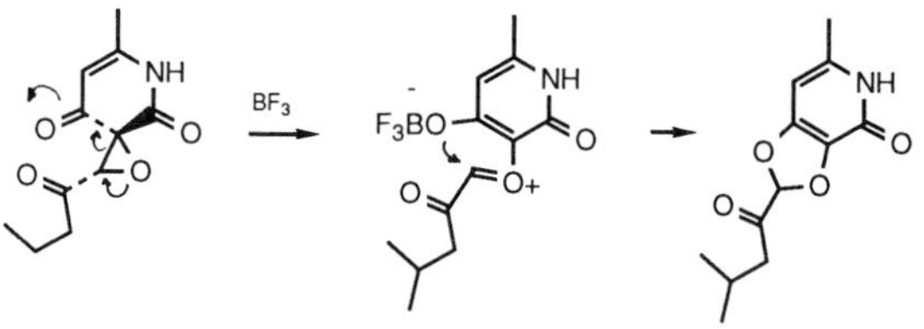

Pyrolysis of α-vinylglycidic esters results in the production of dihydrofurans [Hudlicky, 1989a]. Fragmentation of the epoxide ring is aided by the ester group which stabilizes the incipient carbanions.

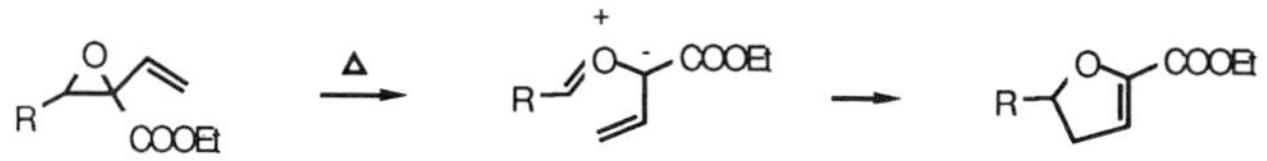

The 1,3-dipolar adduct of a 1,4-benzodiazepin-2-one 4-oxide with ethyl propiolate underwent a rearrangement involving rupture of an aziridine ring [J.P. Freeman, 1982].

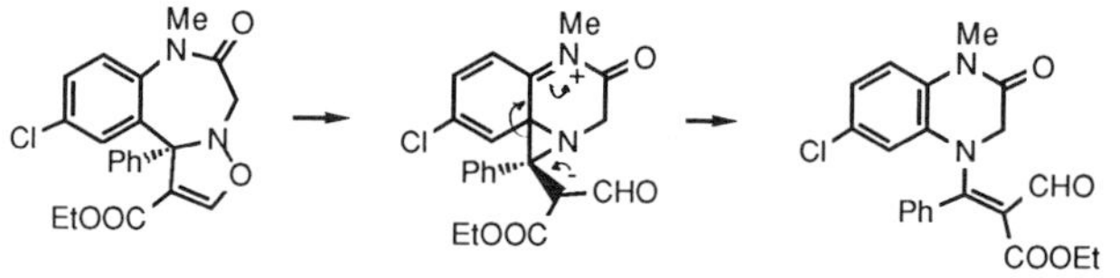

Controlled thermal decomposition of 2-aroylaziridines in solutions led to dihydropyrazines [Lown, 1972] as a result of dimerization of the *N*-alkylphenacylamine intermediates. These latter substances may have originated from the 1,3-dipoles via fragmentation.

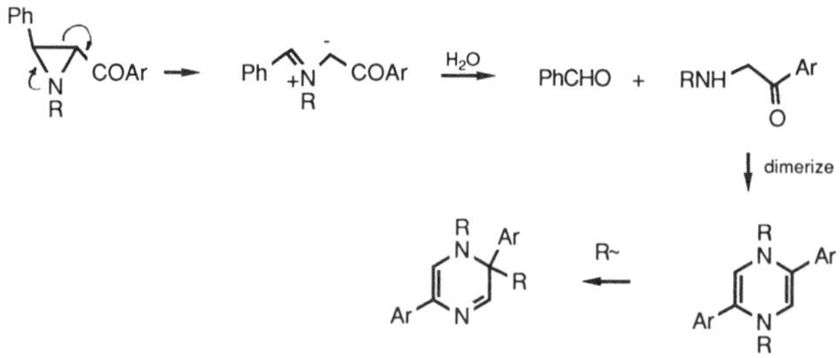

### 7.2.3. Cleavage of Four-Membered Rings

As expected, 2-acylcyclobutanols and derivatives behave similarly to the cyclopropanols. Interestingly, the four-membered ring of the photocycloadducts from 4-oxy-2-quinolone and 3-hexyne opens in different manners according to the nature of the oxy substituent and the reaction conditions [Naito, 1983]. The 4-acetoxy congener affords a benazocine derivative on exposure to base, while the methoxy compound suffers cleavage of the exocyclic C—C bond connecting C-3 of the quinoline nucleus.

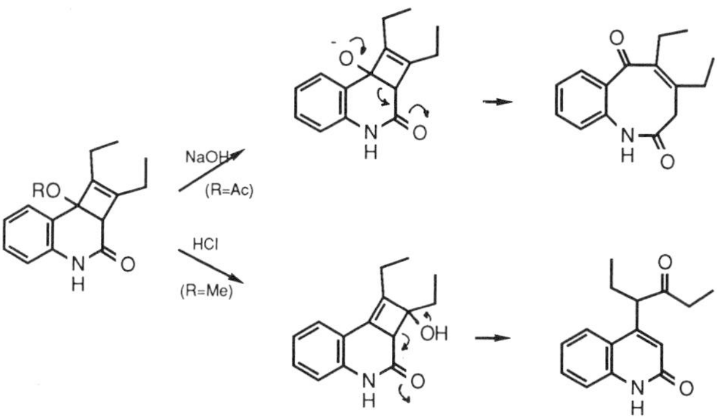

NaOH (R=Ac)
HCl (R=Me)

3-Methyl benzofuran-4,5-dione is a potentially useful dienophile for synthesis of the tanshinones. An excellent method for the preparation of this quinone is by photocycloaddition, retro-aldol fission, and oxidation [J. Lee, 1987].

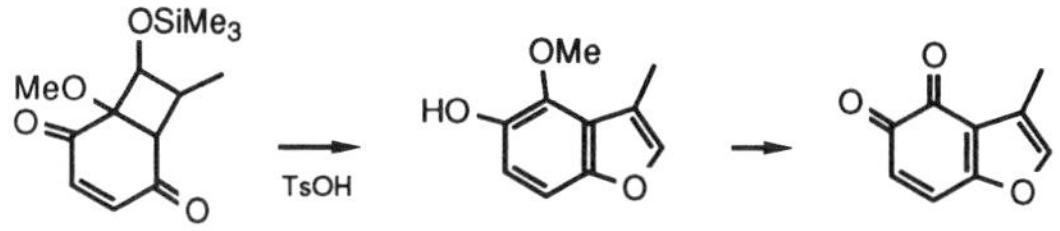

TsOH

Synthesis of (−)-quadrone from (−)-10-camphorsulfonic acid [H.J. Liu, 1988] entailed photocycloaddition of a diquinane enone with allene as a method for reductive $\beta$-alkylation. The steric bias and thermodynamic control inherent to the diquinane system ensured the desired configuration of the sidechain to be introduced at the ring juncture. The direct formation of a keto ester by ozonolysis and reductive work-up was driven by strain relief.

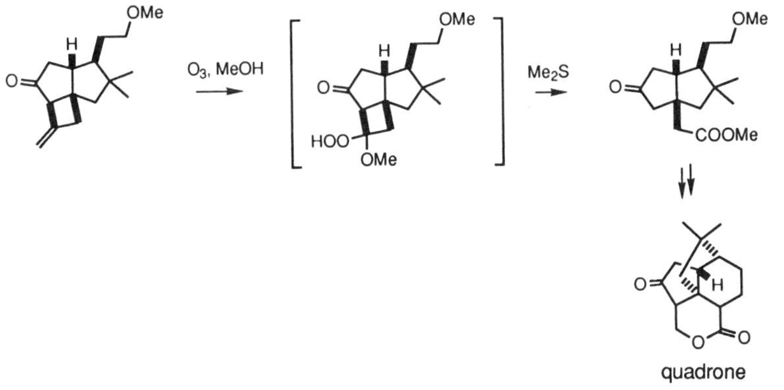

$O_3$, MeOH
$Me_2S$

quadrone

A synthesis of ishwarane employs a photocycloaddition to introduce the three-carbon subunit onto an octalone system [R.B. Kelly, 1970, 1971]. An oxidoreduction sequence followed by a retro-aldol/aldol tandem serves to accomplish the first stage of the ring construction.

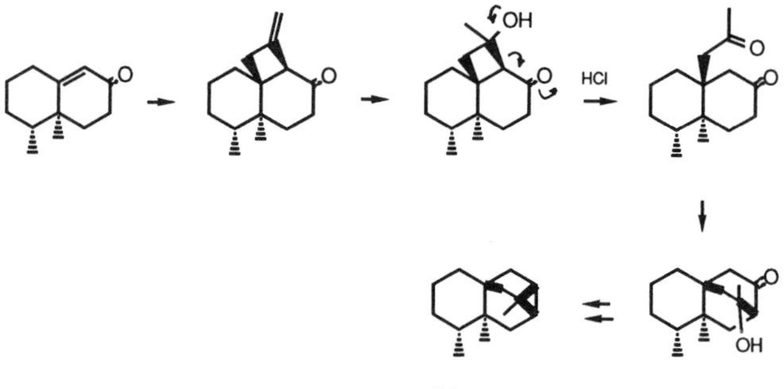

ishwarane

The synthesis of chasmanine [Tsai, 1977] is an elaborate process that features rearrangement of a bicyclo[2.2.2]octane to the [3.2.1]counterpart. Assemblage of the [2.2.2] system relied on the enone–allene photocycloaddition, degradation of the exocyclic methylene group and a tandem retro-aldol/aldol reaction sequence.

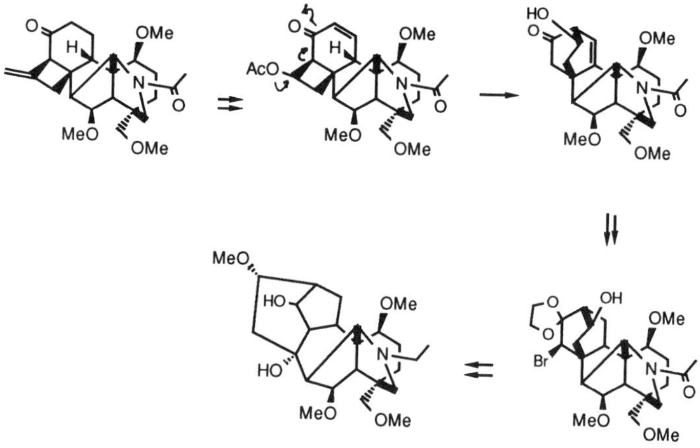

chasmanine

Work originally directed toward synthesis of lycopodine which ended up in the 12-epimer [Wiesner, 1968a] is nevertheless elegant in its approach. The key step is an intramolecular photocycloaddition followed by retro-aldol fission of the four-membered ring.

A synthesis of diterpenes such as maritimol [Bettolo, 1983] exploits the indirect introduction of a two-carbon chain at C-9 by means of the photocycloaddition protocol. The chain is required for elaboration of the bridged system.

Allene–enone photoadducts give 1,5-diketones on reaction with mercuric perchlorate in acetone [Duc, 1981]. Evidently, hydratomercuration of the double bond initiates a ring scission. Demercuration of the diketones may be followed by aldolization.

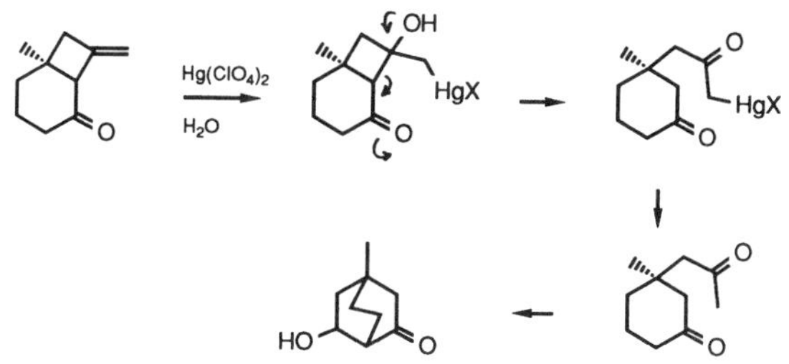

A synthetic approach to cyathin-$A_3$ [Ayer, 1981] is based on the manipulation of a photocycloadduct via epoxidation of the methylenecyclobutane moiety and benzenethiolysis. The latter step exposes a strained $\beta$-ketol for fragmentation.

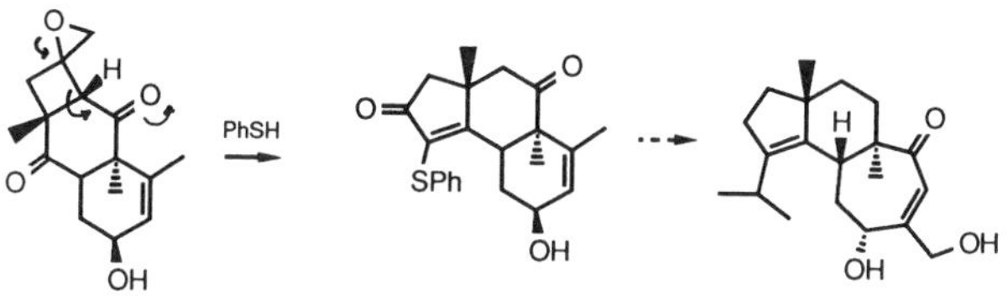

cyathin $A_3$

One of the two regioisomers produced from photocycloaddition of 3,6-dimethyl-2-cyclohexenone with 3-methyl-1,2-cyclopentanedione was found to be a good precursor of norketotrichodiene [Yamakawa, 1976].

In the synthesis of an aromatic intermediate of songorine [Wiesner, 1973] the photocycloaddition of vinyl acetate to a tetracyclic enone took an unusual course in that the regiochemistry is opposite to that commonly observed. This result has probably facilitated the experimental work, but the anomaly remains unexplained.

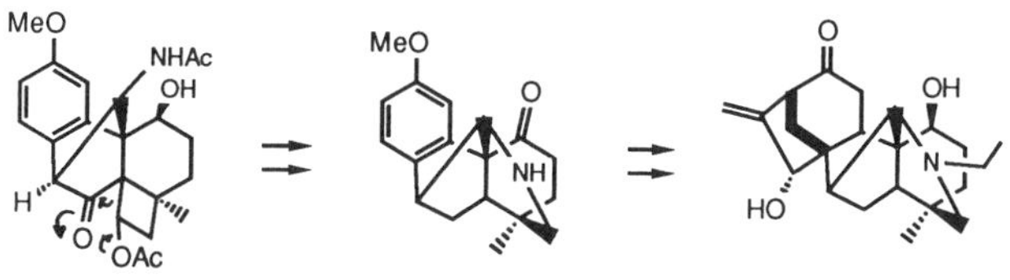

songorine

1-Succinimido-5-chloro-2-pentanone has been obtained from a carbocyclic analog of 1-oxapenm in acidic solution [Agathocleous, 1986]. Transannular retro-aldol-type fragmentation is likely the first step of the transformation.

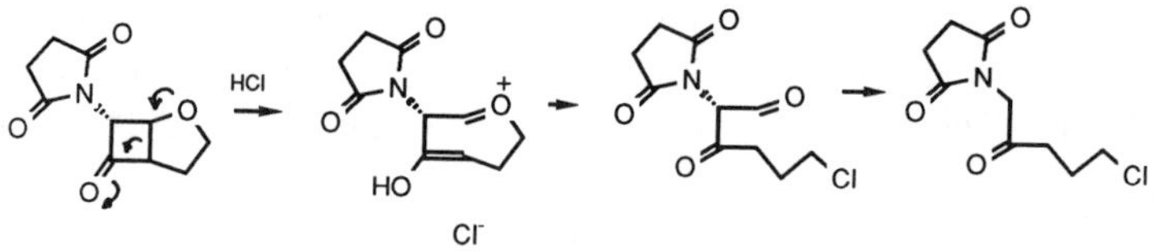

A silyl group transfer probably initiated the in situ fragmentation of the cycloadducts of dichloroketene and silyl enol ethers [Krepski, 1978].

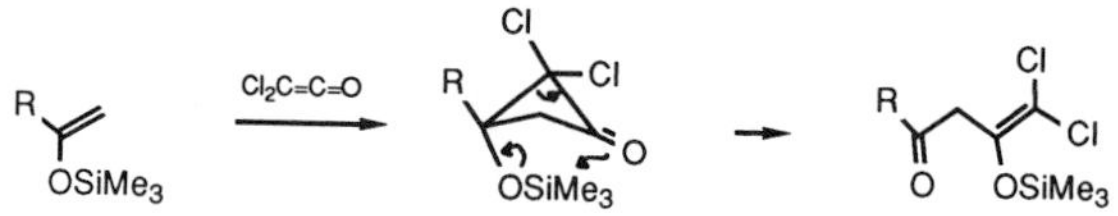

Crucial information for the structural determination of annotinine was gained by the identification of a julolidine derivative generated from dehydrogenation of a $C_{14}$-acid [Wiesner, 1958]. The degradation involved two fragmentation steps: in the cleavage of the four-membered ring, and the assisted opening of the $\gamma$-lactone.

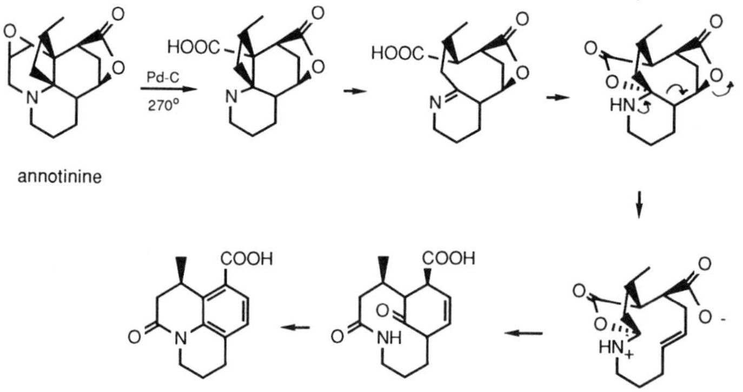

### 7.2.4. The deMayo Reaction

The elaboration of 1,5-dicarbonyl compounds by photocycloaddition of enolizable 1,3-dicarbonyl substrates or derivatives with alkenes and effecting retro-aldol fission of the adducts is commonly known as the deMayo reaction. The retro-aldol step is spontaneous when unprotected 1,3-dicarbonyl compounds are used [deMayo, 1962, 1964; Casals, 1976]. The value of such process has been amply demonstrated, for example in the synthesis of loganin [Büchi, 1973a; Partridge, 1973], hydroxyloganin [Tietze, 1974], chrysomelidial [Takeshita, 1982], (−)-sarracenin [S.W. Baldwin, 1982b], and hirsutene [Disanayaka, 1985].

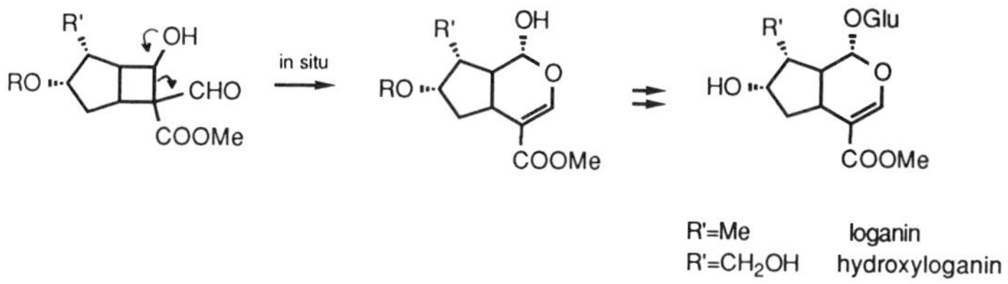

R'=Me loganin
R'=$CH_2OH$ hydroxyloganin

The original work involves photocycloaddition of the enolacetates of 1,3-diketones and the adducts were saponified to achieve the retro-aldol fission [Challand, 1967]. This sequence is the basis of a synthesis of $\beta$-himachalene, after which many other applications have appeared. These include approaches to hirsutene [Tatsuta, 1979], stipitatonic acid [Lange, 1967], a precursor of fusicoccin H [Grayson, 1984] and model studies of taxane synthesis [Berkowitz, 1985; Kojima, 1985; Cervantes, 1986; Kaczmarek, 1986]. The use of enol derivatives other than the acetates has also been explored.

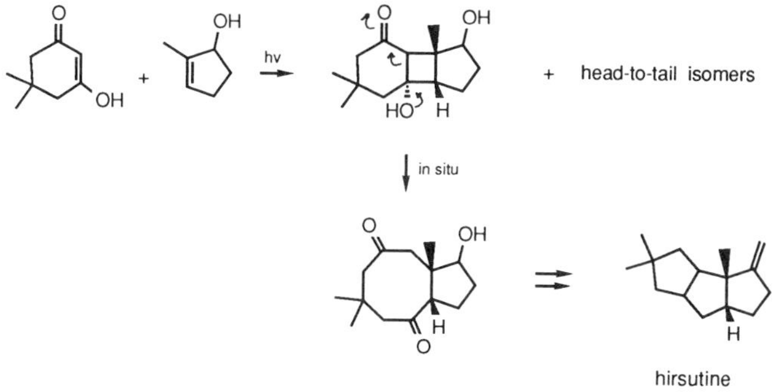

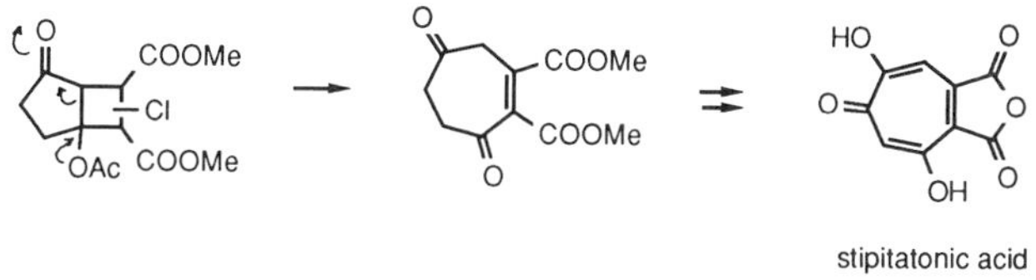

hirsutine

stipitatonic acid

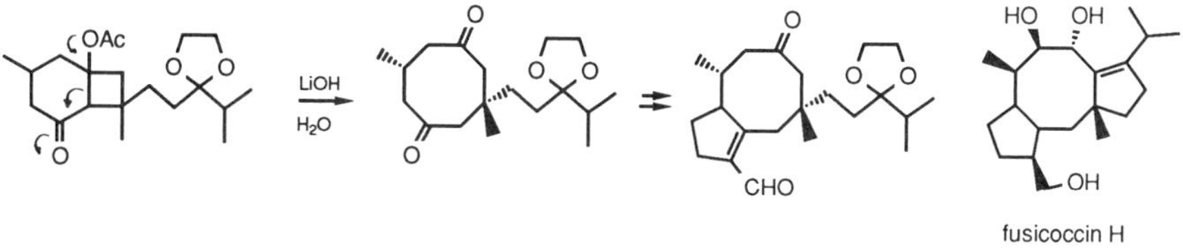

fusicoccin H

Different fates of the two photocycloadducts of dimedone enolacetate and diketene on their reaction with dimethylamine have been observed [T. Kato, 1978a]. One of these adducts underwent lactone aminolysis only, whereas its epimer gave a cyclooctanedione. The latter isomer was identified as that containing *cis* oxygen substituents on the contiguous quaternary carbon atoms.

Acyl transfer among these oxygen atoms is instrumental to the retro-aldol process.

An extension of the deMayo reaction is the use of enol derivatives of $\beta$-keto esters as photoaddends [Liu, 1982]. In such cases 1,5-keto esters are produced.

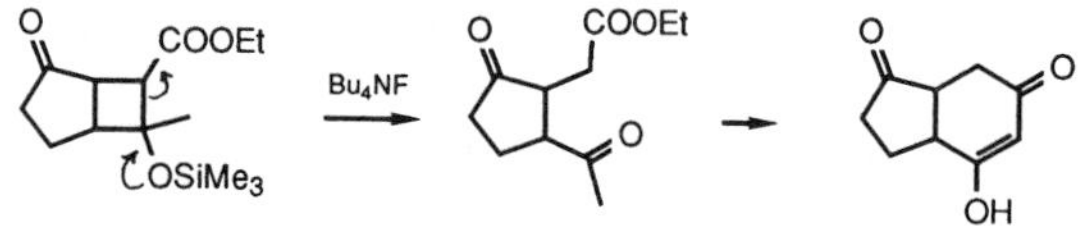

The most significant modification of the deMayo reaction in a synthetic context is the intramolecular version. An elegant accomplishment is a synthesis of longifolene and sativene [Oppolzer, 1984]. A tactical adjustment of this approach is the replacement of the acetate by a mixed carbonate with benzyl alcohol such that unveiling of the ketol and retroaldolization would not engender the formation of a bridged cyclopentanone (by aldol condensation).

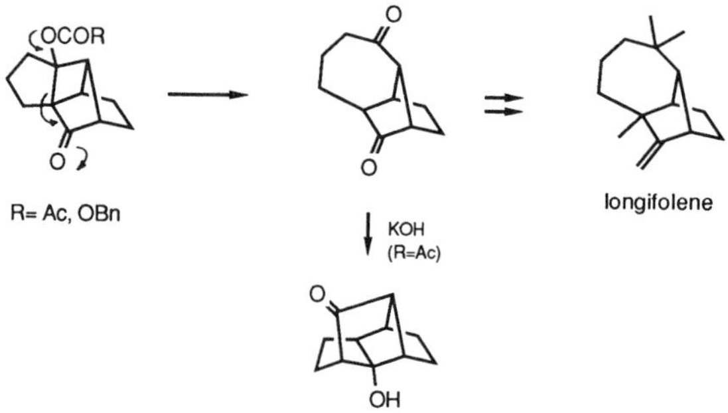

Among other intramolecular deMayo reactions in service to synthesis are found in the approaches to $\beta$-bulnesene [Oppolzer, 1980b], daucene [J. Seto, 1985], and the 11-epimer of precapnelladiene [A.M. Birch, 1983]. Many more applications are expected in the future as intricate polycyclic systems may be readily obtained by careful design and execution of this reaction [Oppolzer, 1979], although an approach to cytochalasin-C [Musser, 1982] was thwarted by the inability to disengage the donor oxygen from a $\beta$-sulfonylethyl chain.

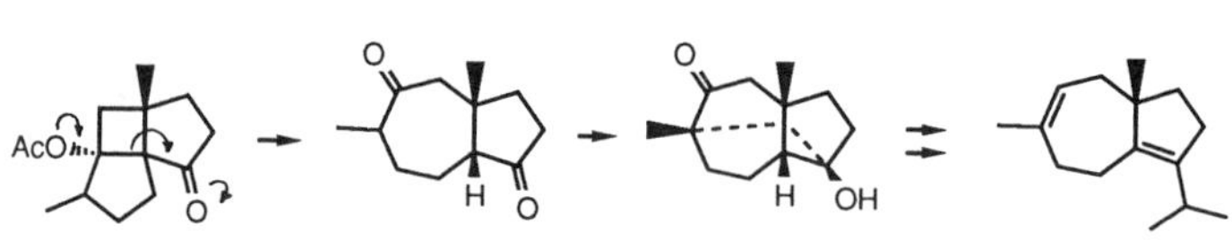

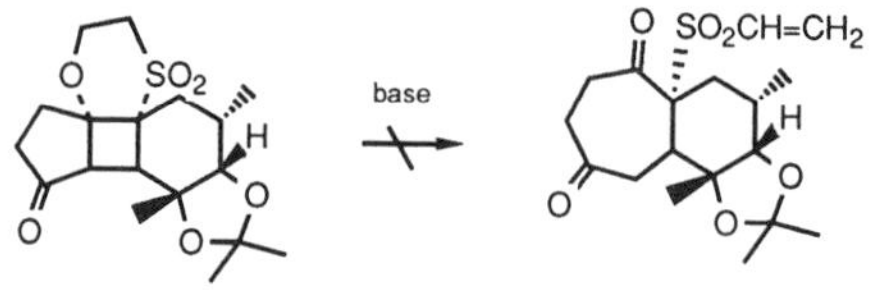

New addends for the deMayo reactions are 2,2-dimethyl-1,3(4*H*)-dioxin-4-one, 2,2-dimethyl-3(2*H*)-furanone and their derivatives [S.W. Baldwin, 1981]. This variation, either intermolecular or intramolecular, has been employed in syntheses of (+)-elemol from α-terpineol [S.W. Baldwin, 1985], grandisol [Landmesser, 1980], the purported structure of genipic acid [S.W. Baldwin, 1980], (−)-acorenone [S.W. Baldwin, 1982b], occidentalol [S.W. Baldwin, 1982d], a model of prostaglandin intermediate [M. Sato, 1984], (−)-histrionicotoxin from (+)-glutamic acid [Winkler, 1986a, 1989], and the carboxyclic framework of the ingenane diterpenes [Winkler, 1987b].

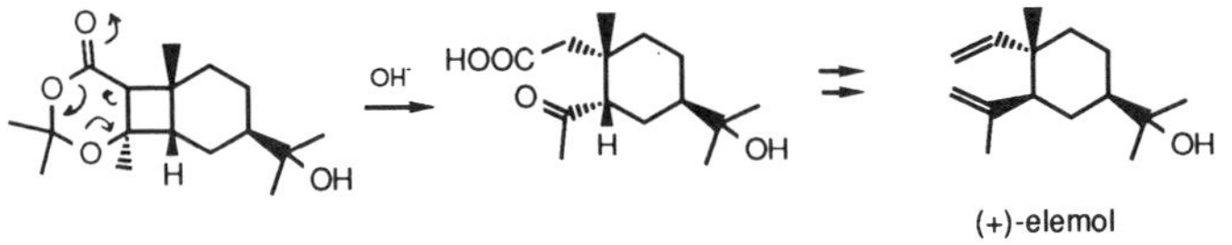

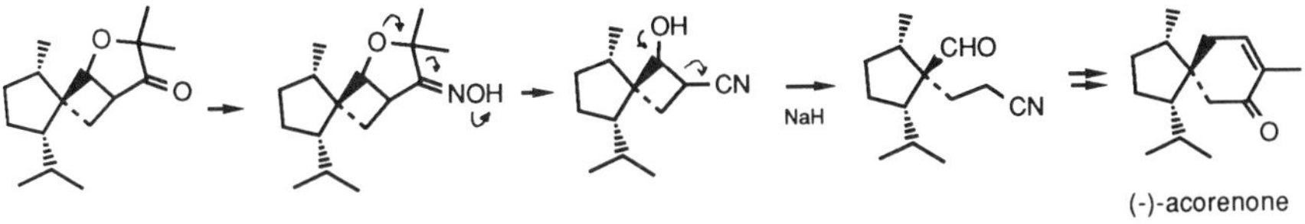

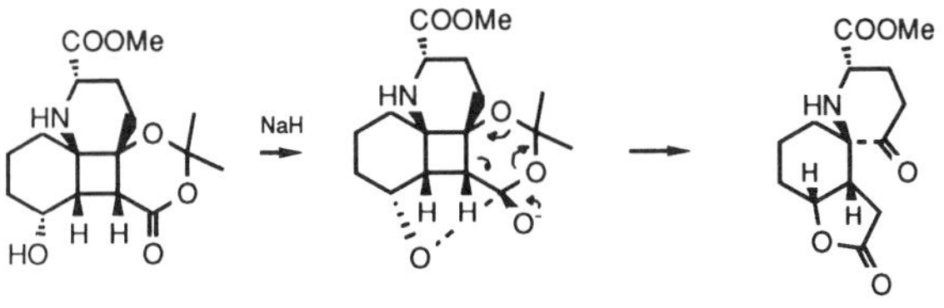

(+)-elemol

(-)-acorenone

Assembly of the tricyclic nucleus of the taxanes by a similar approach [Winkler, 1986c] was, however, met with an unexpected rearrangement instead of the desired fragmentation. In connection with this study it has also been found that intramolecular dioxenone photocycloaddition is strongly dependent on the orientation of the chromophore [Winkler, 1986d, 1989a]. The structures of the adducts were confirmed by their fragmentation behavior.

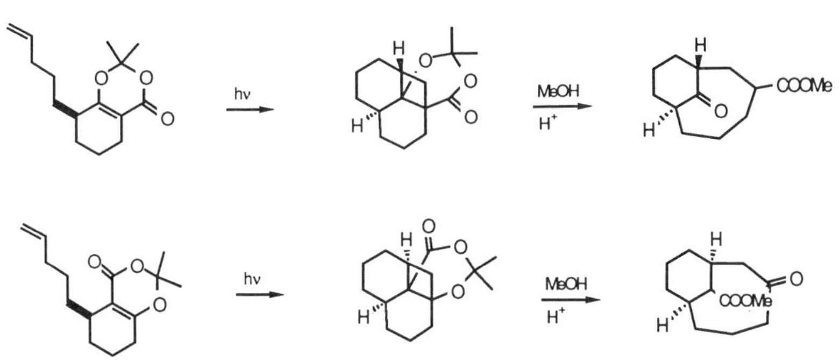

The general approach is amenable to producing optically active compounds. A synthesis of (+)-grandisol [Demuth, 1986] based on this method is interesting because it shows the preference of a twist-boat conformation of the enone ring; the more exposed face would be closer to the isopropyl group of the menthane moiety.

HCOOH

$H_2O$ acetone

(+)-grandisol

## 7.3. FRAGMENTATION OF β-HYDROXY ESTERS AND NITRILES

The fragmentation of β-hydroxy esters and nitriles is completely analogous to the retro-aldol fission. Only several representative examples are mentioned here.

Cleavage of a hydroxybicyclo[3.2.0]heptenecarboxylic ester is the key step of a prostaglandin synthesis [S. Goldstein, 1981]. The ring expansion of cyclic β-keto esters on reaction with (*Z*)-propenyl phenyl selenone [Sugawara, 1985] likely involves the formation and fragmentation of hydroxy ester anions.

$NaBH_4$; NaOMe

Corey's lactone for prostanoids

$PhSeO_2$; KH

Desilylative $C \rightarrow N$ transacetylation has been effected in a β-lactam system [Kametani, 1986]. A penem could be the intermediate.

F⁻

A convenient synthesis of β-ribofuranosylmalonic esters, which are useful intermediates for *C*-nucleosides, was based on high-pressure Diels–Alder reaction of furan and dialkyl (acetoxymethylene)malonates which was followed by dihydroxylation and fragmentation [Katagiri, 1988].

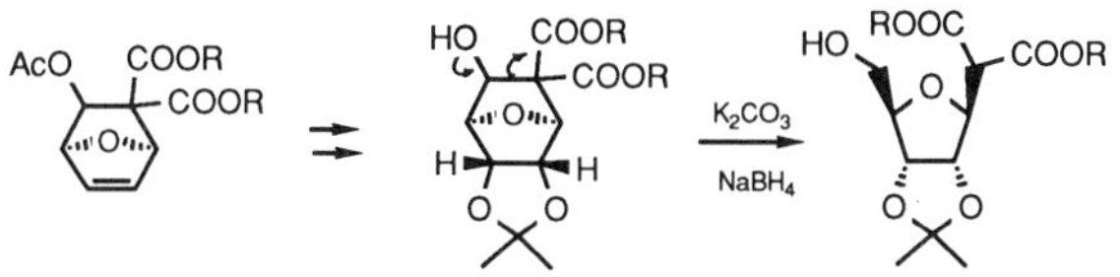

Unexpected products from aldol condensation of ethyl 2-oxo-1-(2-oxopropyl)-1-cyclopentanecarboxylate include a 1,3-cyclooctanedione and two isomeric 2-acetylcyclohexanones [Tsuzuki, 1977]. The latter two compounds are derived from the former by alternative modes of aldolization and subsequent fissions.

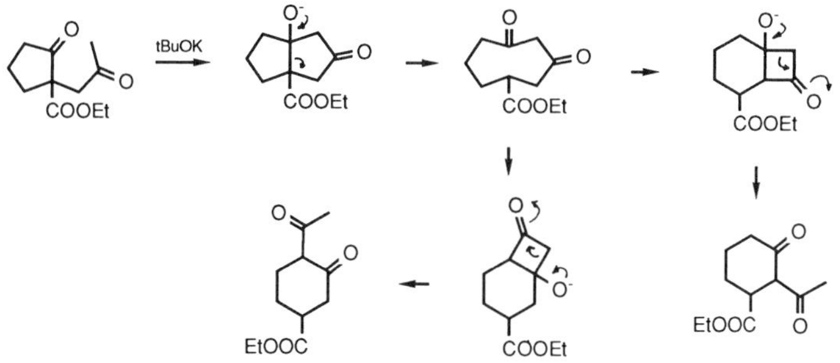

Heating of ξ-pyrromycinone results in a dehydration product and one from ring cleavage [Brockmann, 1959]. Formation of the latter species represents a reverse step of biosynthesis.

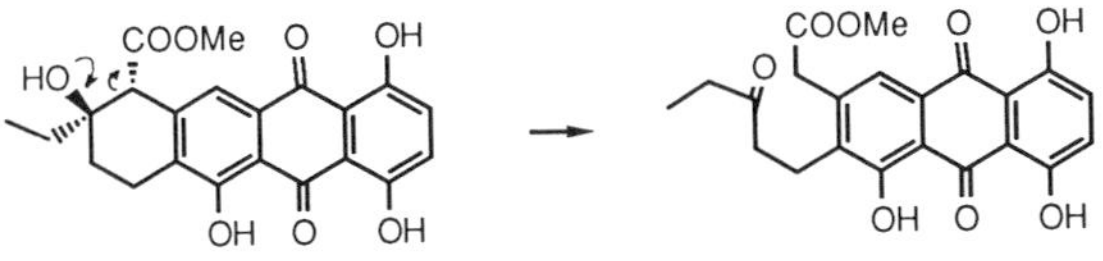

The Diels–Alder approach to 4-substituted 2-cyclohexenones [A.J. Birch, 1966] has found many applications. The popularity of this method is due to the availability of many methoxycyclohexadienes which can be prepared by Birch reduction of the corresponding anisoles. In situ conjugation of the dienes prior to the Diels–Alder reaction further simplifies the operation. In the synthesis of anticapsin [Souchet, 1987] the adduct was treated with acid. As expected, a cyclohexenone was generated, but an intramolecular Michael addition followed immediately.

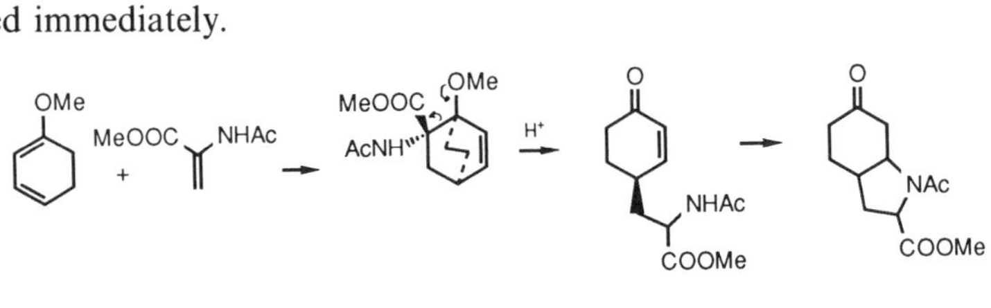

An access to the ketolactam intermediate for synthesis of pumiliotoxin-C [Ibuka, 1976] embodies a fragmentation of a bridged hydroxy nitrile followed by hydration of the nitrile function and Michael addition of the resulting amide to the enone system, all in one operation.

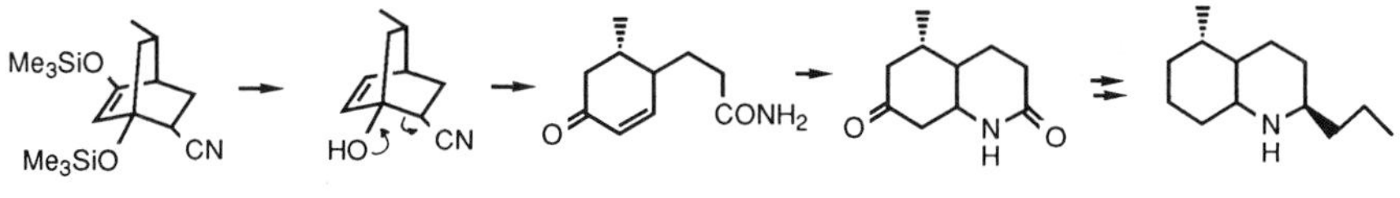

pumiliotoxin C

In an eriolanin synthesis [M.R. Roberts, 1981], stereocontrol of the tertiary carbon center in the sidechain was invested in a bicyclo[2.2.1]hepten-1-ol. Fragmentation of the latter compound resulted in a cyclohexenone which was systematically elaborated into the target molecule. It should be noted that the bicycloheptenol used here was obtained by a reflexive Michael addition.

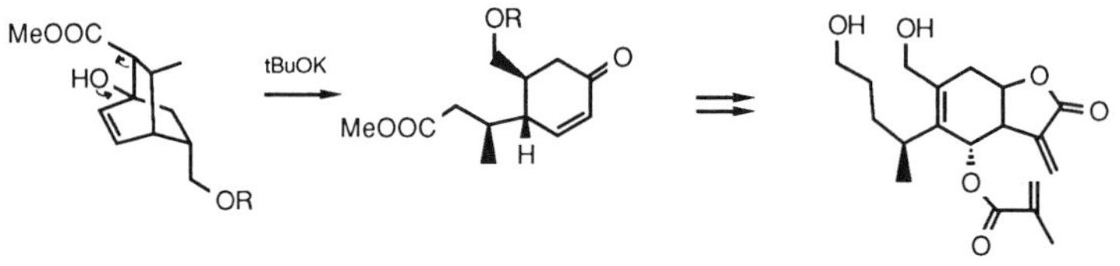

eriolanin

Another tandem reaction sequence started by fragmentation was reported [Scheffer, 1980]. There was a 1,4-acyl migration interposed between the fragmentation and the Michael addition.

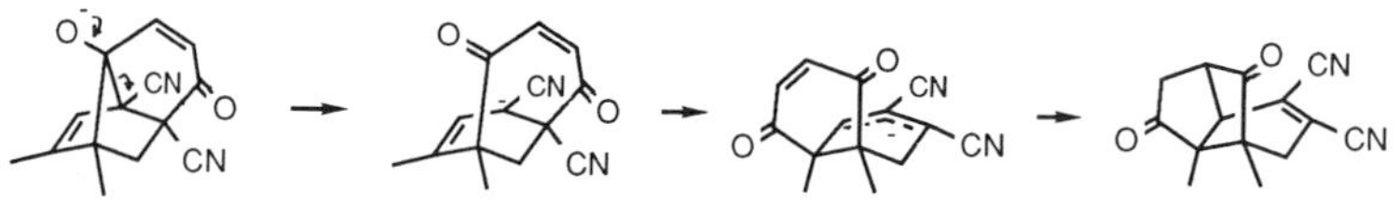

Solvolysis of 2,2-dimethyl-5-(1-bromomethylethylidene)-1,3-dioxane-4,6-dione under mild alkaline conditions affords the 2-acetonyl derivative of Meldrum's acid [Hunter, 1980]. This transformation can be envisaged to involve hydration, 1,3-elimination and fragmentation of the spirocyclic cyclopropanol.

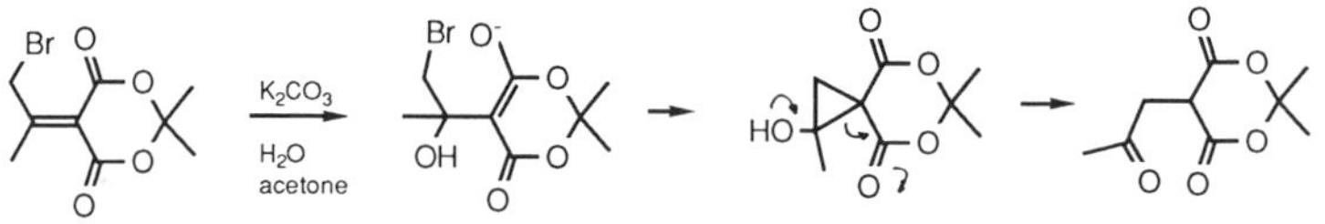

Tenulin is an aldol tautomer of isotenulin [Barton, 1956; Herz, 1962]. Thus conversion of tenulin to isotenulin can be effected by treatment with a base.

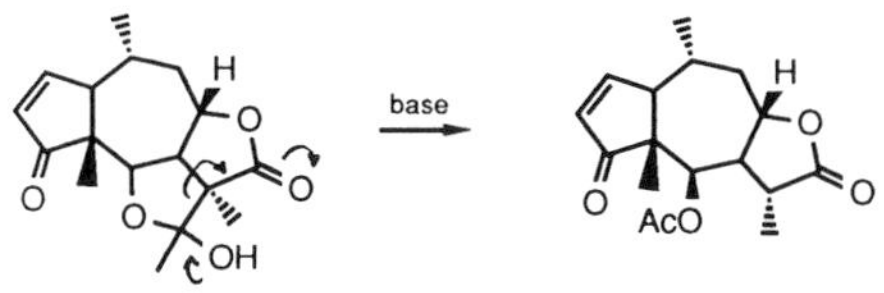

The epimerization of the A-ring hydroxyl of the gibberellins, for example $GA_1$ methyl ester, by dilute alkali involves seco intermediates [Cross, 1961; MacMillan, 1967].

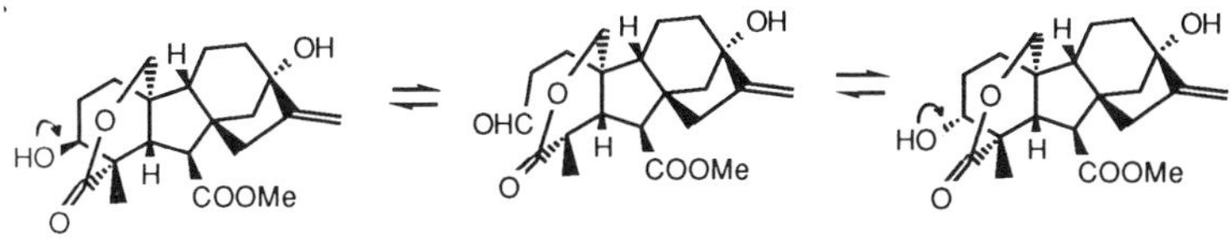

A Cope rearrangement, retro-Prins and retro-Michael reactions are induced on pyrolysis of glauconic acid [Barton, 1965]. The presence of the anhydride group definitely facilitates the retro-Prins reaction.

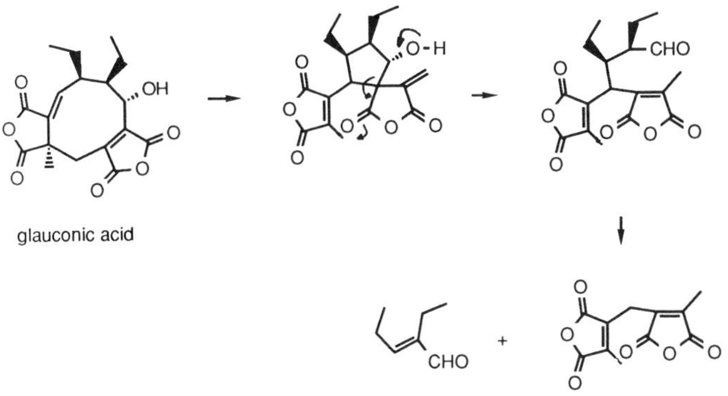

glauconic acid

One acyclic example may be cited concerning the fragmentation of β-hydroxy esters. Thus the cleavage of β-hydroxy-α-phenylsulfenylcarboxylic esters is extremely facile after deprotonation with lithium diethylamide [Hoye, 1980].

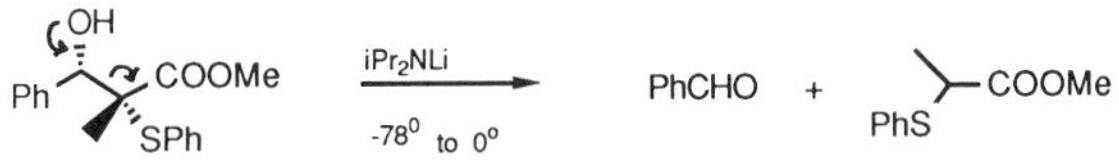

## 7.4. REDUCTIVE FRAGMENTATION OF ε-ENONE ESTERS

Relatively few cases of this fragmentation are known. The most prominent example concerns with a synthesis of dodecahedrane [Paquette, 1979]. Dissolved metal reduction of a heptacyclic dienedione destroyed the ring system because the radical anions found a fragmentation pathway by virtue of the ester acceptors.

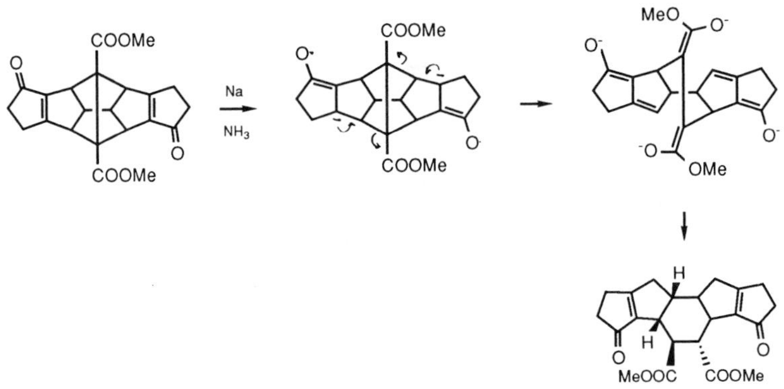

## 7.5. FRAGMENTATION OF β-HYDROXY SULFONYL, NITRO, AND NITROSO UNITS

Hydroxyl compounds containing an acceptor group at the β-position fragment readily because of the existence of a conjoint chain of atoms. The sulfones are readily available by various methods; therefore, they have found extensive use. One such service is in a synthesis of multifidene [Burks, 1984]. The hydroxy sulfone derived from cyclopentadiene and dichloroketene via [2+2]cycloaddition and further transformations can be directly transformed into a vinylcyclopentene during a Wittig methylenation. Apparently fragmentation was a prerequisite to unveil the aldehyde functionality.

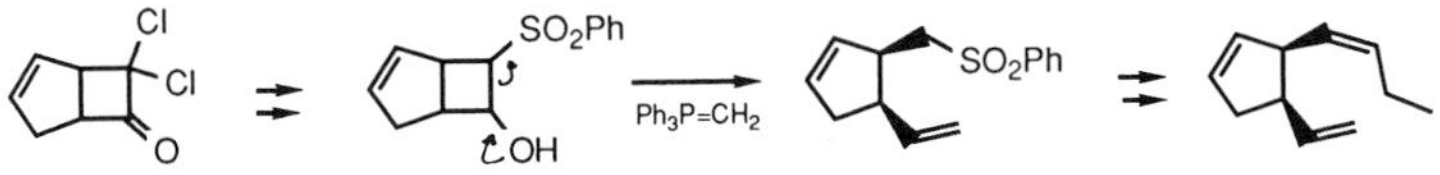

multifidene

In a model for constructing the taxane skeleton the fragmentation of aβ-hydroxy sulfone was studied [Trost, 1982]. Another application is described in a synthesis of muscone [Trost, 1980b] in which the focus is the formation and fragmentation of a bicyclic β-sulfonyl alkoxide ion.

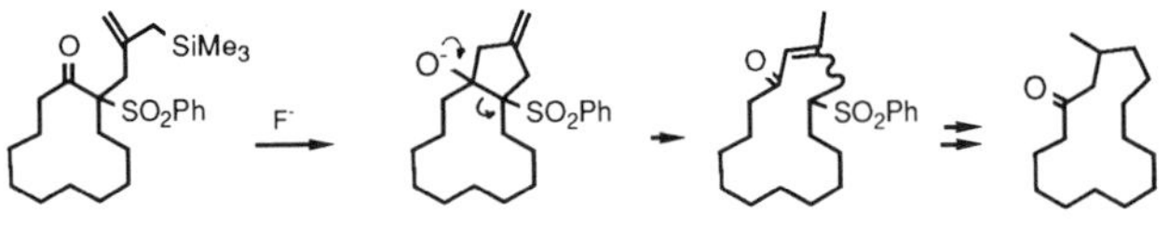

muscone

2-Chlorotropones which are further substituted with a nitro group at either C-5 or C-7 have been converted into *m*-hydroxybenzaldehydes [Forbes, 1968]. It is conceivable that the reaction proceeds via hydration, fragmentation of the norcaradienols, and oxidation. Both the hydration and the fragmentation steps are governed by the nitro group.

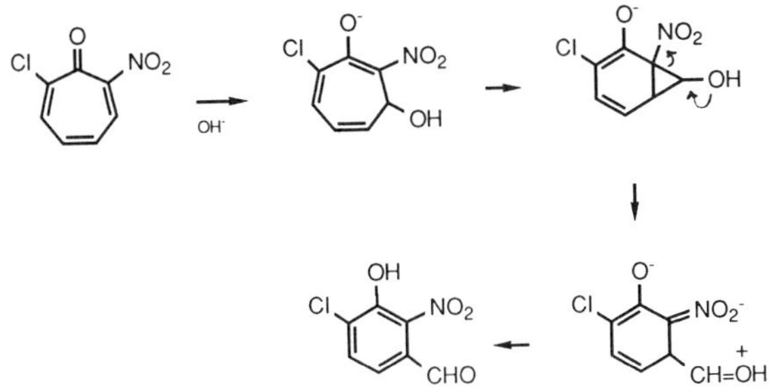

A variant of the cyclohexenone synthesis based on the Diels–Alder/ fragmentation protocol is that using a nitroalkene as dienophile [Clarke, 1988].

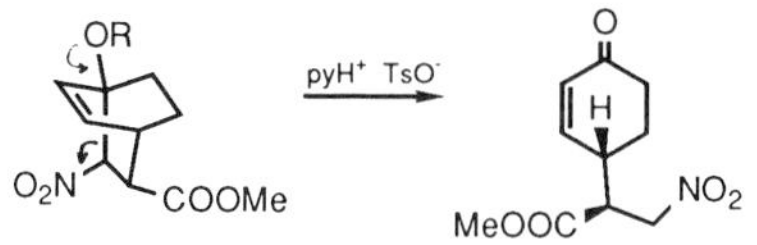

Nitroglycosides are readily epimerized to furnish the more stable products [Baer, 1971], undoubtedly via the open-chain isomers.

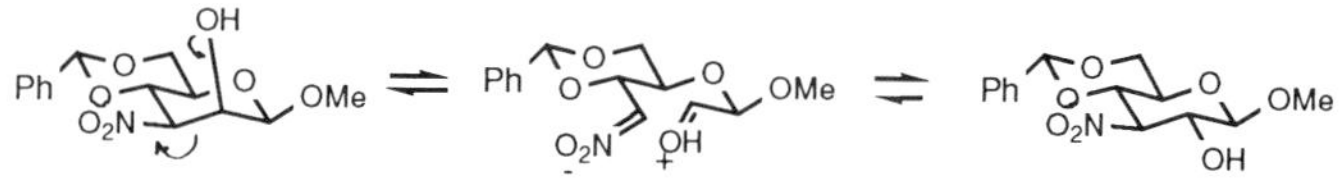

Chain elongation based on Michael reaction of α-nitroketones to enones may be further modified. As shown in the following equations, complete relocation of functionalities may be performed [Lorenzi-Riatsch, 1981]. It is also interesting to note the steric effect on the aldol condensation.

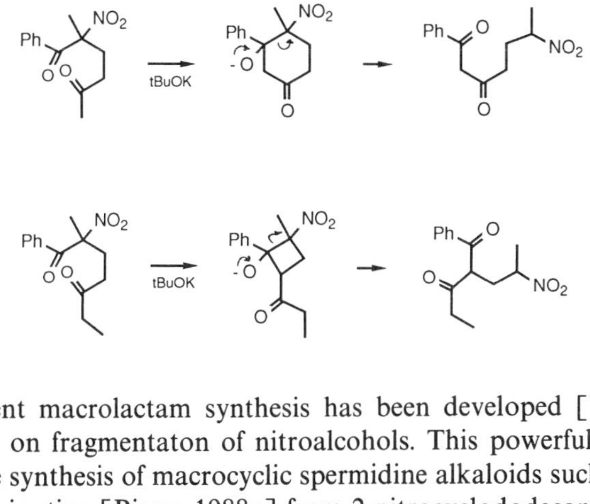

An excellent macrolactam synthesis has been developed [Wälchli, 1985] which hinges on fragmentaton of nitroalcohols. This powerful technique has permitted the synthesis of macrocyclic spermidine alkaloids such as inandenin-10-ol and oncinotine [Bienz, 1988a] from 2-nitrocyclododecanone.

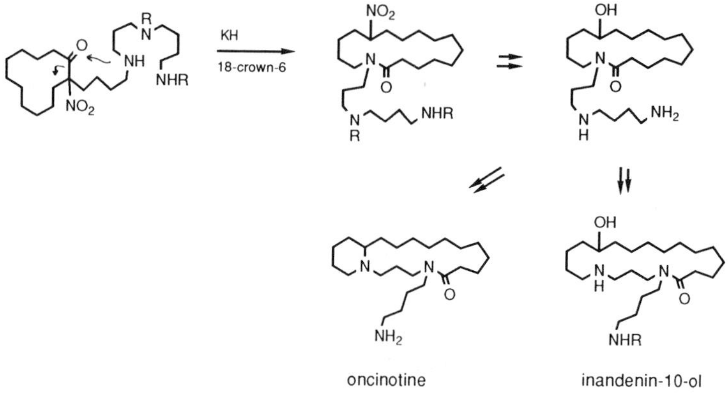

oncinotine

inandenin-10-ol

Macrocyclic ketones including muscone have also been synthesized [Bienz, 1988b] by effecting the annulation step with an enamine sidechain.

An old procedure for preparation of *C*-nitro hydrazones [E.C.S. Jones, 1930] consists of condensation of α-hydroxymethyl nitroalkanes with a diazonium salt. Formaldehyde is eliminated from the adducts.

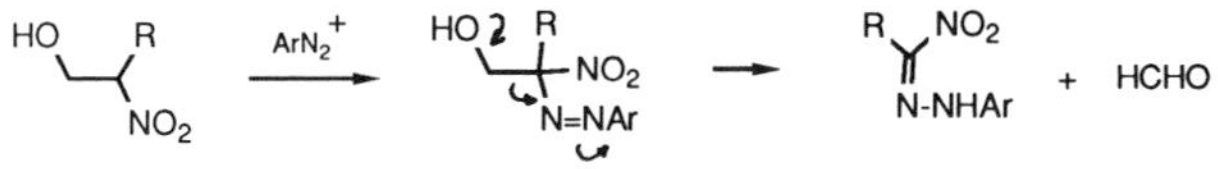

It is reasonable to expect a facile fragmentation of *β*-nitro amines if the elimination pathway is suppressed. Directly related to this fragmentation is the equilibration of the toxic fungal components cyclopiamine-A and -B in polar solvents [Bond, 1979].

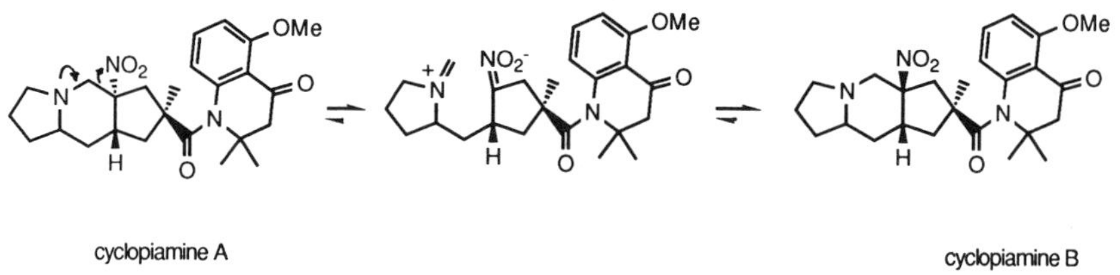

cyclopiamine A

cyclopiamine B

(ω-3)-Oximino macrolides [Mahajan, 1972], and in principle many other regioisomers, can be prepared from bicyclic enol ethers by nitrosation under hydrolytic conditions. As shown in a simple synthesis of phoracantholide-I [Mahajan, 1981], in situ fragmentation constitutes the key operation.

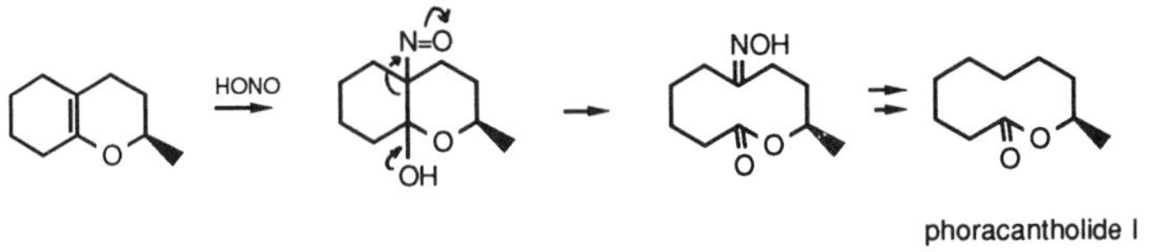

phoracantholide I

Loss of the hydroxymethyl sidechain occurs when *cis*-2,5-dimethyl-2-hydroxymethyl-5-(2-hydroxyhexyl)tetrahydropyrrole-1-oxyl is exposed to air [Keana, 1985].

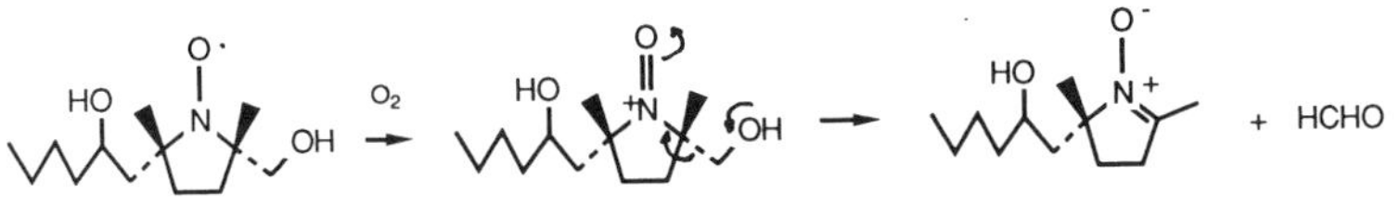

Anodic oxidation of *N*-alkyl-$\beta$-hydroxy hydroxylamines in buffer solutions results in a mixture of nitroso compounds, oximes, and aldehydes [Ozaki, 1984]. The latter compounds are fragmentation products. The relative importance of the fragmentation pathway and disproportionation depends on the molecular structure.

The oxidative cleavage of styrenes by diazonium ions is intriguing. This reaction apparently proceeds via anti-Markovnikov addition followed by fragmentation [Quilico, 1928; Ainley, 1937].

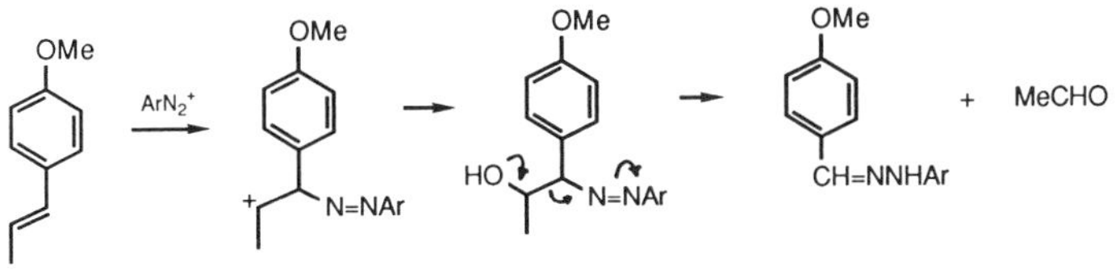

Fragmentation instead of *O*-acetylation occurs when the spiroannulated pyrazolene-borneol system is exposed to acetic anhydride [Snatzke, 1971].

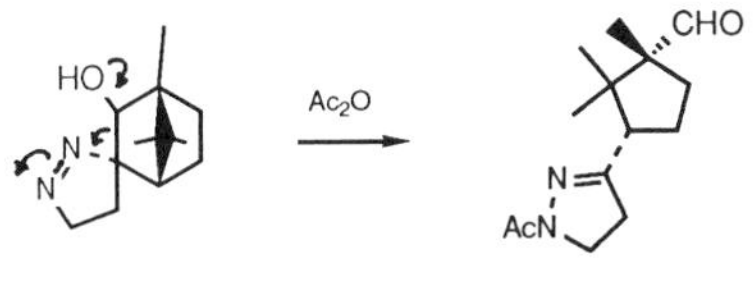

## 7.6. RETRO-MANNICH REACTIONS

### 7.6.1. N—C—C—C=O Arrays

The Mannich reaction is reversible. It means that fragmentation of the Mannich reaction products occurs readily, as any **d2a=d** system would. For special synthetic operations $\beta$-aminoketones may be acquired by other methods and retro-Mannich degradation may then be induced.

The retro-Mannich reaction of aminocyclopropyl ketones is as facile as the corresponding retro-aldol process. The Hantzsch acid gives two major products on heating in diglyme [Biellmann, 1971]. The pyrrole derivative may have been formed via a bicyclic intermediate.

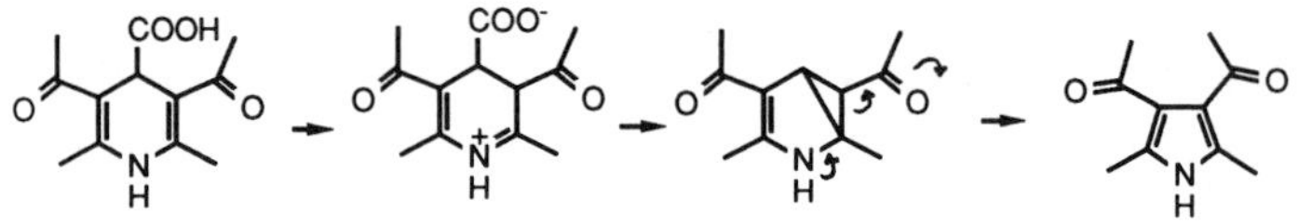

Methylation of 2-(diethylamino)-3-ethoxycarbonyl-5-phenyl-3*H*-azepine leads to a cyclopropanodihydropyridine [Streef, 1987]. The reaction has been considered to proceed via an azahomobenzvalene intermediate.

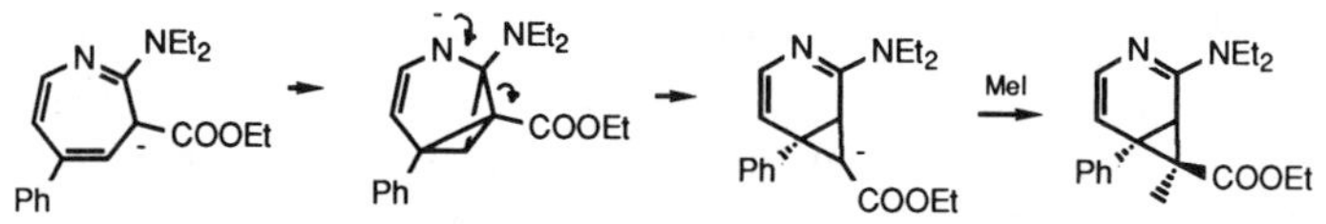

Scopinone is remarkably unstable. In distilled water at room temperature it is degraded into *m*-hydroxybenzaldehyde in substantial yields within four hours. Both acidic and basic reagents bring about this transformation in much higher yields, and scopinone methobromide is even more susceptible to this degradation [Meinwald, 1959]. The penultimate step of the aromatization is a retro-Mannich fragmentation.

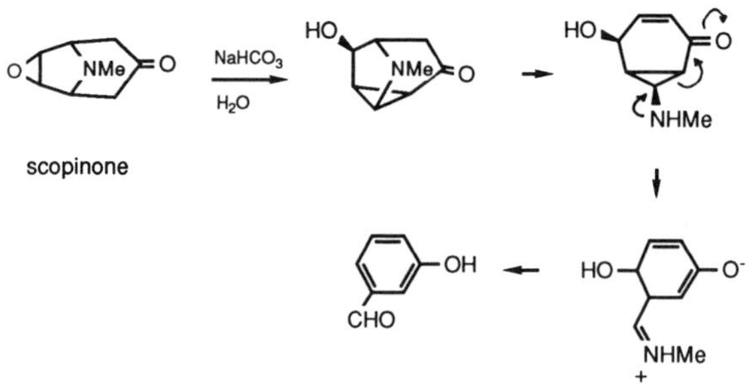

scopinone

The photoinduced reaction of α-bromoacetone enolate with 6-iodo-9-ethylpurine in the presence of ammonia gives 1-acetonyl-9-ethylhypoxanthine [Nair, 1985]. A mechanism hypothesizes the intermediacy of a 2-acetylaziridinium ion which is attacked by ammonia. The ensuing 2-acetyl-3-aminoaziridine then undergoes fragmentation.

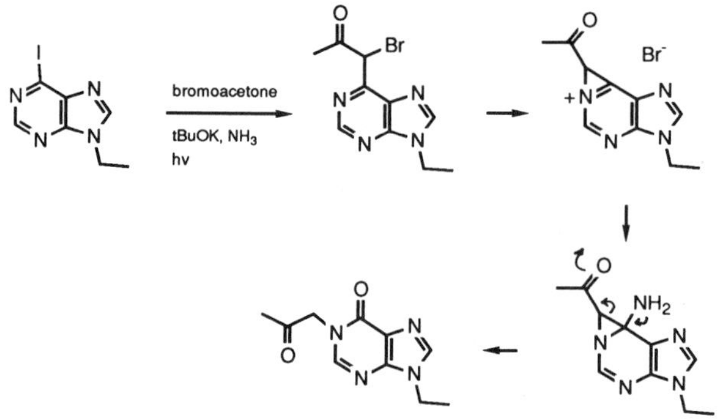

2-Pyrrolines are obtained by hydrogenation of nitrocyclopropyl ketones over Raney nickel [L.I. Smith, 1951]. Fragmentation of the aminocyclopropyl ketones is a crucial step preceding the heterocyclization.

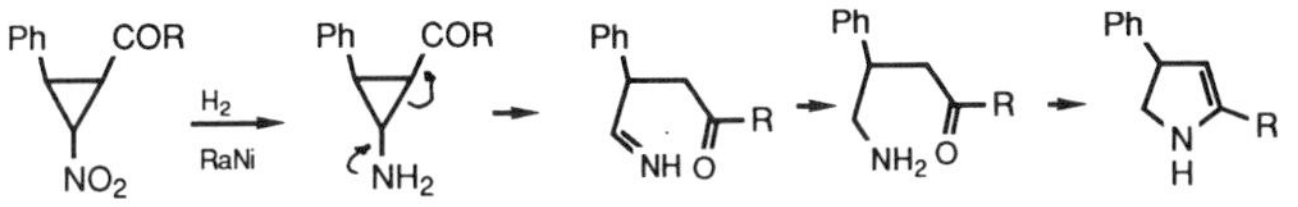

The major thermolysis products of azidocephems in the presence of methanol are ring expanded lactim ethers [Spry, 1984]. Azirenes, the probable intermediates of these reactions, would be attacked by methanol and a **d2a=d** fragmentation then follows.

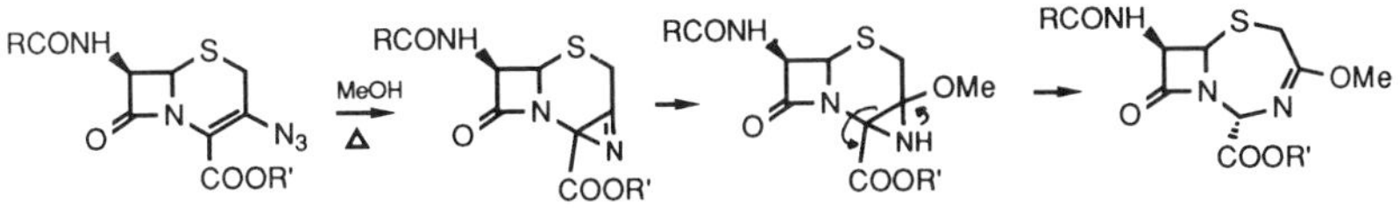

The cyano group is an acceptor of comparable ability to the carbonyl in stabilizing an adjacent carbanion. Therefore it is not surprising that fission of 2-aminocyclopropyl cyanides occurs readily [M. Ikeda, 1977].

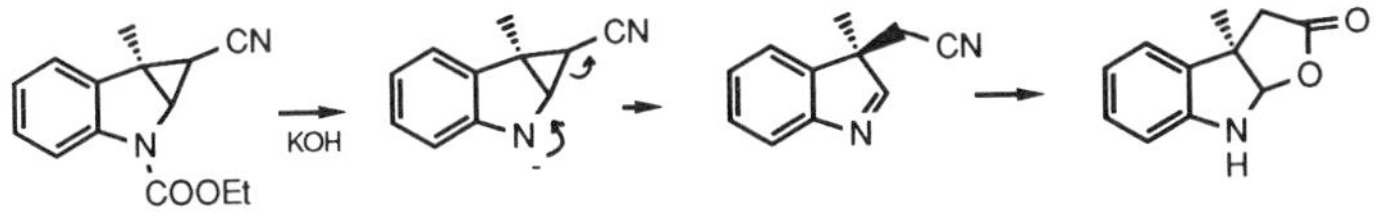

Neoxyberberine acetone has a bridged ring subunit. Its deacetoxy derivative splits off the acetone bridge on acid treatment [Naruto, 1975], as the retro-Mannich reaction is favored by aromatization of the heterocycle.

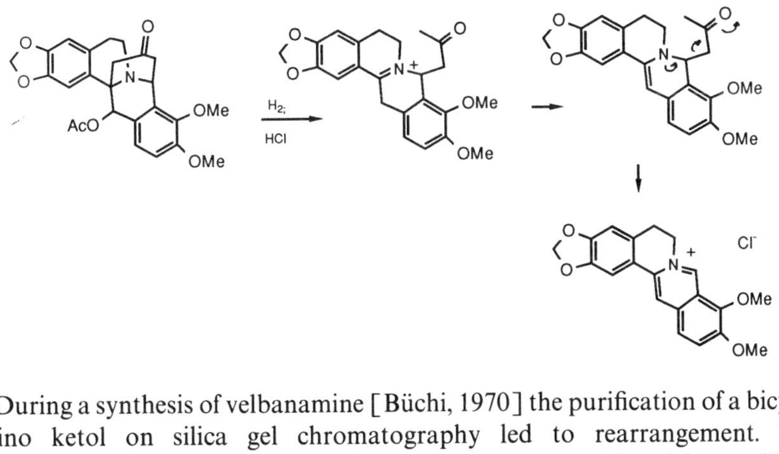

During a synthesis of velbanamine [Büchi, 1970] the purification of a bicyclic amino ketol on silica gel chromatography led to rearrangement. This rearrangement involves fragmentation and a vinylogous Mannich reaction in an alternate site. Of course the destruction of the bridged ring system relieves much strain from the molecule.

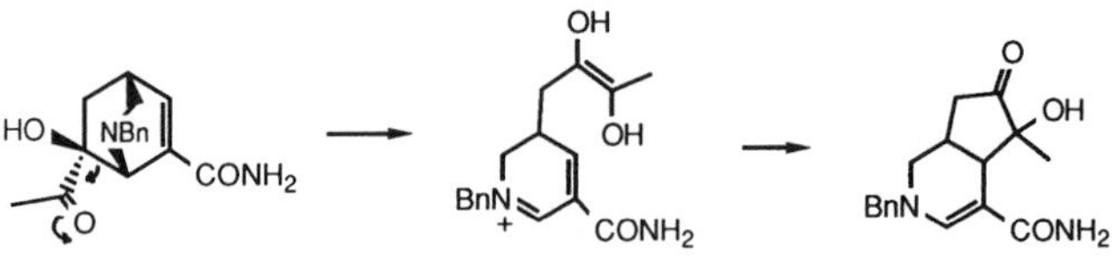

Two reaction pathways, that is retro-Michael and retro-Mannich reactions, are available to $\beta$-amino ketones. Both types of reactions can occur concurrently, as exemplified here [Michne, 1976].

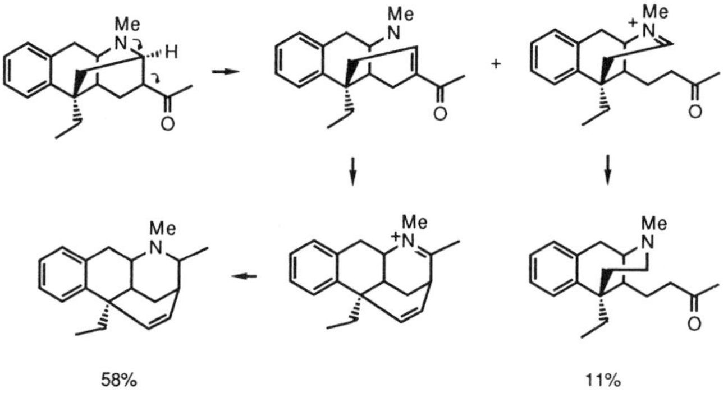

58%

11%

Fragmentation of the indolenine carbamate has been considered as the first step in the polymerization of 1,2,3,4-tetrahydro-3-oxocarbazole by carbobenzyloxy azide [Bailey, 1980].

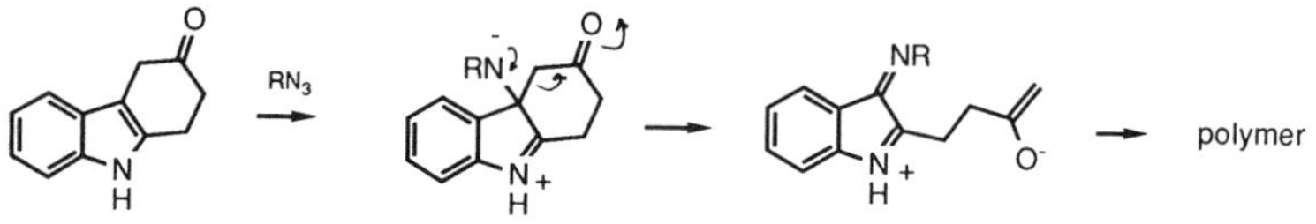

An interesting method for synthesis of certain heterocycles which also incorporate elements of dimedone in a seco form is shown now [S. Miyano, 1983; Matsuo, 1985]. A sequence of intramolecular Michael addition and retro-Mannich reaction accounts for the observation.

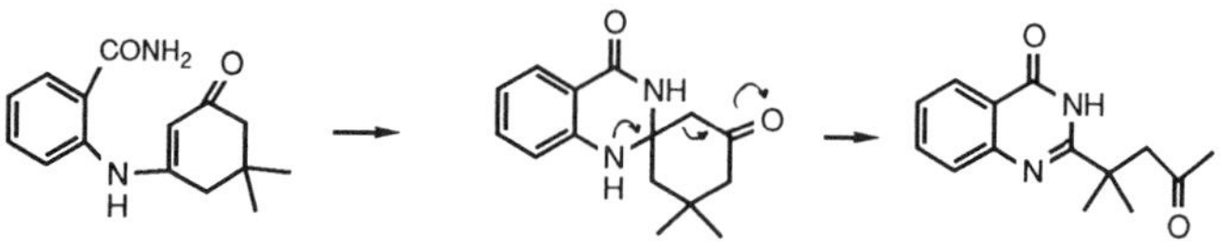

The purple Stenhouse salts undergo cyclization in pyridine, but under acetylation conditions cycloreversion of the initial products by the retro-Mannich reaction occurs [Lewis, 1977]. The fascinating reaction sequence is obscured by the reversible process.

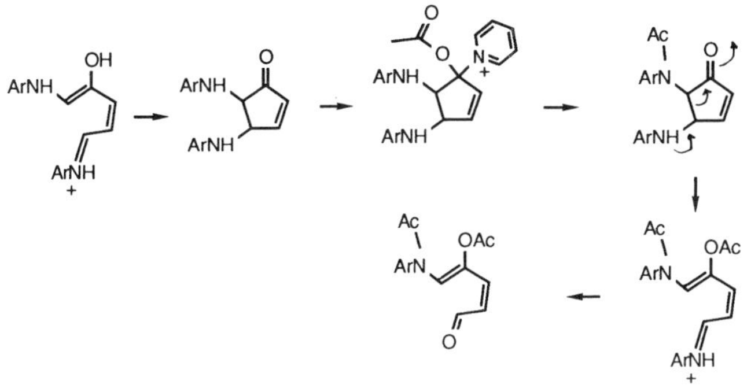

A delightful discovery during a synthesis of emetine [Openshaw, 1963] is concerned with deracemization of a tricyclic ketone. Thus, the (−)-ketone is obtainable on treatment of the racemate with (−)-10-camphorsulfonic acid in refluxing ethyl acetate, and decomposition of the salt. The selective crystallization of the desired diastereoisomer drives the equilibrium towards it via a retro-Mannich/Mannich sequence of the (+)-isomer.

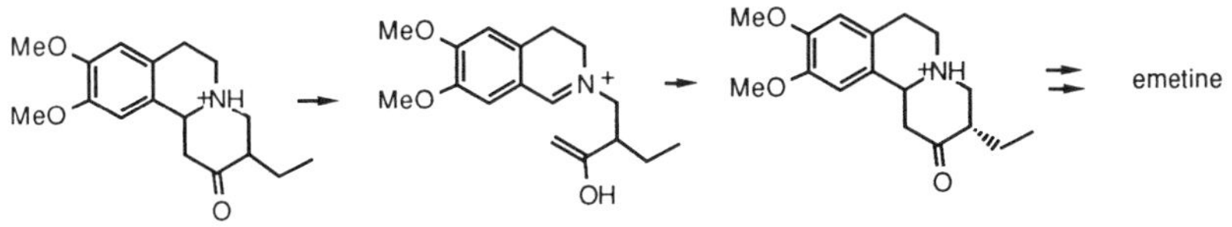

emetine

Undoubtedly the stereospecific cyclization of *N*-(3,4-dimethoxybenzylidene)-3,3-(ethylenedioxy)butylamine [Rubiralta, 1987] arises from equilibration of the product via fragmentation and ring closure. It should be noted that the desired stereoisomer is a *trans*-diequatorial 2,5-disubstituted piperidine derivative.

An enantioselective synthesis of vincamine [Oppolzer, 1977] also profits from the ready equilibration by the reversible Mannich reaction manifold. The required *cis*-isomer separates and is resolved by addition of (+)-malic acid. The other isomers can be recycled.

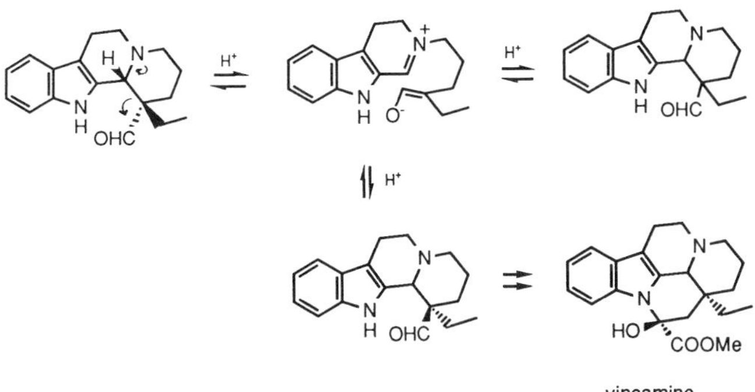

vincamine

Intervention of the retro-Mannich process can also complicate synthetic operations. One such untoward reaction forced the abandonment of a short route to (−)-vindoline [Feldman, 1987].

A transformation product of tabersonine (Zn/HOAc; $LiAlH_4$) undergoes an abnormal Oppenauer oxidation [Mauperin, 1971]. There is a retro-Mannich fission and inversion of configuration at the quinolizidine ring juncture.

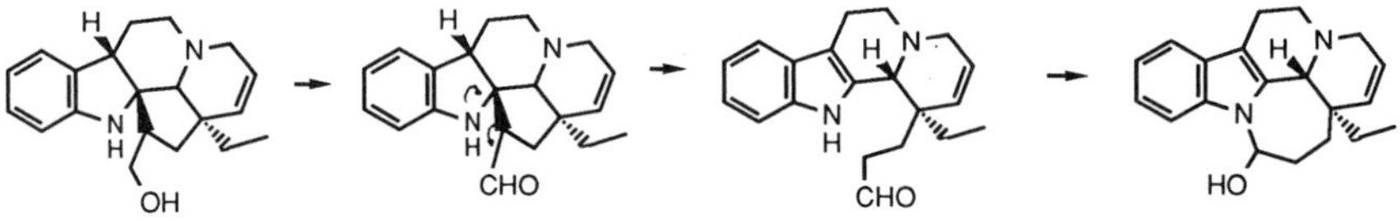

Borohydride reduction of a C-seco ketoester derived from vincadifformine leads to a tricyclic product [Hugel, 1981a], because of the intervention of a fragmentation of the spirocyclic system.

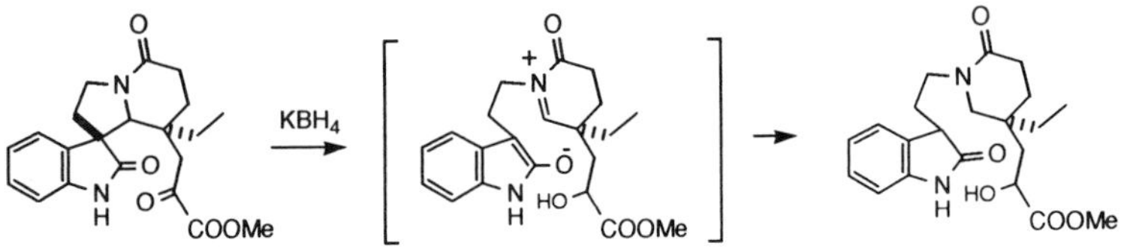

The destruction of the ring system of benzylpenicillin potassium by 1,2-diphenylethylenediamine has been observed [Ogawa, 1988]. Opening of the β-lactam, release of penicillamine by a hetertocycle exchange process hasten the retro-Mannich reaction.

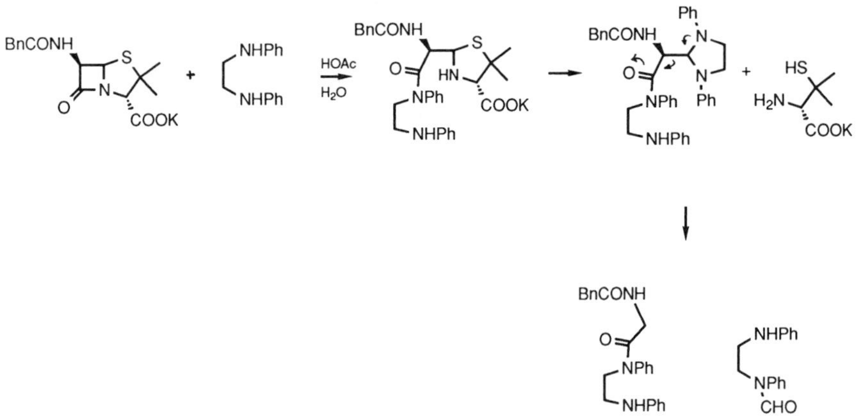

The synthetic approach to hetidine can obviously be simplified by its correlation with an atisine-type diterpene [Wang, 1986] through the Mannich/retro-Mannich interconversion.

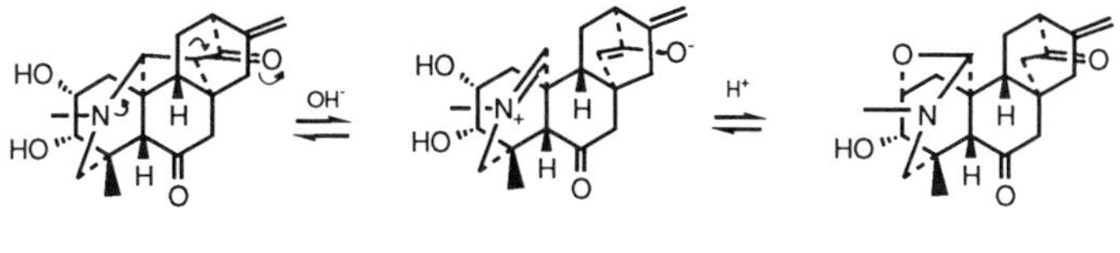

hetidine

Certain 3-aminocyclobutanones may be obtained from the reaction of enamines with ketenes. However, these strained molecules undergo thermal fragmentation very readily [Opitz, 1963]. Note that they contain a **d2a=d** circuit.

A short route to mesembrine [Winkler, 1988] is based on intramolecular photocycloaddition of $\beta$-aminoenone to alkene. The adduct is strained and fragmentable by a retro-Mannich reaction. Completion of the synthesis requires only *N*-methylation and Mannich cyclization. A formal total synthesis of (−)-vindorosine from a tryptophan derivative has also been achieved [Winkler, 1990].

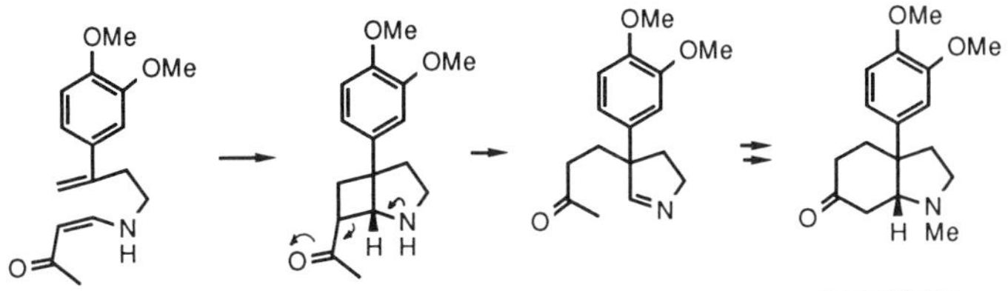

mesembrine

This variant of the deMayo reaction has been exploited in the preparation of 1,5-cyclooctanediones, particularly in connection with synthesis of the taxane ring system [Tamura, 1972, 1975; Schell, 1984; Swindell, 1984]. The intermolecular version has also been reported [Tsuda, 1986].

An interesting approach to the *cis*-octahydroquinoline skeleton is via an intramolecular Diels–Alder reaction of a 2-alkenyl-1,2-dihydropyridine, followed by hydrogenation and retro-Mannich degradation [Comins, 1983]. The Diels–Alder reaction ensures the *cis* juncture of the ring system.

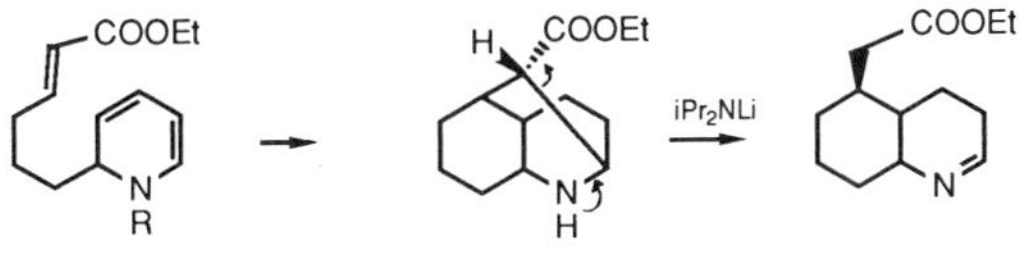

2-Azatricyclo[4.4.0.0$^{2,8}$]dec-9-en-7-ones are usually stable after they are formed by a photochemical reaction of triazolines. However, the presence of an ester substituent at C-6 greatly facilitates the retro-Mannich fission [Schultz, 1987].

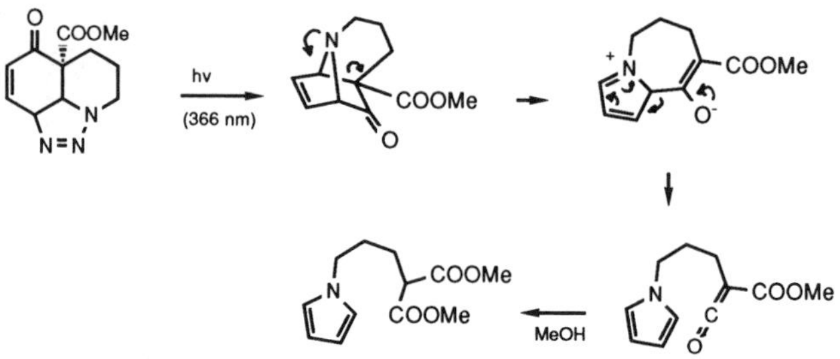

A rare case of fragmentation of $\beta$-iminoketone [Bach, 1986] pertains to an attempted hydrolysis of the corresponding oxime under reductive conditions in the context of synthesizing histrionicotoxin.

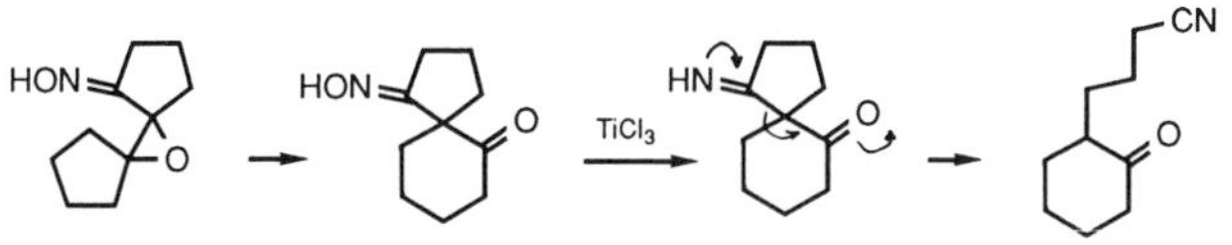

As acceptor the nitroso group is as effective as a carbonyl. Accordingly, the aza-analog of the retro-Mannich reaction is a possibility under proper circumstances [Lockhart, 1978].

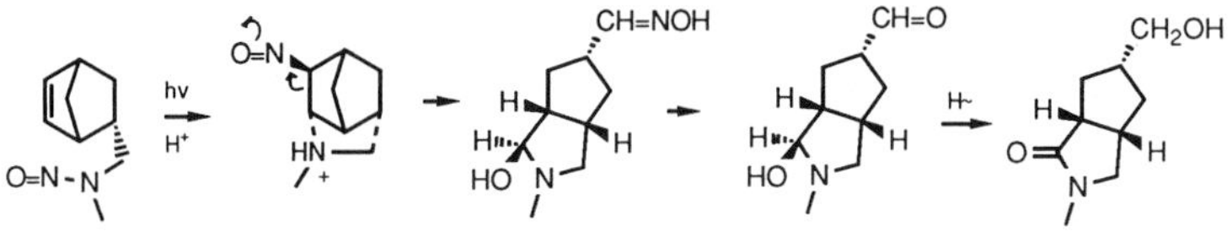

### 7.6.2. N—C—C—C=N Arrays

Fragmentation of this system is a modified retro-Mannich reaction and finds most of its examples in complex indole alkaloids, both during chemical transformations and synthesis. For example, in a synthesis of aspidospermine [Stork, 1963] via Fischer indolization of a perhydrolololidine in which the tertiary carbon atom common to all three rings is not in the desired configuration, a hidden thermodynamically controlled fragmentation–recyclization corrected this misalignment.

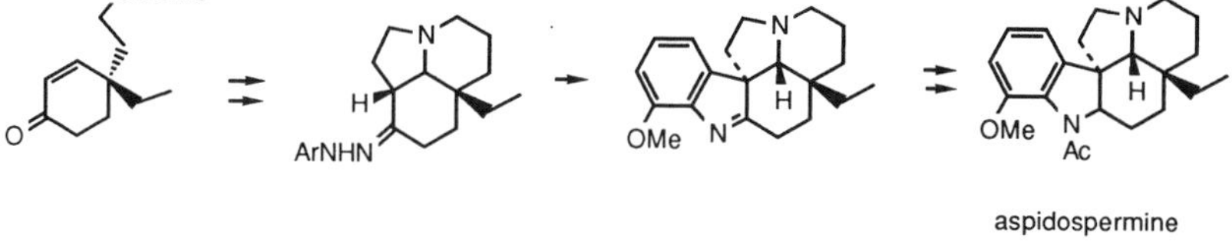

aspidospermine

(−)-Vincadifformine racemizes to the extent of 99% [Takano, 1989] after microwave irradiation in dimethylformamide for 20 minutes. Formation of the chanoiminium ion and further depolarization through complete ring opening to a dihydrosecodine intermediate accounts for the observation.

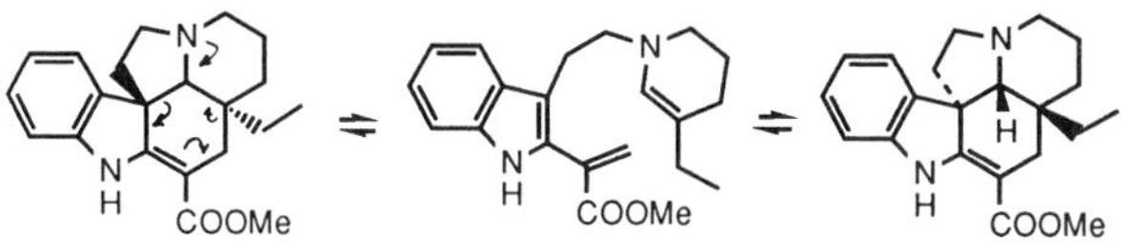

(+)-vincadifformine (-)-vincadifformine

The apparent epimerization of the ethyl group of 19,20-dihydroakuammicine [Scott, 1974] is the result of C-ring cleavage and reclosure. Actually, epimerization occurs at the other three asymmetric carbon atoms.

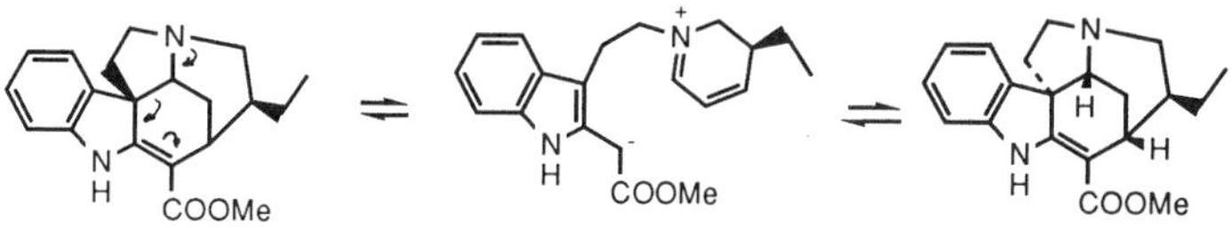

Akuammicine itself gives 2-hydroxycarbazole [P.N. Edwards, 1961] among other products on heating with methanol. Aromatization of the fragmentation product now renders a transformation of the ring system irreversible.

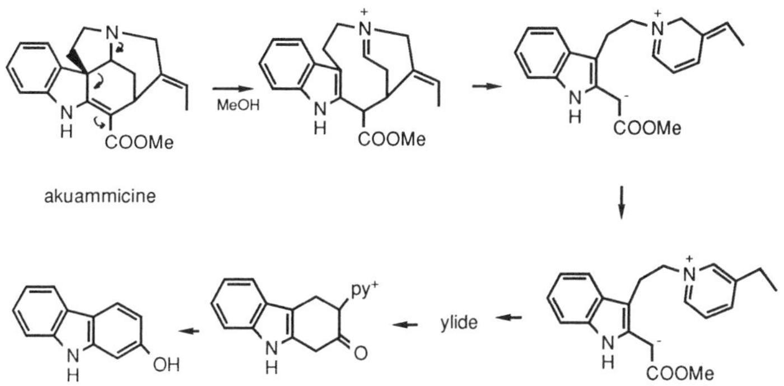

akuammicine

The fragmented species may be arrested by reduction. Thus, borohydride reduction of apopseudoakuammigine [Joule, 1962] leads to a tetracyclic indole derivative.

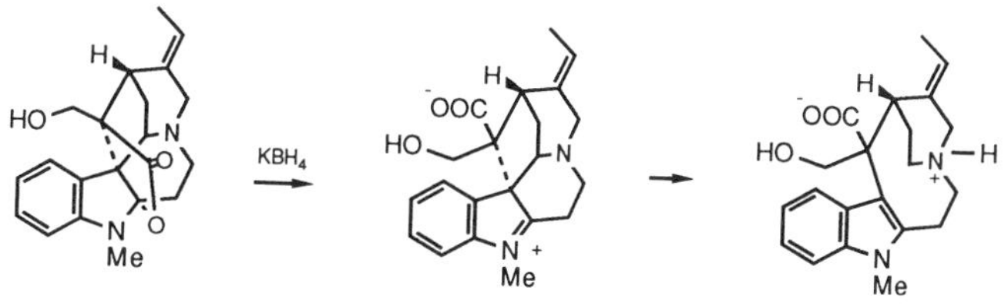

Involved in the conversion of precondylocarpine to andranginine [Andriamilisoa, 1975] are two fragmentations which convert the molecule into a tricyclic compound. The latter undergoes recyclization by the Diels–Alder process.

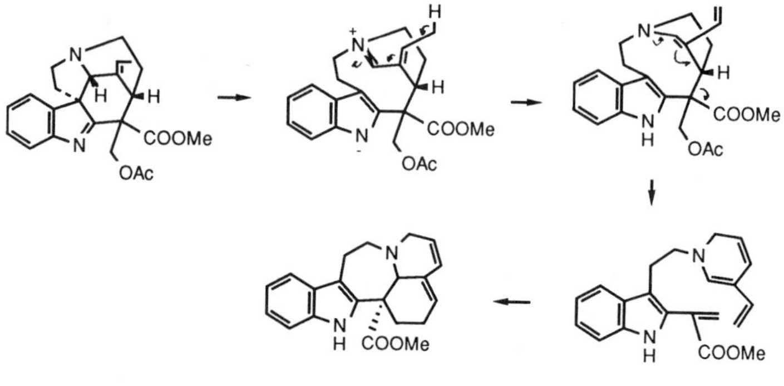

andranginine

The transformation of 19,20-dihydropreakuammicine to the skeleton of ngouniensine [Massiot, 1982] follows a similar pathway except the recyclization is a Mannich-type reaction.

Certain amount of tubifoline is generated by heating condyfoline [Schumann, 1963]. This behavior of condyfoline entails a transformation involving CD-ring scission, C=N double bond migration, and recyclization.

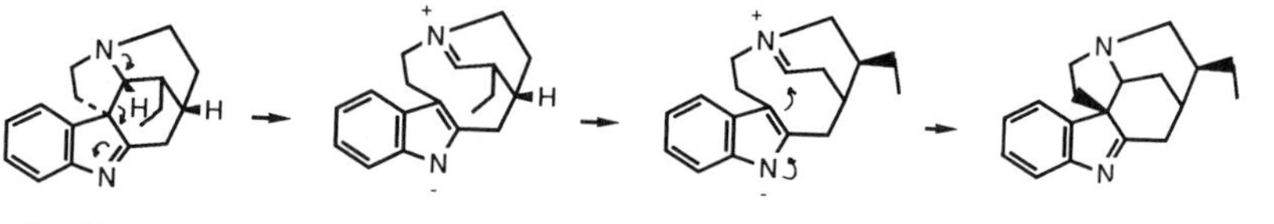

Condyfoline tubifoline

In anhydrous acetic acid, the chloroindolenine derived from vincadifformine is transformed into a hexacyclic base [Lewin, 1982], apparently according to the following mechanism.

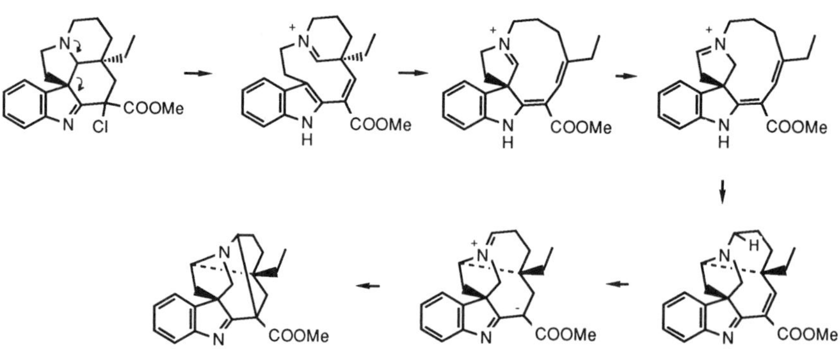

Isomerization of the aspidosperma to the eburna alkaloid skeleton may be effected in several ways. The crucial and likely the first step is the fragmentation [Hugel, 1981; Magnus, 1986].

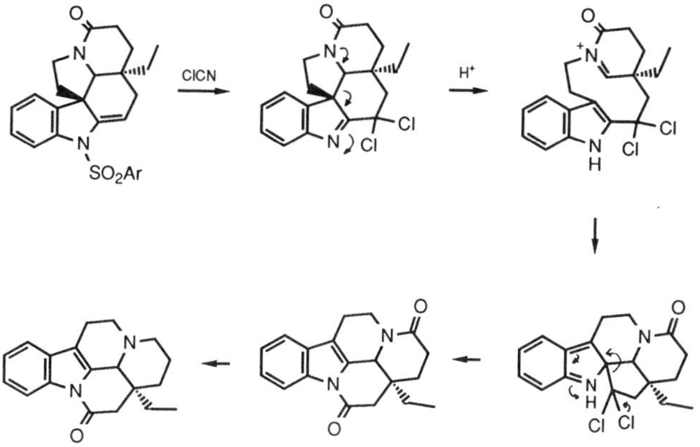

ClCN
H+

eburnamonine

A double fragmentation ensues when aspidospermidine 16-oxime is submitted to Beckmann rearrangement conditions [Hugel, 1974]. This is followed by reclosure to an E-seco-eburnamonine base.

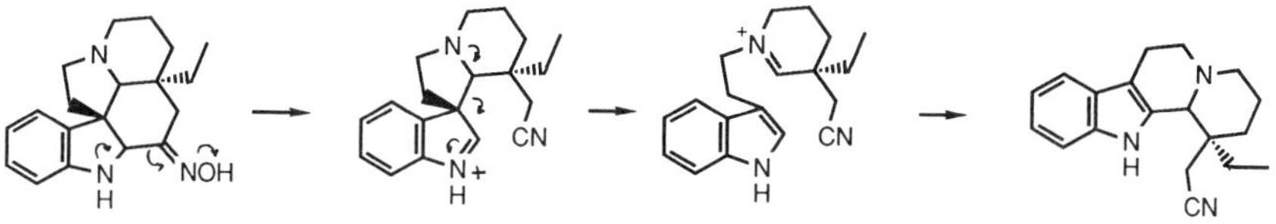

Of particular interest is a simple synthesis of strempeliopine [Hajicek, 1981] from an aspidosperma base by a similar isomerization process.

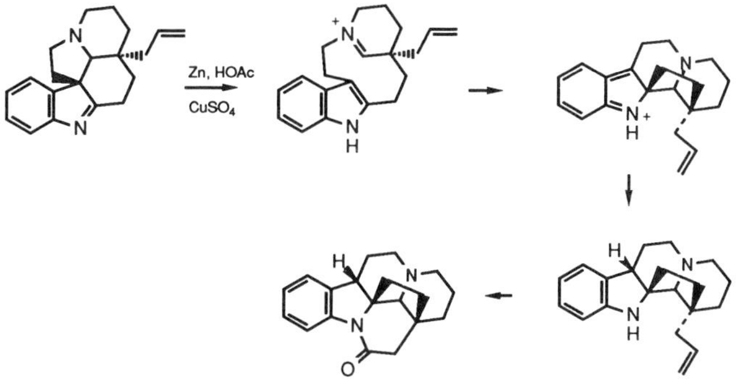

Zn, HOAc
$CuSO_4$

strempeliopine

It is also possible to manipulate the eburna skeleton back to the aspidosperma family. Thus an E-seco congener of eburnamonine (a hydroxy lactam) has been found to undergo intramolecular alkylation and isomerization on acid treatment to give a product whose hydride reduction results in aspidospermidine [Harley-Mason, 1965].

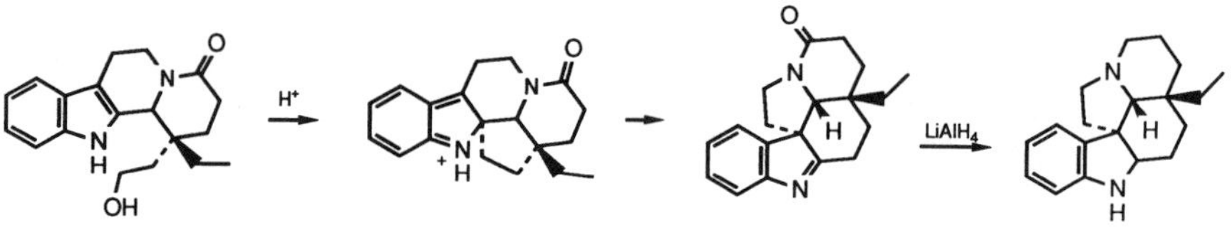

A superb scheme for synthesis of many aspidosperma alkaloids [Kuehne, 1981, 1983] depends on the formation of bridged indoloazepines and their fragmentations. The strategy may be illustrated by a route to ibophyllidine.

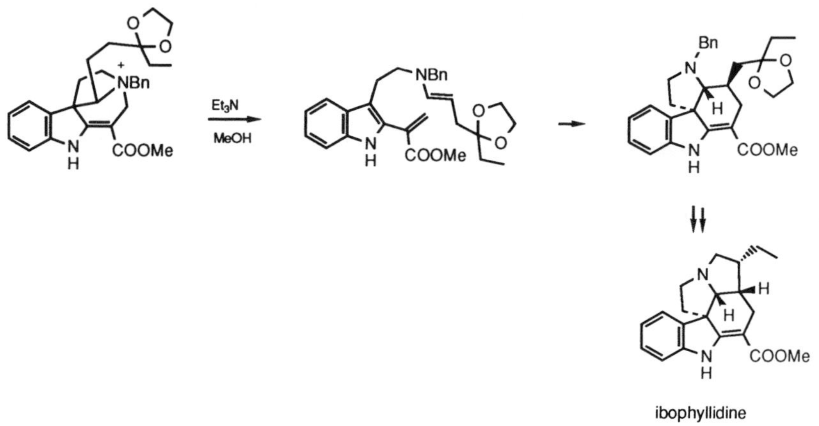

ibophyllidine

Oxidative degradation of 16-epi-19(*S*)-vindolinine [Atta-ur-Rahman, 1983] creates an indolenine chromophore which is susceptible to fragmentation, resulting in a tricyclic formamide.

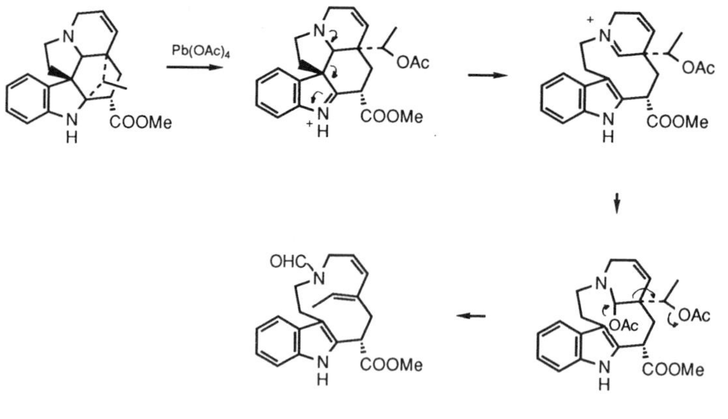

The formation of cleavamine from catharanthine [Gorman, 1963] under reducing conditions may be envisaged as an indolenine-induced fragmentation with subsequent loss of the ester group.

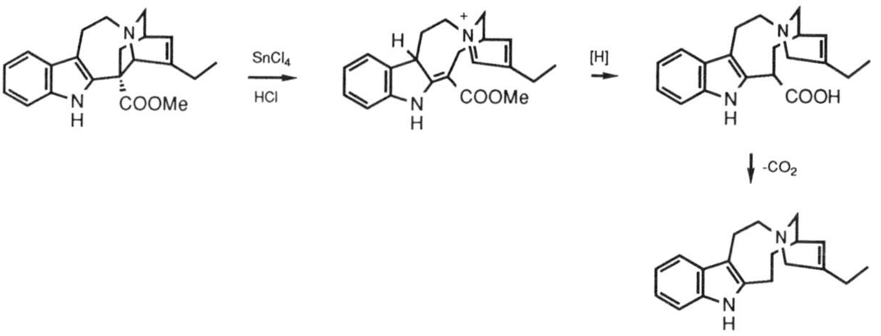

In simpler structures such as *N*-methyltetrahydro-γ-carboline various electrophiles have been used to cause the isomerization to the tetrahydropyrimido[1,6-*a*]indole derivative [Ebnother, 1969; Bhandari, 1971; Bailey, 1972; Nagai, 1979; Hershenson, 1980]. The latter ring system can actually be assembled by the Fischer indole synthesis of 1,3-disubstituted 4-piperidinone [Cattanach, 1968].

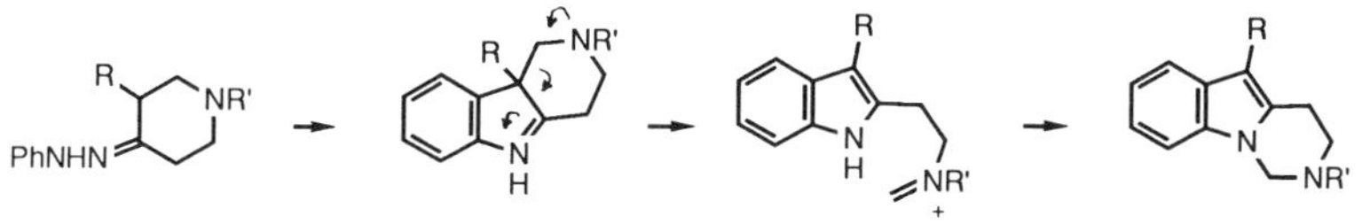

The formation of a 4-oxo-1,2,3,4-tetrahydro-β-carboline from a 3-(cyanomethyl)aminomethylindole by acid treatment [MacKay, 1984] is readily interpreted in terms of two cyclization processes interposed by a fragmentation step. However, the possibility that the latter two steps may actually be a 1,2-rearrangement cannot be excluded.

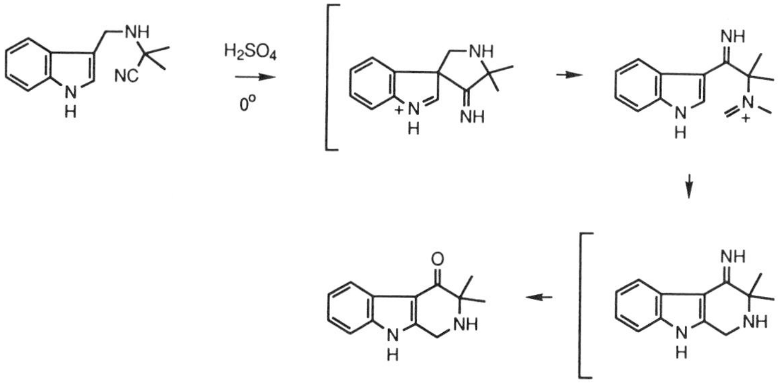

The Vilsmeier–Haack reaction of *N*-benzyl-1,2,3,4-tetrahydrocarbazole to give the 1-formyl derivative [Y. Murakami, 1983] may proceed by reaction at the ring juncture initially, but this substituent falls off on hydration as the molecule returns to an aromatic state.

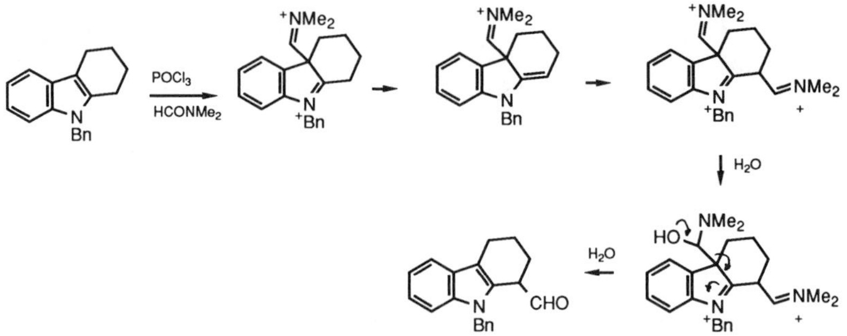

A few selected reactions related to the retro-Mannich fission are now mentioned. These include the conversion of an amino dicyanoperhydroindan to a spirocyclic keto aminoacrylonitrile [Berkowitz, 1976] and the reaction of a Diels–Alder adduct of tetracyanoethylene and a conjugate tetraene with sodium methoxide [Pfoertner, 1975].

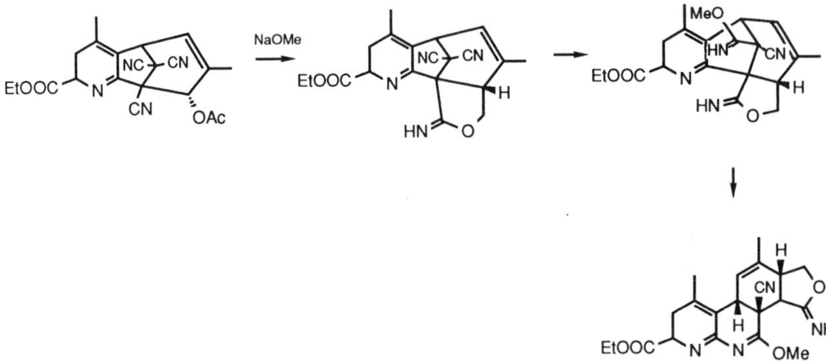

More remote analog of the retro-Mannich reaction is the diazoalkane synthesis [Sekiya, 1976] by alkaline decomposition of *N*-(benzamidomethyl)-*N*-nitrosamines.

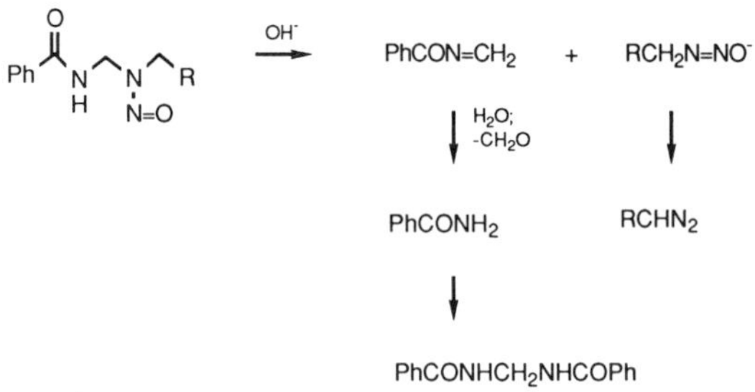

## 7.7. FRAGMENTATION OF THE O—C—C—C=N SYSTEMS

The O—C—C—C=N array differs from the fragmentable system discussed above by an interchange of heteroatoms. Since both heteroatoms are donors, the fragmentation susceptibility for the two systems is similar. Thus a retrobiosynthetic process is implied in the degradation of the *Rauwolfia perakensis* alkaloid RP-7 by alkali [Kiang, 1966]. Vellosimine has been obtained.

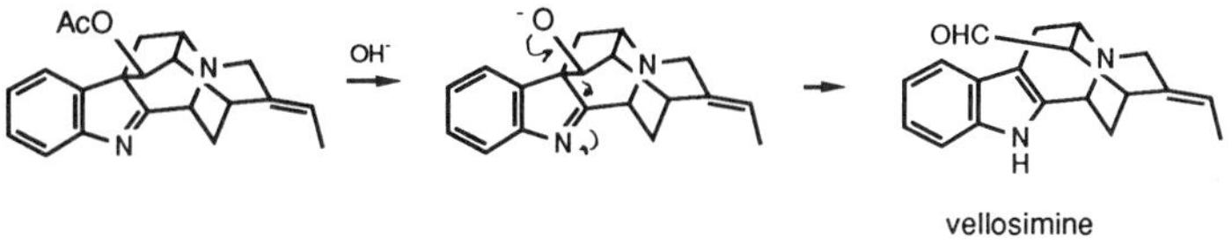

vellosimine

The following Bischler–Napieralski cyclization of a $\beta$-ethylenedioxy amide is attended by scission of the $C_{\alpha}$—$C_{\beta}$ bond [Bosch, 1983a]. Such fragmentation has also been observed during ketalization or thioketalization of 3-acylindoles [K.M. Smith, 1983b] and it has been applied to a partial synthesis of dehydrocoproporphyrin [K.M. Smith, 1983a].

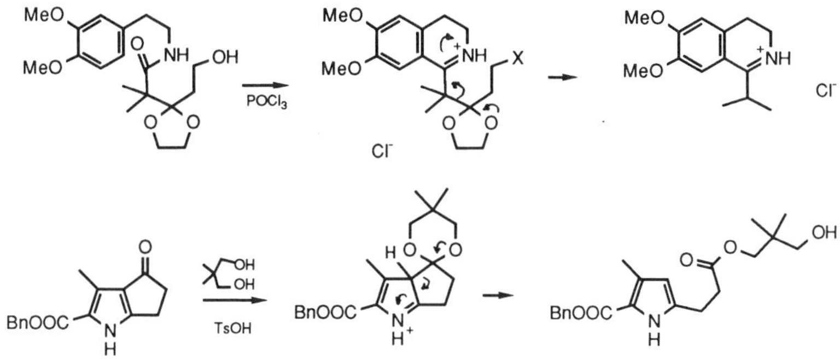

Under the Bischler–Napieralski conditions 3-[2-(propionylamino)-1,1-(ethylenedithio)ethyl]indole gives 3-propionylindole [Murakami, 1987] as a result of reaction at C-3 and fragmentation of the ensuing spirocyclic indolenine.

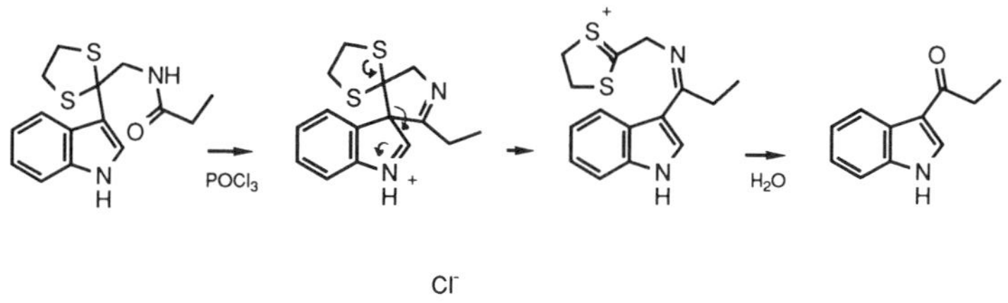

A synthesis of (−)-rhazinilam from (+)-1,2-dehydroaspidospermidine [Ratcliffe, 1973] is accomplished by peracid oxidation and treatment with ferrous sulfate. The crucial intermediate contains the **O—C—C—C=N**$^+$ segment.

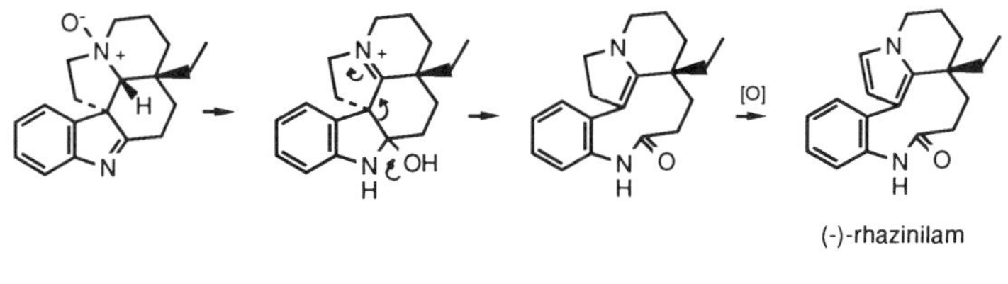

(-)-rhazinilam

During degradative studies of himandridine [Mander, 1967] the complex cage structure was unraveled by dehydrogenation of a B-aromatic compound. The C—C bond cleavage also indicates a 1,3-relationship of the amino alcohol system.

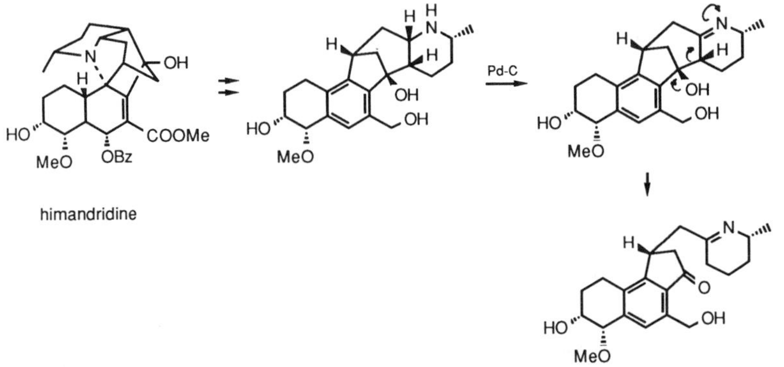

himandridine

Beckmann fragmentation is enhanced considerably by the presence of a donor group at the a position, which enables a push–pull mechanism to operate. Many synthesis have been designed with this fragmentation to performed required C—C bond cleavage. Among these is the indirect inversion of the C-4 configuration of resin acids, for example from isopimaric acid to its epimer [Tresca, 1973] via a hydroxycyclobutanone oxime.

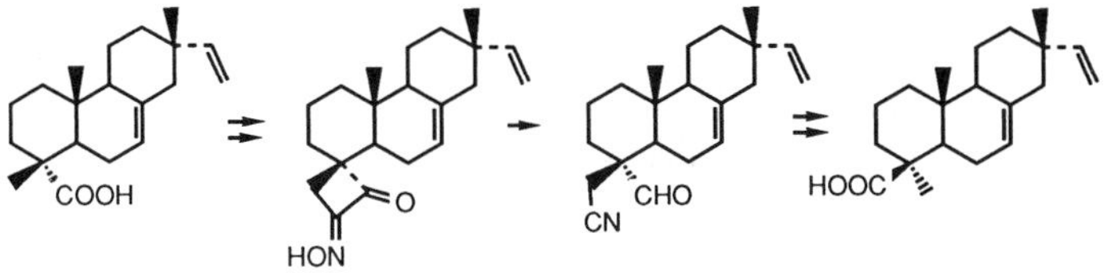

More popular is the use of 1-methoxybicyclo[2.2.2]oct-5-en-2-one oximes as starting materials for synthesis of complex molecules, which include trichodermin [Colvin, 1973], mesembrine, and the morphinan skeleton [Kametani, 1987], and the acyltetronic acid portion of tetronomycin [Hori, 1989].

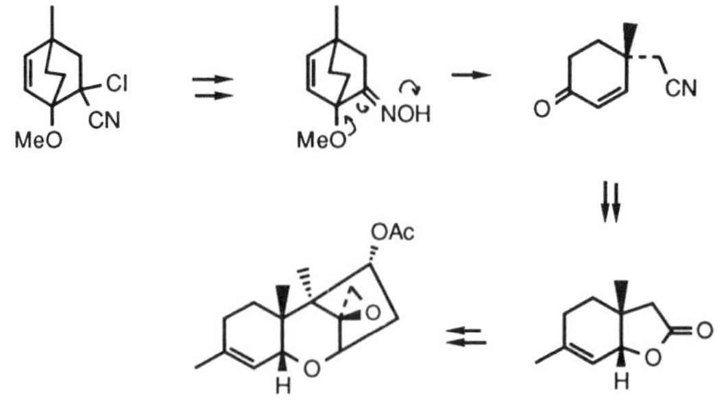

trichodermin

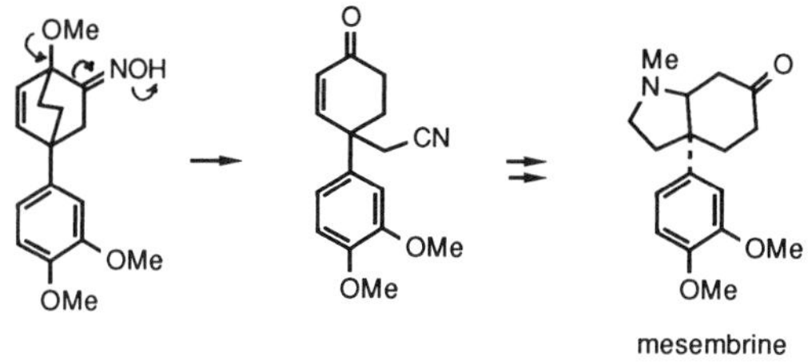

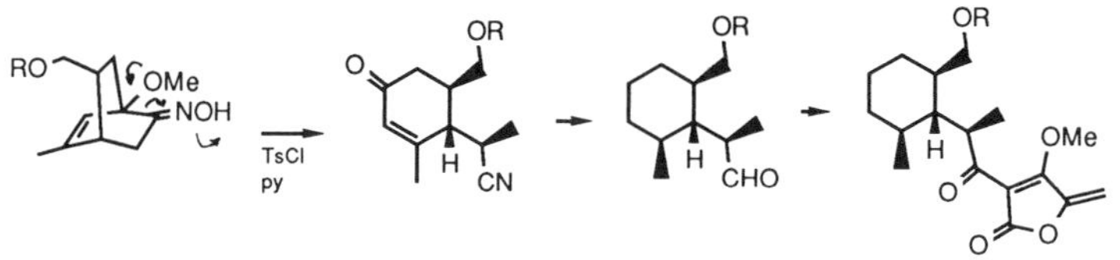

5-Isoxazolones which are readily obtained from *β*-keto esters undergo *N*-nitrosation/fragmentation to give alkynes [Boivin, 1991]. Actually there are two fragmentation reactions, the first one results in cleavage of the heterocycle, and the second step is initiated by decarboxylation.

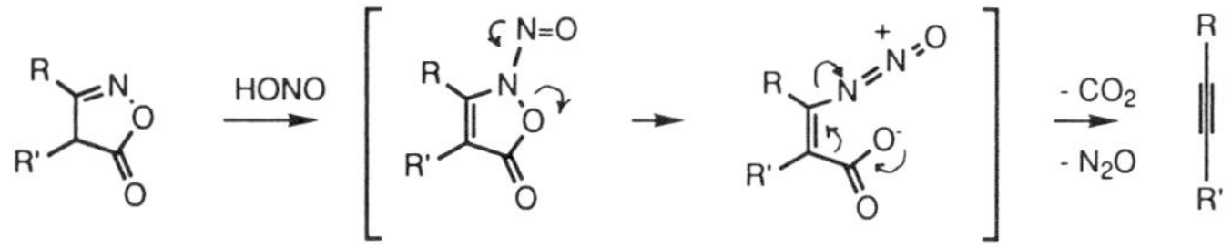

# 8

# d3d FRAGMENTATIONS

The systems in which two donor atoms are separated by three other atoms comprise the most impressive numbers of varieties and examples among fragmentation reactions. A permutation of donors and intervening atoms is possible, and the intervening atoms are not limited to carbon. Fragmentation reactions of a three-carbon chain which are linked to two effectors (donors) have been referred to in the chemical literature as the Grob fragmentation.

## 8.1. O—C—C—C—O SYSTEMS

### 8.1.1. HO—C—C—C—OR Arrays

In these systems the nucleofugal OR may be a hydroxyl, ether, or ester. The general reactivity pattern applies when diols undergo fragmentation, that is priority of ionization follows the trend of tert. > sec. > prim. A few examples are shown now.

Cu-bronze, Δ; + $CH_2O$ + $H_2O$

[Barton, 1961]

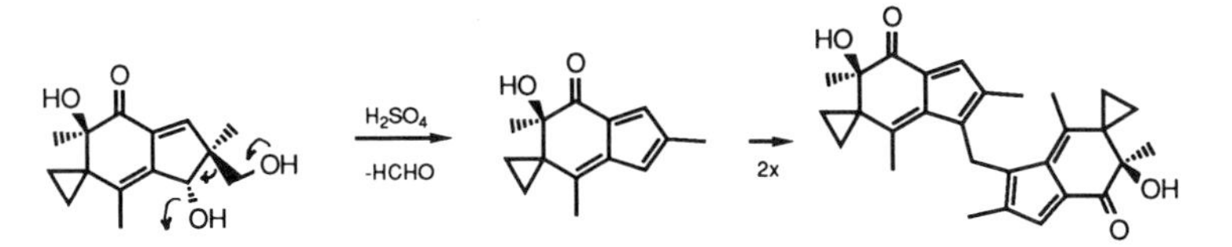

$H_2SO_4$

-HCHO

2x

[Weinreb, 1971]

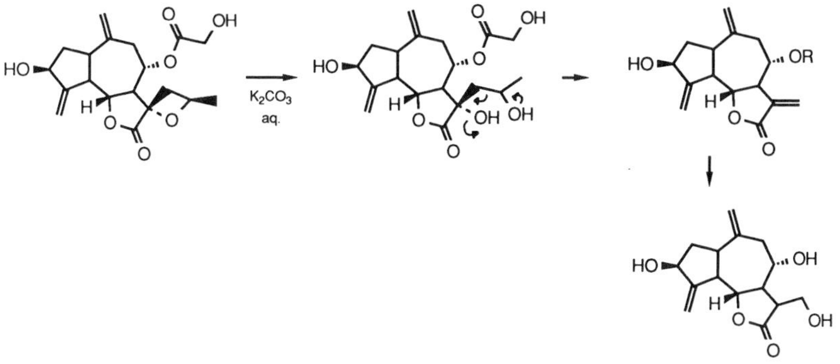

$K_2CO_3$ aq.

[Collado, 1986]

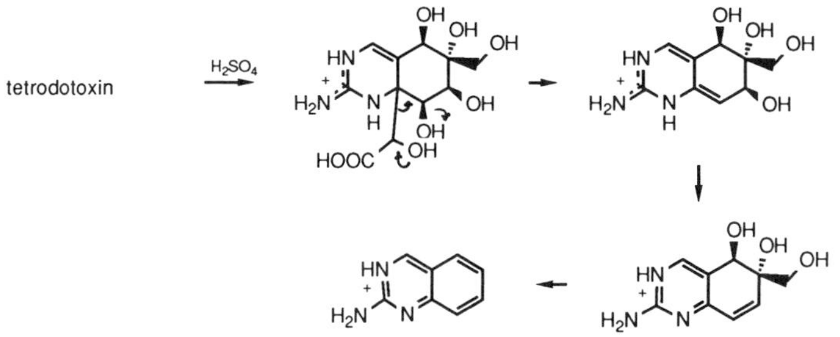

tetrodotoxin

$H_2SO_4$

[Goto, 1965]

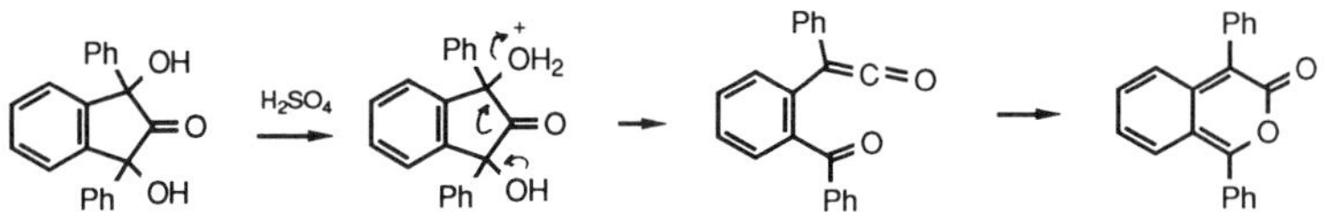

$H_2SO_4$

[Holland, 1970]

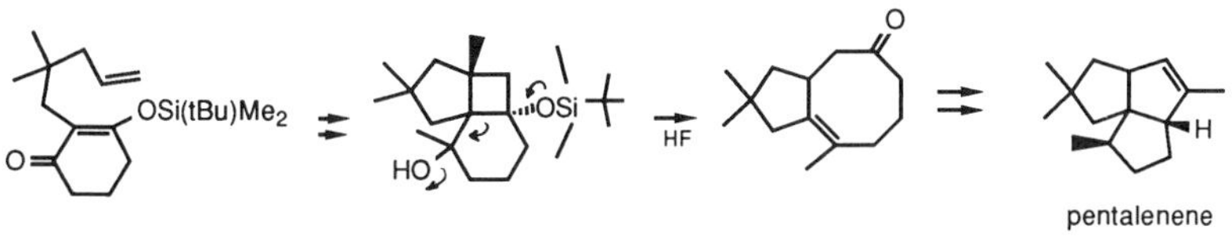

OSi(tBu)$Me_2$

HF

pentalenene

[Pattenden, 1984]

For steric reasons, for example Bredt's Rule, etc., exceptions are known:

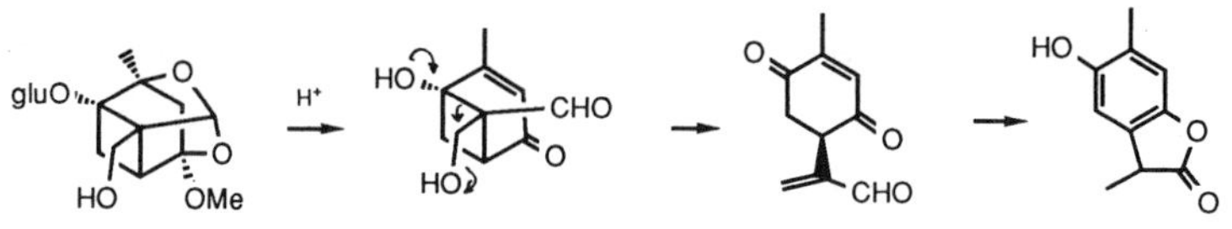

[Aimi, 1969]

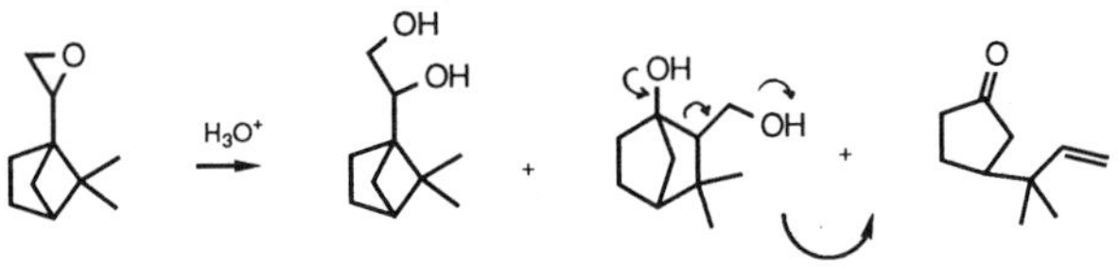

[Sasaki, 1970]

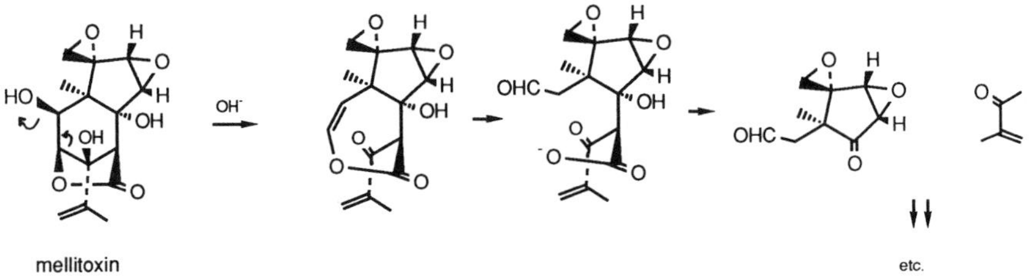

mellitoxin

etc.

[Hodges, 1964]

The $\beta$-glucosidase-catalyzed hydrolysis of salicortin [Clausen, 1990] leads to the loss of glucose and carbon dioxide. From the structure of the product it can be deduced that the transformation consists of fragmentation and a Michael reaction.

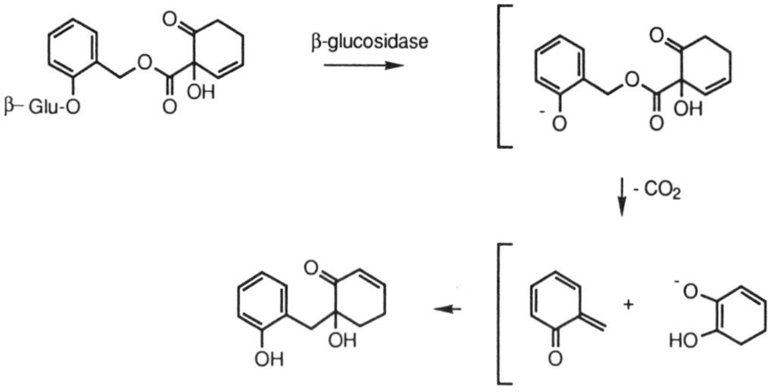

β-glucosidase

$-CO_2$

Workup of the reductive dimerization of 1-phenyl-2-(1,3-thiolan-2-yl)ethanone induces a fragmentation [Gammill, 1980]. Note that benzaldehyde (not benzyl alcohol) is one of the products.

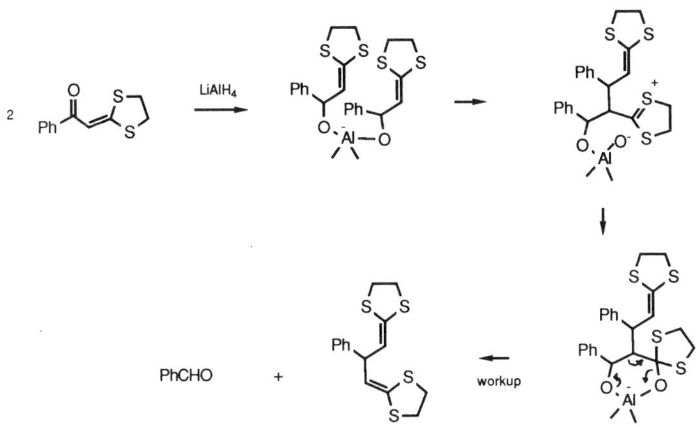

During a synthesis of (+)-compactin [Grieco, 1983] from a Diels–Alder adduct of methyl 7-oxabicyclo[2.2.1]hept-2-enecarboxylate and a diene, the generation of the 8′a-hydroxyl group and the 4′a,5′-double bond enlisted a fragmentation procedure.

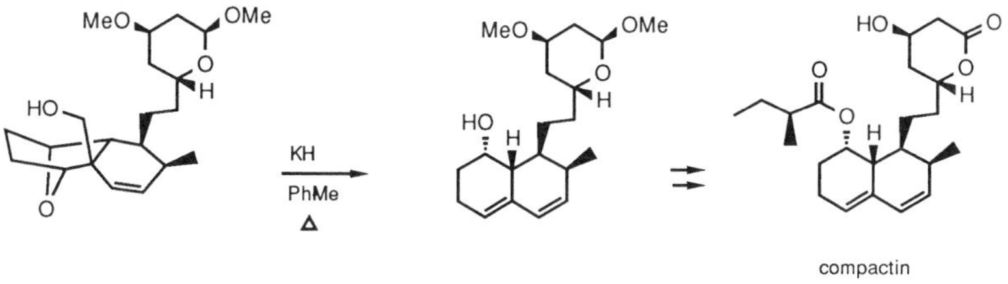

compactin

If a 1,3-diol subunit is a cyclopropanol, opening of the small ring invariably results upon its contact with an acid. Such cyclopropanols may be generated in situ [Read, 1973; Sato, 1986] and they would undergo fragmentation spontaneously. Under nonacidic conditions their isolation may be achieved [T. Sato, 1985].

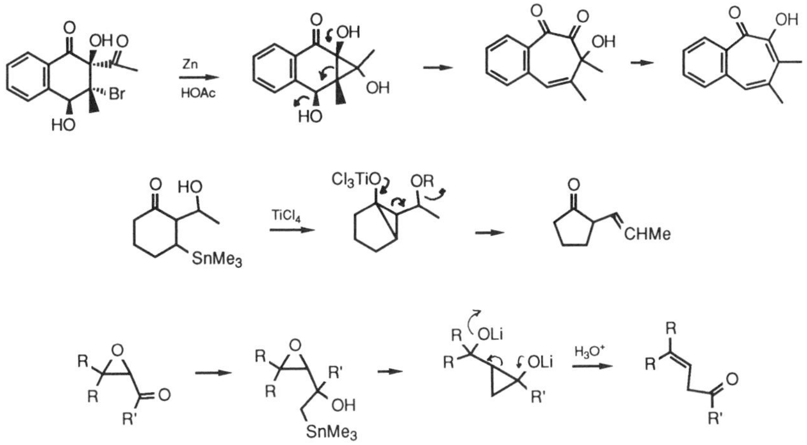

Of special interest is the following sulfonyl diol derivatives [Pohmakotr, 1985], the α-hydroxyl supersedes the sulfonyl as the leaving group.

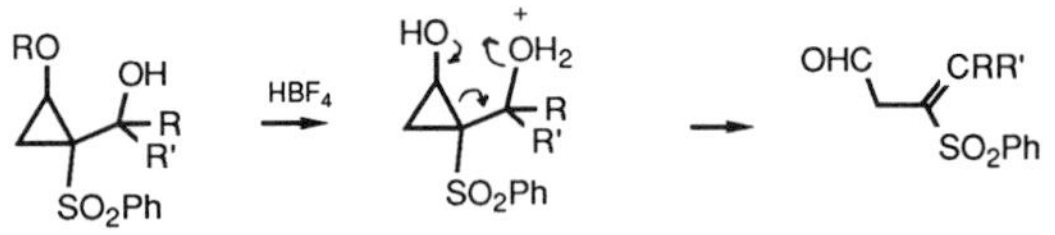

Certain 1,3-diketones produce cyclopropanediols when submitted to the Clemmensen reduction [Cusack, 1965; Wenkert, 1965; Kaplan, 1967]. These compounds fragment immediately.

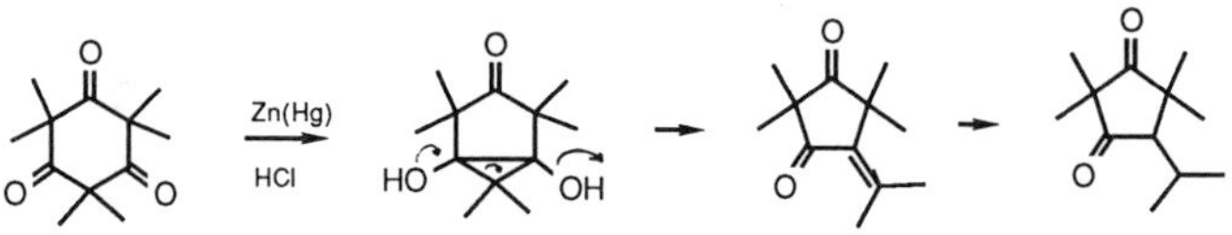

1,2-Cyclopropanol derivatives generated by other methods are of course not immune to fragmentation [M. Sato, 1977; Horowitz, 1952; Wolff, 1984].

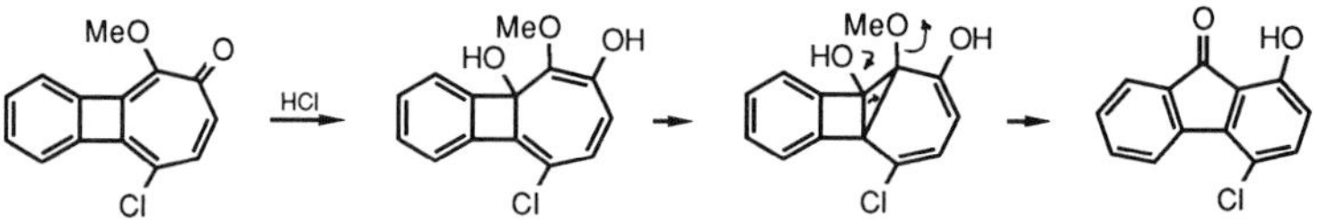

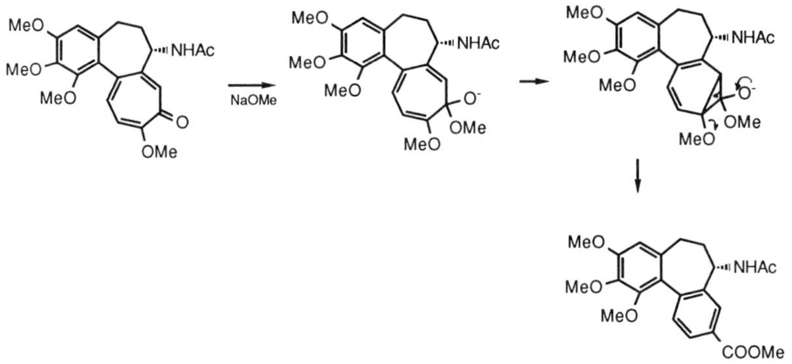

allocolchicine

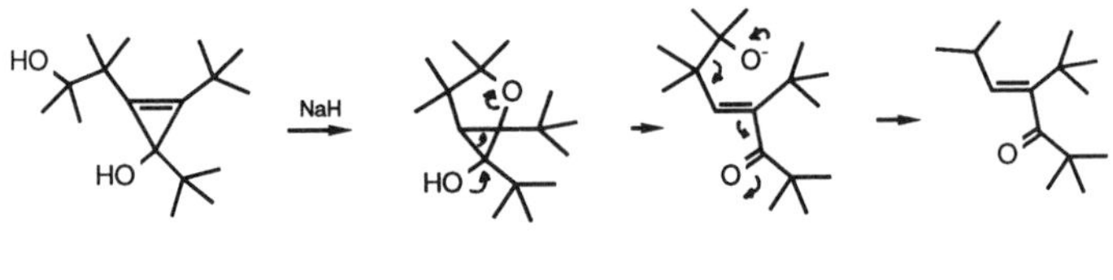

Cyclopropanone hydrates are obtainable from hydrogenolysis of the appropriate benzodioxepines. When an α-hydroxyalkyl group is present in these substrates the products are β,γ-unsaturated acids [Dowd, 1985a].

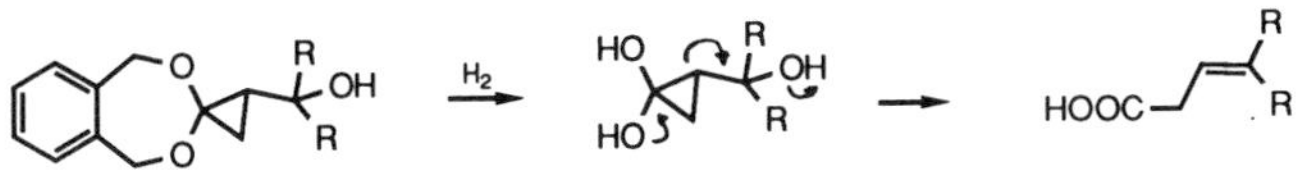

Four-membered rings are still highly strained, and the facile fragmentation of 1,3-cyclobutanediols is nothing extraordinary [Clark, 1979; Tobe, 1985].

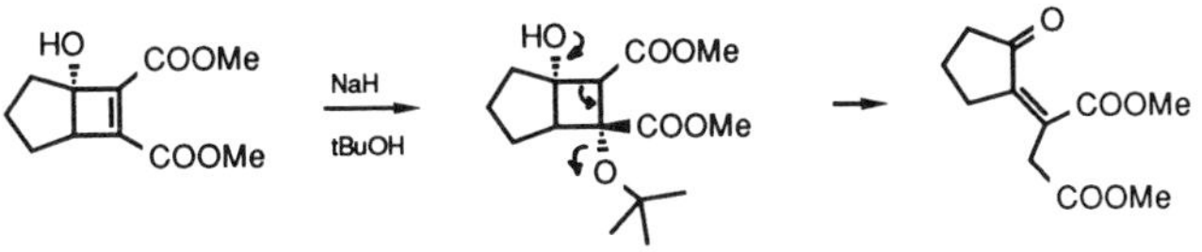

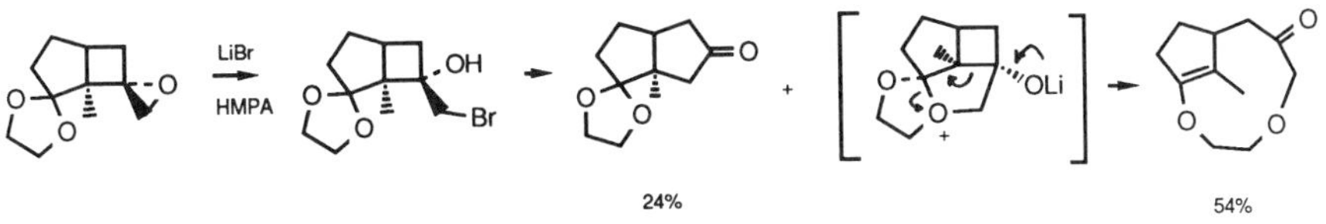

Epoxides of allylic and homoallylic alcohols are prone to disintegrate such that two carbonyl groups or a carbonyl and an allylic alcohol emerge. Isomerization of a photolysate of 1-acetyl-1,2-epoxy-2,2,6-trimethylcyclohexane to a 1,4-diketone [Müller, 1975] is particularly favorable on account of strain release upon cleavage of a three- and a four-membered ring.

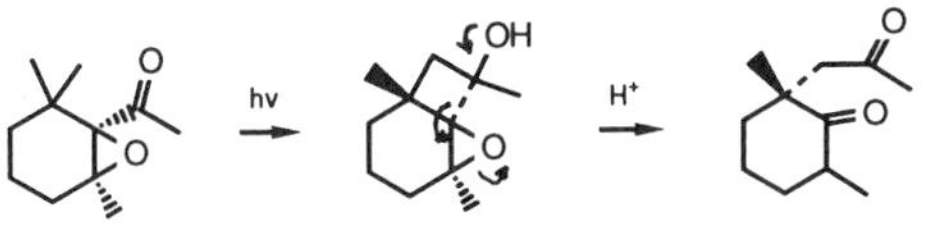

The fragmentation has been incorporated in a nice synthesis of the prostaglandins [Trost, 1980a].

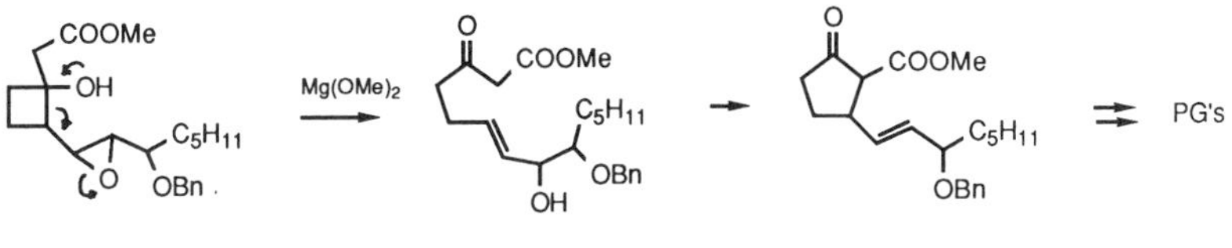

Loss of a two-carbon sidechain during C-ring aromatization of a *cis*-AB pregnane triol epoxide [Campbell, 1978] occurred to some extent. This type of fragmentation has also been encountered in a synthetic study of guaiane sesquiterpenes [Marshall, 1971c]. The intention was a $S_N2$ opening of an epoxide by methylmagnesium bromide to establish stereospecifically the secondary methyl group in the five-membered ring.

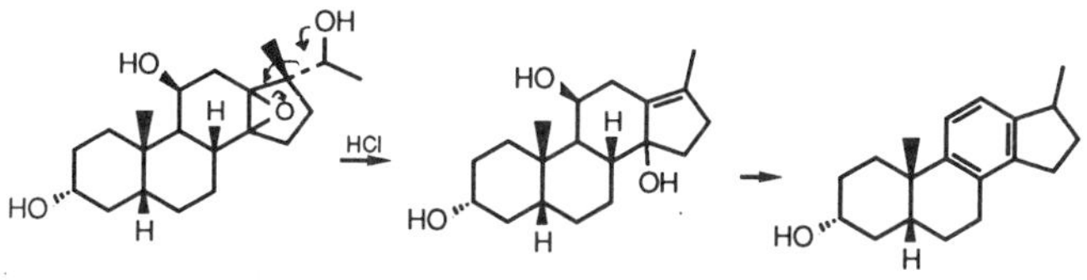

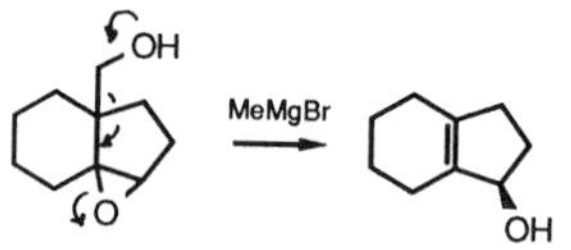

The structure of morolic acid [Barton, 1951] was corroborated by its conversion to an epoxy diol and, eventually, norolean-16,18-dienyl acetate.

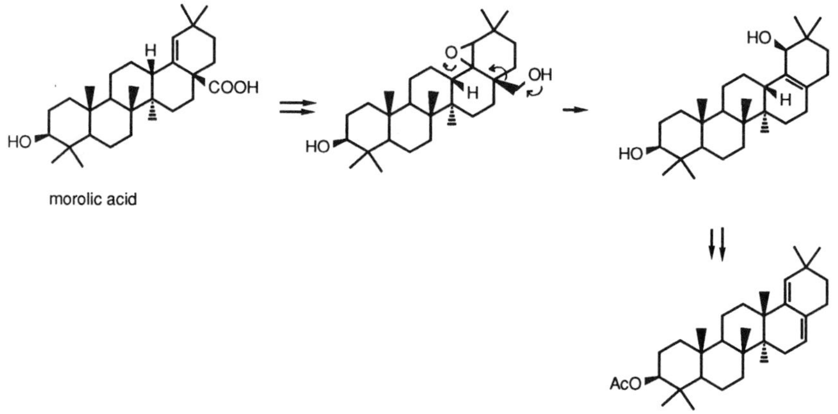

morolic acid

The diol obtained from hydrogenation of fumagillol fragments on treatment with lithhium aluminum hydride [Tarbell, 1961]. This reaction helped interrelate many of the functional groups in the molecule.

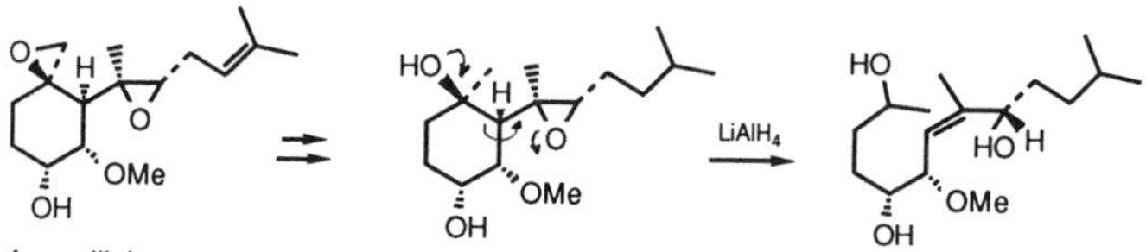

fumagillol

An interesting observation concerns the access of an optically active AB-synthon for daunomycinone [Rama Rao, 1984] by Sharpless epoxidation and reduction. At a higher reaction temperature (−15°C versus −50°C) the epoxide undergoes fragmentation to give a $\beta$-tetralone.

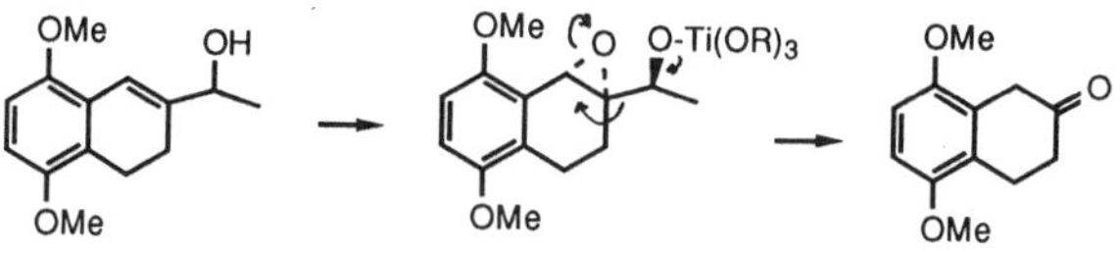

Treatment of the fish poison callicarpone with sodium carbonate leads to two phenolic products. Loss of acetone from the molecule accompanies the aromatization in one of them [Kawazu, 1966]. This reaction clarifies the nature of the various functional groups and their distribution.

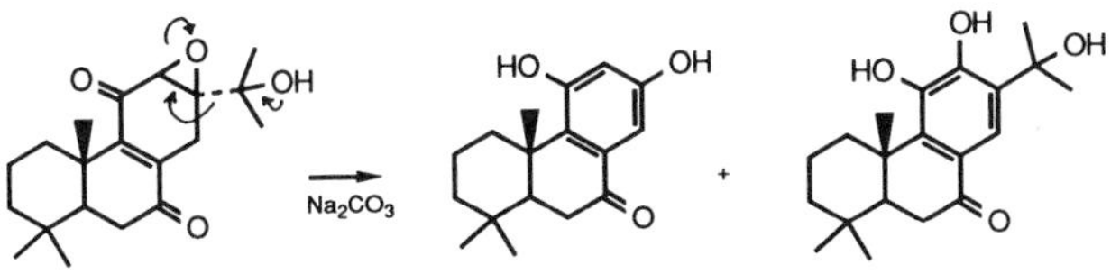

An example of the epoxy ether fragmentation is the degradation of a chromone oxide into a chromonol by acid [Donnelly, 1979a].

Chromenone derivatives isolated from *Diospyros montana* Roxb. could be artifacts formed from diosquinone [Lillie, 1976]. Ethanol present in the chloroform used for extraction might have caused a fragmentation. Heterocyclization of the resulting phenolic ketoaldehyde is most favorable.

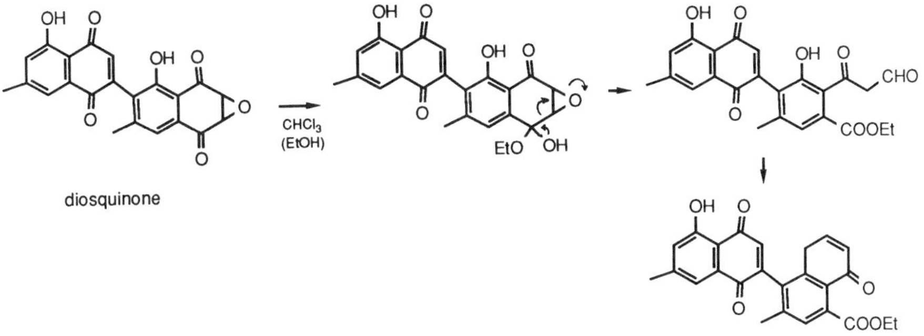

diosquinone

Epoxides of homoallylic alcohols which are placed within a rigid framework show distinct steric requirements for fragmentation. It seems that in the norbornane skeleton a preference for *syn*-coplanar arrangement of the oxygen functions is indicated [Holton, 1984; Waddell, 1985]. This phenomenon boded well for the eventual application of the reaction in a successful synthesis of *ent*-taxusin [Holton, 1988].

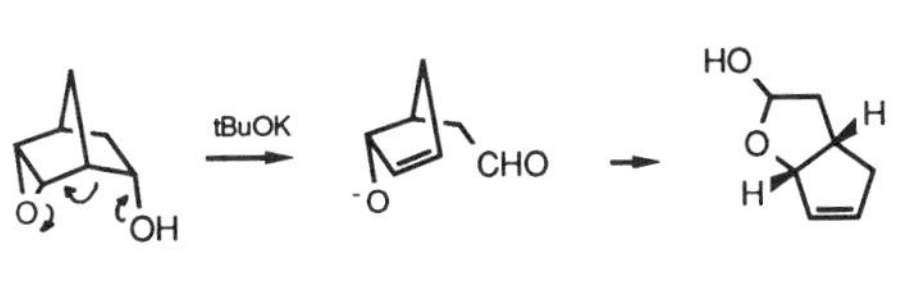

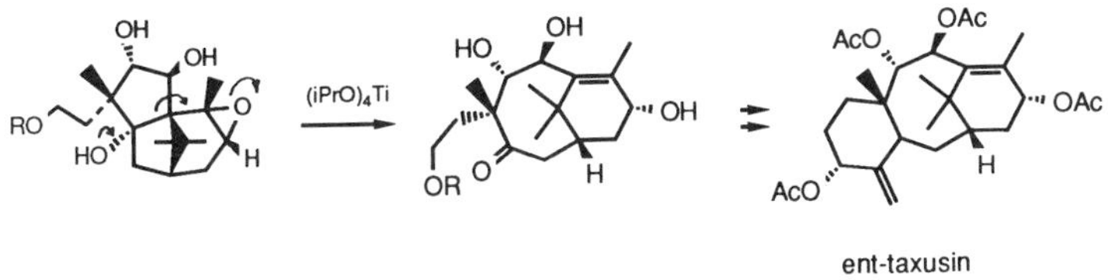

ent-taxusin

The loss of 3-furancarboxaldehyde from the Meerwein–Pondorff reduction product of limonin, on its reaction with potassium hydroxide [Arigoni, 1960] was taken as evidence for the location of the furan ring and the ketone group, as the released carboxyl formed a new $\delta$-lactone.

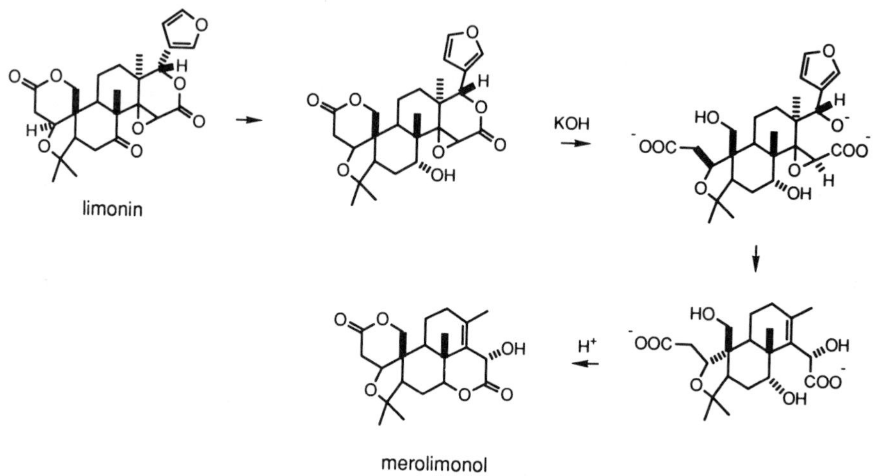

limonin

merolimonol

Of the two bridged ring 1,3-dioxy systems which fragment as shown below [Böhm, 1989; Rice, 1975], the first has synthetic implications because the compound is easy to prepare, and the product, containing three different functional groups with defined stereochemistry, is expected to be an excellent building block for many complex molecules.

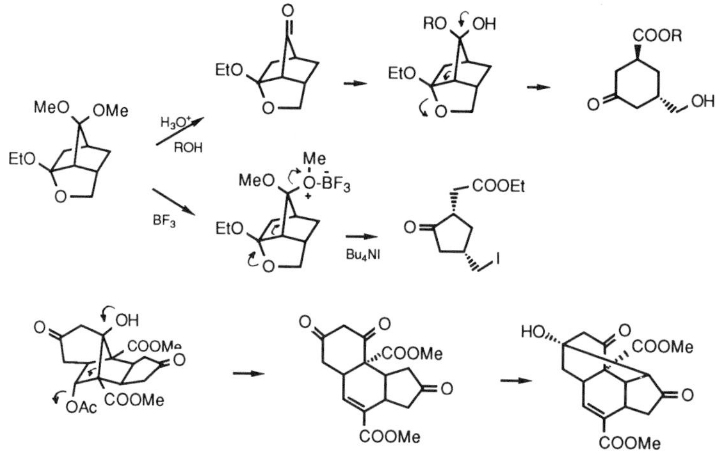

Certain nitrotrifluoroacetoxylation products of 3-acetylated pyranoid glycals tend to give aliphatic nitroalkenes on hydrolytic workup [Zajac, 1986]. The fragmentation is particularly favorable when the 3-acetoxy group is equatorial, because the stereochemistry fulfills the conformational requirements for a concerted process. In other cases, a stepwise mechanism operates involving **d2a = d** fragmentation and elimination.

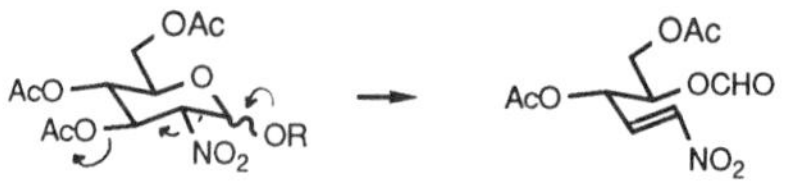

### 8.1.2. RO—C—C—C—OH Arrays

In most instances fragmentation of the RO—C—C—C—OH system has been observed for those with tertiary hydroxyl groups as nucleofuges. Also steric constraints often prohibit the reversal of roles for the two donors.

This fragmentation which is a modified version of a cyclohexenone synthesis has found numerous utility in the syntheses of terpenes such as nootkatone [Dastur, 1974], glutinosone [Murai, 1980], $\beta$-vetivone [Murai, 1981], an azapropellane system [Uff, 1980; cf. Crabb, 1977].

MeO HCOOH MeO OH OCHO O nootkatone

MeO NMe OH R O NHMe R O R NHMe

An unusual route to spirocyclic compounds [Nagumo, 1988] has been reported. A method for manipulating two identical ketone groups of linear triquinanes [Mehta, 1983] also depends on the fragmentation of the tetracyclic aldol tautomers.

O $BF_3 \cdot OEt_2$ EG OH O OH

H H O HH O MeO O MeO HO TsOH H H HH O

The Grignard reaction of cyclobutanones which are fused to a tetrahydrofuran or tetrahydropyran ring [Kimbrough, 1967] differs from that of the carbocyclic analogs, as the latter undergo only enolization and ring opening. The importance of the ethereal oxygen atom in governing the outcome of such reaction is undeniable.

O Ph Ph OH Ph Ph O Ph $H^+$ Ph Ph O Ph Ph Ph O Ph

By means of the fragmentation and other necessary transformations a synthesis of edulinine [Naito, 1983] has been achieved from an $N$-methyl-4-methoxy-2-quinolone.

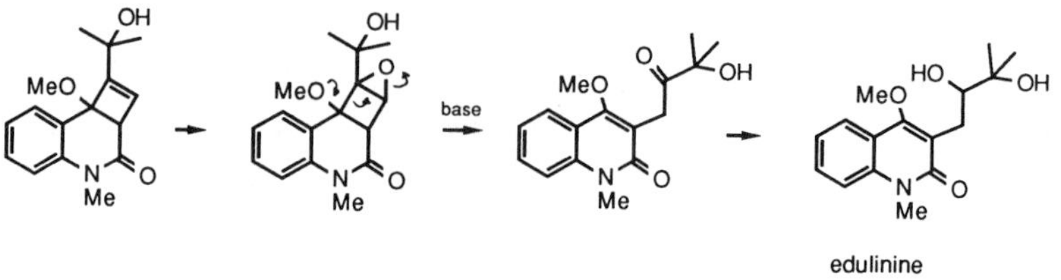

edulinine

Again, the most fragile molecules belonging to the 1,3-diol monoether family are those interposed by a cyclopropane ring. $\beta,\gamma$-Unsaturated aldehydes may be prepared by taking advantage of the fragmentation of such substances [Corey, 1975; S.F. Campbell, 1975; Adams, 1986].

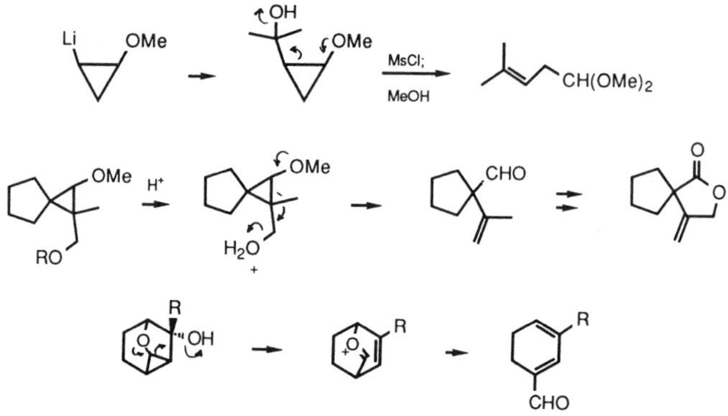

A spiroketal intermediate for monensin synthesis [Ireland, 1985] has been devised. The fragmentation is the crucial step (although the leaving group is an ether oxygen).

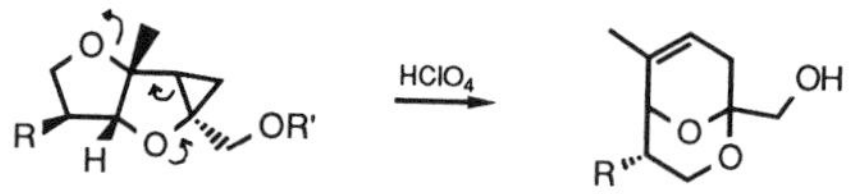

The production of the thermodynamically more stable *cis*-lactone from thermolysis of 1,2-dihydro-6-methoxybenzocyclobuten-1-yl 3-butenoate [Kametani, 1981] must have involved a sequence of fragmentation and reclosure of a *trans*-annexed lactone ring. Accordingly, the presence of the methoxy substituent is essential.

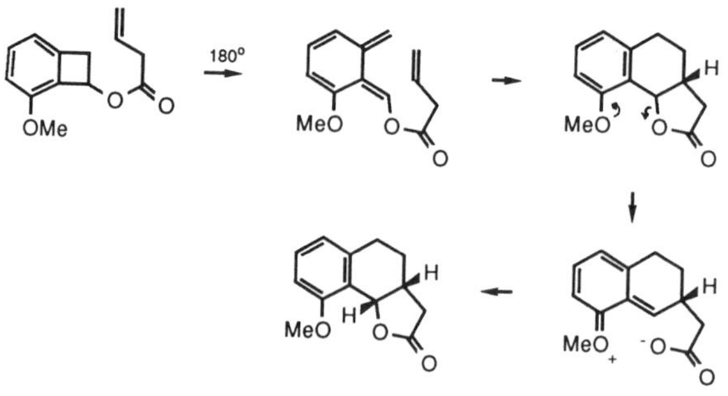

Methylenation of cyclopropanochromanone with dimethylsulfonium methylide was found to be less than straightforward [Bennett, 1979]. The expected epoxide product underwent fragmentation.

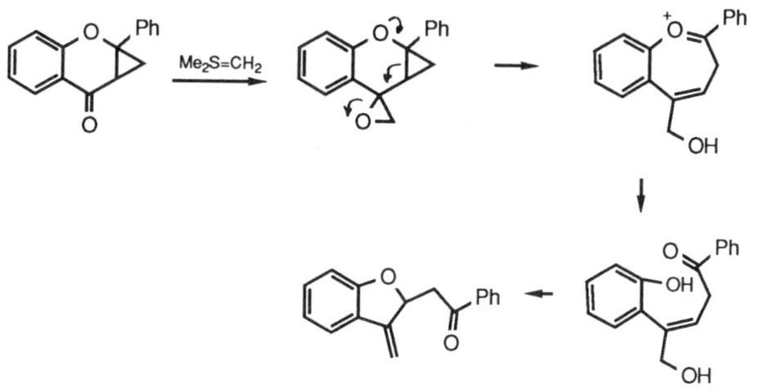

Another example of 1,3-diether cleavage is the isomerization of terpinolene diepoxide into 2,2,4-trimethyl-4-vinyl-3-oxacyclohexanone [Ho, 1983b] on heating with alumina in toluene. This reaction is an ethanolog of tetramethylallene diepoxide [Crandall, 1968].

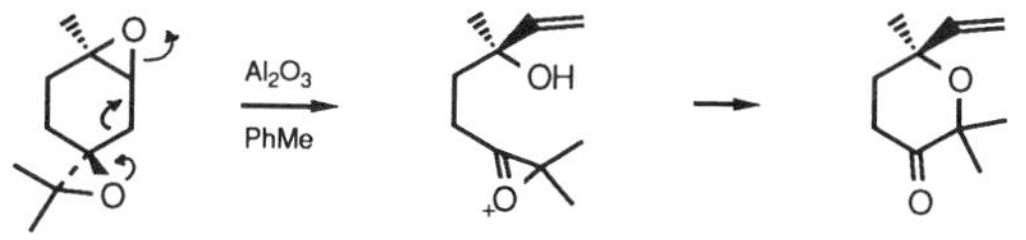

The major product of $BF_3$-catalyzed rearrangement of 16-hydroxy-12-oxopodocarp-9(11)-ene α-epoxide is a tetralin derivative [Nakano, 1981]. The fragmentation that initiated the rearrangement may have involved cleavage of the C-9/C-11 bond because much steric strain of the molecule would be dissipated.

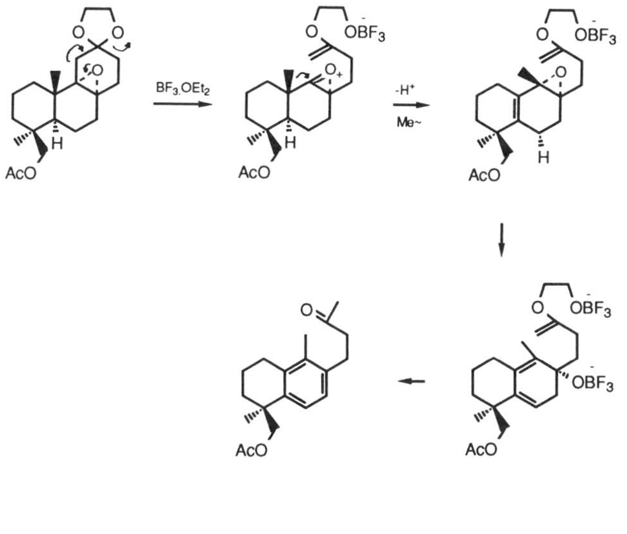

### 8.1.3. O—C—C—C—$OSO_2R$ Arrays

The 1,3-diol monosulfonates are probably the most widely employed substrates in controlled fragmentation reactions. The base-promoted fragmentation is stereospecific such that the configuration of double bond to be created is predetermined by the stereochemistry of the C—C—$OSO_2$ segment.

3-Hydroxy-2,2,4,4-tetramethylcyclobutyl tosylates [Wilcox, 1963] and 3-mercaptocyclobutyl tosylates [Block, 1986] undergo ring fission on treatment with base. Generally the *trans*-isomers react faster.

A practical synthesis of 2-deoxy-D-ribose from glucose [Hardegger, 1957; D.C. Smith, 1957] is based on the fragmentation of the 3-mesylate.

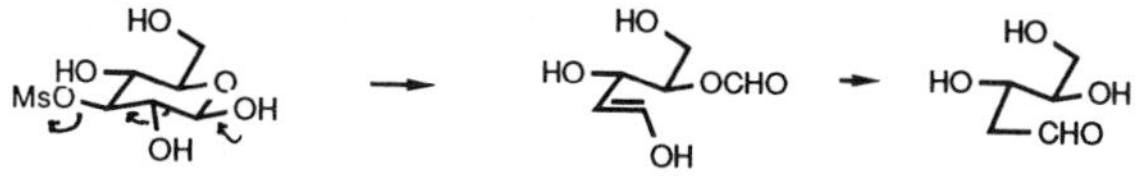

Ring opening of 6-epihydroxyloganin aglycone methyl ether monotosylate [Kinast, 1976], the formation of A-seco aldehydes from certain triterpenes [Y. Tsuda, 1975] which provides a reliable chemical method for the determination of relative configuration of ring-A substituents, and the indirect butenylation of (−)-dihydrocarvone during synthesis of *ent*-valeranone [Marshall, 1968a] are a few of the applications of the **d3d** fragmentation.

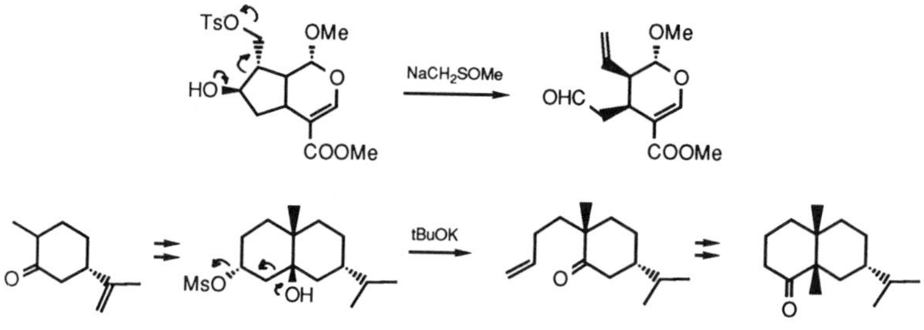

The fragmentation products may be further transformed into less recognizable compounds if participatory functional groups are present in proximity to the hydroxy sulfonate. The secondary reaction could be a Conia reaction [D.S. Brown, 1980], an intramolecular Michael addition [Barton, 1961], or some other processes.

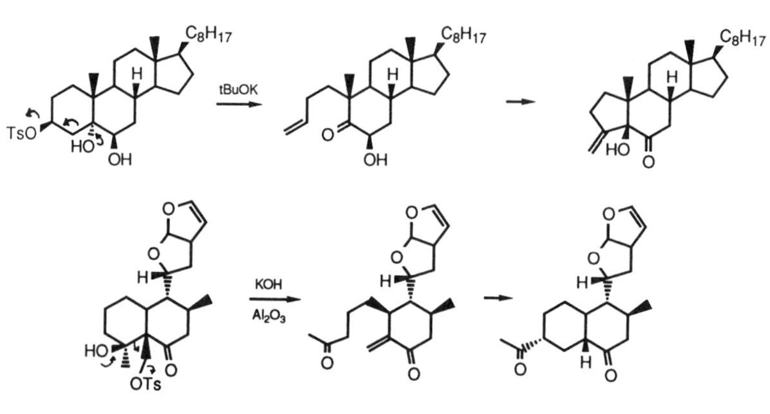

A synthesis of the cecropia juvenile hormone JH-I [Zurflüh, 1968] features two prominent fragmentation steps. Each of these reactions (and the attendant preparation of their precursors) provides a full configurational control of the alkene product.

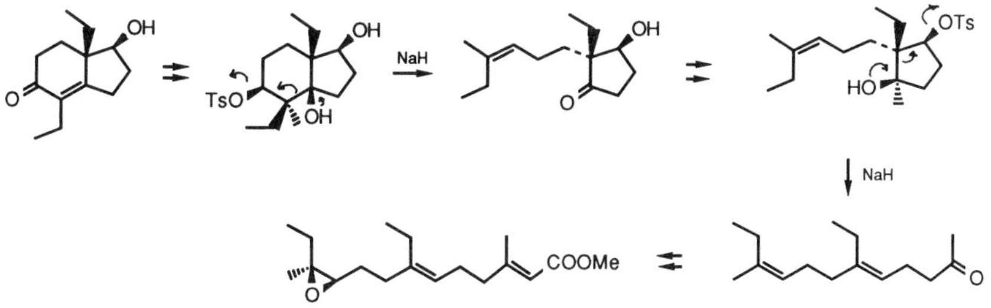

Peripheral fragmentation performed on appropriately substituted decalins has significant impact on the success in synthesis of sesquiterpenes such as geijerone [M. Kato, 1979], and saussurea lactone [Ando, 1978]. Of particular interest is the latter work in which the fragmentable entity is generated by isomerization of an epoxy mesylate, and under this Meerwein–Pondorff–Verley reduction conditions the unsaturated aldehyde product is better preserved.

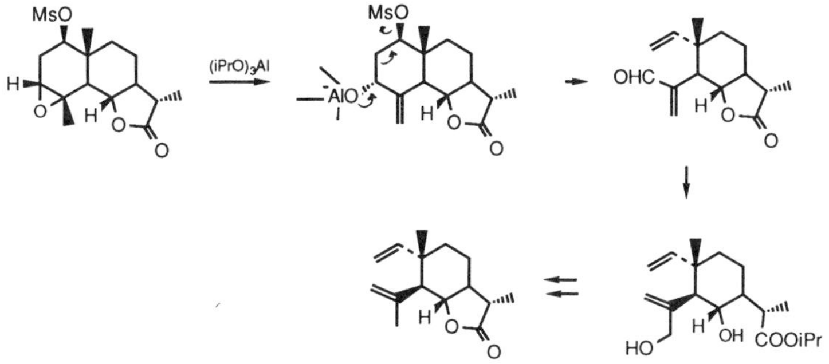

Another peripheral fragmentation provides a benzocyclobutene with a remote vinyl pendant that its thermolysis readily gives a tetracyclic compound containing five of the chiral centers of chenodeoxycholic acid [Kametani, 1981a].

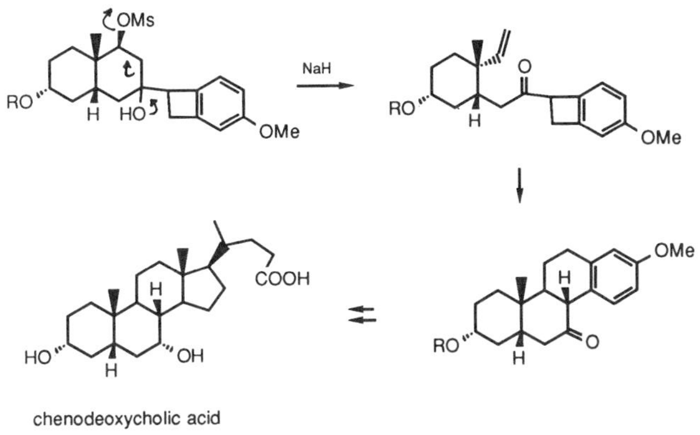

chenodeoxycholic acid

Cleavage of bridged ring systems by this method is the key operation whereby the spirocyclic framework of hinesol [Marshall, 1969b] is revealed. In a synthesis of atisine [Nagata, 1963, 1967] a bicyclo[3.2.1]octane is broken in order to construct the subunit with a [2.2.2] skeleton.

HO OMs tBuOK O OH

hinesol

OAc Ms N H H OMs KOH Ms N H H O O N H H OH

atisine

Reversing the roles of the two oxygen atoms in the bicyclo[3.2.1]octane necessarily changes the product from a 3-vinylcyclohexanone into a 3-methylene-cyclohexanecarboxaldehyde, and this has been demonstrated in a steroid derivative [Nagata, 1968].

Reductive fragmentation of $\beta$-keto tosylates embedded in a bicyclo[2.2.2]-octane skeleton [W. Kraus, 1969] is facile. It is remarkable that the corresponding ketals also fragment, although with an efficiency highly dependent on the medium homogeneity. In a heterogeneous reaction the lithium ion of $LiAlH_4$ is poorly solvated and its coordination with the sulfonyl oxygen atom(s) would help fragmentation.

O OTs $LiAlH_4$ O⁻ OTs CHO $CH_2OH$

O O OTs $LiAlH_4$ O O O S O LiO Ar O O+ O O

An unusual synthesis of phoracantholide-I [Nagumo, 1991] involves a **d3d** fragmentation to generate the macrolide skeleton. The donor group is an oxygen atom of a ketal function.

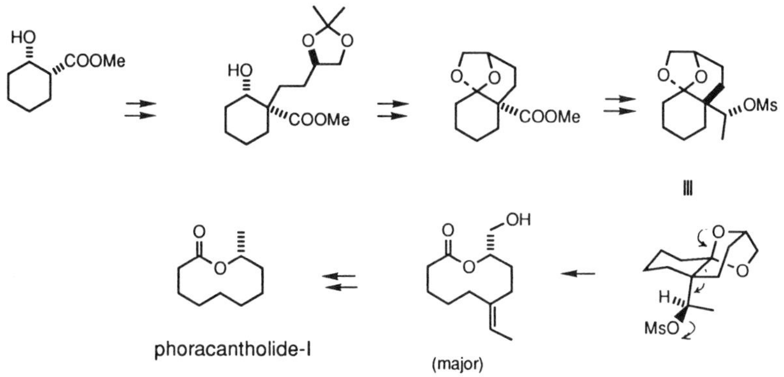

Acetolysis of 7-ethylenedioxo-*endo*-2-norbornyl tosylate proceeds by a fragmentative route, resulting in a 3-cyclohexenecarboxylic ester [Gassman, 1973].

Delightful syntheses of bazzanene [Kodama, 1980] and trichodermol [Still, 1980] focus on the elaboration of 1-oxybicyclo[2.2.2]octenes and partial deannulation at a later stage. Stereocontrol is available from such intermediates.

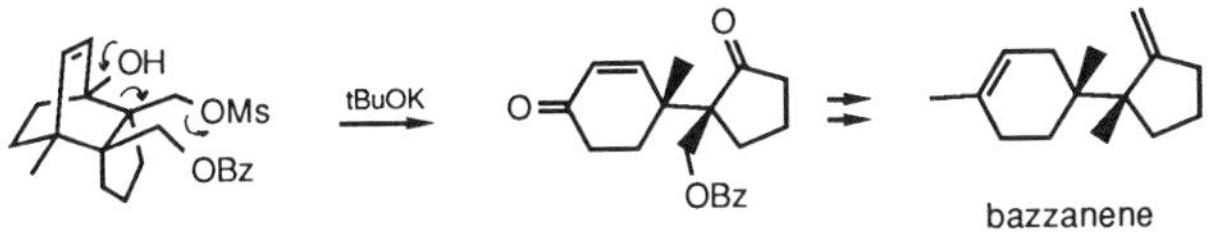

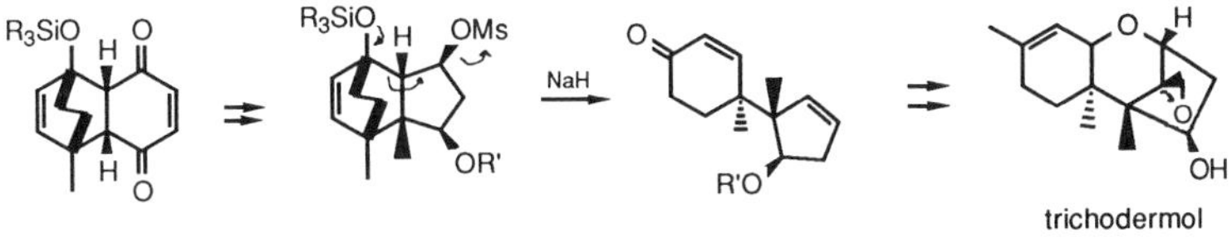

A synthesis of subergorgic acid [Iwata, 1988] was conducted via an intramolecular alkylation of a spirocyclic ketone. The required substrate containing a two-carbon pendant was obtained from a fragmentation route.

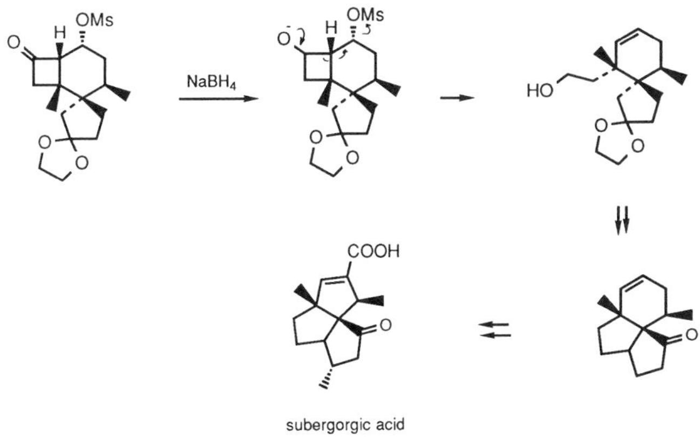

Fragmentation became an obligatory step in a synthesis of longifolene [McMurry, 1972] because a conjugate methyl transfer operation also effected an aldolization as a result of nonbonding interaction alleviation.

The unwanted (?) aldol step actually served to differentiate more readily the two carbonyl groups. Analogous situations arose in a synthesis of daucene [J. Seto, 1985] and the framework assembly of nakafuran-9 [H. Seto, 1983].

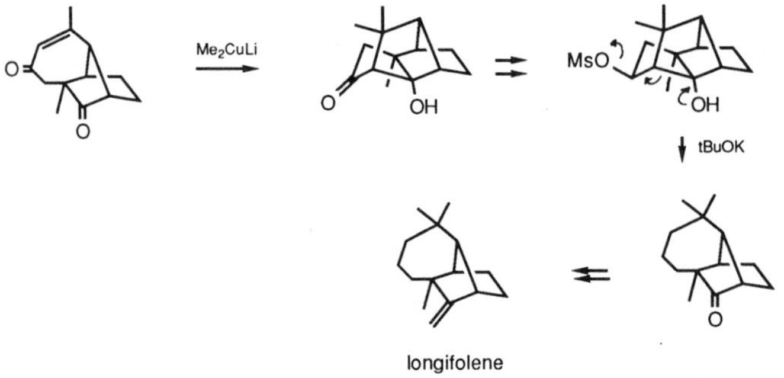

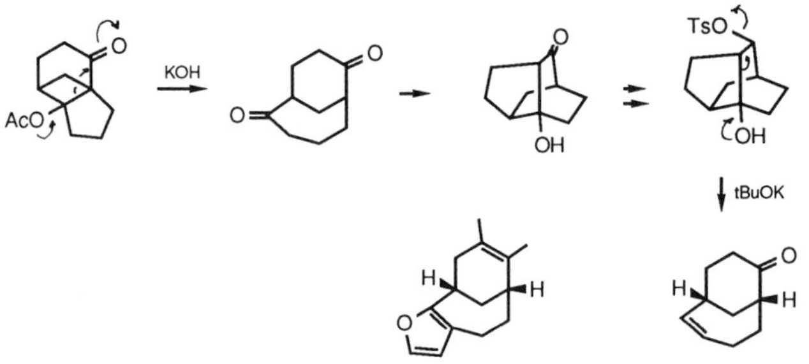

Intercyclic fragmentation is an important strategy for the synthesis of macrocycles. As methods for the introduction of asymmetry along the peripheries of common rings are well developed, synthesis of complex molecules containing large rings would be greatly simplified by coupling the common ring assembly with the necessary fragmentation.

The magnificent approach to caryophyllene and isocaryophyllene [Corey, 1964a] exemplifies the solution to stereoselective access of (*E*)- and (*Z*)-cyclononenes. Regioisomers of the cyclononenones can be obtained from isomeric hydrindediol monotosylates [Caine, 1985].

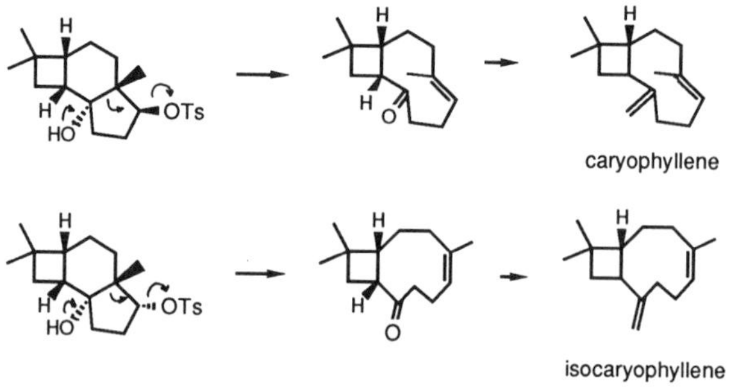

4-Cycloheptenones, especially those fused to another ring with the C-2/C-3 bond, are readily acquired by a sequence of photocycloaddition of 3-acyloxy-2-cyclopentenones with alkenes, reduction, sulfonylation, and fragmentation. Since the pioneering use in a synthesis of $\beta$-himachalene [Challand, 1967] this fragmentation route has enjoyed great popularity in applying to many other natural product syntheses including $\beta$-bulnesene [Oppolzer, 1980b]. (For variations in substrate preparation and conditions for fragmentation, see: Sampath, 1987; Tietze, 1980.)

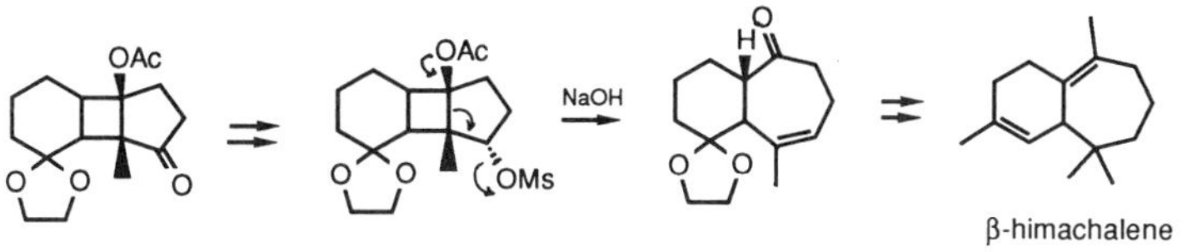

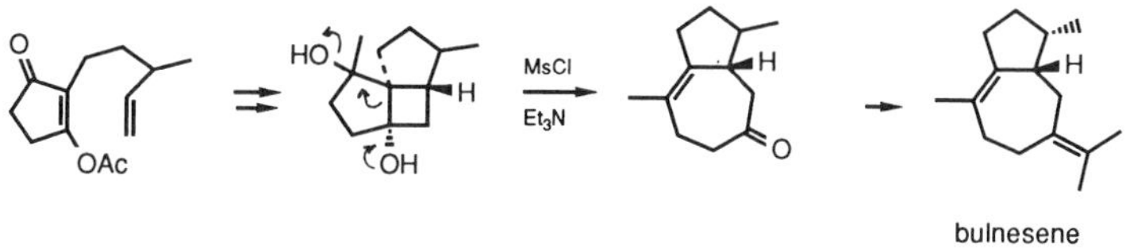

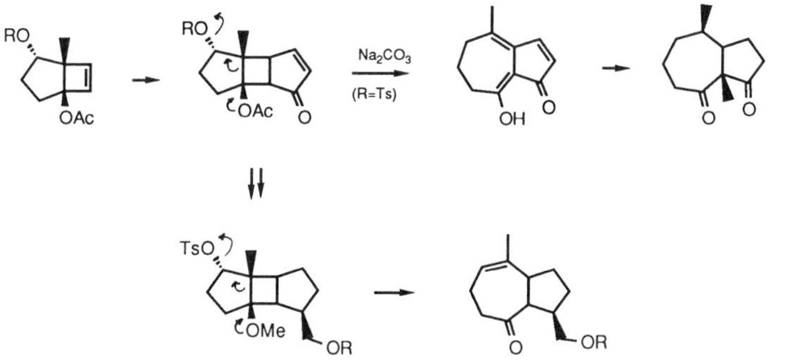

OAc OAc NaOH OMs

β-himachalene

OAc MsCl $Et_3N$ OH

bulnesene

RO OAc RO $Na_2CO_3$ (R=Ts) OAc OH

TsO OMe OR OR

A slight alteration of the photocycloaddends permits the access to 4-methylenecycloheptanones, as shown in a synthesis of dehydrokessane [H.J. Liu, 1977].

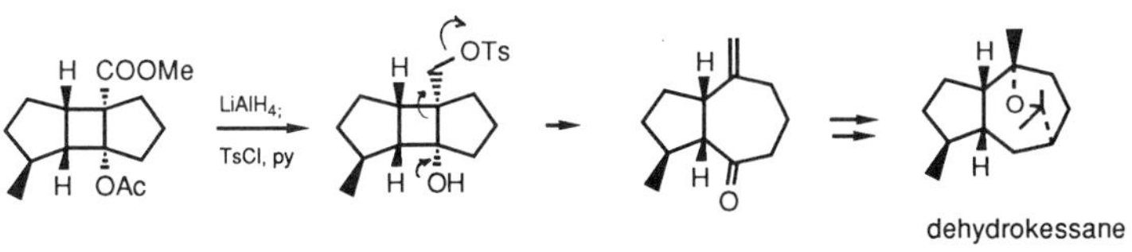

H COOMe H OAc $LiAlH_4$; TsCl, py H OTs H OH

dehydrokessane

The hydrazulenone skeleton has also been secured from a doubly bridged system [H. Seto, 1985a].

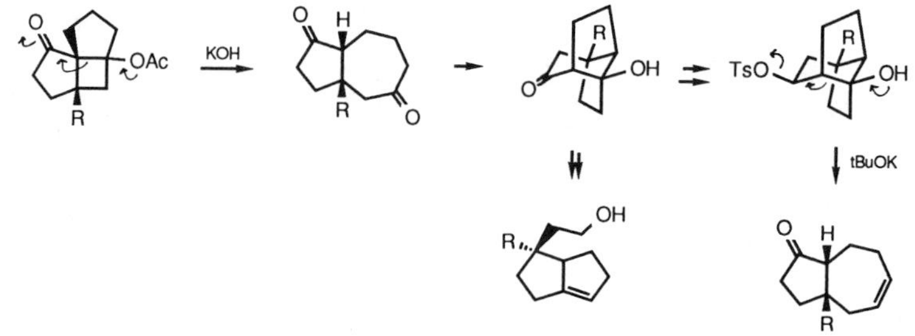

The fragmentation approach to 5-cyclodecenones has been actively investigated in the context of terpene synthesis. The stereospecific C—C bond scission [Wharton, 1965] provides an exceptionally concise route to the trienone intermediate for periplanone-B [Cauwberghs, 1988] which is a sex attractant of the American cockroach. A different decalindiol derivative affords the tricyclic nucleus of copaene [Heathcock, 1972] via photocycloaddition of a cyclodecadienone. A total synthesis of the longipinenes [M. Miyashita, 1974] is also based on the same strategy.

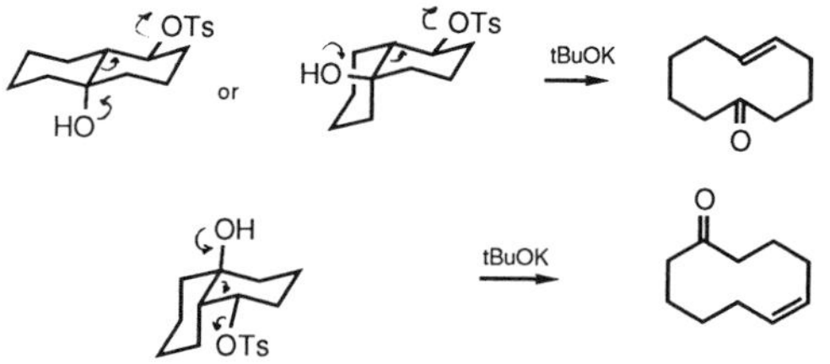

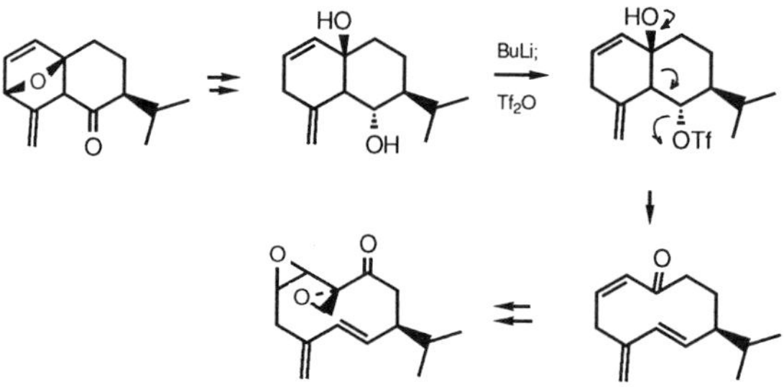

periplanone B

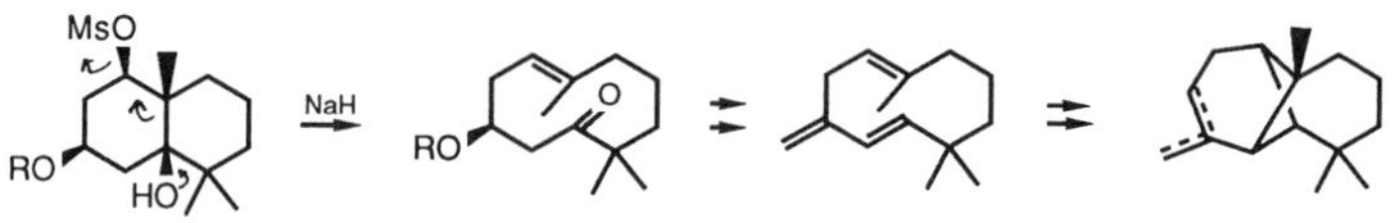

α/β-longipinene

In an exploration of hydrazulene synthesis leading to 4-deoxydamsin [Marshall, 1975b] the cleavage of an octalin system was effected. Subjecting the properly modified cyclodecadienone to solvolytic conditions indeed resulted in a desired bicyclic product.

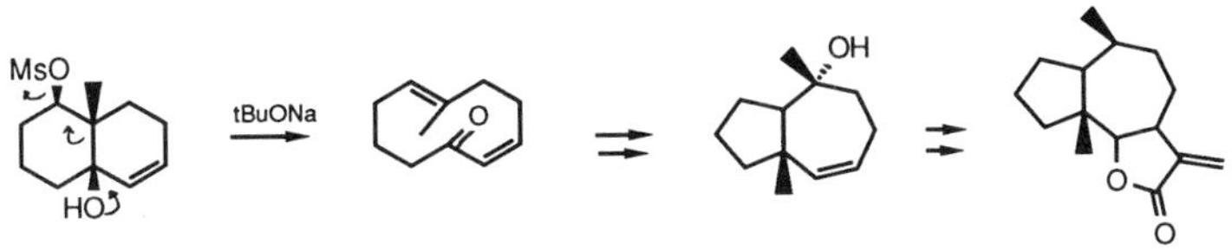

The 15-membered ring ketone, commercially known as exaltone, has been prepared from the inexpensive cyclododecanone via bicyclic intermediates [Ohloff, 1967] and using a fragmentation protocol.

Synthesis of larger ring compounds containing a carbonyl pendant is feasible by fragmentation of proper bridged substrates [Marshall, 1967]. The indisputable value of the operation can be judged by reports of its involvement such as in the synthesis of rudmollin [Wender, 1986]. (Note that the relative locations of the OH and $OSO_2R$ groups are extremely important, see, for example, [Duthaler, 1976].)

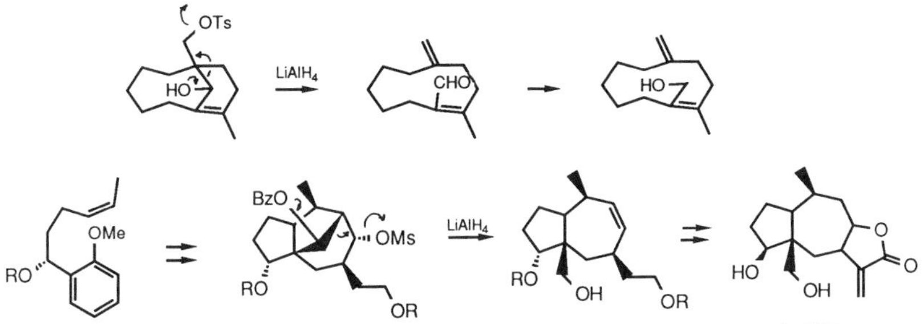

rudmollin

Even more complicated ring skeletons may be unraveled by fragmentation routes. The assembly of sanadaol [Nagaoka, 1987, 1988] and the bicyclo[5.3.1]-undecane portion of taxane diterpenes [Nagaoka, 1984] are impressive examples.

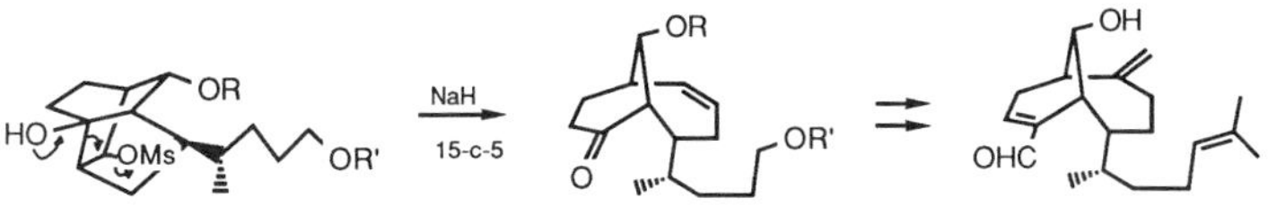

(+)-sanadaol

A route to zizaene [Barker, 1983] has been realized despite drawbacks in regiocontrol and stereocontrol at various stages. An intramolecular photocyclo-addition and fragmentation are the key steps. (For a model study toward the same objective: see, [Oppolzer, 1980a].)

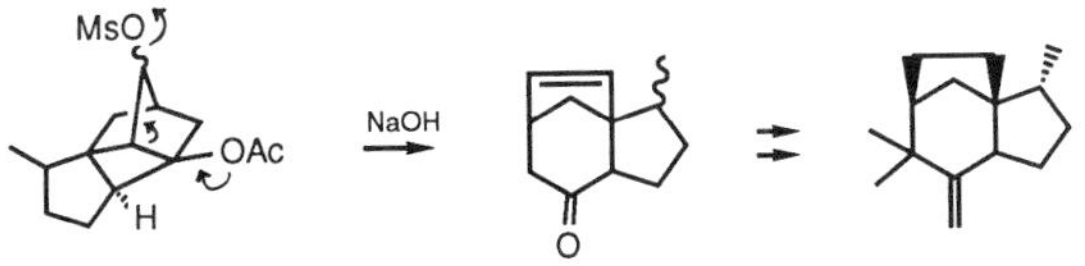

zizaene

The presence of fused cyclobutane rings in goverdhane suggests an approach from a diene, and a retrosynthetic analysis of this subgoal points to a fragmentation scheme for its creation [Marchand, 1989].

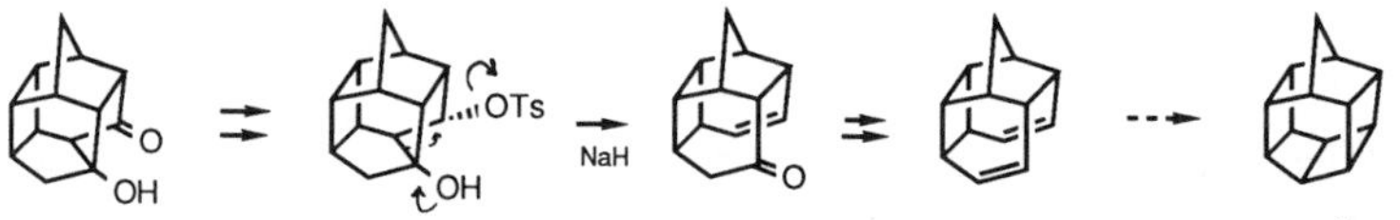

goverdhane

A ring fission study [Klunder, 1985] of certain 1,3-disubstituted caged molecules indicates a dependence of the leaving group. A computational analysis of the two types of homocubyl derivatives has also been accomplished [Ivanov, 1985].

A rare case of fragmentative elimination of enol tosylates to provide acetylenic ketones was incorporated in a synthesis of hybridalactone [Corey, 1984]. The electronic requirements for the fragmentation are the same as other 1,3-diol monosulfonates.

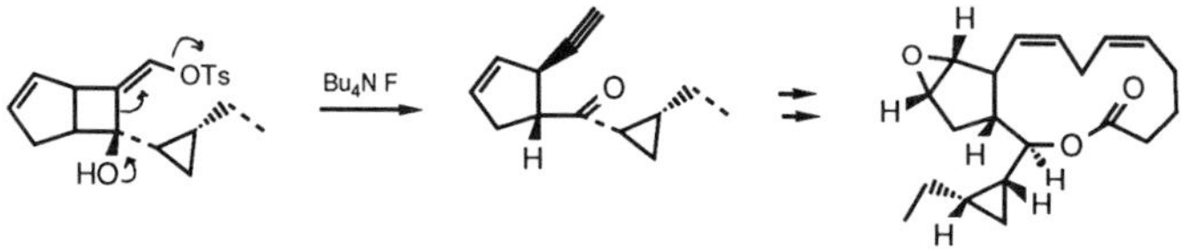

hybridalactone

Several examples in which ether-assisted fragmentation figures prominently have been reported. These are a synthesis of phoracantholide [M. Ikeda, 1983], heteroexpansion of the A-ring of cholesterol [Coxon, 1968], and that of an epoxycyclooctanol [Cooper, 1971].

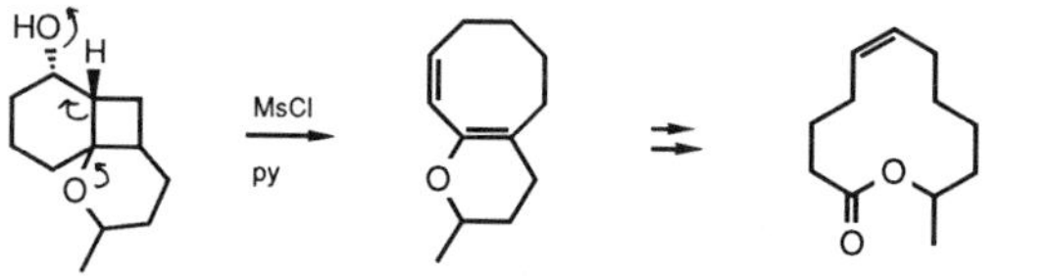

phoracantholide

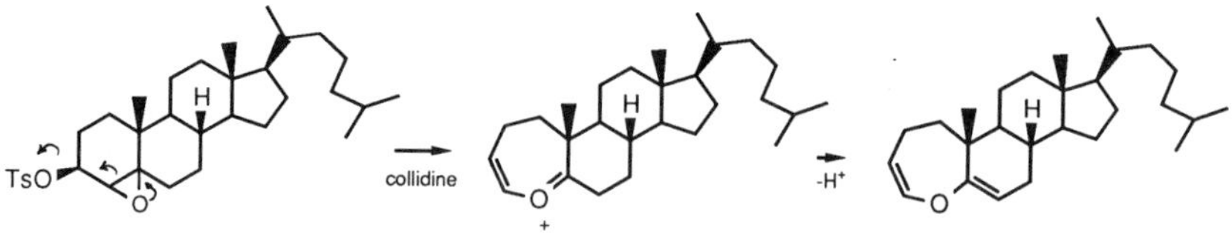

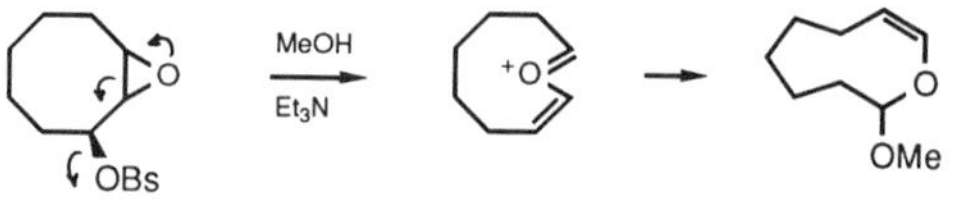

The solvolytic generation of 1,5-cyclopentadecanedione [Gray, 1977] from an epoxy tosylate is, however, most probably the result of a fragmentation of the tosyloxy diol arising from hydration of the epoxide.

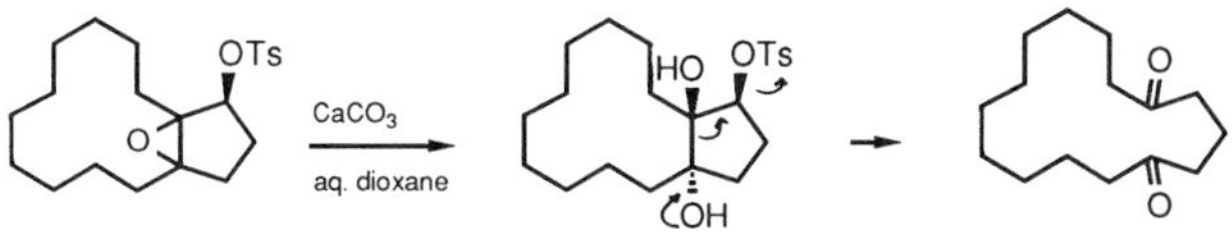

(2,5-Cycloheptadien-1-yl)methanol emerged from hydride reduction of a 2-tosyloxycyclobutanone which is *trans* annexed to a cyclohexene [Casadevall, 1978]. The reagent apparently caused migration of the tosyloxy group with the formation of a cyclopropanone. Rapid reduction and fragmentation ensued and the reaction was terminated by further reduction.

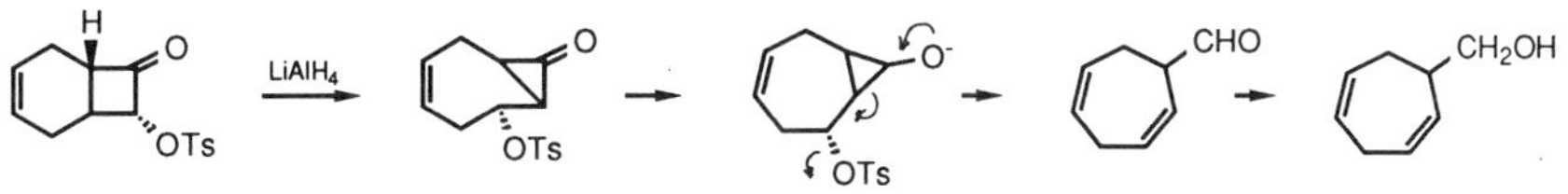

It has been suggested [S.W. Baldwin, 1972] that the generation of two cyclobutanones from base treatment of 2-methyl-2-tosyloxymethylcyclohexanone is by fragmentation into a ketene intermediate followed by a [2 + 2]-cycloaddition.

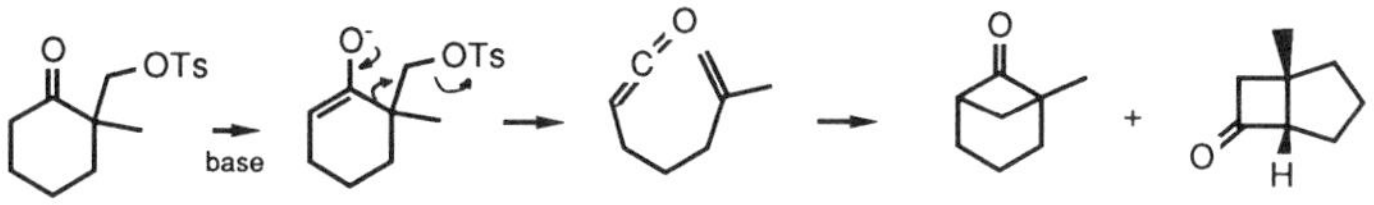

## 8.2. O—C—C—C—Hal SYSTEMS

### 8.2.1. Halocyclopropanol and Derivatives

Halocyclopropanols are among the most readily fragmented (by donor-assisted electrocycloreversion process) because of their inherent strain. As they are also easily prepared from halocarbenes and enol derivatives, the combined steps constitute a valuable method for homologation of ketones, especially cycloalkanones. It is significant that the chemistry of tropones has received a great impetus on account of their accessibility by this route. Convenient syntheses of nezukone [A.J. Birch, 1968], γ-thujaplicin [Macdonald, 1978b], and 7-chloro-2,3-benzotropone [Parham, 1961] may be mentioned. A fascinating transformation of hydroquinone dimethyl ether into a benzocyclobutenone [A.J. Birch, 1964] was also based on this chemistry.

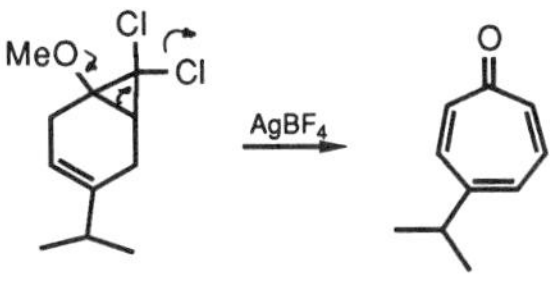

nezukone

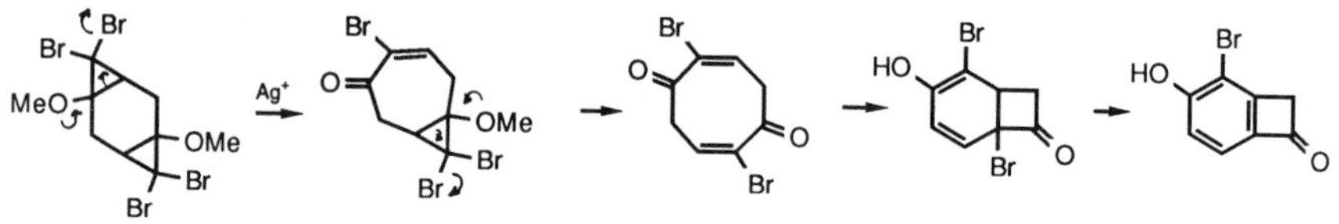

Many variants of this ring expansion scheme have been developed for the synthesis of macrocyclic ketones [Parham 1967; Stork, 1975; Hiyama, 1974] because of the economic implications of such fragrant substances. The expediency of this method also presents itself as the preferred route to a cycloheptenone template in a synthesis of parthenin [Heathcock, 1982].

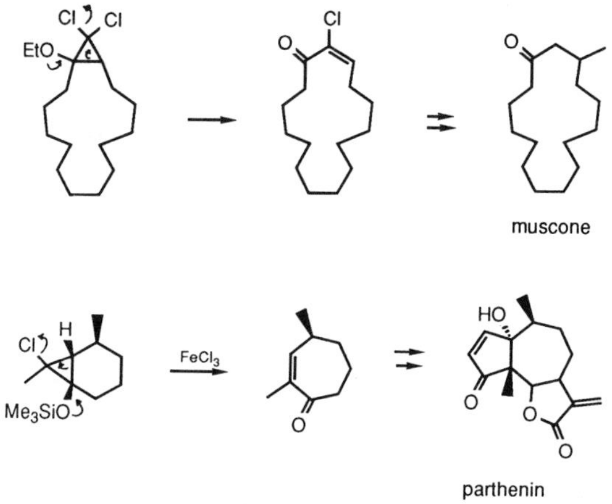

muscone

parthenin

The debromination of 2,2,6,6-tetrakis(bromomethyl)cyclohexanone with zinc dust [Leriverend, 1966] furnishes a dispirocyclic ketone, a spiro[2.6]nonanone, and in the presence of sodium ethylenediaminetetraacetate, a small amount of hexahydro-3a,6a-dimethyl-2(1*H*)-pentalenone [Krapcho, 1968; Mulder, 1980]. The bicyclic products may have arisen from cyclopropanol intermediates.

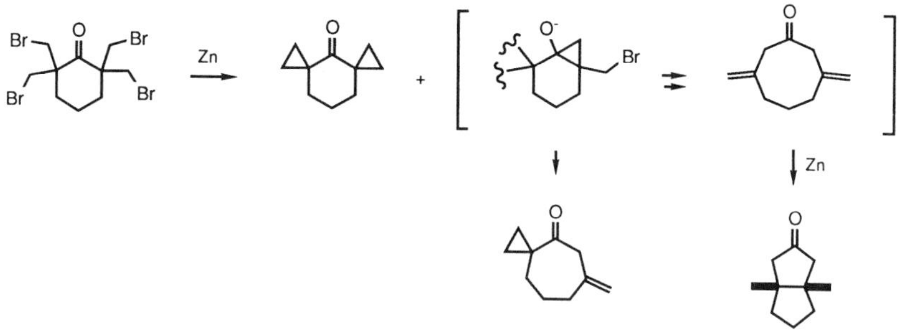

When the donor oxygen atom of the substrate is part of a ring the reaction causes an analogous expansion of the heterocycle [Schweizer, 1960]. A practical synthesis of pyrenophorin [Hirao, 1986] has exploited this opportunity.

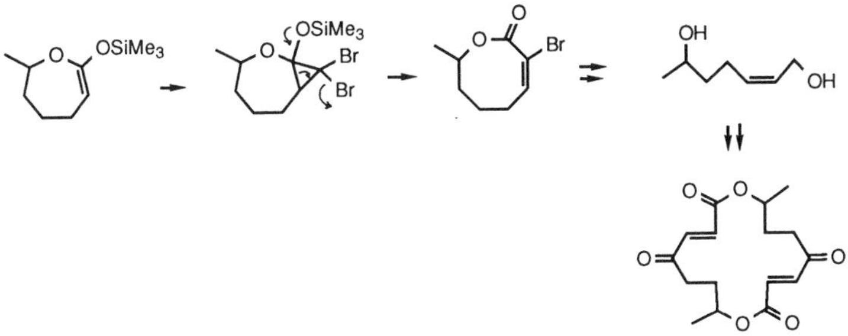

pyrenophorin

The dibromocarbene adduct of 1-acetoxycyclopentene is decomposed in situ to provide 2-bromo-2-cyclohexenone [Davis, 1966]. It should be noted that the homologation is not limited to ketones, as esters [McElvain, 1959; Slougui, 1982, 1985, 1987] and aldehydes [Blanco, 1981; Hiyama, 1978] are also readily obtained from the corresponding enol derivatives.

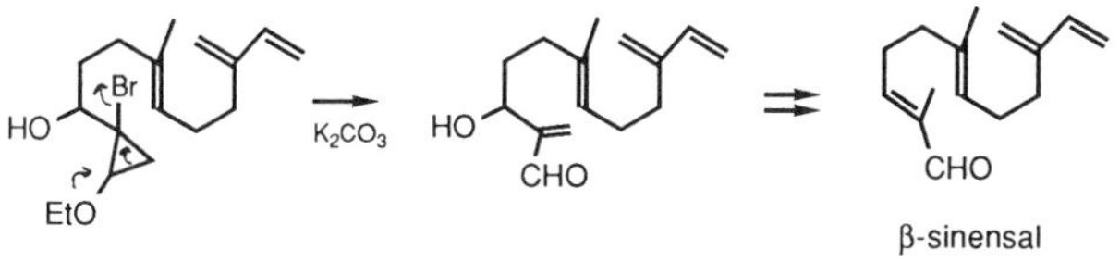

β-sinensal

The decomposition of these cyclopropanol derivatives in the presence of proper reagents often directly furnishes other types of products. Pyrazoles and pyrimidines [Parham, 1967, 1968, 1969] have been secured.

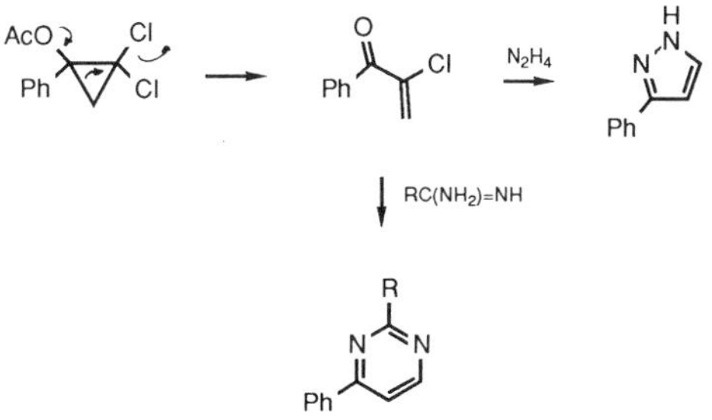

The stronger C—F bond entails the selective cleavage of the other C—Hal bond of 1-fluoro-1-halocyclopropanes [Molines, 1985]. Noteworthy is the solvolytic ring opening of *gem*-difluorocyclopropanol derivatives [Crabbé, 1973; Y. Kobayashi, 1979] which also leads to saturated α,α-difluoroalkylcarbonyl products. The diminished importance of the dehalogenative pathway is due to the C—F bond strength and the ability of the fluorine substituents to stabilize the incipient carbanions by a polarity alternation mechanism.

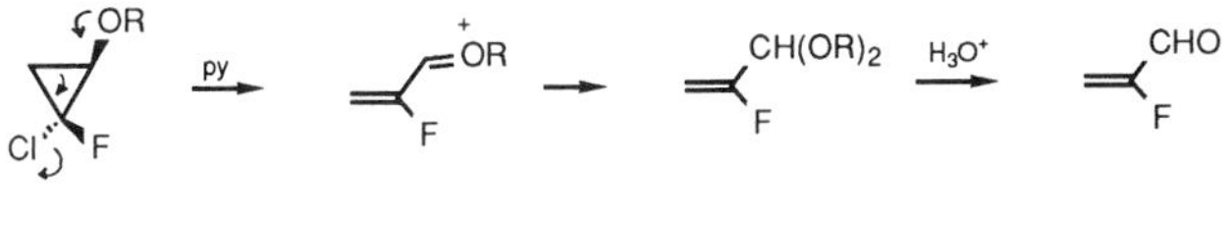

Favorskii rearrangement of unstrained α-haloketones usually proceeds via cyclopropanone intermediates. In the case of α,α-dichloroketones the ring opening step is accompanied by the expulsion of the second chlorine atom. More interesting reactions are those involving the dichloro derivatives of β-diketone monoenol ethers [Verhé, 1977].

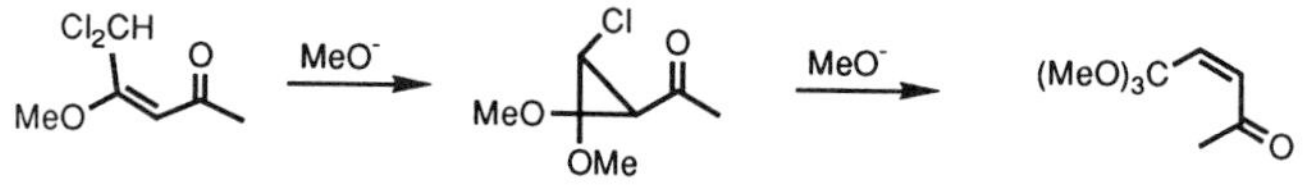

The isomerization of 2-bromotropone ethyleneketal to 2-bromoethyl benzoate [J.E. Baldwin, 1969] occurs by way of fragmentation of the norcaradiene tautomer.

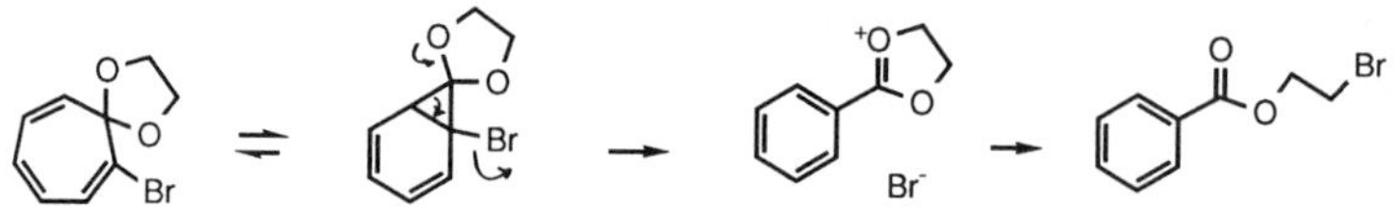

Involvement of 7-norcaradienol intermediates satisfactorily explains the ring contraction of various 2-substituted tropones [Forbes, 1967; Biggi, 1973].

The high strain of the methylenecyclopropene system contributes to the facility in the eliminative ring fission of a bromo derivative [Billups, 1976] which follows hydration of the exocyclic double bond.

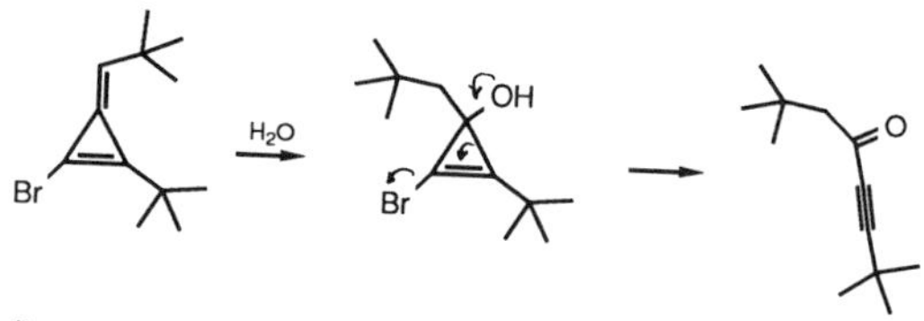

### 8.2.2. Others

*exo,exo*-6-Acyloxy-8-chloro-*cis*-bicyclo[4.2.0]octan-7-ones have been converted to methyl 2-oxocyclohexaneacetate [Hassner, 1986] under the Favorskii rearrangement conditions. Ketene formation via fragmentation of the tertiary alkoxide ion is the most reasonable account of the event. The fragmentation is dependent of conformational factors, it being forbidden when the cyclohexane moiety is part of a rigid framework such as ring-A of the steroids.

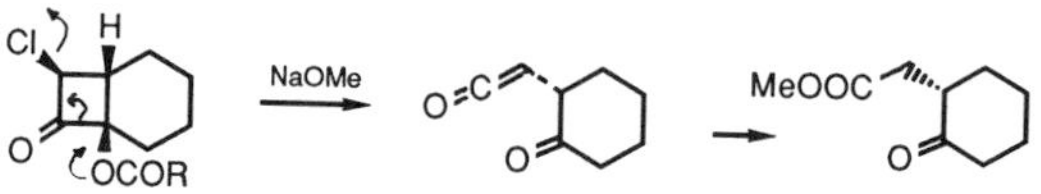

The synthesis of Ormosia alkaloids [H.J. Liu, 1969] featured fragmentation of a photocycloadduct to complete the indirect alkylation at the α-position of the original enone as a step toward elaboration of the hydroquinoline portion of the target molecules.

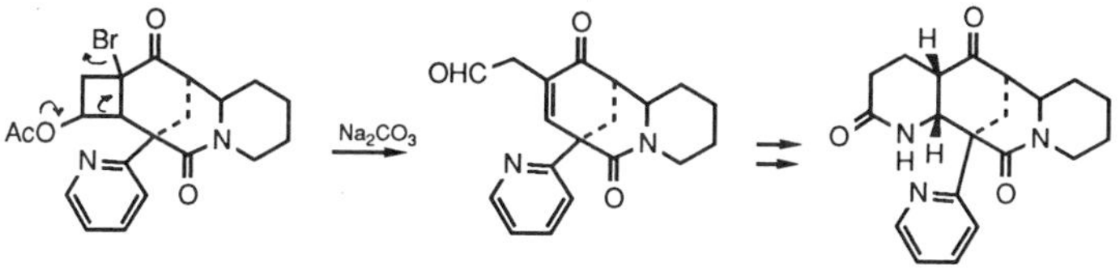

The mechanism for tropolone genesis from dichloroketene–cyclopentadiene adduct has been clarified [P.D. Bartlett, 1970]. The ether formation from a spiroannulated analog [Tsunetsugu, 1983] further indicates the entering position of the oxygen nucleophile and its role in the fragmentation pathway.

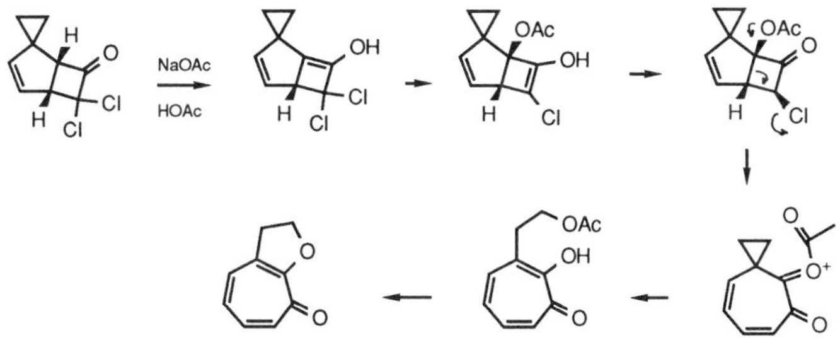

In an enantioselective approach to heteroyohimbine alkaloids [Flecker, 1984a,b] using an iridoid compound as starting material it is crucial to transform the five-membered ring into two functional chains. Accordingly, the necessary steps consist of $\beta$-iodo acetate formation and fragmentation to give a latent secologanin derivative.

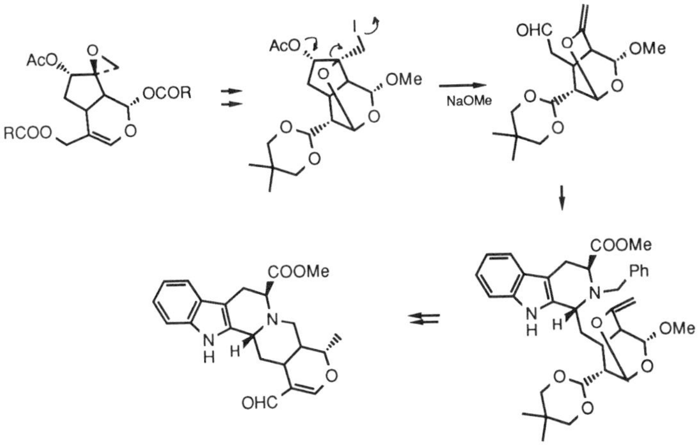

The monoacetate of spirolaurenone glycol does not revert to the diol on saponification [M. Suzuki, 1980]. Instead, there is a loss of the bromine atom as the result of a fragmentation.

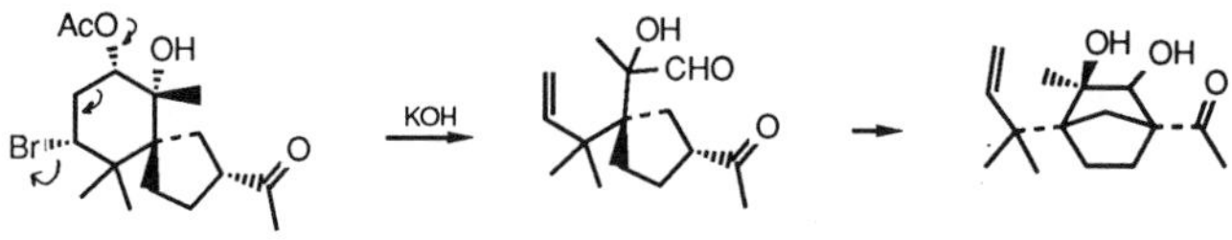

The following equations depict the despanning of bridged systems.

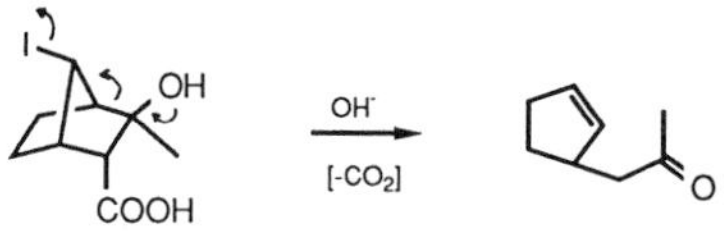

[Beckmann, 1965]

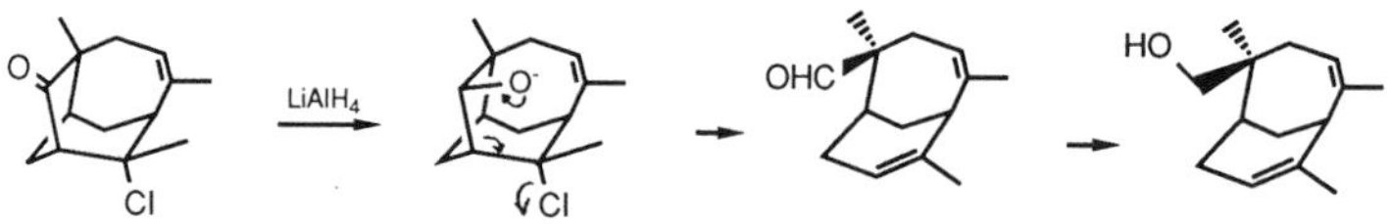

[Frater, 1976]

[W. Fischer, 1975]

It may be added that the chloroadamantane fragmentation is more facile with an oxygen donor than with a sulfur donor [Grob, 1975b].

An extensive modification/destruction of the five-membered ring of verrucarol ensues when the major dichlorochemiacetal derived from its treatment with hypochlorite is further reacted with alkali [Burrows, 1986].

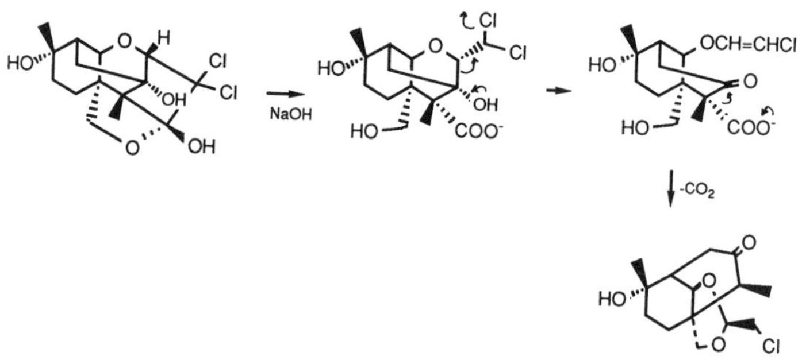

From reaction of 19-iodotabersonine with DBU [Diatta, 1976] a minor macrocyclic $N_b$-formyl diene has been obtained in addition to the expected vinyl compound. A small amount of water apparently intercepted the chano-iminium ion and triggered a deiodinative fragmentation.

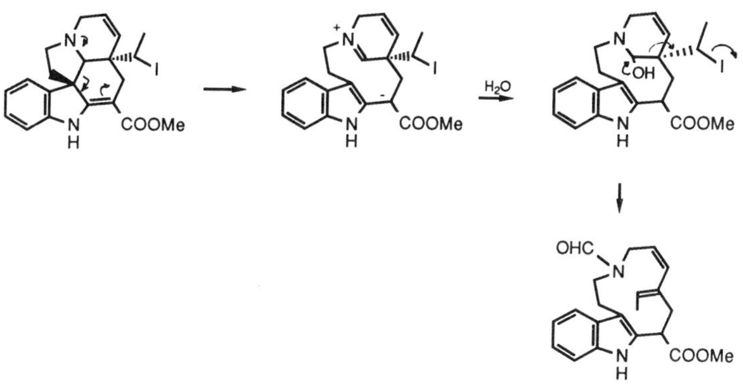

Base-catalyzed cleavage of the D-ring of 20-chloro-5α-pregnane-3β,16β-diol is stereospecific [Adam, 1967] due to a concerted reaction. *N*-Chloroveratramine loses the heterocycle when treated with methoxide ion [R.W. Franck, 1964; cf. Adam, 1971]. An azomethane must be the intermediate.

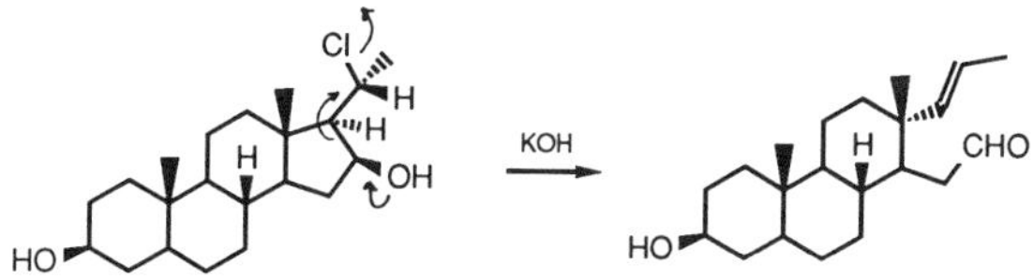

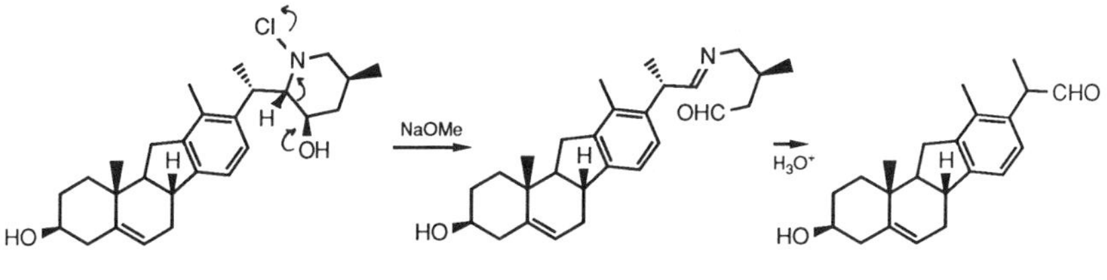

γ'-Chlorinated propargyl alcohols are labile to base [Hauptmann, 1977]. They undergo a **d3d** fragmentation.

## 8.3. O—C—C—C—N SYSTEMS

Fragmentation is often the favored reaction pathway for γ-hydroxy ammonium salts. Thus, the second stage Hofmann degradation of 19,20-dehydroervatamine [Knox, 1971] results in a tricyclic ketone by the extrusion of formaldehyde. Double bond migration of the primary product is expected.

When the intervening carbon chain of the O—C—C—C—N segment is part of a ring system, the fragmentation gives rise to an unsaturated carbonyl compound [Tietze, 1979; Mannien, 1967]. An impressive array of fragmentation reactions during Hofmann degradation of delphonine [Wiesner, 1960] has been witnessed.

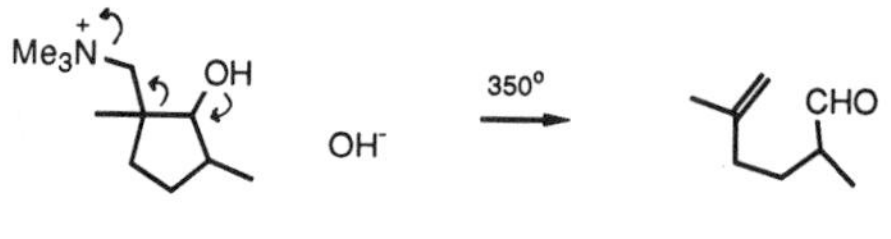

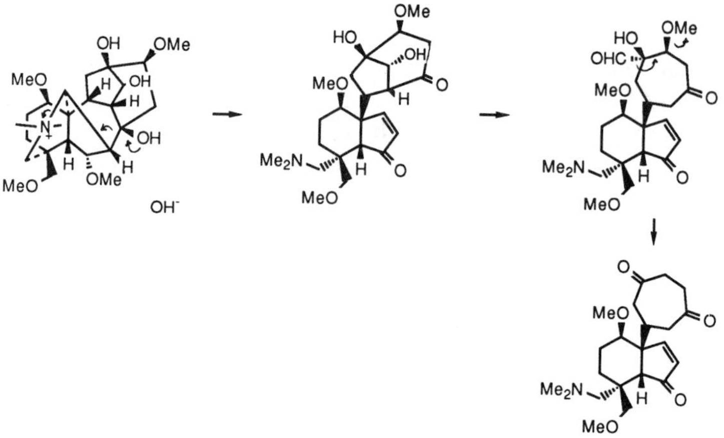

Terramycin is a very fragile molecule. Thus, its treatment with iodomethane is sufficient to cause an extensive degradation resulting in a tetralone [Conover, 1956]. This reaction is initiated by fragmentation of the $\gamma$-hydroxy ammonium salt followed by a retro-aldol fission.

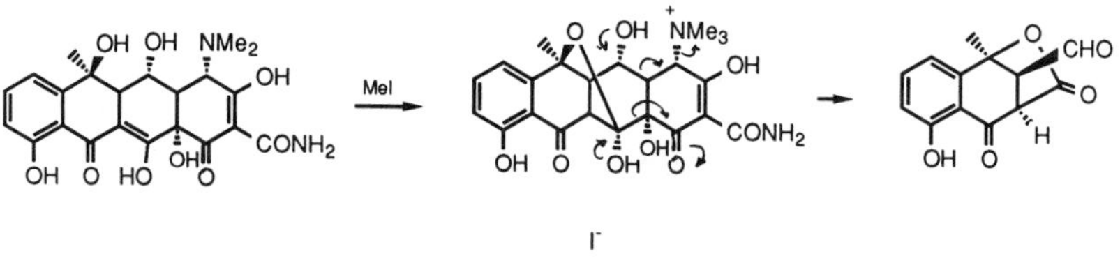

An example of molecular rearrangement intervened by the generation and fragmentation of an O—C—C—C—N segment is found in the chemistry of voaenamine epoxide [Ohkata, 1975; Milenkov, 1987; Morita, 1976].

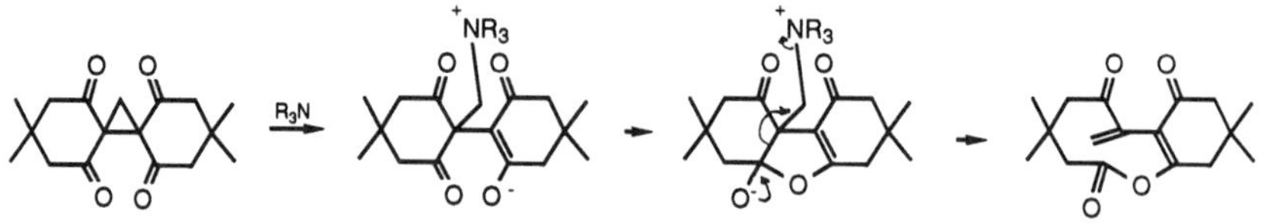

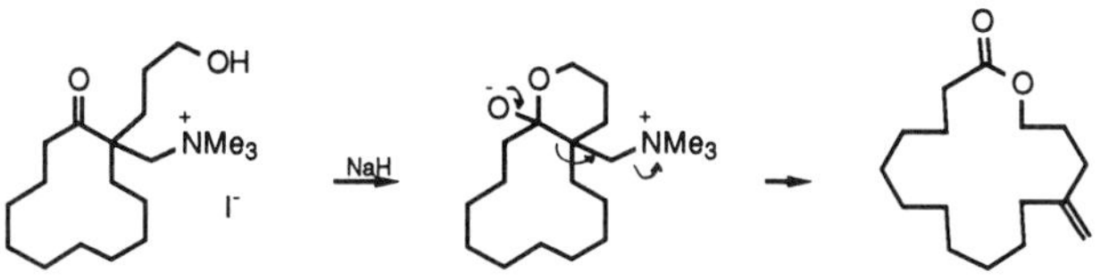

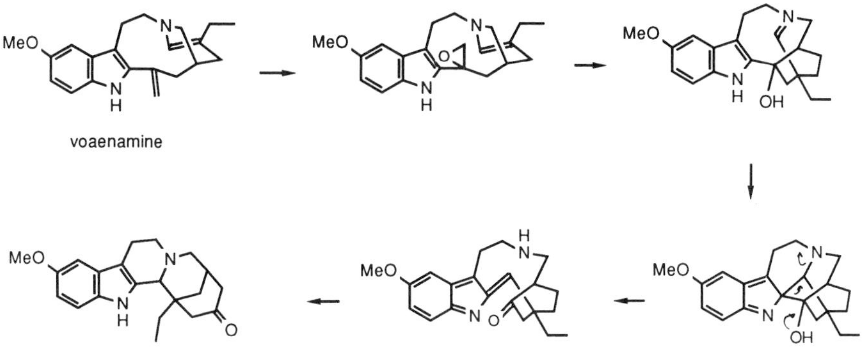

voaenamine

The structure of the Hofmann base derived from lycopodine methiodide has been clarified [Chin-You, 1971]. The mechanism for its formation could involve a retro-Michael elimination followed by hydride transfer and Mannich reaction.

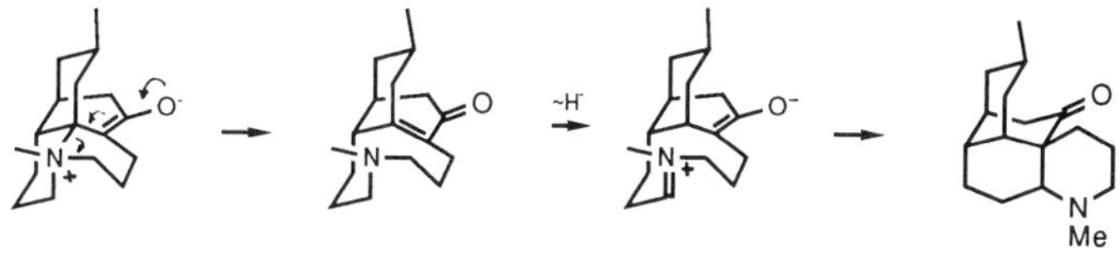

Tetradethioverticillin-A is susceptible to base degradation [Minato, 1973] by virtue of its embodiment of fragmentable sequences. The products of this degradation are biindol-3-yl and 1,6-dimethyl-3-formylpiperazine-2,5-dione.

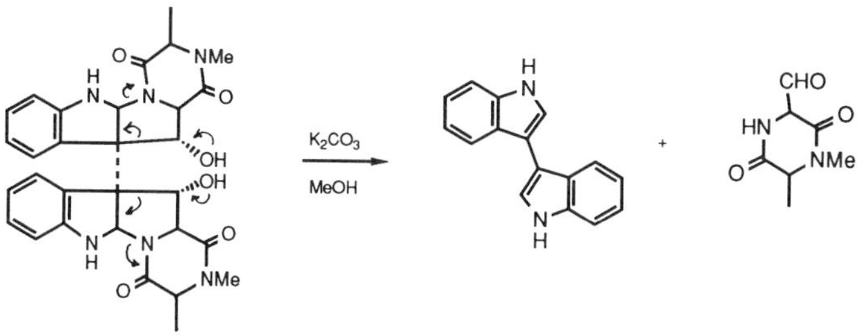

Abnormal products of a Pschorr reaction of 8-amino-1,2,3,4-tetrahydro-1-phenethylisoquinolines includes aromatic aldehydes arising from benzylic alcohols [Kametani, 1970b].

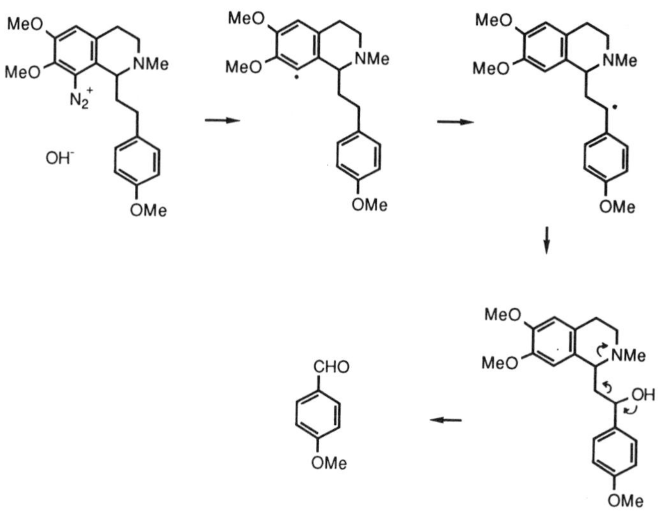

Nonaromatic diazonium ions are very unstable and their ejection of dinitrogen is spontaneous. In the last stage of the photolytic reaction of 2,2,4,4-tetramethyl-1,3-cyclobutanedione ditosylhydrazone salt in methanol the nitrogen expulsion is likely a fragmentation which is instigated by an ethereal oxygen [P.K. Freeman, 1969].

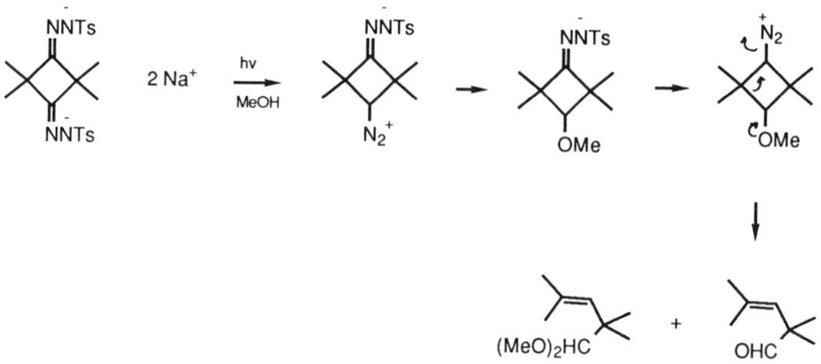

Diazotized 7-aminotropolones undergo rearrangement to salicyclic acid derivatives [Nozoe, 1951]. Such transformation is quite general with 2,7-disubstituted tropones [Magid, 1968] in which the substituents are donors.

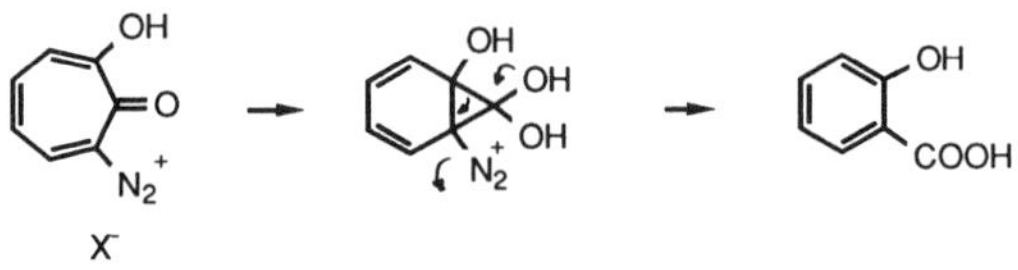

Replacement of the hydroxyalkyl chain at C-2 of substituted indoles by an amino group on reaction with an azide reagent [Bahadur, 1979] is readily rationalized by a fragmentation mechanism of a diazonium ion.

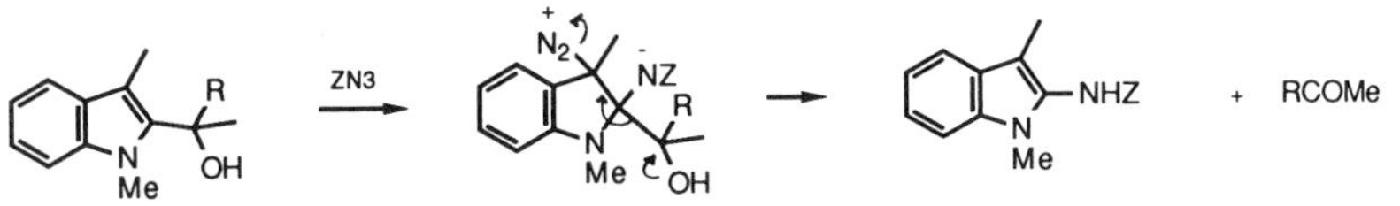

## 8.4. O—C—C—C—X SYSTEMS

This section covers some examples in which the leaving groups are outside those mentioned above. Many of these are sulfur-centered.

The role of a nucleofugal donor can be played by a phenylthio group in $\gamma$-hydroxy dithioacetals when it is properly activated [Semmelhack, 1977].

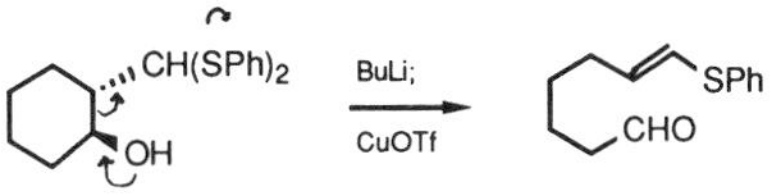

An electron transfer quenching mechanism has been proposed for the photochemical reaction of 3,3,6,6-tetramethyl-1-thiacheptane-4,5-dione [Johnson, 1972, 1975]. Elimination of isobutene from a sulfonium oxide zwitterion is considered to be the first step of the process.

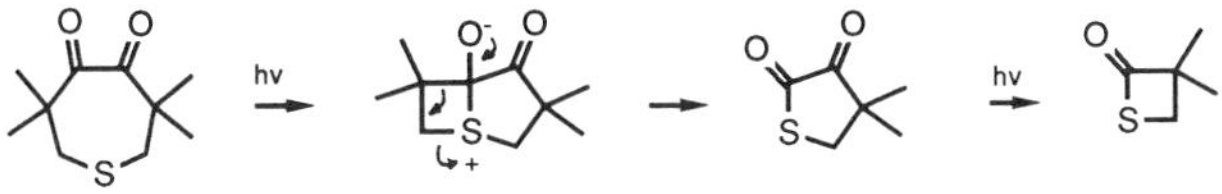

Homologation of ketones can be accomplished by reaction with [(phenylsulfonyl)methylene]dilithium [Eisch, 1985]. Vinylic sulfones and selenones in which the $\gamma$ position is substituted by a hydroxyl group are liable to fragment, often initiated by Michael addition with the nucleophile [Shimizu, 1984].

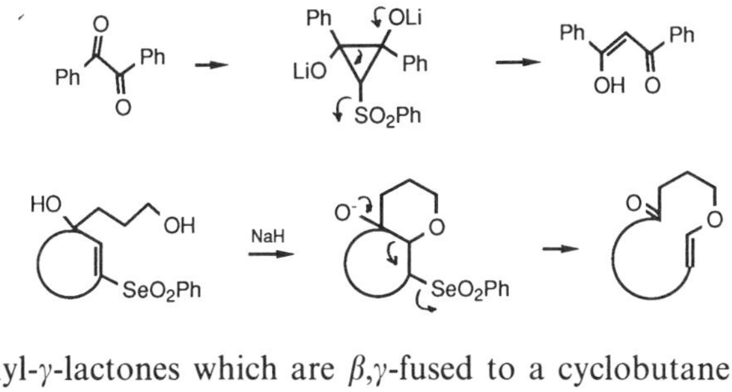

$\alpha$-Sulfonyl-$\gamma$-lactones which are $\beta,\gamma$-fused to a cyclobutane and $\alpha,\gamma$-bridged to another carbocycle would fragment on treatment with an alkoxide ion [Kende, 1989]. Depending on the length of the bridge, either unsaturated ketones or bicyclic compounds may be produced. Thus, the very strained fragmentation products from a propylene-bridged analog gains stability by undergoing an intramolecular Michael addition or a Michael-initiated aldol condensation.

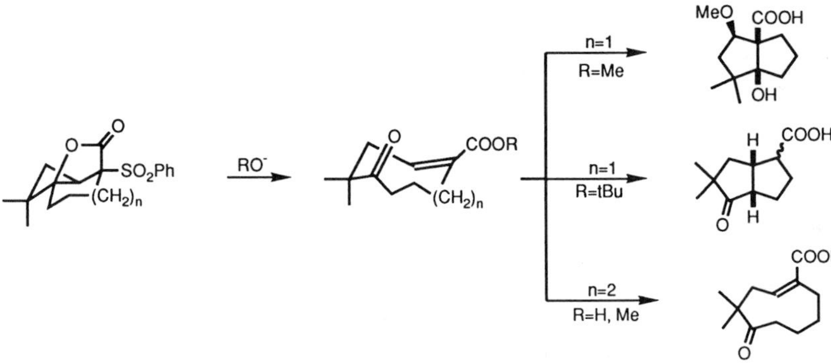

n=1
R=Me

n=1
R=tBu

n=2
R=H, Me

β-Sulfenyl alcohols may be caused to fragment by oxidation, for example using lead tetraacetate [Trost, 1978b]. This reaction has been employed in a synthesis of (−)-α-cadinol [Caine, 1977].

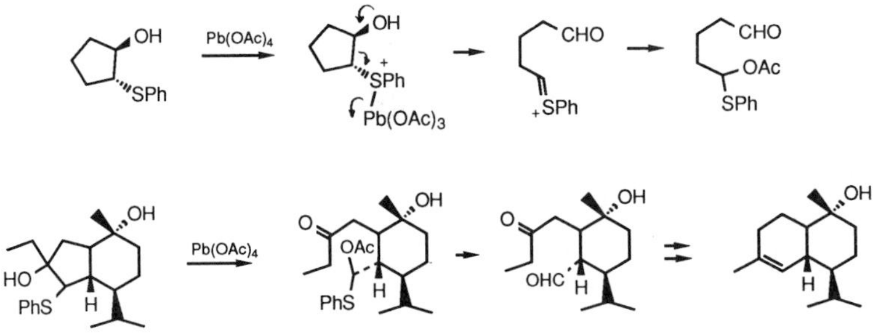

(-)-α-cadinol

A fragmentation pathway has been developed for the elaboration of the furyl dihydropyrone system which is characteristic of the limonoids [Emmer, 1981].

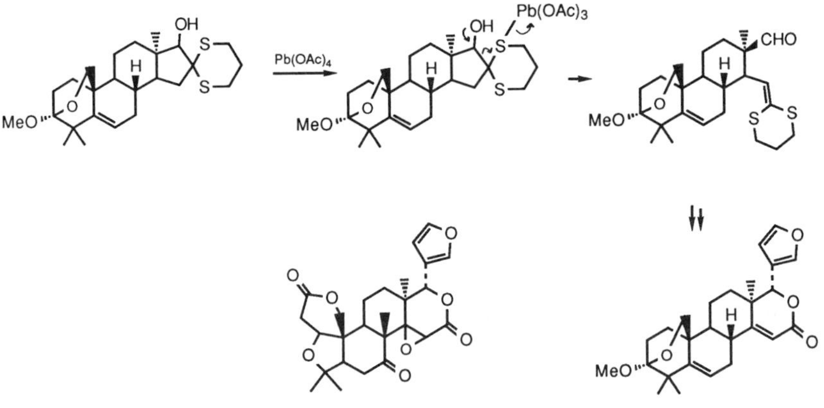

limonin

An undesired product from desulfurization of a cyano alcohol during synthesis of the phyllocladane nucleus is a naphthalene derivative [Shishido,

1983]. Destruction of the bridged ring system is due to the presence of a γ-hydroxy nitrile function.

## 8.5. O—C—C—O—X AND O—C—C—N—X SYSTEMS

As seen in many other fragmentable systems the nature of the intervening atoms is of minimal consequence. Thus, β-peroxy alcohols behave similarly to 1,3-diols with respect to fragmentation. Epoxidation of alkenes under conditions which favor epoxide opening by the carboxylic acid coproduct often results in C=C bond cleavage, recent examples being the oxidative ring fission of triisopropylsiloxycyclohexenes [Magnus, 1990]. It is possible that the fragmentation product undergoes further reactions such as the Baeyer–Villiger oxidation [Bernhard, 1977].

Autoxidation of 8-methoxyberberine phenolbetaine in wet THF results in an oxidative $C \rightarrow N$ transacylation [Moniot, 1977]. This reaction presumably involves cleavage of the pyridinium nucleus with formation of a fragmentable 2-hydroxy-3-hydroperoxyaziridine.

Photosensitized oxygenation of $\beta$-ionol in basic media gives rise to dihydroactinidiolide [Isoe, 1968]. Elimination of the hydroxyethyl group by a fragmentation of the endoperoxide is conceivably a crucial step of the reaction pathway.

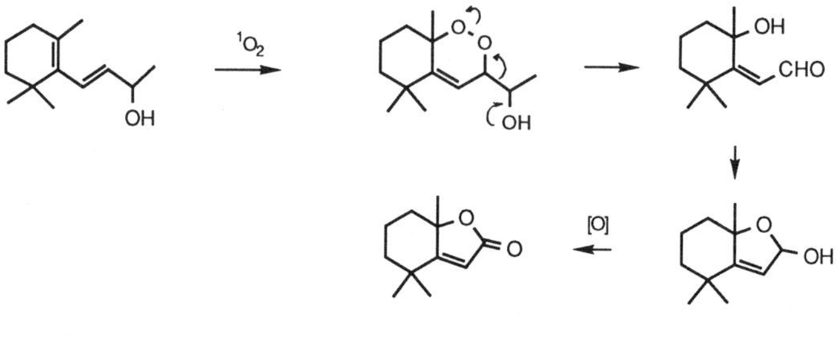

dihydroactinidiolide

Baeyer–Villiger oxidation of nonenolizable $\alpha$-oxy ketones has been shown to afford keto acids [Madge, 1980].

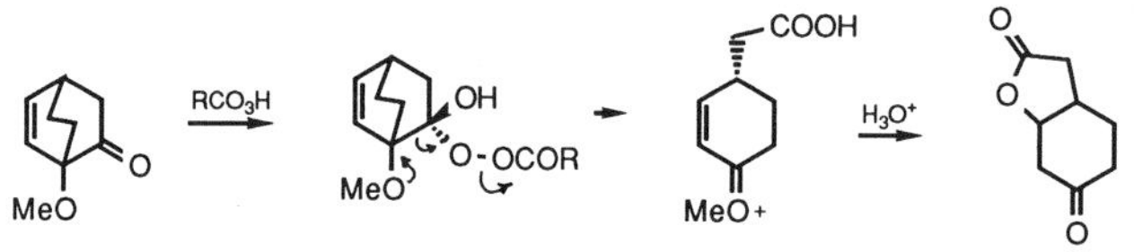

Ring-C contraction of colchicine [Iorio, 1984] and colchiceine [Cech, 1949] may be effected by hydrogen peroxide. The latter reaction is particularly noteworthy as many tropolones undergo ring cleavage to give muconic acid derivatives.

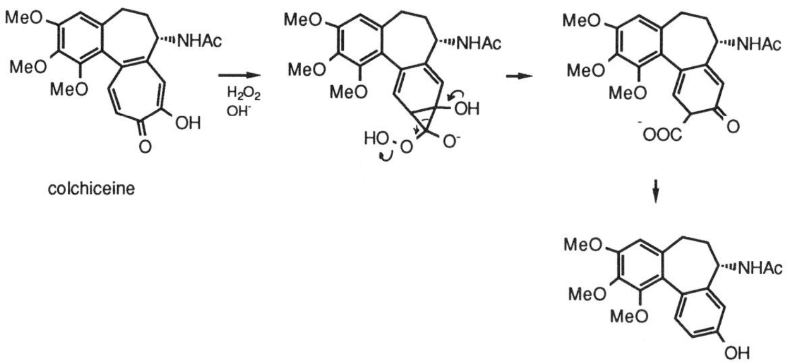

colchiceine

The reaction of dimedone with singlet oxygen in the presence of fluoride ion leads to 5,5-dimethyl-1,2,3-cyclohexanetrione [Wasserman, 1985b]. The fluoride ion increases dramatically the rate and efficiency of this transformation by hydrogen-bonding with the enol.

It is interesting to note that the trione actually condenses with dimedone in an aldolization fashion, but the dimeric species would also intercept singlet oxygen to give a fragmentable hydroperoxide. In other words, the final product acts as a catalyst for the oxidation.

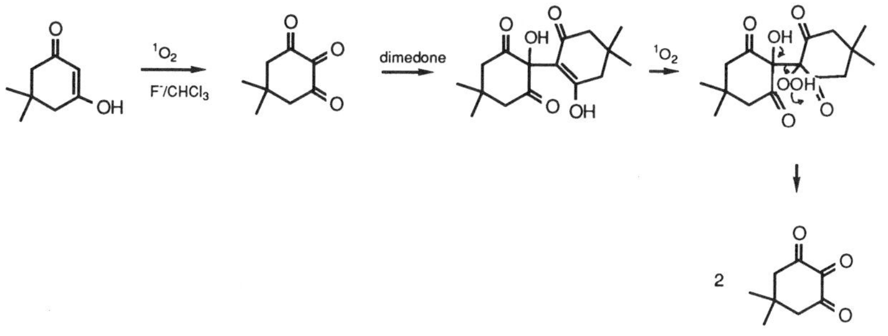

Cyclic α-diketones undergo degradation into aldehydo- or keto-acids (cf. degradation by hydrogen peroxide [Weitz, 1921]).

With alkaline hydrogen peroxide aliphatic β-diketones react to give three carboxylic acids [W. Cocker, 1975].

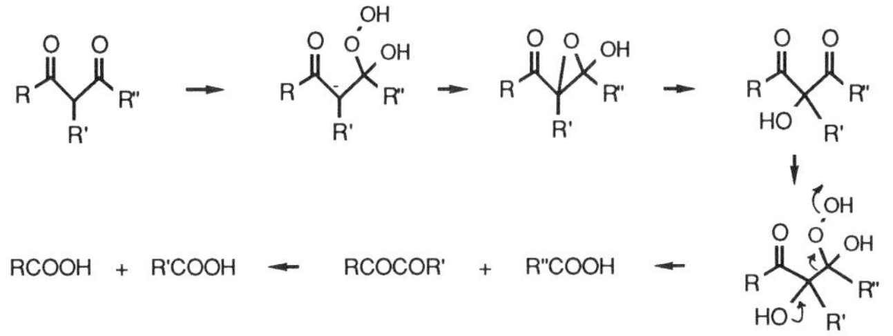

α-Hydroperoxy ketones decompose in the presence of base via cyclic intermediates to afford acids and carbonyl compounds [Doering, 1954].

β-Hydroxy-α-azohydroperoxides formed by autoxidation of the corresponding hydrazones are very sensitive to acids, bases, and reducing agents [Utaka, 1982]. The $C_\alpha$—$C_\beta$ bond cleavage occurs readily and the resulting keto acyldiazene reacts further with nucleophiles present.

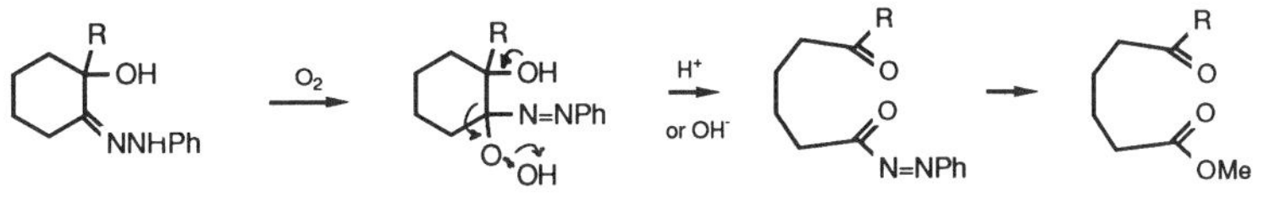

A familiar reaction of indoles is the oxidative ring scission [Witkop, 1951; Deyrup, 1969]. Many different reagents have been used to accomplish this transformation, including peracids.

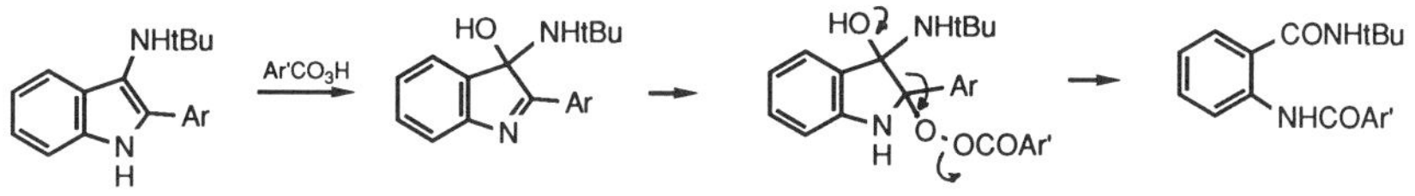

Salicyaldehyde is among the reaction products of tropone with superoxide anion in dimethyl sulfoxide [S. Kobayashi, 1979]. The solvent plays an important role in this reaction, as no reaction having been observed in benzene, acetonitrile, or dimethylformamide.

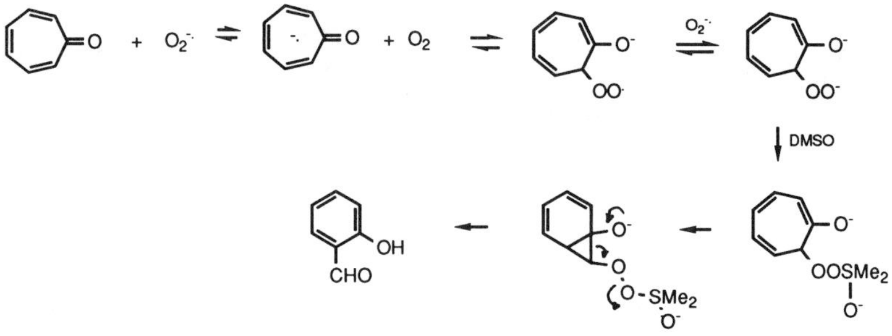

Either $\gamma$- or $\delta$-lactones are obtained by chromic acid oxidation of $\gamma$- and $\delta$-hydroxy alkenes, respectively [Schlecht, 1985]. A proposed mechanism for this oxidative cleavage depicts a fragmentation that generates formaldehyde, a Cr(III) species, and a cyclic oxonium ion.

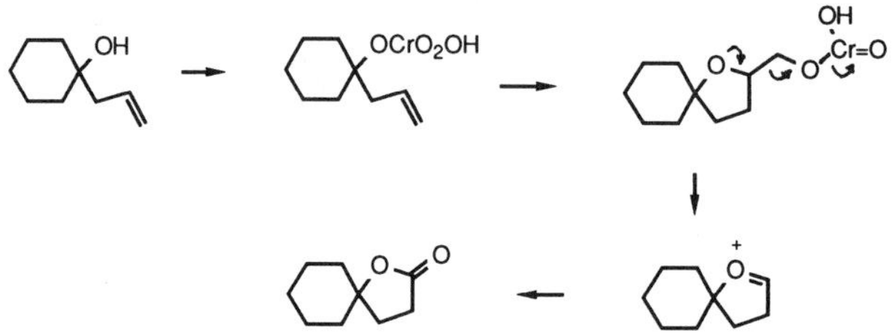

The well-known procedure for *vic*-glycol cleavage by lead tetraacetate [Criegee, 1933] has been incorporated into a scheme for macrolide synthesis [Wakamatsu, 1977]. Contrapolarization of one of the oxygen atoms by Pb(IV) allows for fragmentation to take place.

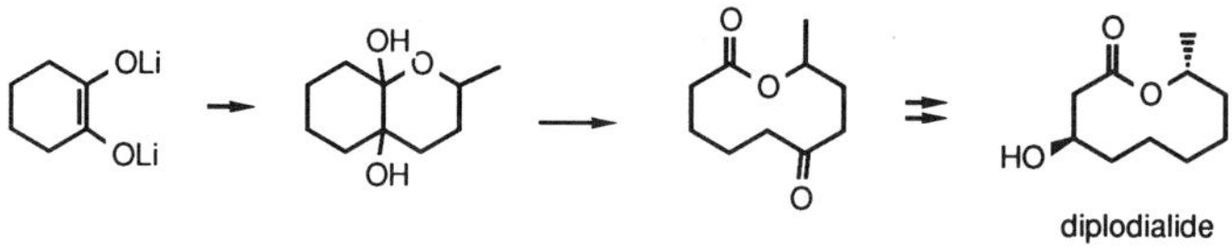

diplodialide

A route to 1,3-diketones is based on cleavage on modified substrates, for example those obtained by cyclopropanation of enediol derivatives [Ito, 1977].

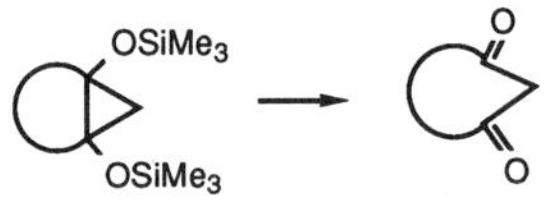

$\alpha$-Hydroxy acids undergo autoxidation in the presence of Cu(I) ion. It has been proposed that Cu(III) carboxylates are the fragmenting species [Toussaint, 1984].

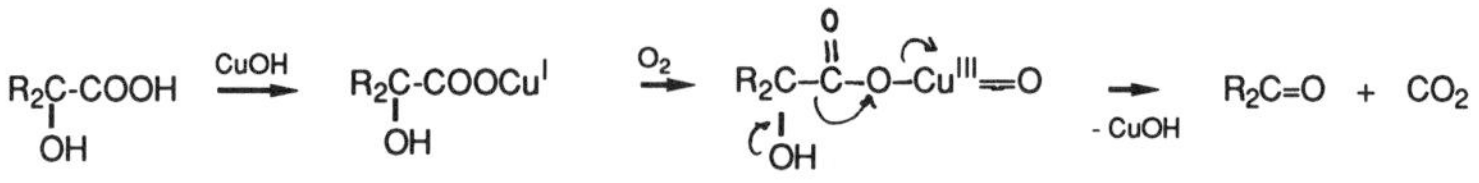

Hofmann degradation of α-hydroxy amides gives aldehydes. The fragmentation is presumably concerted because detection of α-hydroxy isocyanates failed [C.L. Stevens, 1956].

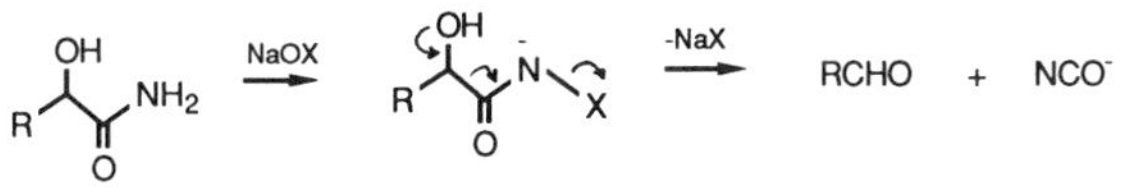

*N,N′*-Dimethyl-*N′*-(2-hydroxy-2-phenylethyl)hydrazine needs only contact with acid to decompose into benzaldehyde, formimine, and dimethylamine [Urry, 1964].

The formation of thioxanthone [S. Jones, 1990], albeit in low yield, from an intramolecular Friedel–Crafts reaction indicates the occurrence of a tandem fragmentation.

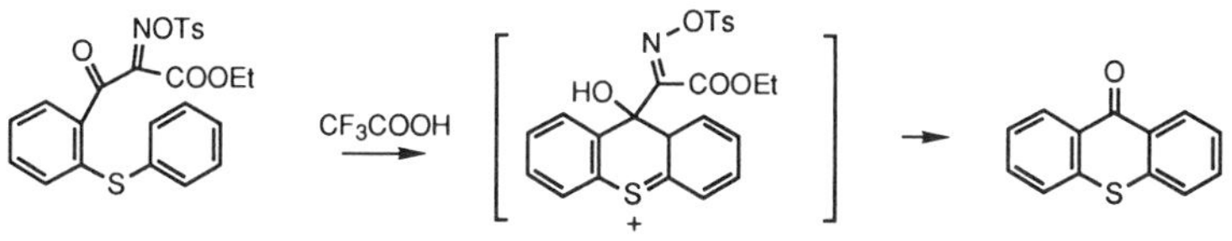

## 8.6. O—C(=O)—C—C—X SYSTEMS

More than one hundred years ago the formation of itaconic anhydride by pyrolysis of citric acid was reported [Anschütz, 1880]. The decarboxylation of β-donor substituted carboxylic acids is so well known that only a few cases of are presented here to highlight the structural requirements.

Decarboxylation of the malonic acid substituted with a butenolide [Jaroszewski, 1982] to give (*E,Z*)-2,4-hexadienedioic acid is most efficient at pH ~3, when it exists as the monoanion.

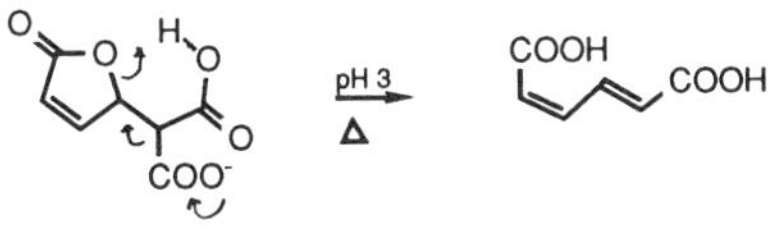

The *endo* isomer of 1,4-dimethyl-2-isobutyl-6,7-dioxabicyclo[2.2.1]heptane-2-carboxylic acid exhibits a considerably faster rate for decarboxylation than the *exo* acid, and both isomers are more reactive than the ethyleneketal of acetoacetic acid [Atkinson, 1977]. The decarboxylation is assisted by strain relief and it proceeds via the zwitterions. Zwitterion formation apparently is favored by the *endo* acid because the oxygen within the longer bridge is more basic.

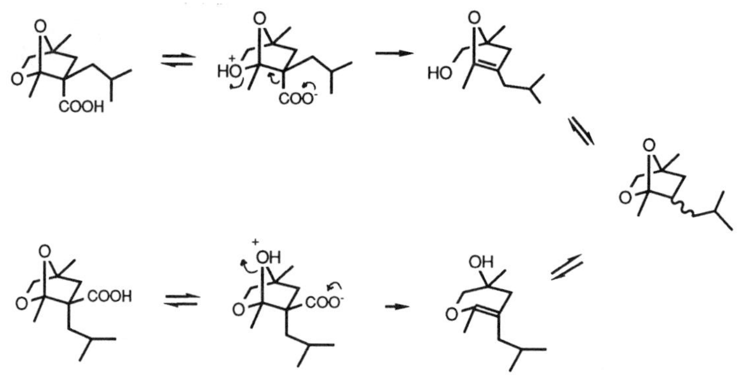

Heating of (2-cyanoethyl-1-hydroxy)cyclohexaneacetic acid effects decarboxylation and simultaneous hydration of the nitrile function [Dolby, 1965]. An imino lactone could be the intermediate.

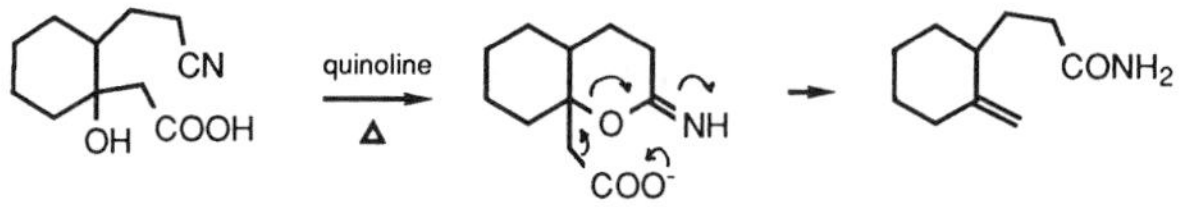

Activation of the hydroxyl group of the $\beta$-hydroxy carboxylic acid facilitates the fragmentation. Dimethylformamide dimethylacetal [Hara, 1975] or a combination of triphenylphosphine and diethyl azodiformate [Mulzer, 1979] are two of the commonly used reagent systems. Olefins are also formed by pyrolysis of $\beta$-lactones [Krapcho, 1974].

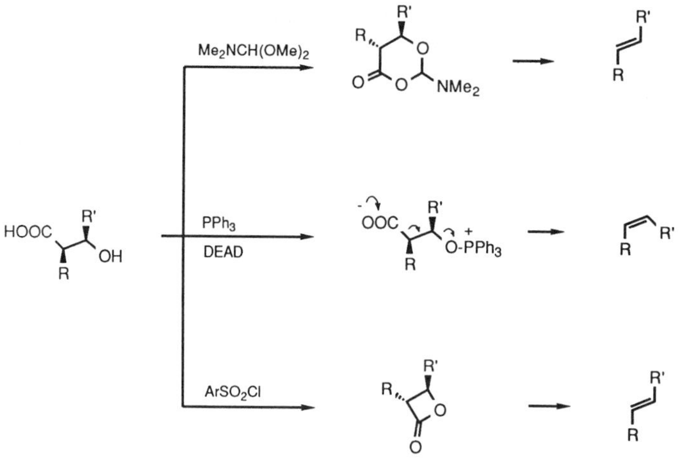

Glycidic acids decompose thermally to afford carbonyl compounds with one less carbon atom. This process constitutes the final step of a classical and indirect method for homologation of ketones and aldehydes via the Darzens glycidic ester condensation [Newman, 1968].

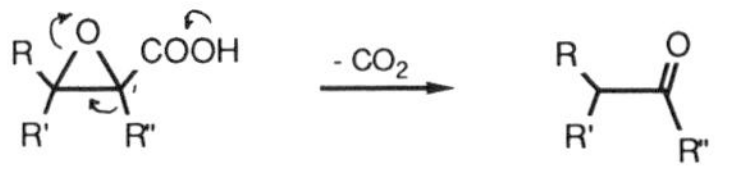

α-Methylene ketones are now accessible from allyl β-ketocarboxylates via acetoxymethylation and Pd(O)-promoted fragmentation [Tsuji, 1986]. Generation of the carboxylate ions under essentially neutral conditions offers great advantages in such synthesis, considering the reactivity of the products.

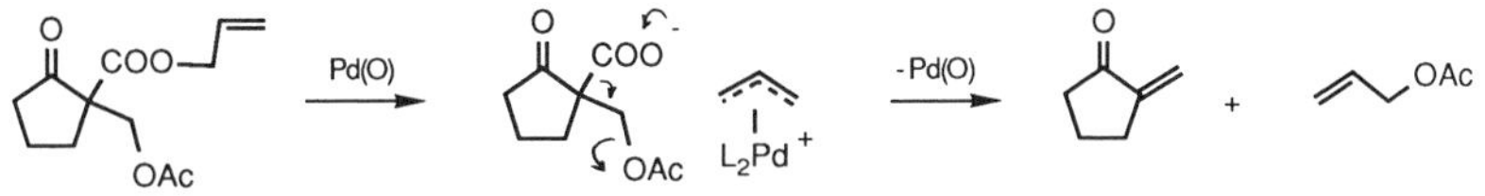

An excellent approach to unsaturated macrolides is based on a decarboxylative double fragmentation [Sternbach, 1979; Shibuya, 1979]. The pivot of this process is a bridged internal ketal which is fused to a cyclohexane moiety. Favorable stereoelectronic alignment of the bonds in the fragmentation involved is noted.

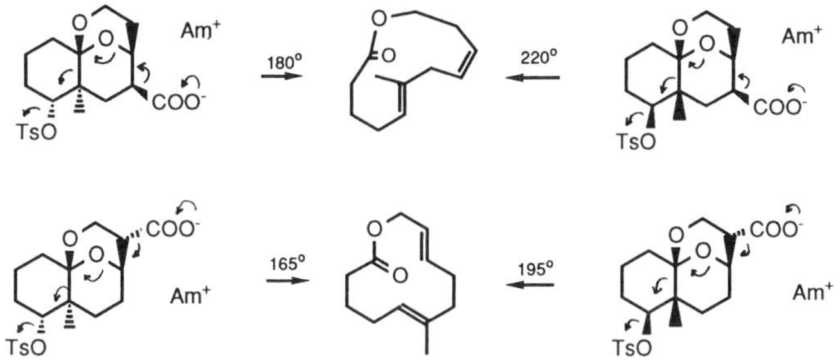

One method for the peripheral unraveling of medium-sized heterocycles is by decarboxylation [Reinecke, 1972].

A convenient synthesis of acrylic acid embodies Mannich reaction of malonic acids and in situ elimination of carbon dioxide and a tertiary amine upon alkylation of the Mannich adduct. In certain variants, alkylation might not be necessary, for example the condensation of cyclohexanone enamines with malonic acid leading to cyclohexylideneacetic acid [Prout, 1973].

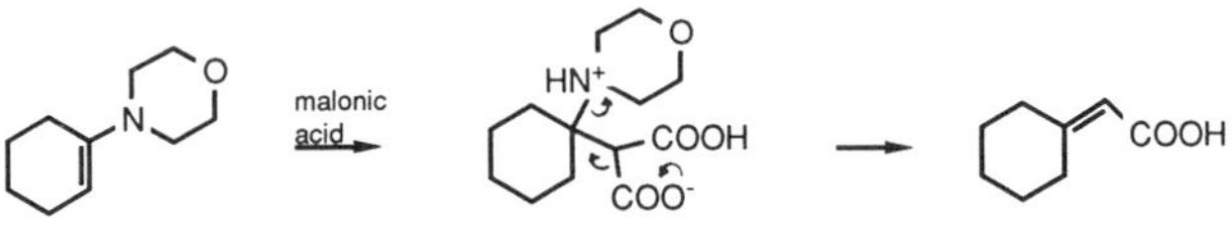

The Mannich reaction of 1,2-benzisoxazole-3-acetic acid affords the allylic amine directly [Uno, 1978].

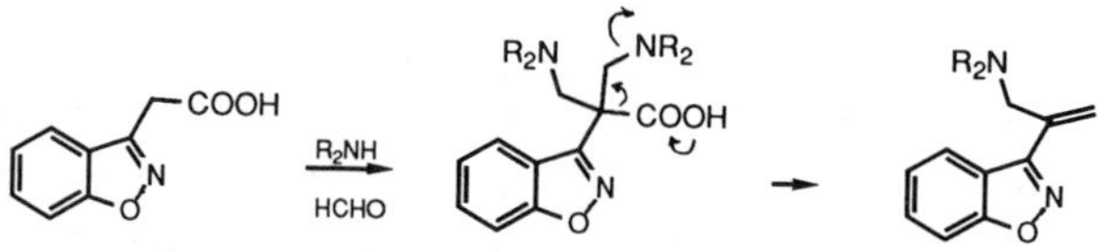

Hofmann degradation of metaphanine methiodide gives a tricyclic cyclopentanone [deWaal, 1966]. There is a ring contraction and decarboxylative C—N bond cleavage.

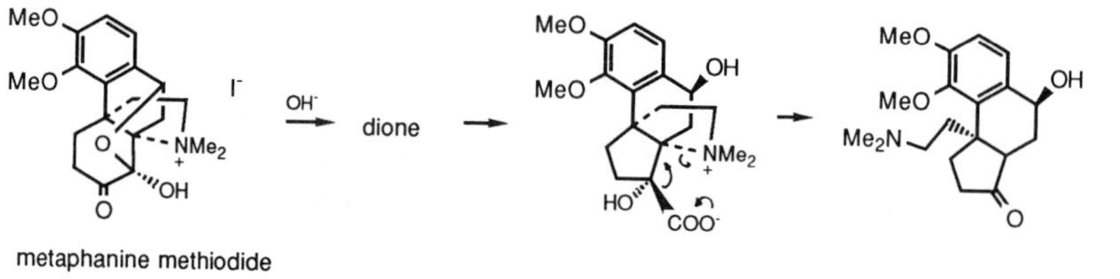

metaphanine methiodide

($\beta$-Halophosphonic acids decompose in a strictly analogous manner as the corresponding carboxylic acids [Conant, 1922; Satterthwait, 1978].)

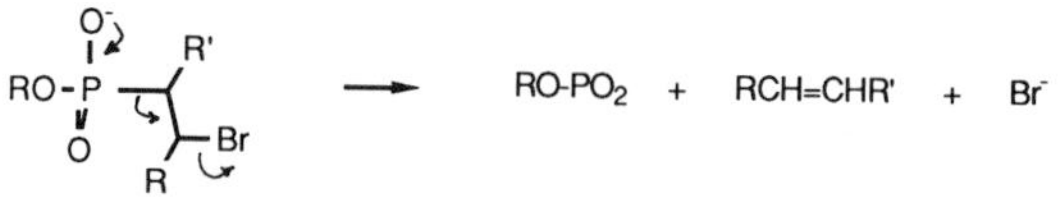

2-Phenyl-1,2,3-triazole-4-carboxylic acid undergoes decarboxylative ring opening in the presence of barium hydroxide [Carman, 1966].

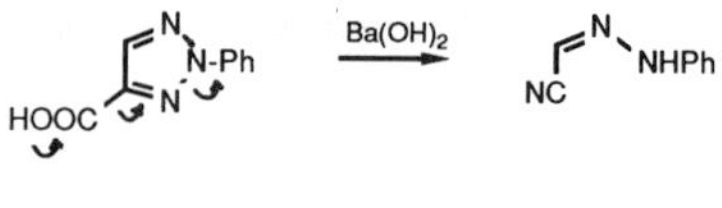

$\alpha$-Amino acids undergo oxidative decarboxylation under various conditions. Hypohalite treatment is the simplest method whereby baikiain has been converted into anatabine [Leete, 1978] and 2-methyltryptophan into 4-acetylquinoline [van Tamelen, 1968]. The latter reaction involves chlorination and heterocyclic opening of the indole-3-acetaldehyde product.

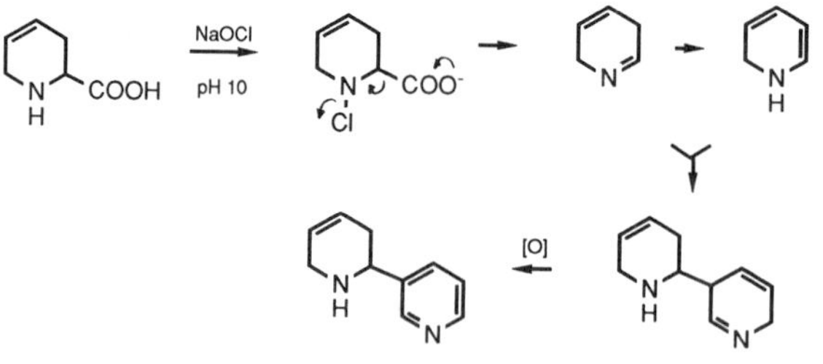

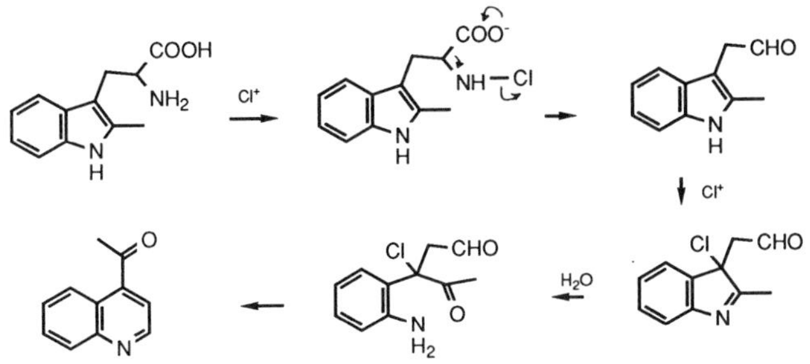

Both dehydration and decarboxylation results when indol-3-ylpyruvic oxime is exposed to dilute sulfuric acid [A. Ahmad, 1960]. This transformation is not entirely unlike the fragmentation of α-(*N*-hydroxycarbamato)oximes [Kiseleva, 1981] in terms of atomic movements.

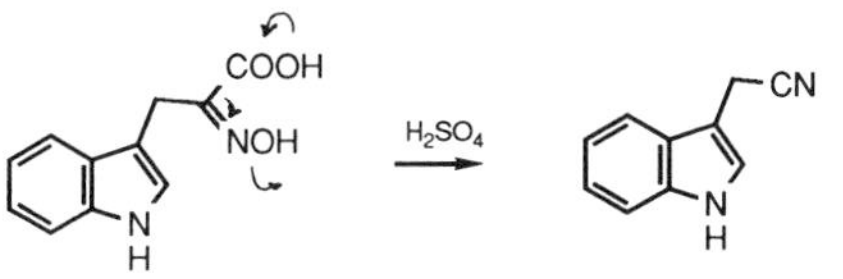

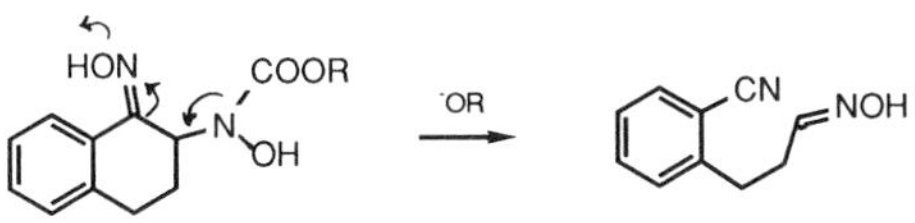

Decarboxylation of other α-donor substituted carboxylic acids can also be effected oxidatively. The oxidant attaching to the α-donor contrapolarizes ($a \rightarrow d$) and set up a **d3d** segment. Synthesis of ketones via α-sulfenyl carboxylic acids has found much practicality, as demonstrated in an approach to juvabione [Trost, 1977b].

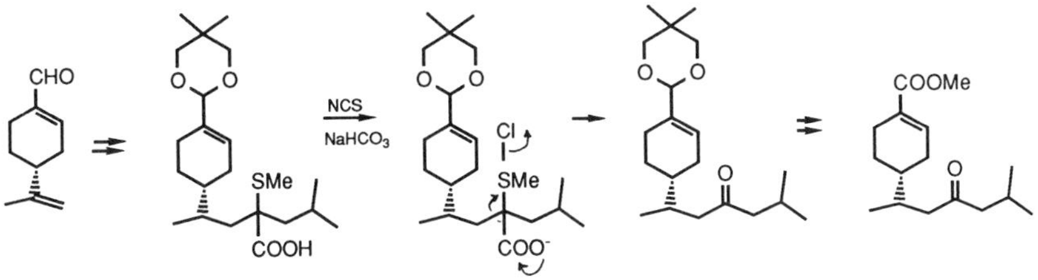

α,β-Unsaturated β-halocarboxylate ions afford acetylenes on heating, the facility of the decarboxylation depends on the nature of the other substituent [Grob, 1964a].

The net reductive removal of an α-mesyloxy group from a lactone by reaction with hydrazine [Paulsen, 1966] seems to proceed via a ketene intermediate, which is then intercepted by hydrazine. A simpler analog of this reaction is the decomposition of α-chloroacetyl hydrazide in alkaline solution [Buyle, 1964].

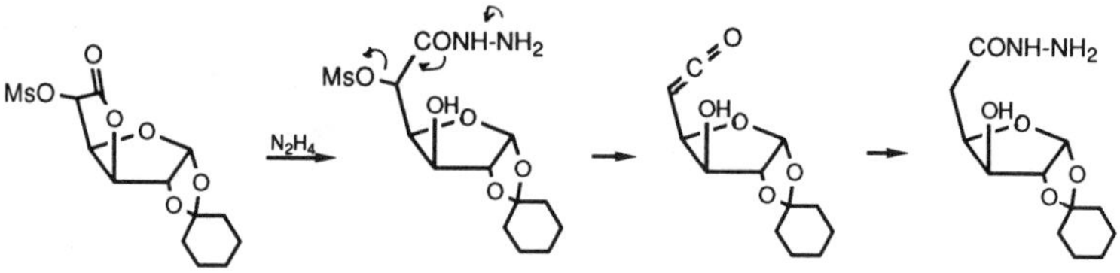

## 8.7. N—C—C—C—X, N—C—C—O—X AND N—C—C—N—X SYSTEMS

The most widely studied subsystems of this category are the cyclopropylamines derived from halocarbene reaction with enamines [Ohno, 1963]. Synthetically the combination of this reaction with enamine formation is complementary to the homologation method based on enol derivatives, with possible regiochemical differences. The reactivity of dienamines has also been scrutinized [DeGraaf, 1973, 1974].

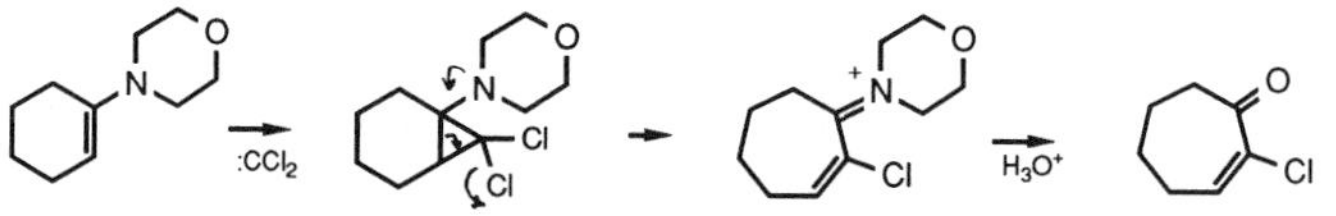

A convenient method for the elaboration of homoaporphine alkaloids [Castro, 1985] is based on the two processes. However, enamides must be employed in the carbene reaction.

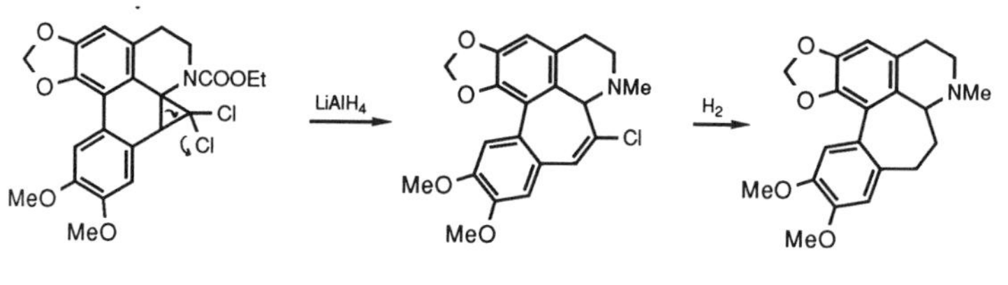

Certain heteroaromatic compounds including indole and pyrrole (and their substituted congeners) contain an enamine moiety and their adduct formation with halocarbenes is quite well known. These adducts often decompose in situ [R.L. Jones, 1969] or during hydrolysis [Closs, 1961]. The typical **d3d** fragmentation is obviously very favorable on account of strain relief and aromatization.

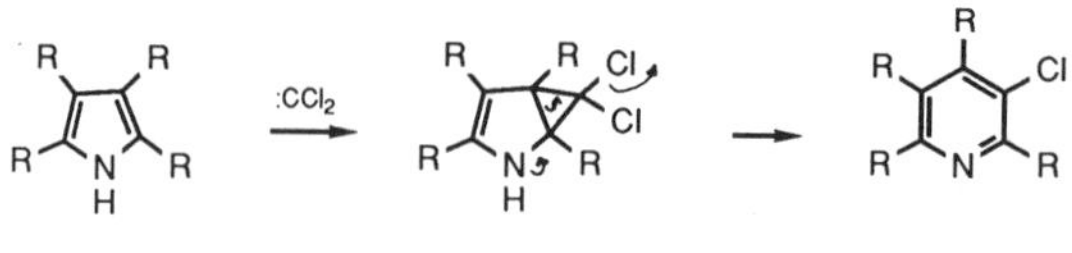

The formal dichlorocarbene adduct of 2,5-dimethylpyrrole has been prepared indirectly [Gambacorta, 1971] and, as expected, it (or more appropriately its conjugate base) undergoes fragmentation to provide 3-chloro-2,6-lutidine.

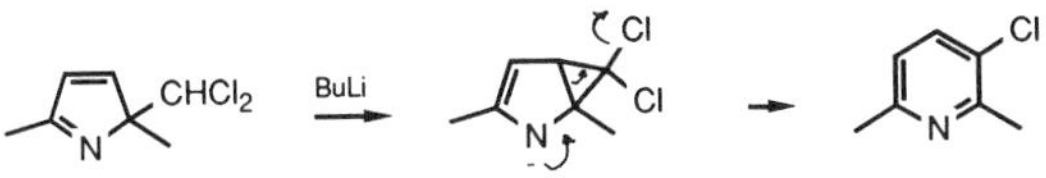

An interesting transformation of a fluorochlorocarbene adduct of *N*,*N*′-dibenzyluracil in methanol [Thiellier, 1975] involves formation of seven-membered heterocycle intermediates.

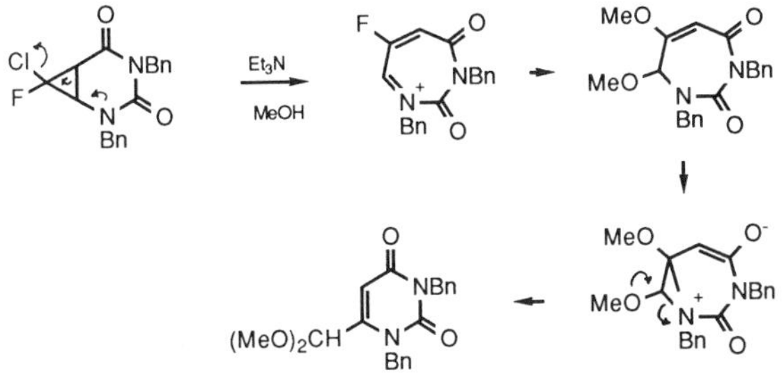

Ring expansion via fragmentation is readily achieved within a nucleoside framework [Marquez, 1984]. Here, an ethereal function acts as the leaving group.

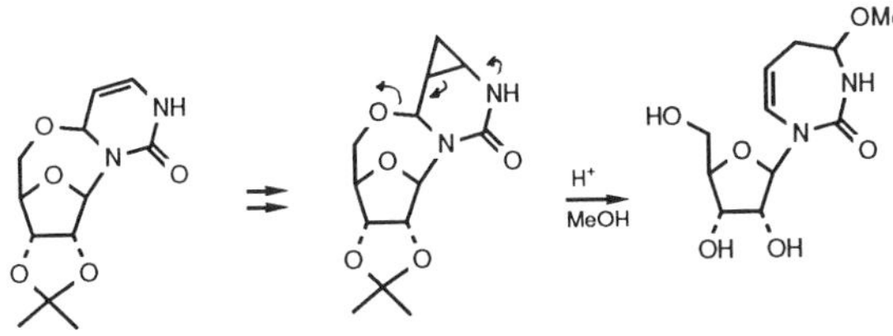

It is rather unusual that a tricyclic bromostyrene is formed in the silver ion-promoted solvolysis of a dibromo[5.3.1]azapropellane [Dhanak, 1987]. The product contains an (*E*)-cyclooctene nucleus with a bulky bromine substituent inside the molecular cavity.

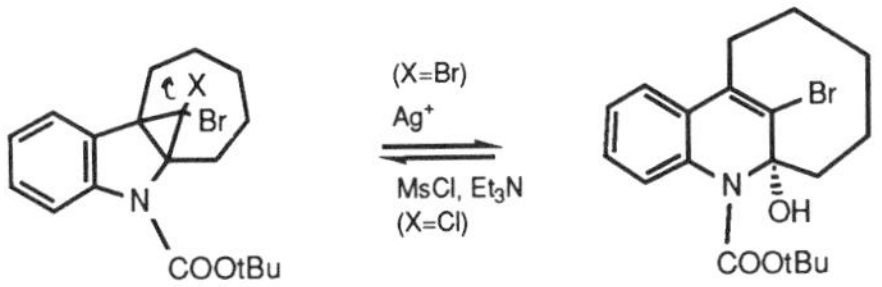

It is possible that the above reaction was not assisted by the nitrogen atom. This nonparticipatory aspect was suggested in the formation of a benzazepine from the dibromocarbene adduct of a 3,4-dihydroisoquinoline [Lantos, 1986].

Conversion of 5-halo-2-methoxytropones to vanillin in liquid ammonia [Kitahara, 1956] may have proceeded via conjugate addition, tautomerization and expulsion of the halide ion.

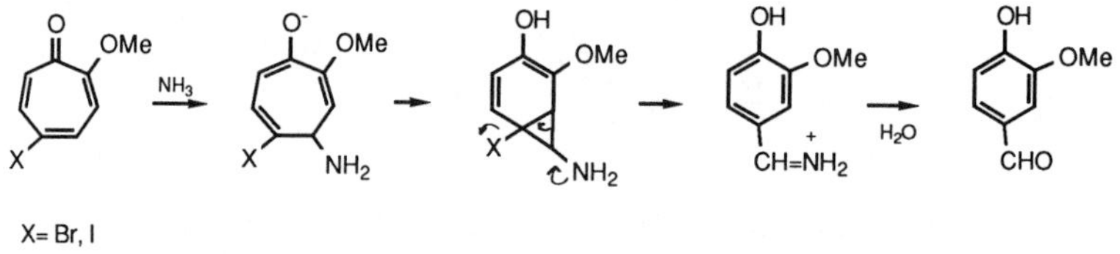

X= Br, I

Solvolysis of 3-azabicyclo[3.3.1]non-9-yl esters shows reactivity differences in which higher rates are associated with the *anti* isomers [House, 1966]. The major product of the *anti* ester is a cyclohexene.

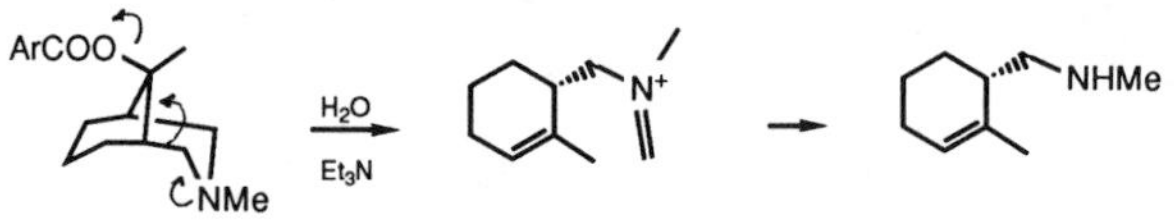

The interference of the Kornblum oxidation of 19-iodotabersonine by water [N. Langlois, 1979] is due to fragmentation.

A minor ketonic product obtained from a double cyclization of a 3-(α-piperidinyl)indole has been converted into a mesylate. Base treatment of the latter compound has resulted in the *trans* → *cis* isomerization of the hydroquinoline moiety [Natsume, 1985]. The driving force for the fragmentation is of thermodynamic nature.

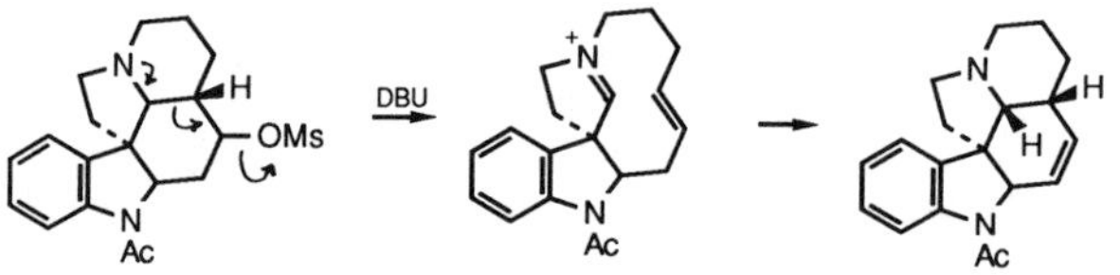

Fragmentation of a γ-amino mesylate and aza-Cope rearrangement are prominent features in a synthesis of isochanoclavine-I [Kiguchi, 1989]. These processes, coupled with reductive photocyclization of enamide, constitute an efficient and stereocontrolled method for entry into the 6,7-secoergoline-type alkaloids.

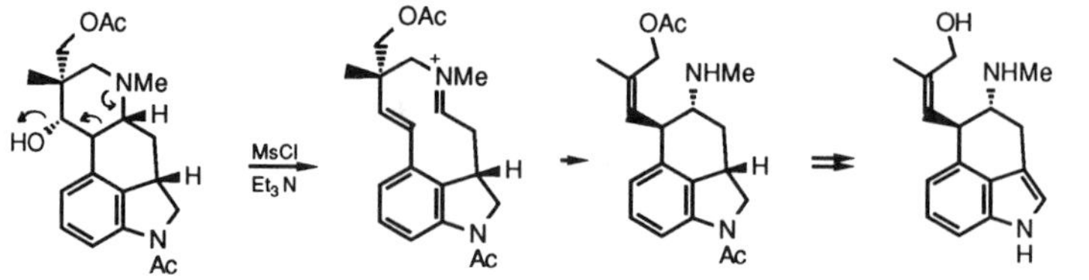

As stated, hydroquinoline derivatives undergo an internal fragmentation, although the cleavage appears to be peripheral [Grob, 1967a]. The facile aza-Cope rearrangement that obscured the true nature of a two-step process has been clarified [Marshall, 1969a] by arresting the macrocyclic iminium ion reductively.

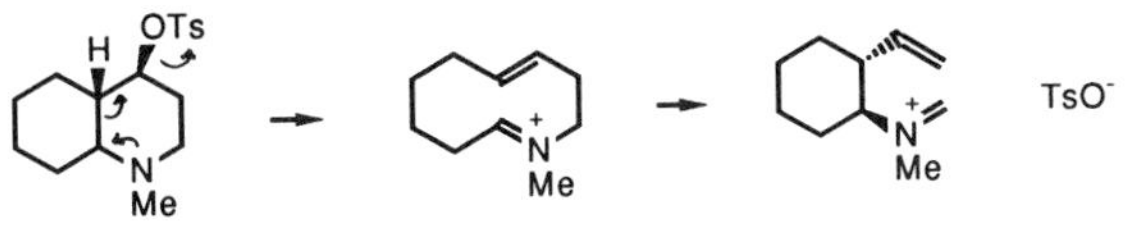

Mechanistic ambiguity does not exist in fragmentation of the decahydroquinoline system(s) [Grob, 1967a; LaBel, 1985] in which the leaving group is extraheterocyclic, that is not in the same ring as the heteroatom.

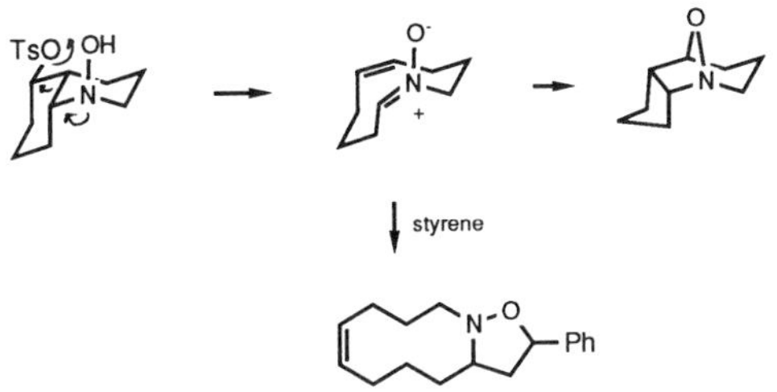

A study on the solvolysis of $\gamma$-amino tosylates which are incorporated into a quinolizidine skeleton reveals a conformational dependence of their reaction courses [Bohlmann, 1965]. Thus, an equatorial tosylate of 13-hydroxylupanine furnishes angustifoline as the major product when treated with a small amount of triethylamine in acetonitrile.

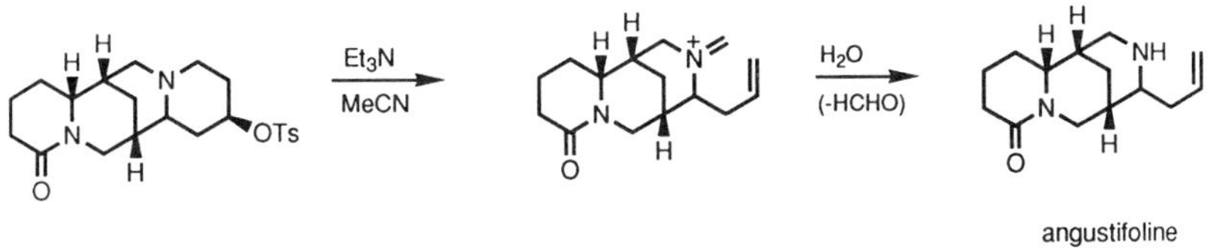

angustifoline

The conessane ring system suffers fragmentation at the C-13/C-18 bond when a 12$\alpha$-mesyloxy group is present [Lukacs, 1970].

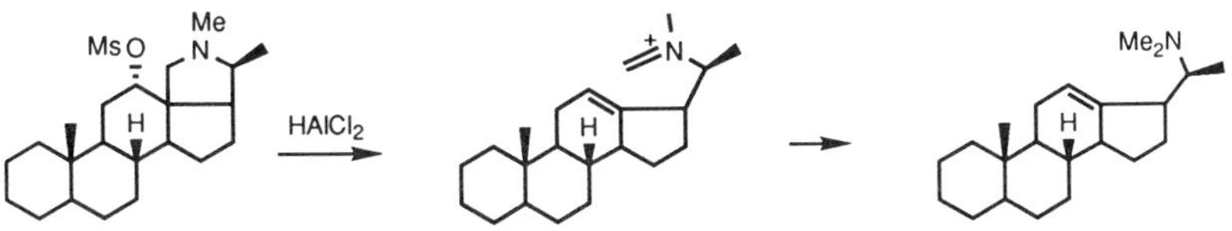

A fragmentation reaction is the key to correlate ajmaline with corynanthe alkaloids [M.F. Bartlett, 1962]. In voacanginol tosylate the nitrogen lone-pair is in a participatory conformation to effect a Grob fragmentation [Morita, 1975].

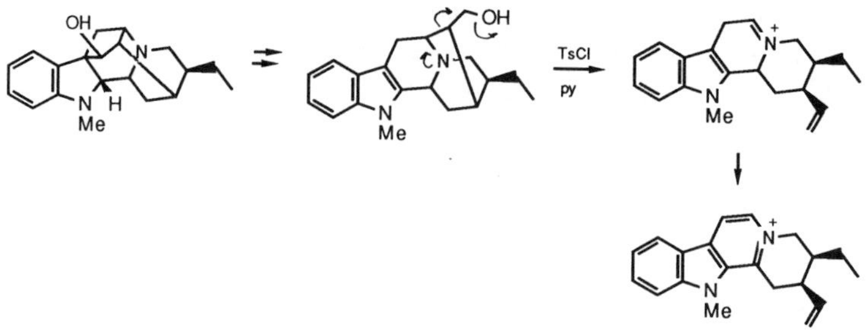

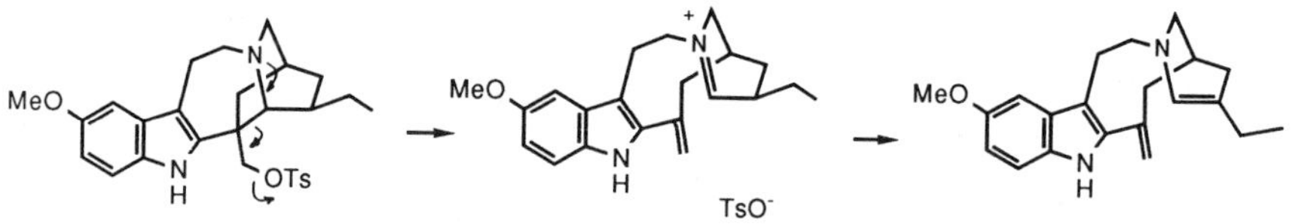

The pyrolytic decomposition of aconitine *N*-oxide results in a nitrone [Wiesner, 1959]. There is a transannular elimination of ethylene and acetic acid, the latter process being a fragmentation.

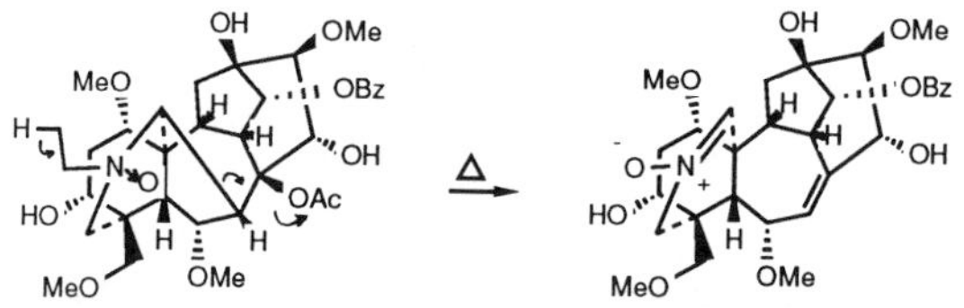

The same fragmentation was thought to be involved in the formation of pyroheteratisine [O.E. Edwards, 1965]. A transacetylation of heteratisine acetate is a prerequisite.

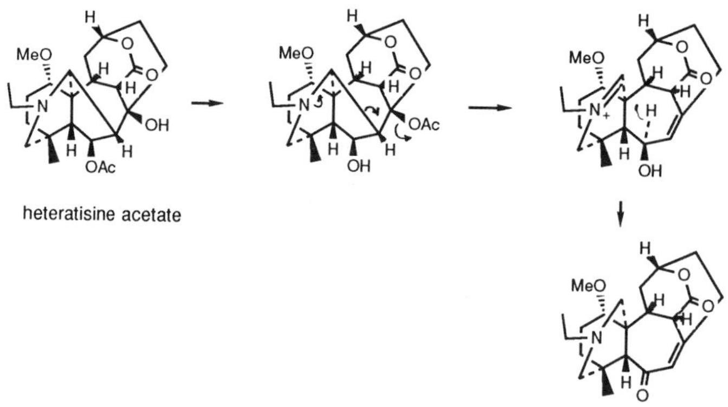

heteratisine acetate

A reductive fragmentation is a crucial operation in a synthesis of daphnilactone-A [Ruggeri, 1989]. This reaction breaks an unwanted C—C bond and sets up the imino–Prins cyclization leading to the complete skeleton of the alkaloid.

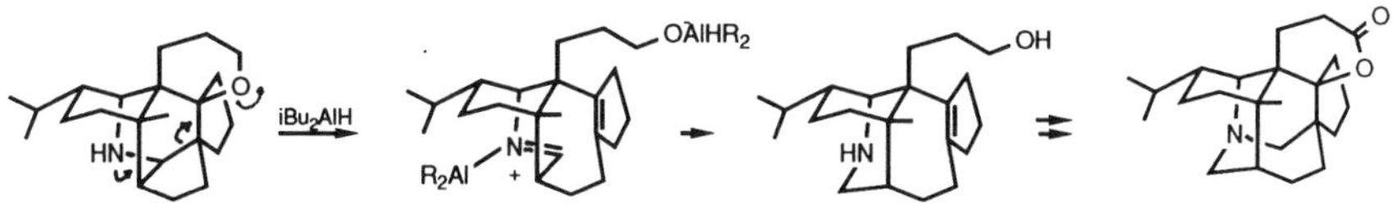

The Diels–Alder adduct of 5,5-dimethoxy-1,2,3,4-tetrachlorocyclopentadiene and methacrylamide does not undergo straightforward saponification [Thompson, 1985]. The observed succinimide formation reflects the strain of the norbornene framework that favors transannular addition of the amide across the double bond. Then the nitrogen atom acts as an effective donor to break up the bridged system.

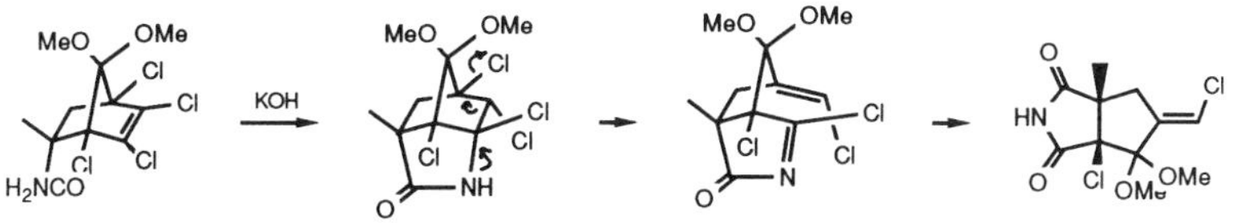

The reaction of dichlorocarbene with 2,2,6,6-tetramethyl-4-piperidone leads to butyrolactams [Lai, 1980; Lind, 1980]. Fragmentation of the dichlorooxirane intermediate is followed by ring closure.

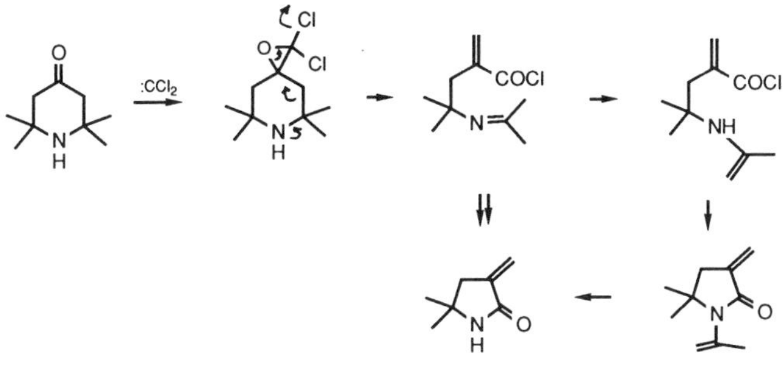

A synthesis of α-isosparteine starts from condensation of anaferine with formaldehyde [Schöpf, 1972]. On treatment with acetic anhydride the intermediate fragments as prelude to a twofold Mannich reaction to give 8-oxo-α-isosparteine. Interestingly, one of the asymmetric centers of *meso*-anaferine adjacent to the nitrogen atom underwent equilibration while those centers of *dl*-anaferine retained their configurations during the reaction.

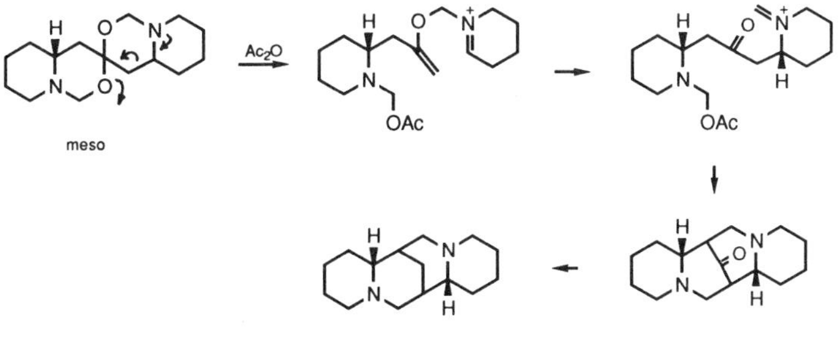

meso

α-isosparteine

Removal of the piperidine ring from jervine has been effected by refluxing in acetic anhydride in the presence of zinc chloride [Fried, 1953]. Fragmentation was triggered by interaction of the ethereal oxygen with the acceptor reagent.

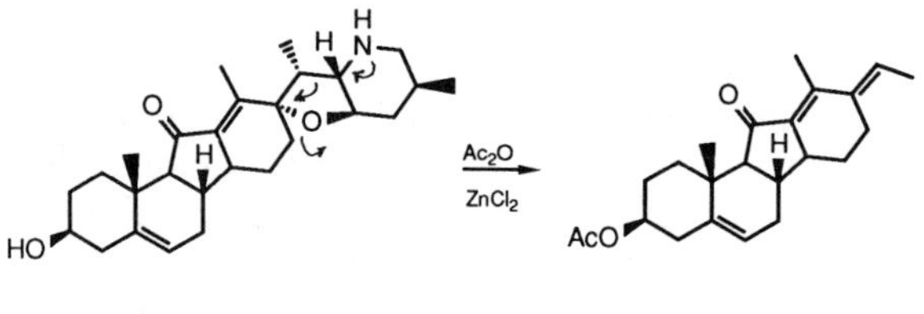

jervine

A scheme for synthesis of velbanamine [Narisada, 1971] entails fragmentation of an intermediate with the iboga skeleton. lead tetraacetate oxidation of an amino alcohol indeed accomplished the desired transformation; however, further reaction of the resulting enamine rendered this approach useless.

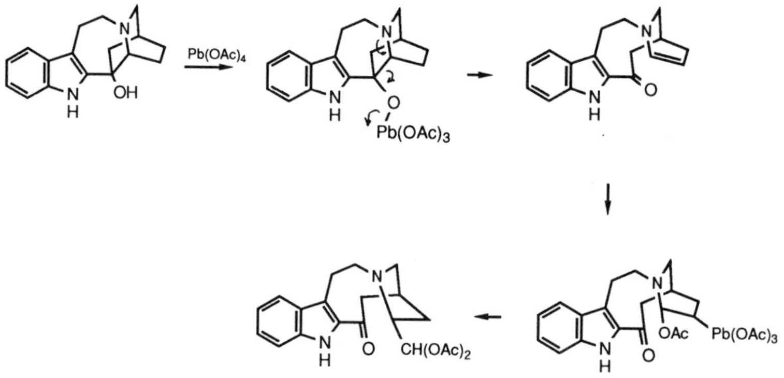

Of certain biosynthetic significance is the conversion of ervitsine to 5,16-dehydromethuenine [Andriantsiferana, 1977] by way of retro-Mannich reaction and recyclization. A similar fragmentation step was involved in syntheses of olivacine [Besselièvre, 1976; Naito, 1981]. The step generated a carbazole from a tetracyclic species.

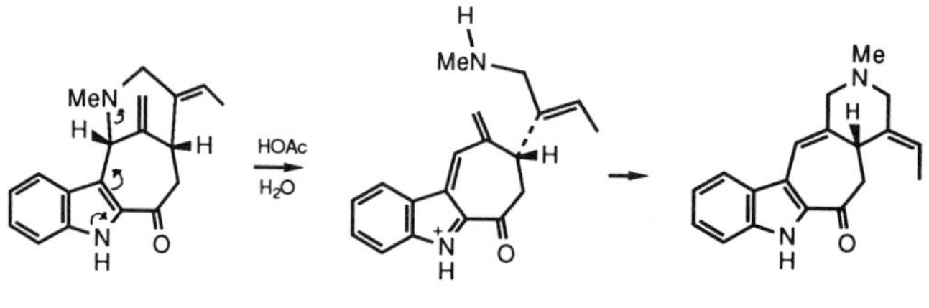

ervitsine

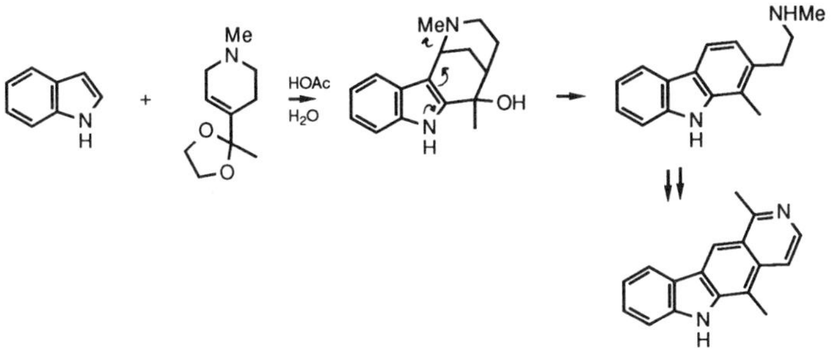

Me N + O O HOAc $H_2O$ MeN OH N H NHMe N H

N

N H

olivacine

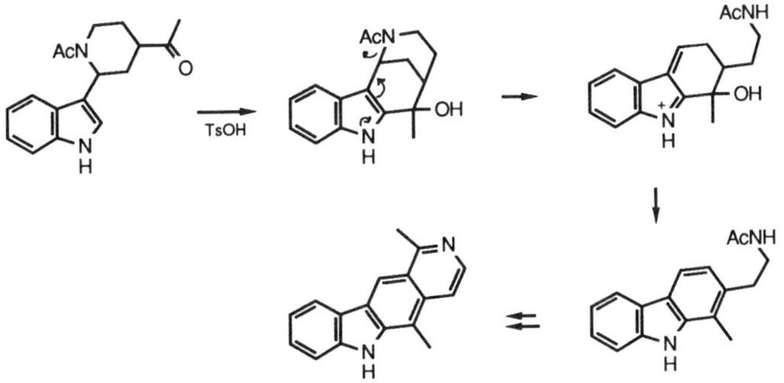

AcN O N H TsOH AcN N H OH AcNH $\overset{+}{N}$ H OH

AcNH N H N N H

The thermal equilibration of indolyltetrahydropyridines [Natsume, 1984] which served as intermediates in synthetic studies of the aspidosperma family proceeded by a retro-Mannich/Mannich reaction sequence. The stability of the isomeric $N_b$-hydroxyethyl derivatives is intriguing.

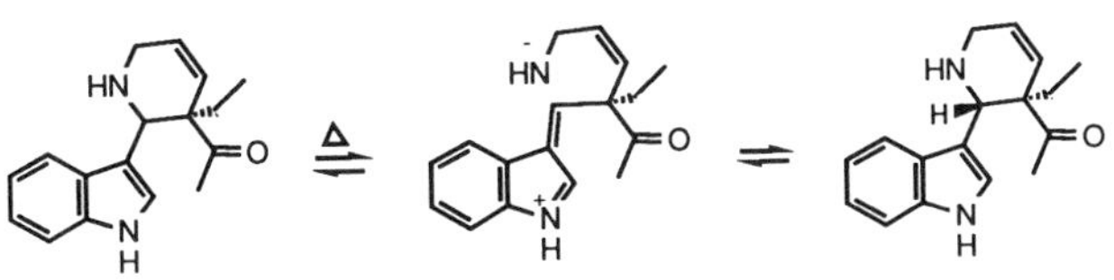

HN O N H Δ HN O $\overset{+}{N}$ H HN H O N H

Sodium borohydride reduction of the gramine obtained from indole, chloroacetaldehyde and a secondary amine leads to a tryptamine [M. Julia, 1973]. The rearrangement apparently involves aziridinium salt formation and fragmentation prior to a reduction.

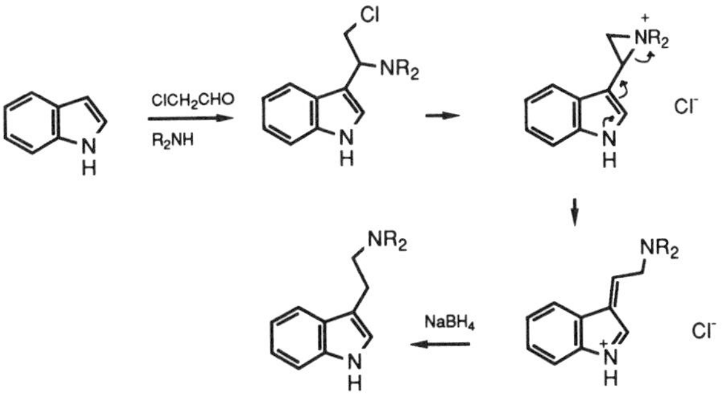

N H $ClCH_2CHO$ $R_2NH$ Cl $NR_2$ N H $\overset{+}{N}R_2$ N H $Cl^-$

$NR_2$ N H $NaBH_4$ $NR_2$ $\overset{+}{N}$ H $Cl^-$

Tetrahydro-$\gamma$-carbolines undergo aza-insertion [Tamura, 1982] which may consist of ylide formation and fragmentation.

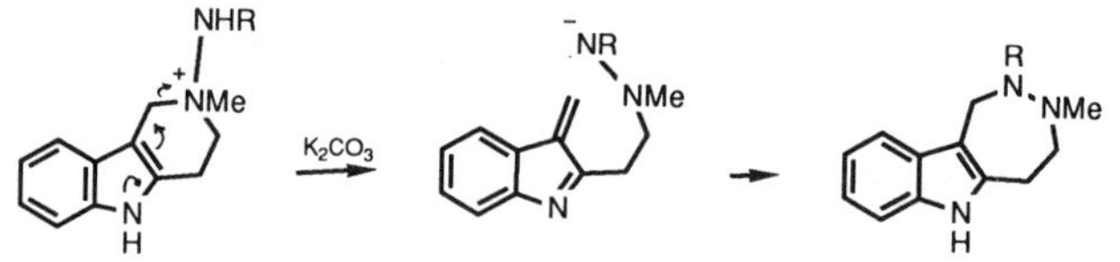

The fascinating isomerization of (19*S*)-vindolinine to the (19*R*)-isomer on merely keeping it in chloroform [Atta-ur-Rahman, 1986] may be explained by the intervention of tetracyclic species.

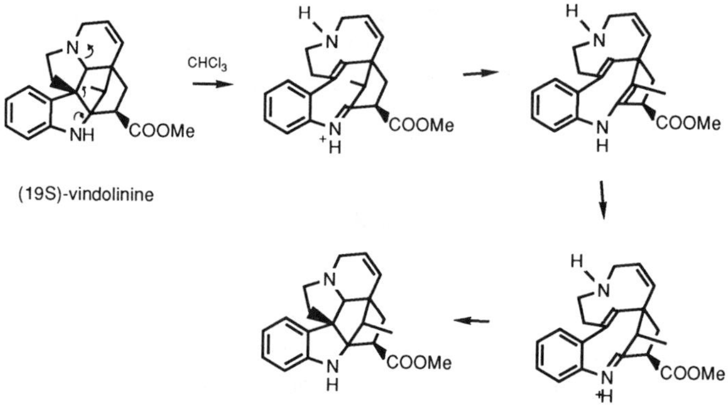

(19S)-vindolinine

(19R)-vindolinine

There have been attempts to construct tetraazasemibullvalenes by a 1,3-dehydrochlorination of bisimidates of the glycouril series [Gompper, 1983]. However, fragmentation has been the only reaction observed. This leads to tetraazacyclooctatetraene derivatives.

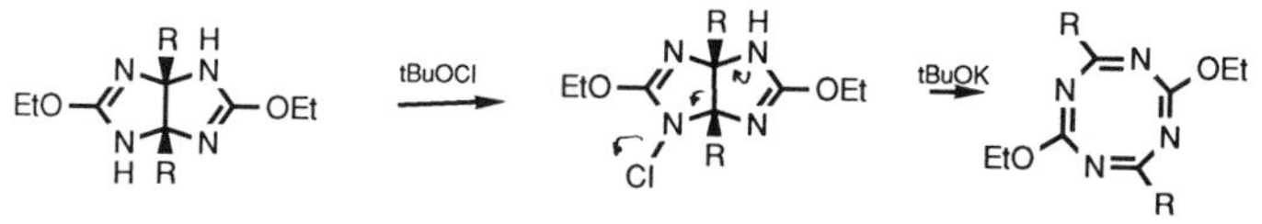

During an Eschweiler–Clarke methylation of polyamines in which the amino groups are separated by propylene groups, unusual fragmentation has been observed [Alder, 1991]. Clear mechanistic information about this process is not available.

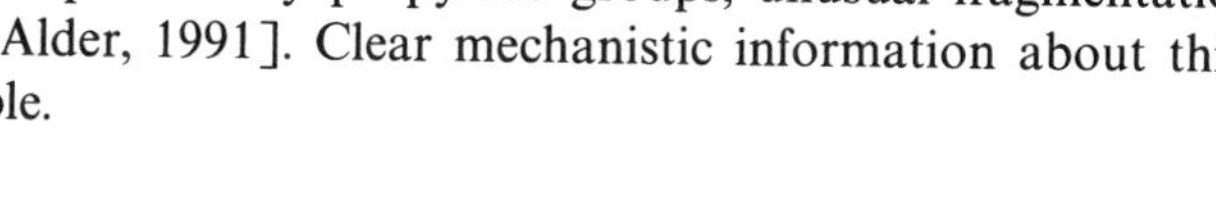

A product from the reaction of *N*,*N*,*N*′,*N*′-tetramethyl-1,3-propenediamine with sulfene is a vinylogous sulfonamide [Paquette, 1966] which is derived from fragmentation of the 3-aminothietane dioxide adduct.

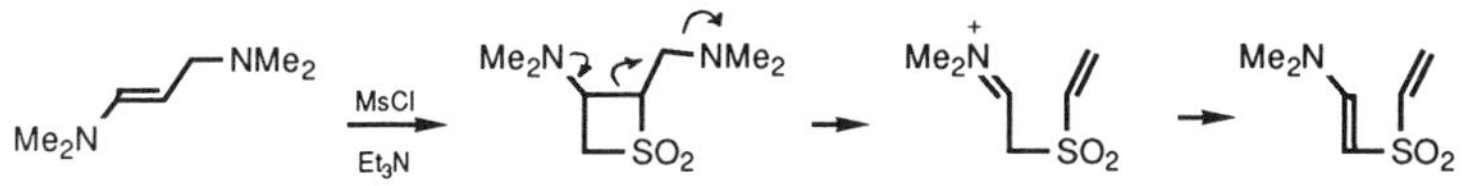

Treatment of 3-phenylacetamido-1-hydroxyazetidin-2-one with dichlorophosphate gives rise to an imidazolidinone [Krook, 1985]. Phosphorylation of the hydroxyl group serves to trigger the $\beta$-lactam ring fragmentation.

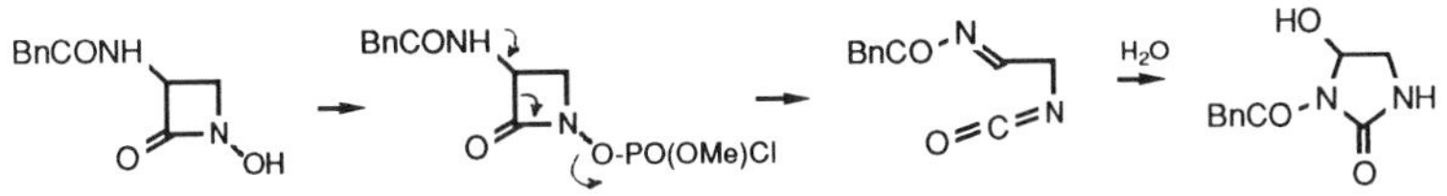

2-Aminocycloalkyl nitrates are produced from photoaddition of *N*-nitrosamines to cycloalkenes in the presence of oxygen [Pillay, 1974]. These compounds are very sensitive to base as they decompose readily into dialdehydes.

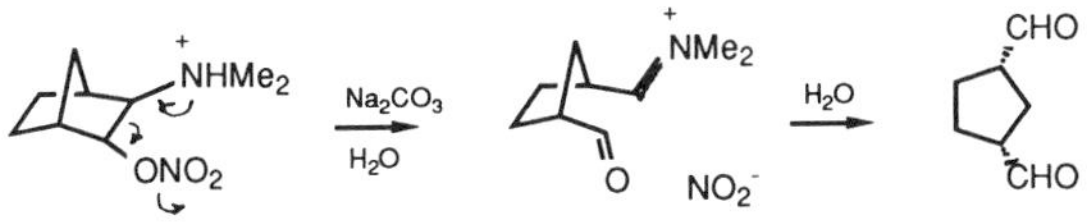

A subtle choice of reaction mechanisms in response to substrate structures and reaction conditions is manifested in the Beckmann rearrangement of quinuclidin-4-yl methyl ketoxime [H.P. Fischer, 1968]. The normal product predominates probably because the nitrilium ion is more stable than the tertiary bridgehead carbenium ion or the 1,4-dimethylene-1-azacyclohexanium cation. The product arising from the last species is only 3%.

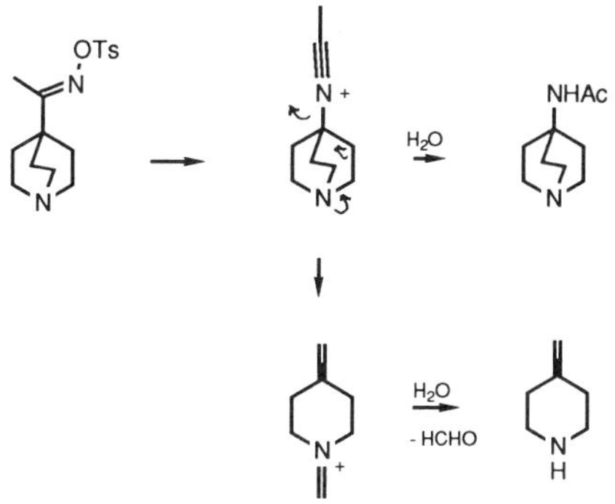

Synthesis of $\alpha$-keto derivatives of ketones, lactones, esters, amides and lactams by reaction of the enamino carbonyl systems with singlet oxygen [Wasserman, 1985a] is expedient. The aminodioxetane decomposes smoothly into a formamide and the desired dicarbonyl compound.

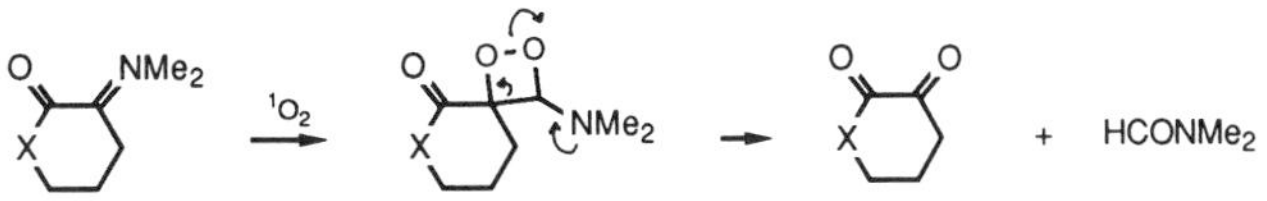

Hydrolysis of *N*-benzoyloxy-1,4-diazabicyclo[2.2.2]octane liberates piperazine, and formaldehyde besides benzoate anion [Huisgen, 1965].

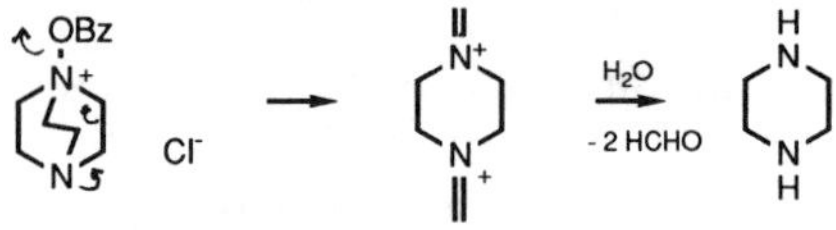

Allophanyl chlorides decompose in the presence of a proton abstractor to give diisocyanates [Sayigh, 1964].

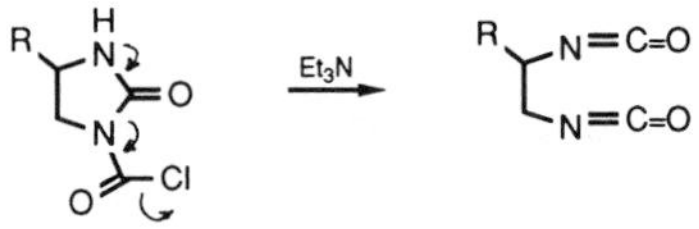

## 8.8. ASSISTED BECKMANN FRAGMENTATIONS

Beckmann fragmentation can be greatly facilitated by the presence of a donor group at the α-position of the oxime, the latter completes a **d3d** segment. It is a versatile synthetic operation because two differentiable functional groups are generated in a controlled fashion.

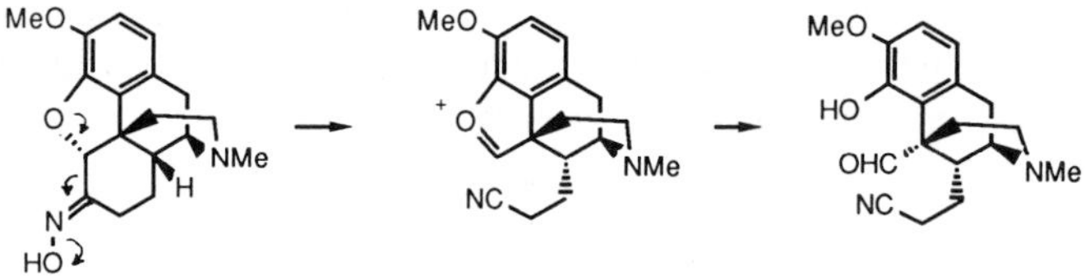

[Schöpf, 1927]

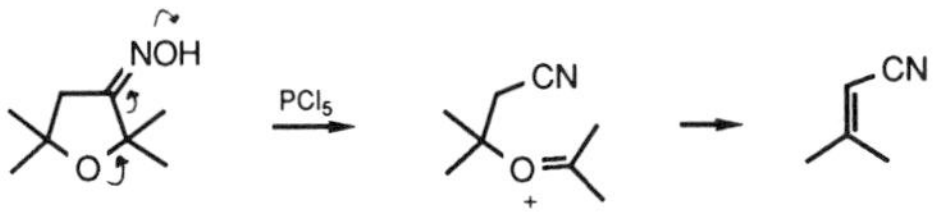

[Hill, 1962]

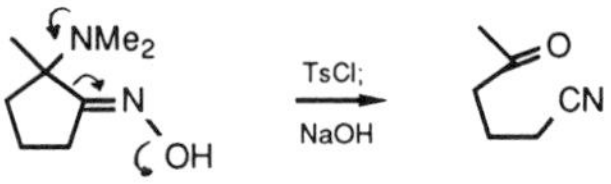

[Lunkwitz, 1968]

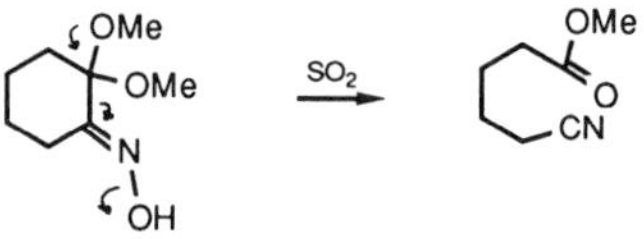

[Rogic, 1975]

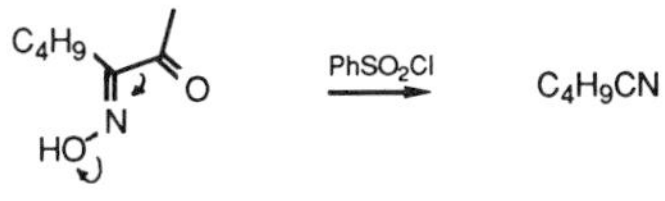

$C_4H_9$ ... $\xrightarrow{PhSO_2Cl}$ $C_4H_9CN$

[Ferris, 1960]

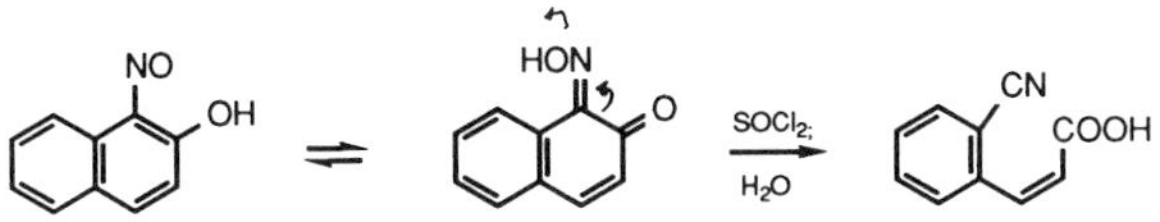

[D. Murakami, 1958]

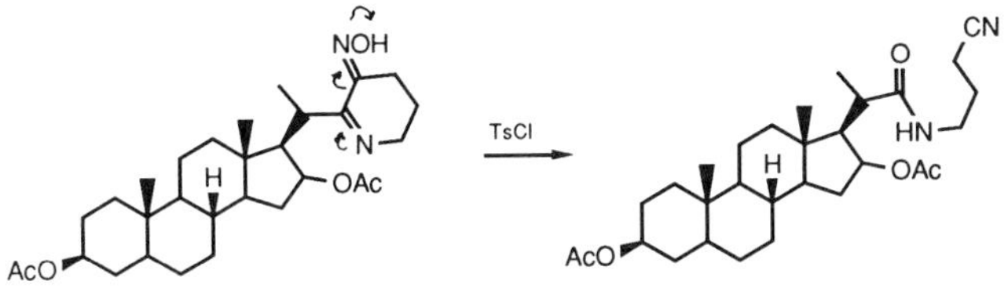

[Y. Sato, 1961]

Molecular strain facilitates the reaction further as shown by an example in which a four-membered ring disintegrates thermally without added reagents [Brook, 1978].

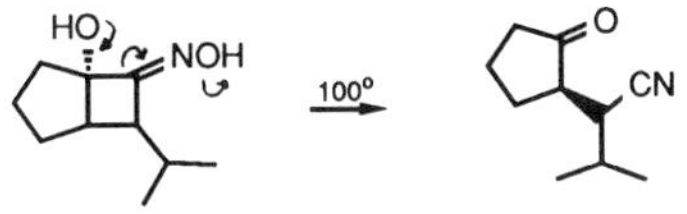

A plan had been delineated for the synthesis of dendrobine [P.J. Connolly, 1985] in which the heterocycle would be created from an α-hydroxy oxime via fragmentation and reductive recyclization. This scheme or a modified route has not yet been realized.

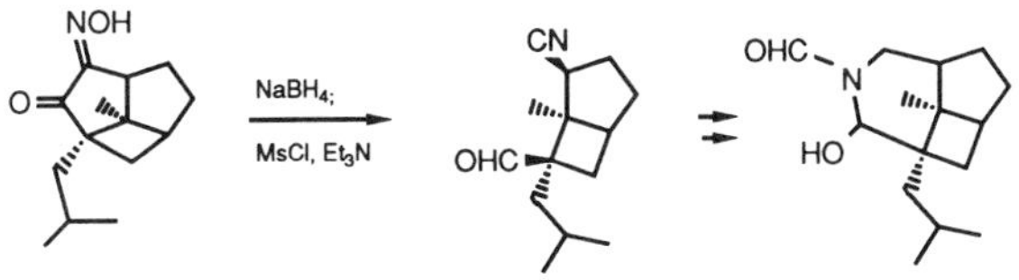

By means of a Beckmann fragmentation the structural relationship of ibogaine, ibogamine, and tabernanthine was established [M.F. Bartlett, 1958].

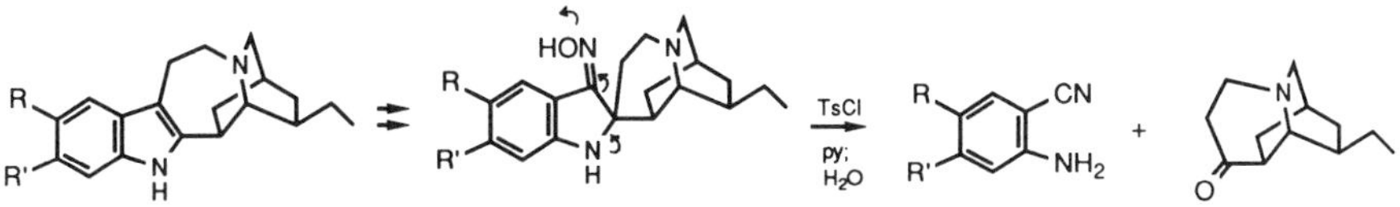

R=R'=H ibogamine
R=OMe, R'=H ibogaine
R=H, R'=OMe tabernanthine

Under the Wolff–Kishner reduction conditions substituted 3-oximino-4-piperidinones afforded 2-pyrrolidones [Bosch, 1983b]. A Beckmann fragmentation follows the reduction (partial or complete) and the resulting iminium nitriles are hydrolyzed and cyclized.

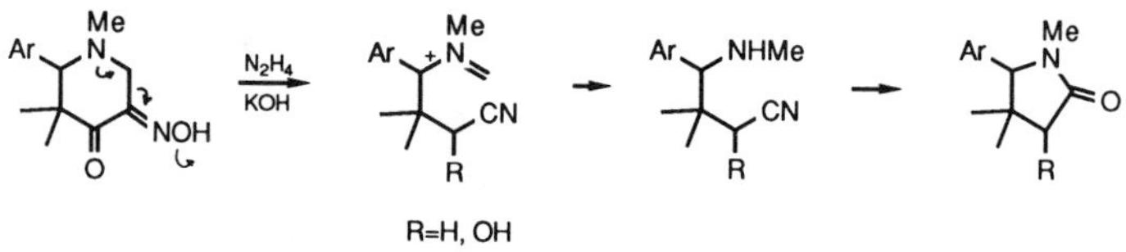

The last stage of the double bond cleavage process involving a combination of an oxidizing agent and trimethylsilyl azide [Zbiral, 1970a,b] resembles the Beckmann fragmentation.

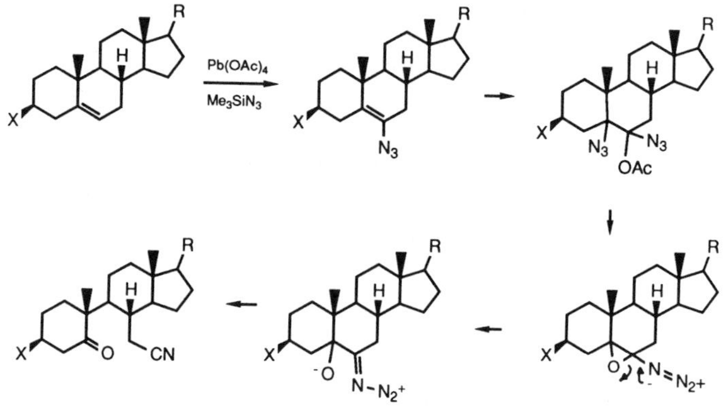

## 8.9. FRAGMENTATION OF OTHER d3d SYSTEMS

Divalent sulfur may behave as a donor, especially when it is conjugated with a good nucleofugal group at the $\gamma$ atom [Little, 1978; Traynelis, 1978].

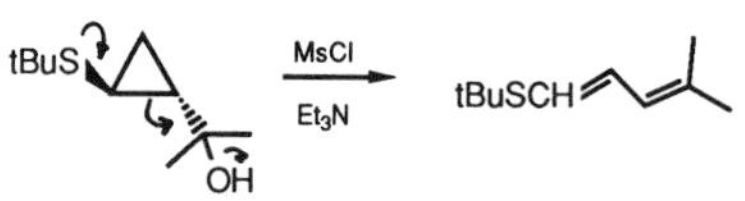

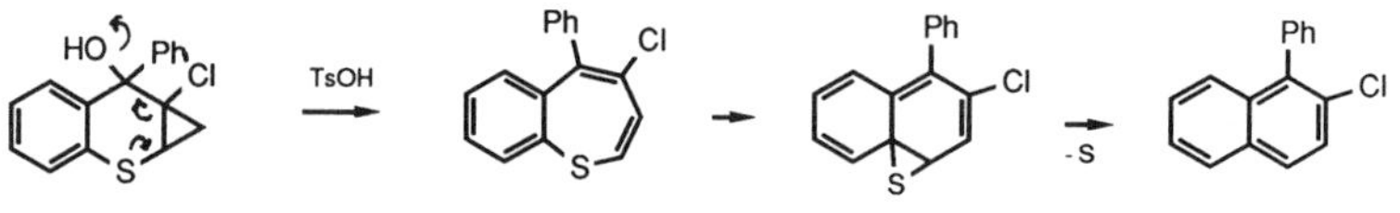

The assisted fragmentation undergone by $\alpha$-sulfenyl ketoximes has found use in a synthesis of corynantheine [Autry, 1968] from yohimbone.

Carbon-bound halogen atoms are also capable of playing the donor role [Heine, 1981]. Interestingly, complete role reversal of the oxygen and chlorine atoms in a dichlorocyclopropanol is attained under oxidative conditions [Macdonald, 1978a; Klehr, 1975; Torii, 1978]. The formation of 2-chlorobenzaldehyde from tropolone on its reaction with thionyl chloride [Abadir, 1952] could also involve chlorine-assisted expulsion of an oxy nucleofuge.

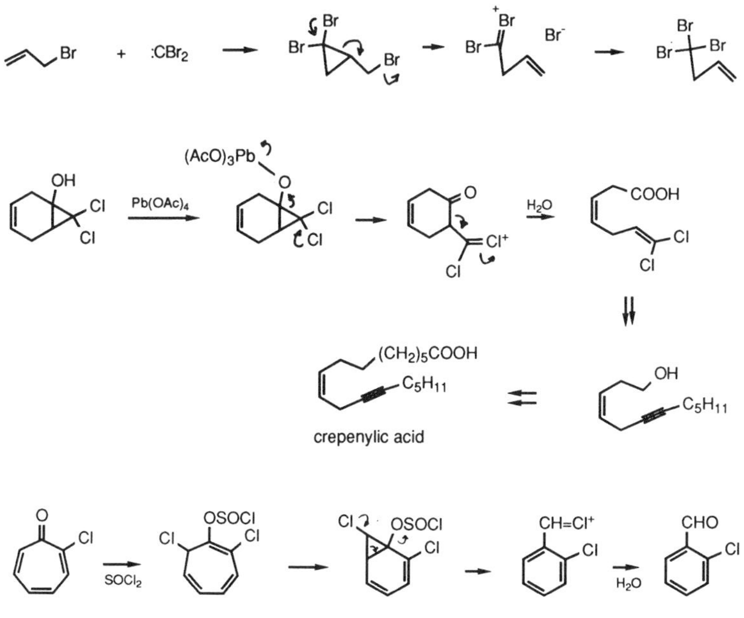

Br + :CBr₂ → … Br⁻ → …

Pb(OAc)₄ … (AcO)₃Pb … H₂O … COOH … Cl … Cl

(CH₂)₅COOH … C₅H₁₁ … crepenylic acid … OH … C₅H₁₁

SOCl₂ … OSOCl … CH=Cl⁺ … H₂O … CHO … Cl

The fluoride ion-induced fragmentation of α-(*N*-siloxy)anilino ketones [Ohno, 1988] is analogous to dioxetane decomposition.

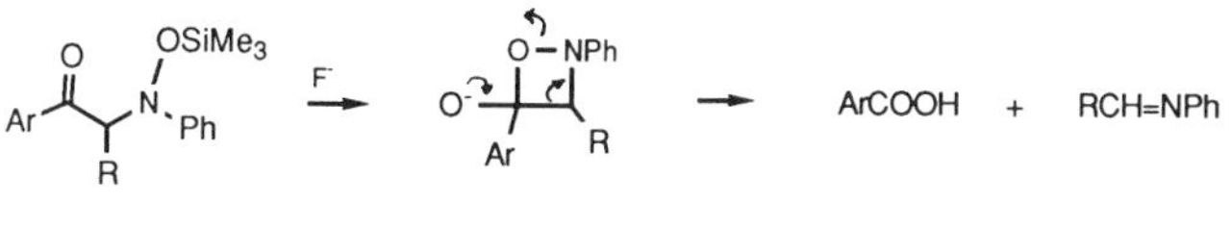

OSiMe₃ … F⁻ … O–NPh … → ArCOOH + RCH=NPh

The best known **d3d** fragmentation processes involving aromatic systems are the benzyne-forming reactions [Stiles, 1960; LeGoff, 1962].

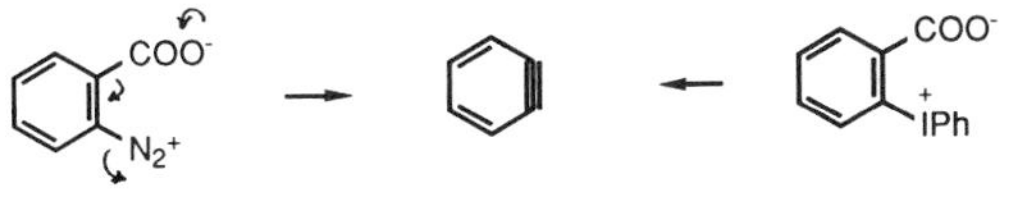

COO⁻ … N₂⁺ → … ← … COO⁻ … IPh

α-Bromobenzyloxyfluorodiazirine decomposes rapidly in the presence of nucleophiles such as alcohols and amines [Moss, 1991]. Two possible reaction pathways involve a **d3d** fragmentation mode after displacement of the bromine atom. However, it is not known whether the fragmentation is preceded by carbene formation.

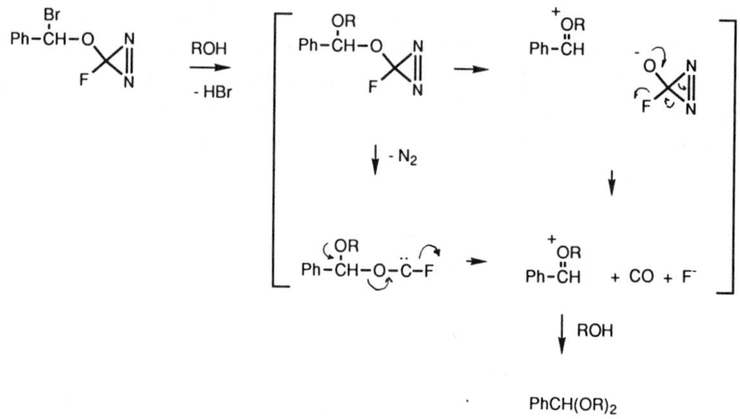

The heterocyclic moiety of dihydrobenzotriazinone is destroyed on its reaction with phosphorus pentachloride [Buckley, 1956]. 4-Chlorobenzotriazine is prone to fragmentation.

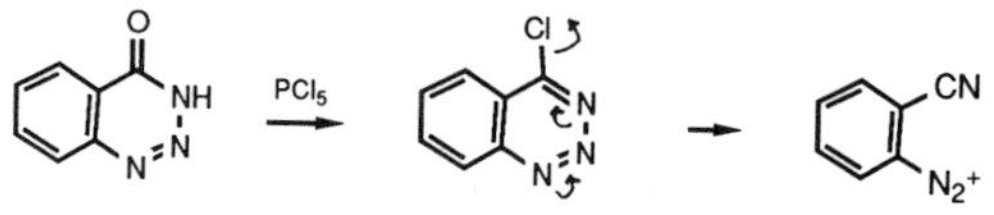

Vilsmeier-type reagents are formylating agents prepared from dimethylformamide and dehydrants such as thionyl chloride and phosgene [Bosshard, 1959]. The formation of *N*,*N*-dimethylformimidoyl chloride derivatives proceeds via a fragmentation pathway.

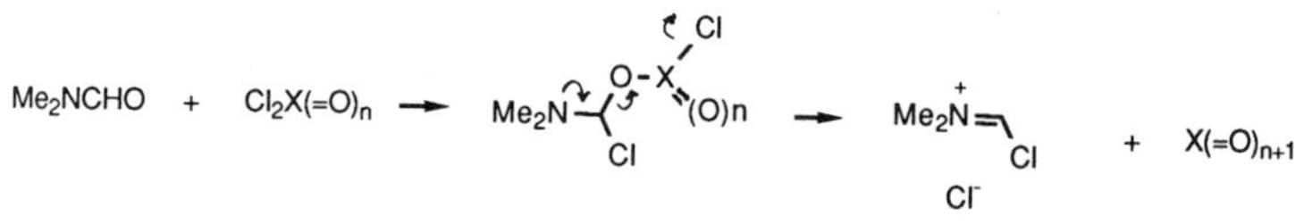

Both the alkaline decomposition of *N*-tosyl-α-amino acid azides to furnish the aldehydes, *p*-toluenesulfonamide, cyanate ion, and nitrogen [Beecham, 1957], and the production of metaphosphoric esters by admixture of diaryl pyrophosphates and carbodiimides [Schofield, 1961; Cramer, 1962] involve a **d3d** fragmentation.

The Wolff–Kishner reduction of α-donor substituted ketones leads to alkene products [Leonard, 1955]. It is possible that the reaction course entails fragmentation of the aza tautomer of the hydrazone. The decomposition of α,β-epoxy hydrazones is known as the Wharton rearrangement [Wharton, 1961].

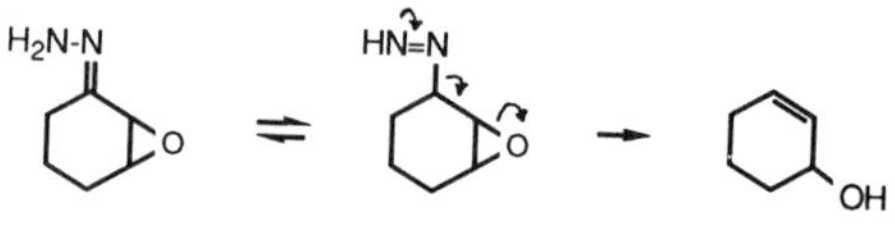

# 9

# FRAGMENTATION OF d4a AND MORE EXTENDED SYSTEMS

## 9.1. d4a SYSTEMS

The few examples of this class of fragmentation deal with ring unraveling from incipient carbocations. Thus, photoisomerization of a cyclopentenone [Hart, 1980] proceeds via bond migration and epoxide opening which generates a $\delta$-oxycarbenium ion. Fragmentation of the latter species affords an opportunity to reach an aromatic product by further deacylation.

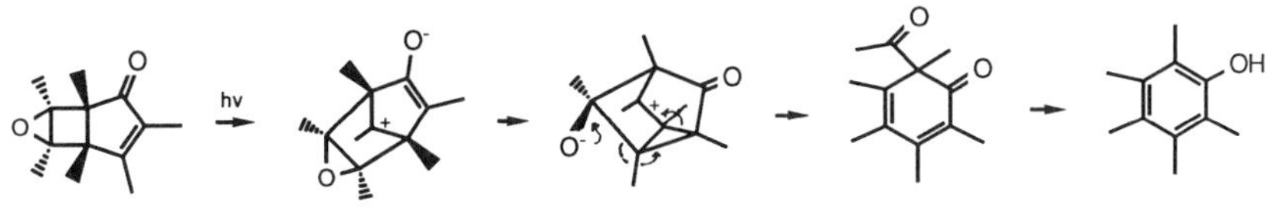

Three products are obtained from 5,5-dimethyl-1-methylsulfonyl-2-phenyl-3,4-epoxycyclopentene by treatment with boron trifluoride etherate [J.A. Miller, 1989]. The most unusual one is an unsaturated aldehyde.

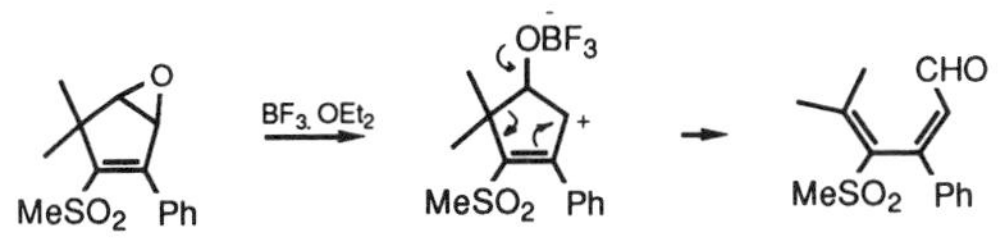

Acid treatment of a secohomocubanedione monoketal generates a pentalenone [Sakkers, 1979]. A series of Wagner–Meerwein rearrangements gives rise to an intermediate in which the cationic center is five bonds away from the ketal oxygen atoms, inducing fragmentation to follow.

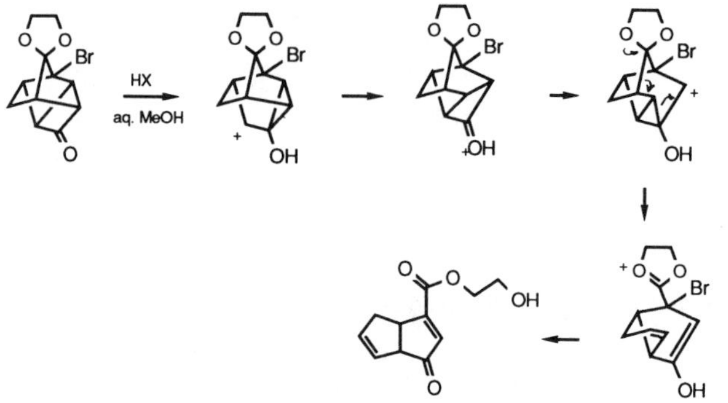

## 9.2. d4a=d SYSTEMS

### 9.2.1. O4C=O and O4C=N Arrays

Two retro-Michael reaction paths are available to the adduct of dimedone and methyl vinyl ketone. Accordingly, heating of the adduct with a secondary amine leads to bisdimedone [Greenhill, 1979].

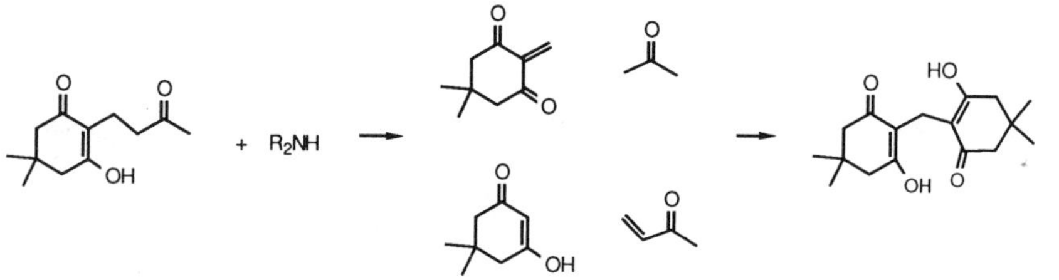

3-Cyclopentene-1,2-dione dimers obtained from $\gamma$-substituted crotonic esters and dimethyl oxalate are very reactive toward nucleophiles. Apparently the dimers are in equilibrium with monomeric species even if the latter cannot be detected [R.T. Brown, 1984]. Regardless the possible occurrence of the reactions in the seco-dimer stage the fragmentation is crucial.

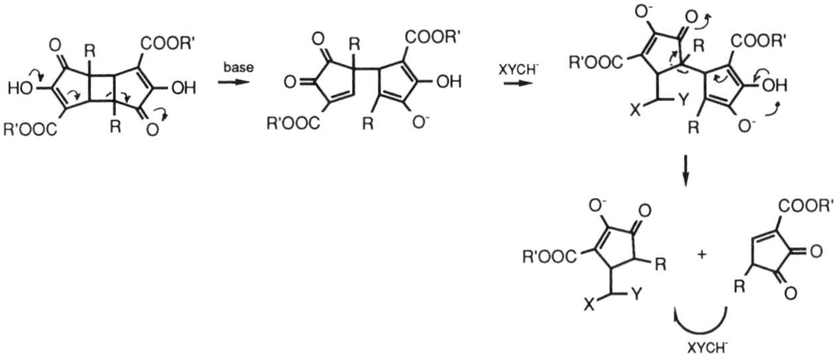

An attempt to assemble the carbon skeleton of seychellene [Jung, 1980a] by a version of intramolecular Michael reaction proves disappointing, the low yield (5%) of the tricyclic diketone is due to a more favorable retro-Michael scission.

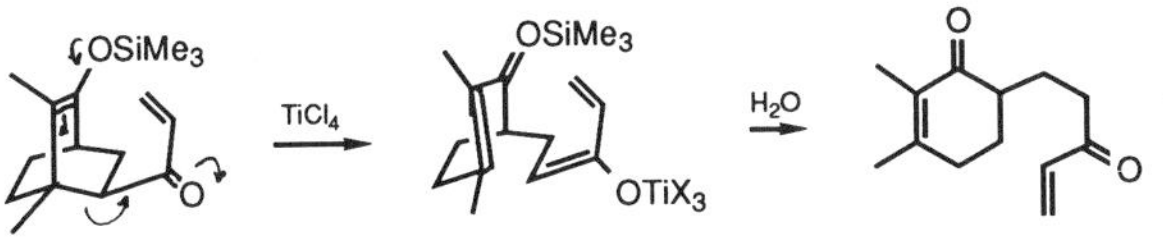

Vinylogous retro-aldol reactions may occur in $\delta$-oxy-$\alpha,\beta$-unsaturated carbonyl compounds, constituting a subclass of the **d4a=d** fragmentation.

The formation of anhydrotaxininol from taxinine on treatment with ethanolic KOH involved a deep-seated rearrangement [Taga, 1964; Y. Yamamoto, 1964]. The central ring of taxinine suffers a vinylogous retro-aldol cleavage, followed by recyclization (aldolization) at an alternative site. Elimination of acetic acid and hydrolysis of the cinnamic ester complete the transformation.

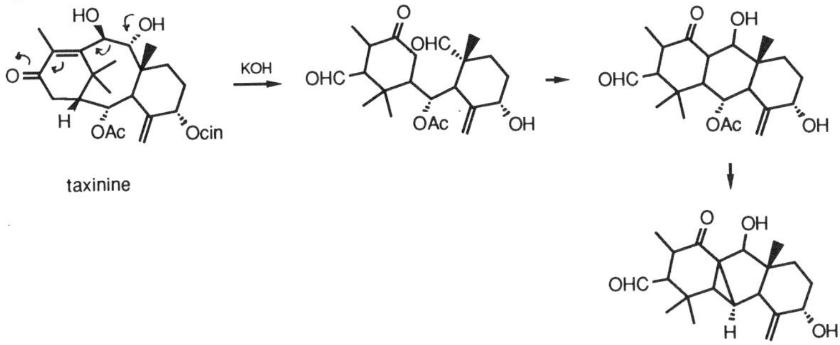

taxinine

$\beta$-Ionone epoxide rearranges in aqueous formic acid to several products of different skeletons [K.L. Stevens, 1975]. An acyclic trione may have formed from the hydrate via a fragmentation initiated by protonation of the enone.

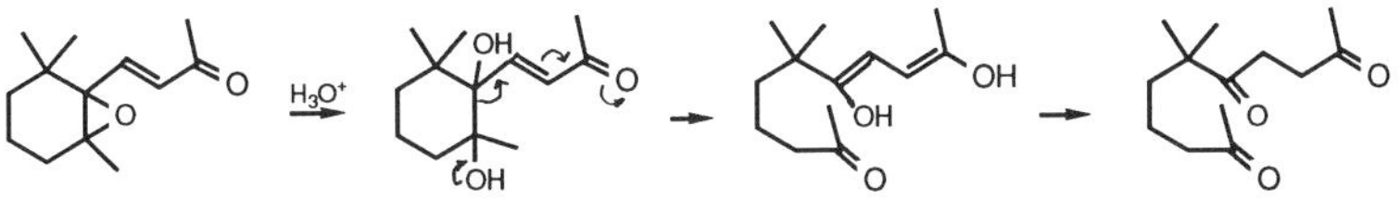

In a synthesis of bilobanone [Büchi, 1969] hydration of the triple bond gives certain amounts of the undesired ketone which undergoes allylic oxidation and expulsion of formaldehyde.

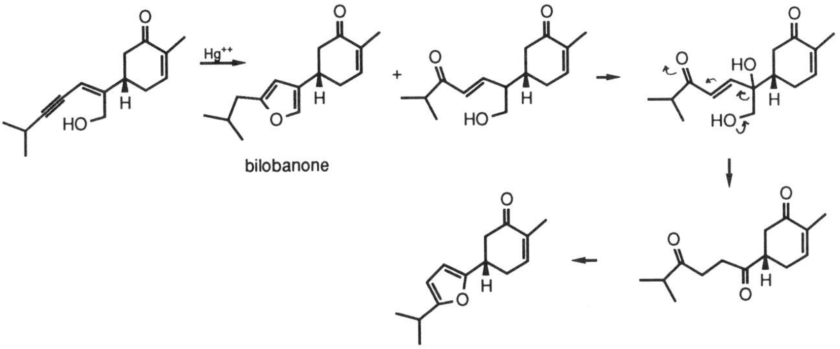

bilobanone

A photoisomer of 4a*H*-pyrosantonin furnishes a bicyclic diketo ester on hydrolysis [Ishikawa, 1973] which may be further degraded to a cyclopentenone.

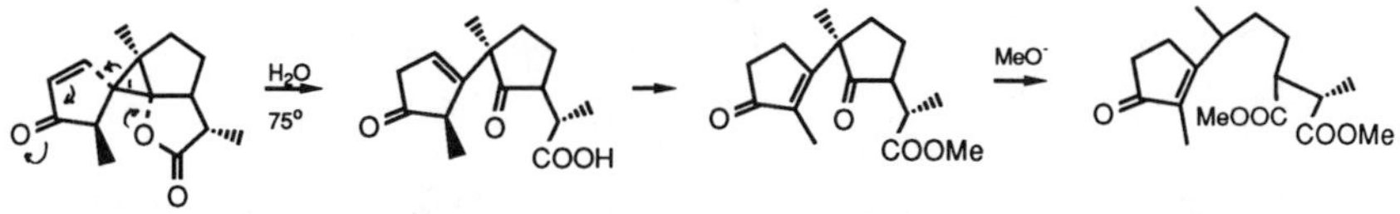

$H_2O$, 75°; COOH; COOMe; MeO⁻; MeOOC; COOMe

Particularly interesting examples of the fragmentation of $\delta$-hydroxy $\alpha,\beta$-unsaturated carbonyl compounds are derivatives of the Wieland–Miescher ketone [J. Cocker, 1957] in which the fragmented species may partake intramolecular reorganization including aldol [Kalyanasundaram, 1980], Michael [Swaminathan, 1970; Dauben, 1977b], and double Michael reactions [Rao, 1982].

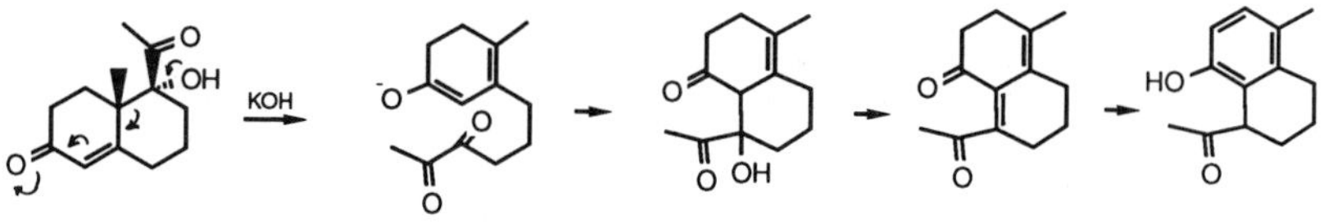

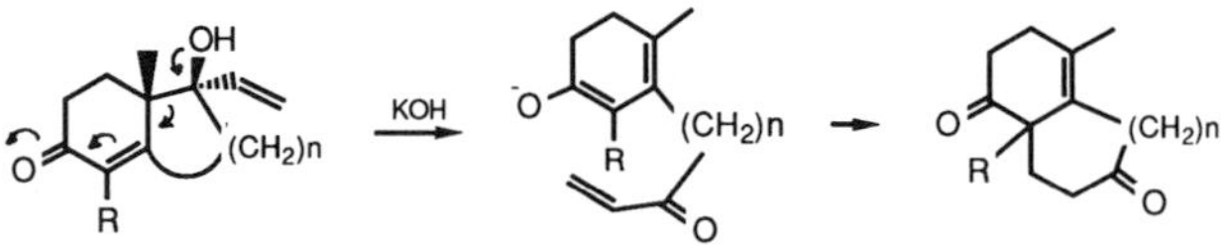

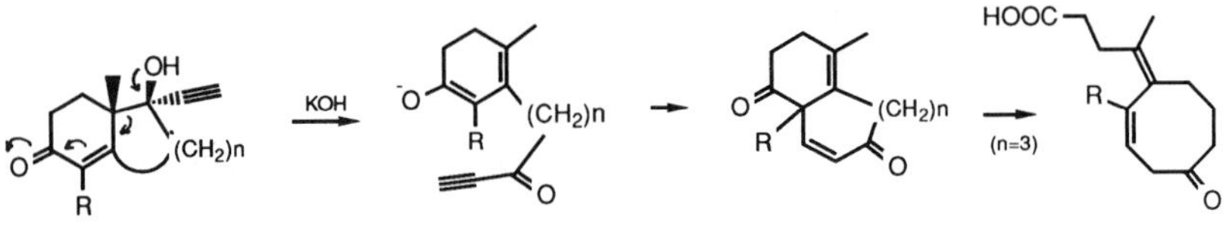

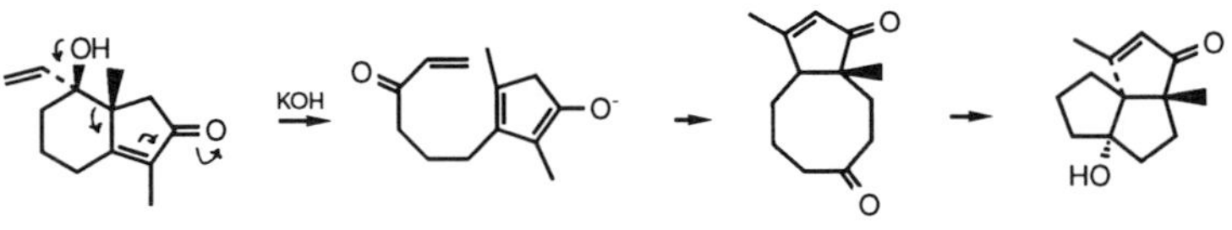

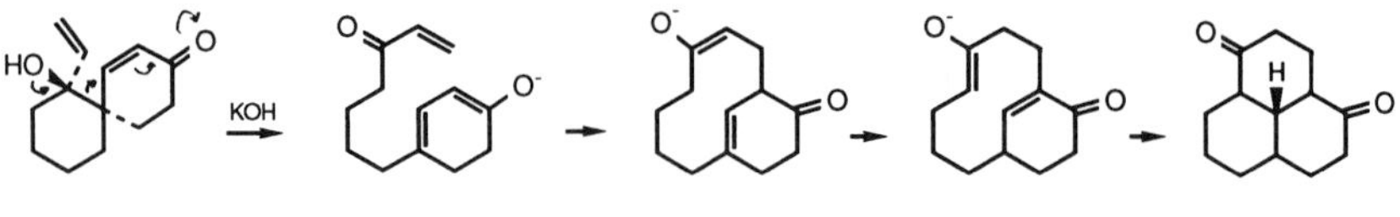

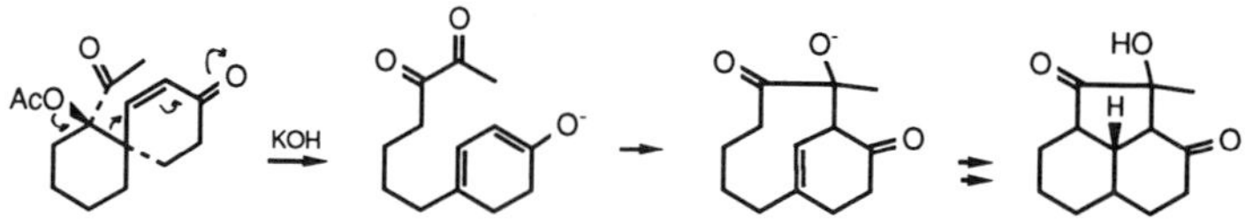

OH; KOH; OH; HO

OH; $(CH_2)n$; R; KOH; $(CH_2)n$; R; $(CH_2)n$; R

OH; $(CH_2)n$; R; KOH; $(CH_2)n$; R; $(CH_2)n$; R; (n=3); HOOC; R

OH; KOH; HO

HO; KOH; H

AcO; KOH; HO; H

An isomeric bicyclic hydroxy ketone has been converted into a bridged tricyclic ketol [Raju, 1980] as a result of succeeding Michael and aldol condensations.

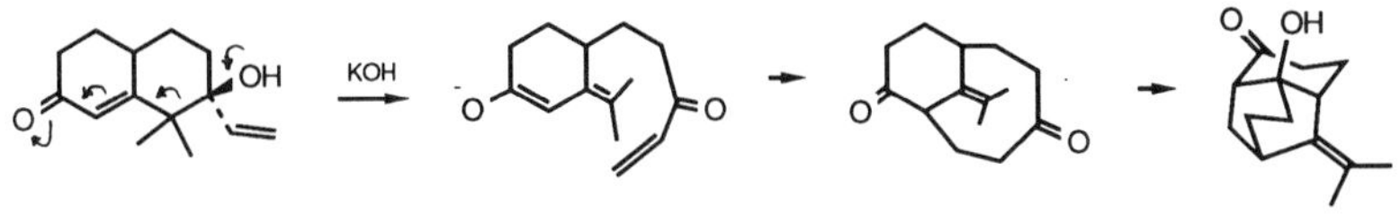

With the cyclohexenone moiety at a higher oxidation level (e.g. as cyclohexadienone [Waring, 1985; MacMillan, 1972] or enedione [Valenta, 1961]) the fragmentation results in phenolic products.

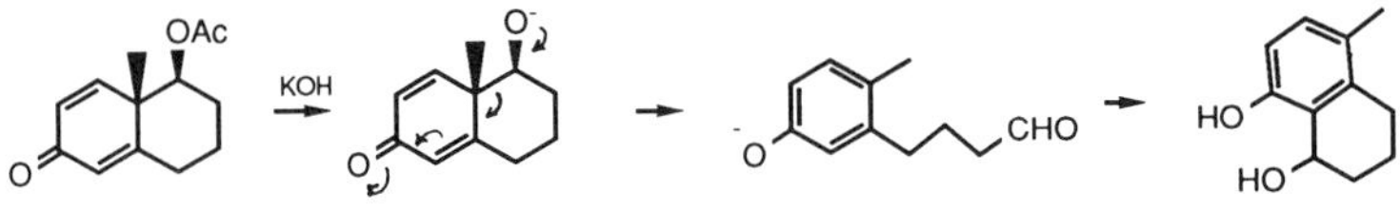

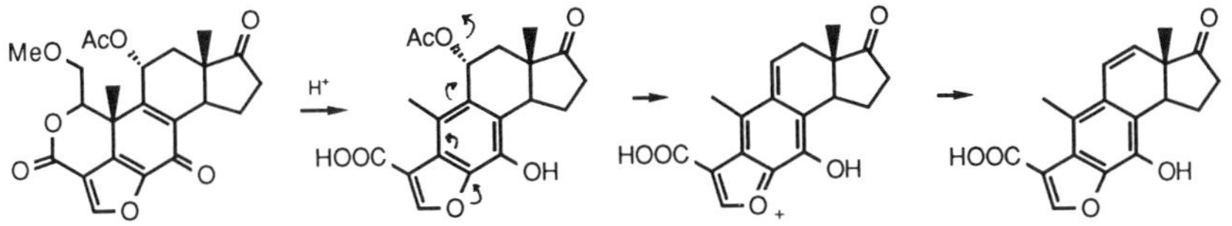

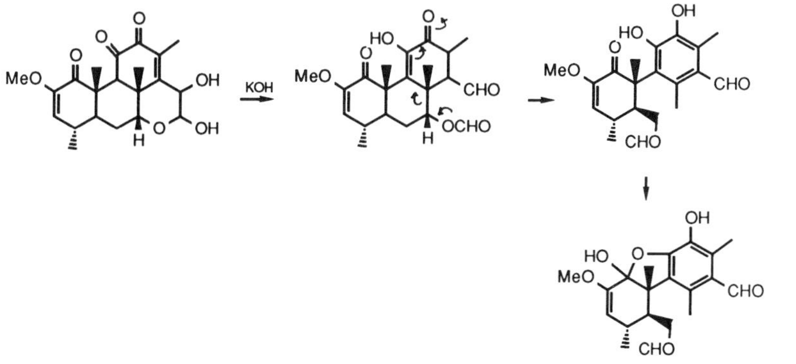

This process must be involved in the enzymatic transformation of androst-4-ene-3,17-dione [Dodson, 1961]. Introduction of a double bond and a C-9 hydroxyl prepares the molecule to fragment. (Analogous fragmentation has been initiated by a carbanion via stepwise reduction of a 11α-xanthate by samarium(II) iodide [Ananthanarayan, 1982].)

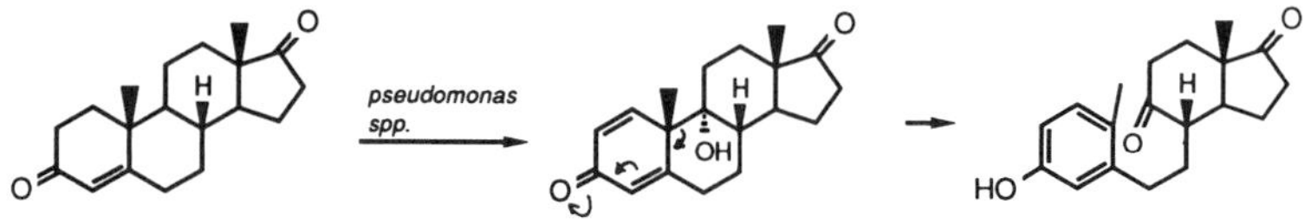

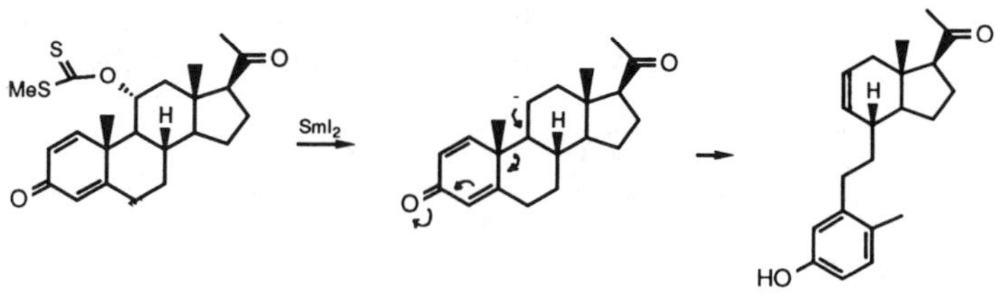

MeS, S, O, H, O, SmI$_2$, H, O, H, O, HO

In the course of a mycinolide-V synthesis [Hoffmann, 1989] the release of an α-phosphonyl ketone from a lactol was complicated by a **d4a**=**d** fragmentation.

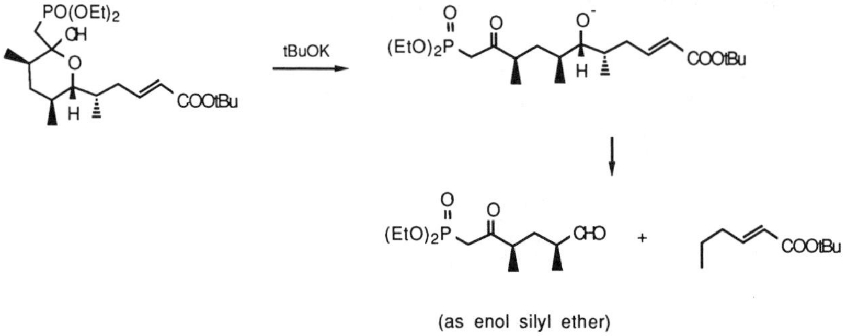

PO(OEt)$_2$, OH, O, H, COOtBu, tBuOK, (EtO)$_2$P, O, O, O$^-$, H, COOtBu, (EtO)$_2$P, O, O, CHO, +, COOtBu

(as enol silyl ether)

The ring contraction reactions of 2-substituted 6-methoxy-α-tetralones [Mincione, 1988] and the dibromides or bromohydrins of the corresponding dihydronaphthalene and benzosuberene [Mandal, 1988] are fascinating. The criticality of the methoxy substituent suggests the involvement of hydride shift, formation of spirocyclic intermediates and a subsequent fragmentation.

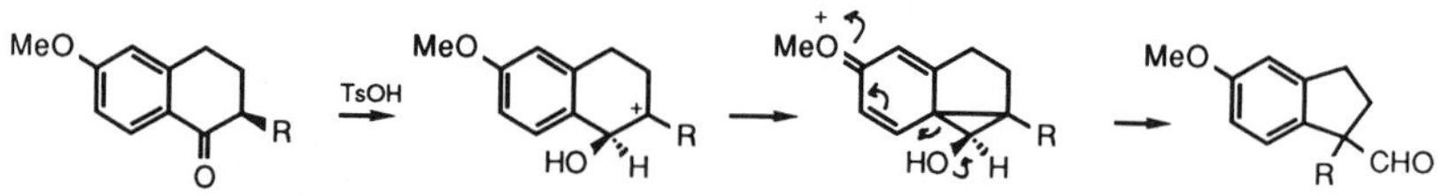

MeO, R, O, TsOH, MeO, +, R, HO, H, MeO, R, HO, H, MeO, R, CHO

During a synthesis of aklavinone [Confalone, 1981] an intramolecular Friedel–Crafts acylation led to an undesired quinone. A spirocyclic species undoubtedly mediated the generation of a rearranged acylium ion via fragmentation.

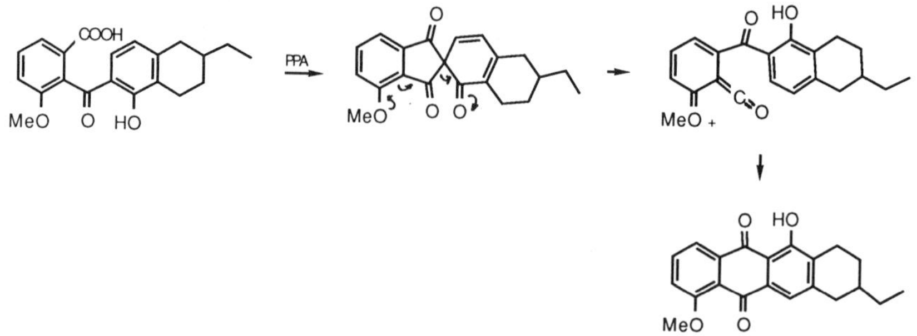

COOH, MeO, O, HO, PPA, O, MeO, O, O, O, HO, MeO, C, O, O, HO, MeO, O

Formation of spirocyclic intermediate by *p*-methoxy participation is also evident in the peracid reaction of an enol ether which served as a model for synthesis of zearalenone [Immer, 1968]. This participation and directed fragmentation completely defeated the approach.

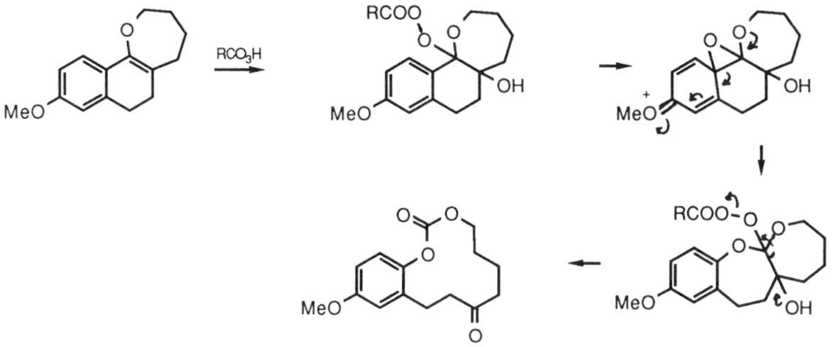

The Diels–Alder reaction of juglone with an α-pyrone forms a 2:1 adduct [Jung, 1984]. Fortunately for the purpose of anthracycline synthesis this adduct may be decomposed to give the desired substance on treatment with aqueous sodium carbonate.

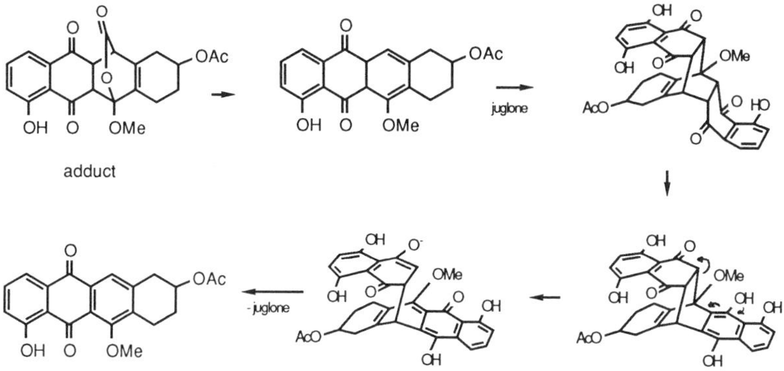

Tannin-protein complexes are degraded by acids [Beart, 1985]. Formation of quinonemethide intermediates from 4,7-flavandiol in acidic solutions has been detected by ultraviolet spectroscopy in the presence of thiophenol [Attwood, 1984].

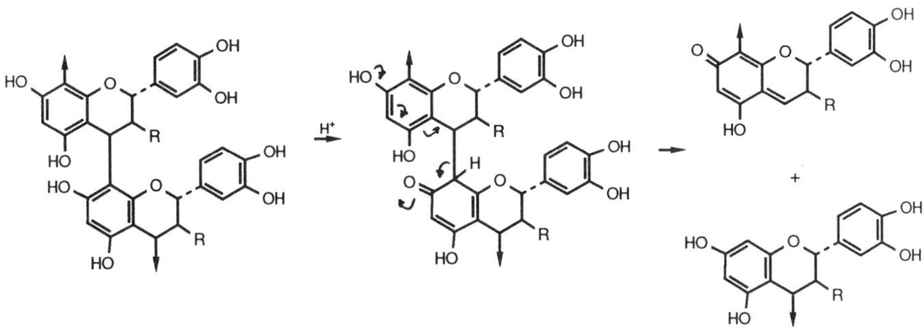

α′-Oxygenated cyclopentenones are members of the **d4a=d** subfamily and several examples of their fragmentation have been documented. It accounts for the generation of the rearranged cyclohexenedione monoketal product from photolysis of 2-methyoxy-4-methyl-4-phenyl-2,5-cyclohexadienone [Matoba, 1987]. A synthesis of the rubrolone chromophore [T.R. Kelly, 1986] also relied on this vinylogous retroaldol reaction.

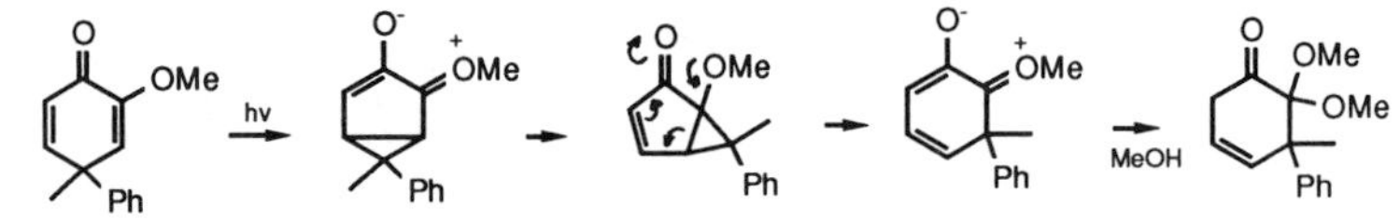

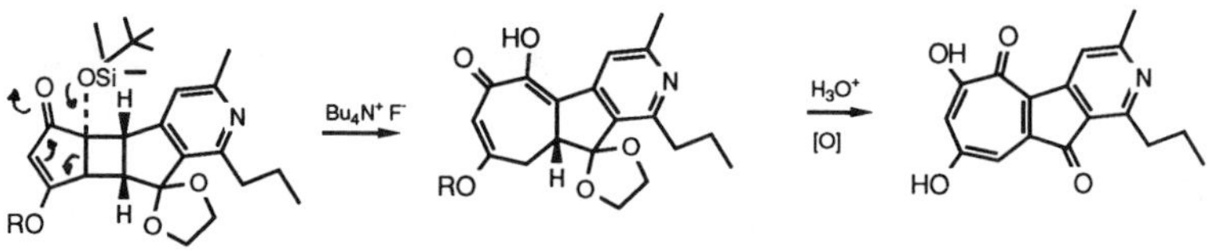

A photocycloadduct of 5-methyl-2,3-dihydrofuran and ethyl 2,3-dioxo-5-phenylpyrroline-4-carboxylate fragments in situ because of the highly strained *trans-syn-cis* stereochemistry [Sano, 1987].

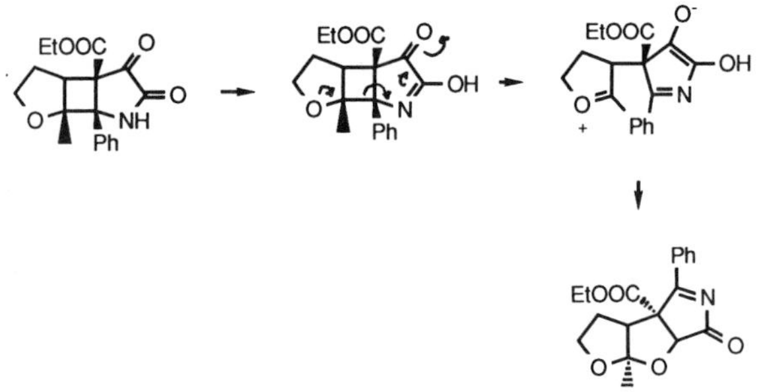

A useful synthesis of 3-formyl-2-cyclohexenones is based on photocycloaddition of methylfuranone with alkenes [Ogino, 1976]. After α-acetoxylation the adducts are treated with a base to induce hydrolysis, fragmentation, and aldolization. The fragmentation is strain-assisted.

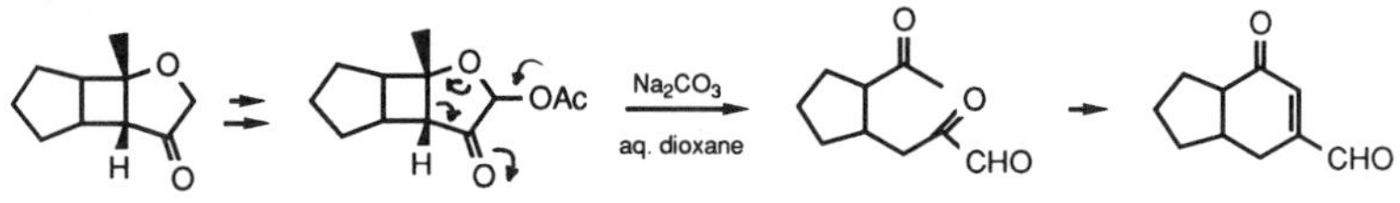

2*H*-Chromenes undergo a Ti(IV)-catalyzed [2 + 2]cycloaddition with 2-alkoxy-1,4-benzoquinones. Exposure of these adducts to protic acids leads to oxygenated pterocarpans [Engler, 1989]. Some of the adducts rearrange in situ which involves fragmentation and recombination steps.

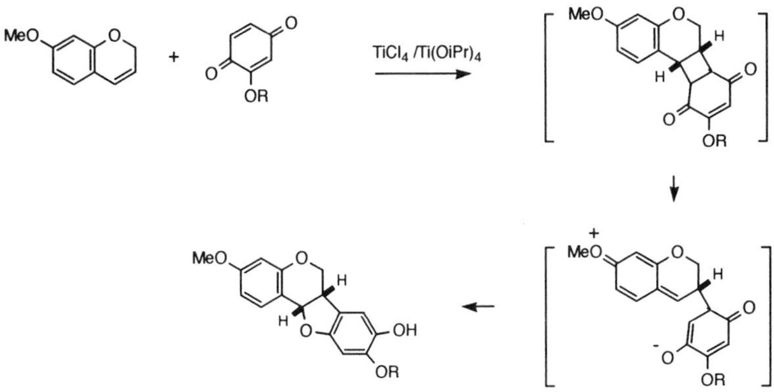

The very congested architecture of lambicin is provided an opportunity to relax through oxidation [Corbella, 1969]. Cleavage of the lactol and the cyclopentane ring becomes feasible in the ketone, and the final product is relieved of three 1,3-diaxial interactions.

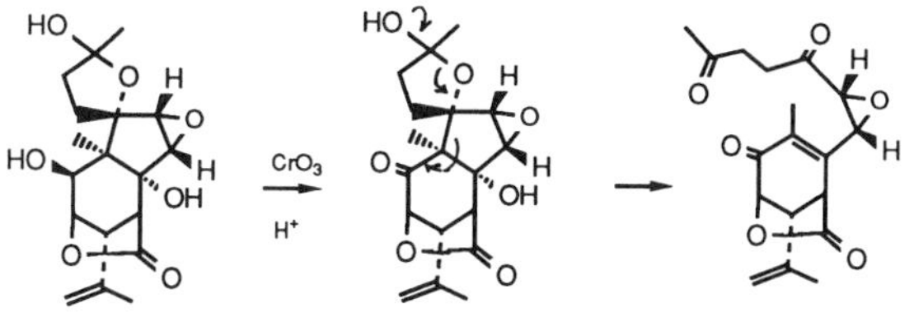

lambicin

Reductive *C*,*O*-dibenzoylation occurs when a nonenolizable α-diketone is treated with benzil in the presence of sodium phenoxide [Rubin, 1979a]. A proposed mechanism indicates formation of a ketal intermediate which fragments.

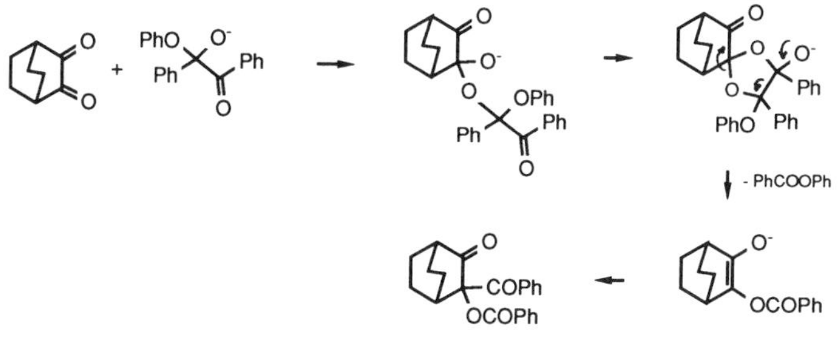

Chromium(II) reduction of a trienetetraone tosylate derived from cucurbitacin-A affords a B-homosteroid [Barton, 1969] due to the presence of carbonyl groups at C-2 and C-11. Consequently the cyclopropyl ketone finds a fragmentation path for strain relief.

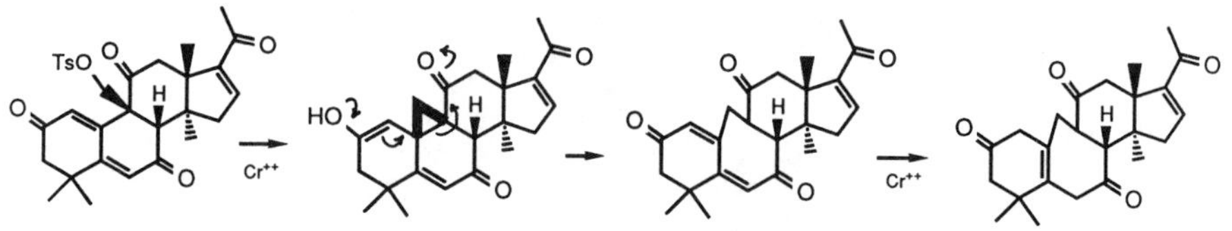

An excellent one-step elaboration of spirocycles entails intramolecular carbenoid reaction of *p*-substituted phenols [Iwata, 1977]. The constitutions of the resulting cyclohexadienols are such that fragmentation is ineluctable. This process has been applied to synthesis of α-chamigrene [Iwata, 1979], a rearranged bromochamigrene [Iwata, 1987], and solavetivone [Iwata, 1981].

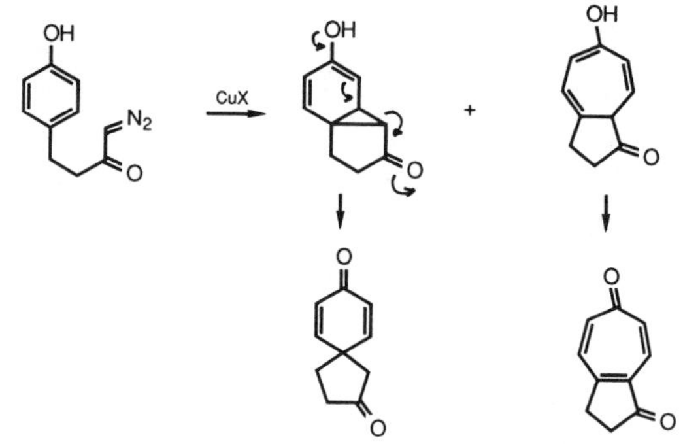

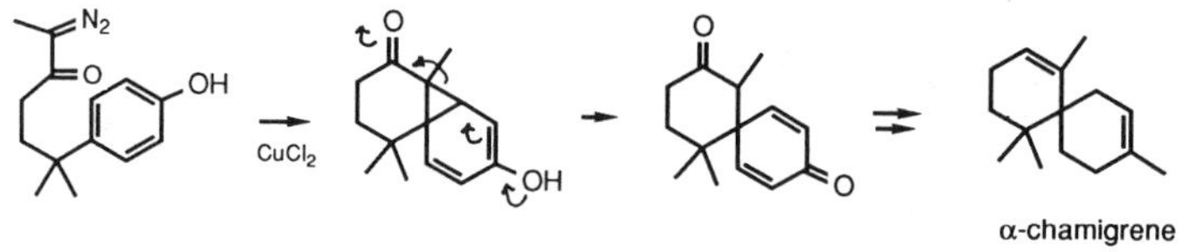

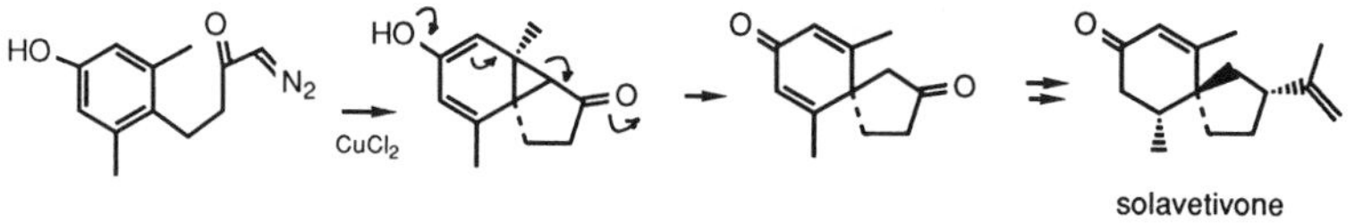

α-chamigrene

solavetivone

The analogous reactions on furan derivatives give rise to very different products. Thus, a cyclopentenone [Nwaji, 1974] and (2*Z*,4*E*)-diene-1,6-aldehydroketones which are useful precursors of 8- and 9-HETEs and (5*S*,12*S*)-diHETE [Adams, 1984a,b] have been acquired by this method.

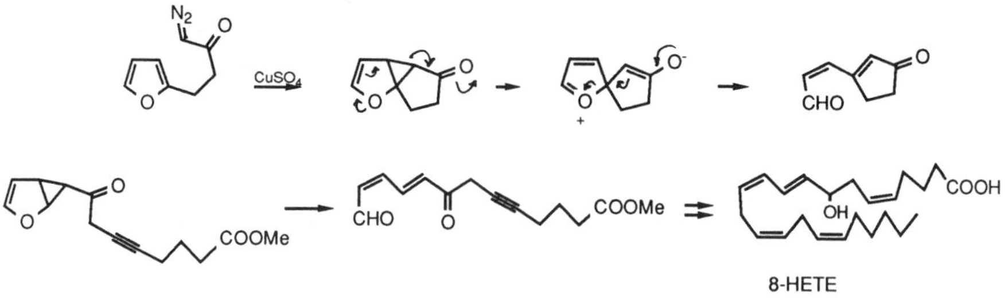

8-HETE

Enols or enol ethers generated in situ that contain the same structural features as the furan adducts are also prone to fragment. Exploitation of this process in synthesis has been frequent, as witnessed in routes to β-vetivone [McCurry, 1971], α-chamigrene [White, 1974], 12,13-epoxytrichothec-9-ene [Masuoka, 1976], and schelhammericine [Y. Tsuda, 1985]. The strictly analogous fragmentation of cyclopropylcarboxylic esters has been found in syntheses of trichodermin [Colvin, 1971] and 9-pupukeanone [Schiehser, 1980].

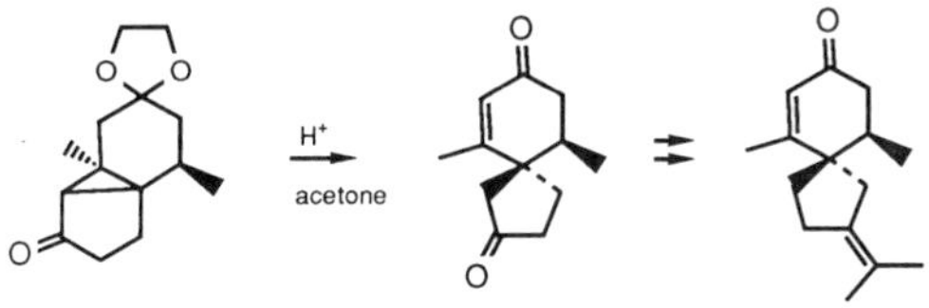

H⁺, acetone

β-vetivone

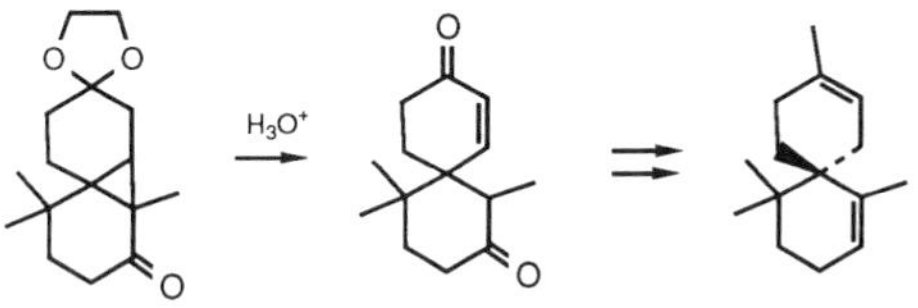

$H_3O^+$

α-chamigrene

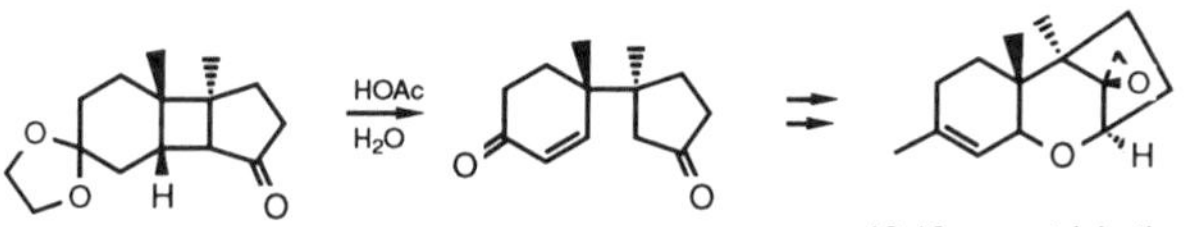

HOAc, $H_2O$

12,13-epoxytrichothec-9-ene

$H_3O^+$

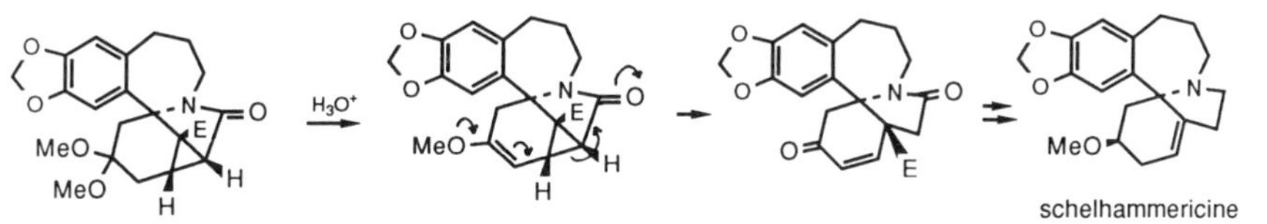

schelhammericine

E= COOMe

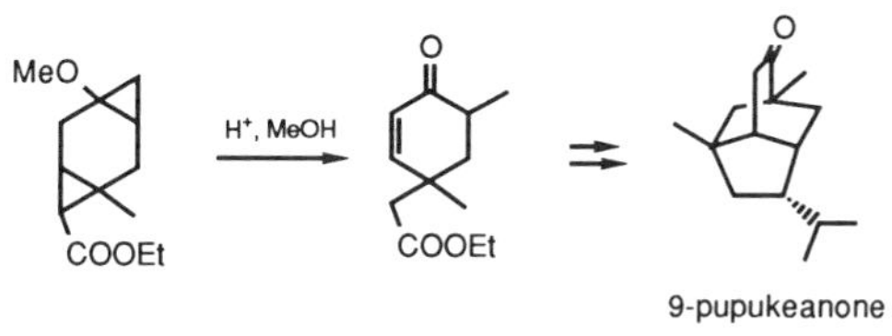

H⁺, MeOH

9-pupukeanone

Other carboxyl supported fragmentations include the generation of trimethyl trichlorohemimellitate by refluxing 5,5-dimethoxy-1,2,3,4-tetrachlorocyclopentadiene with dimethyl acetylenedicarboxylate [Diekmann, 1963], degradation of a cage triketone to a benzocyclobutenetricarboxylic acid [Marchand, 1973], and a process of chain building from small ring precursors [Trost, 1977a].

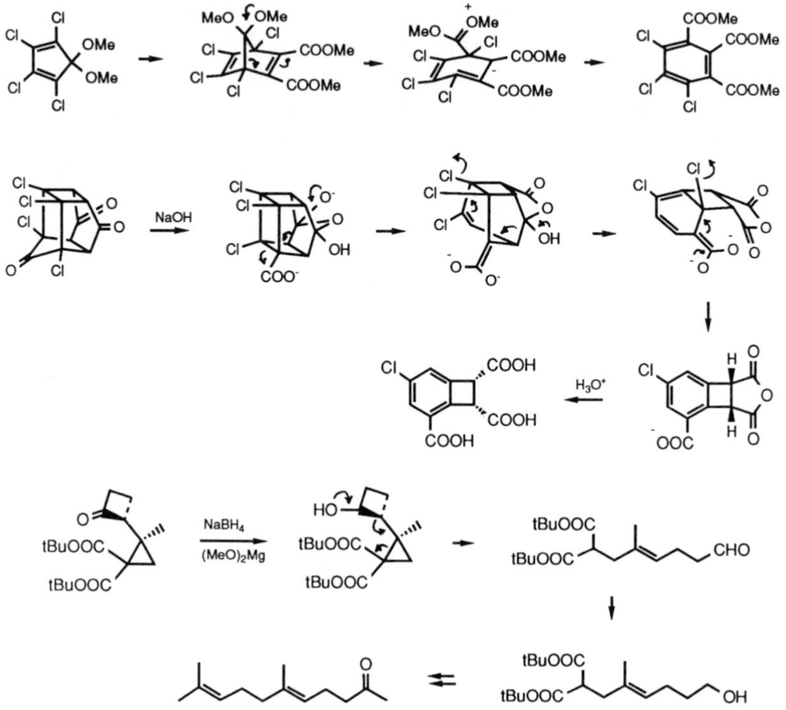

There are several unusual decarboxylation reactions that conform to the **d4a=d** regime. Aflatoxin-$B_1$ follows two parallel degradation pathways on treatment with ammonium hydroxide [Simpson, 1989]. Excision of the alicyclic moiety may have been accomplished by Michael addition, hydrolysis of the lactone, and decarboxylation in the keto form of the phenol. Alternatively, a retro-Mannich reaction could be responsible for the C—C bond cleavage.

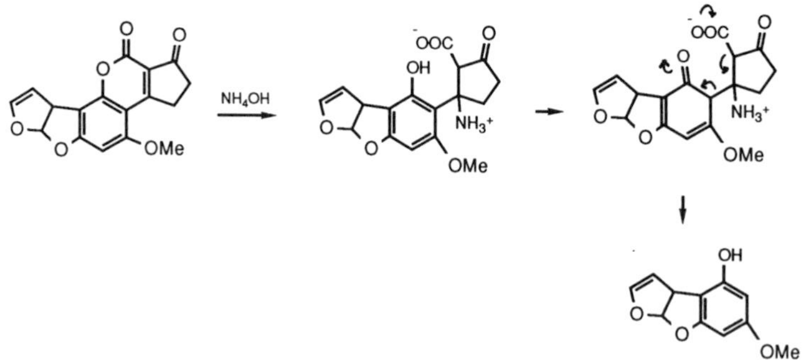

A base-catalyzed deep-seated rearrangement of a perhydroindanedione has been reported [Fayos, 1977]. The overall reaction consists of ring expansion, sulfonyl migration, and decarboxylation.

Many mechanisms that differ in small details in the order of events may be formulated. However, the decarboxylation step is a **d4a=d** process, be it occurring after the ring expansion or concomitant with the fission of the cyclopropane ring in the tricarbocyclic intermediate.

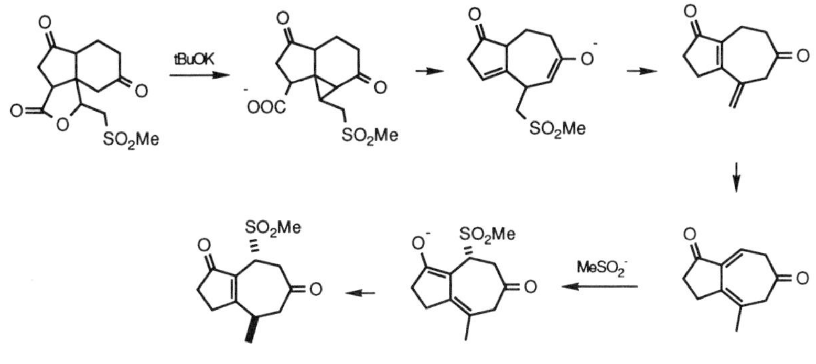

The Strecker degradation of α-amino acids to aldehydes proceeds via decarboxylation. The use of Corey's procedure to degrade a cephalosporin derivative has led to a benzoxazole [Vander Zwan, 1978], due to internal trapping of the imine and oxidation of the dihydrobenzoxazole by the quinone reagent.

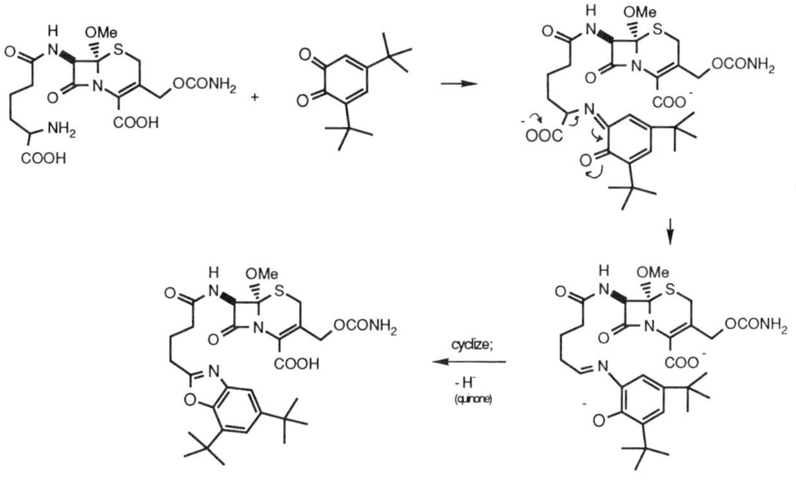

An imidol derived from chilenine undergoes ring closure and decarboxylation to afford nuevamine [Valencia, 1984]. This transformation probably represents the late stage of a catabolic pathway of berberine (via prechilenine and chilenine).

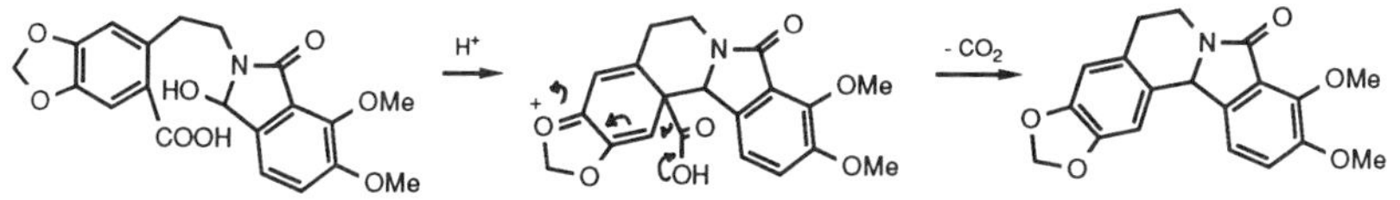

The treatment of 3-(1,2-dimethylindol-3-yl)propanoic acid with trifluoroacetic anhydride results in the formation of much 1,2-dimethyl-3-(trifluoroacetyl)-

indole [Bailey, 1983]. The loss of the sidechain may be explained by fragmentation from the indolenium species.

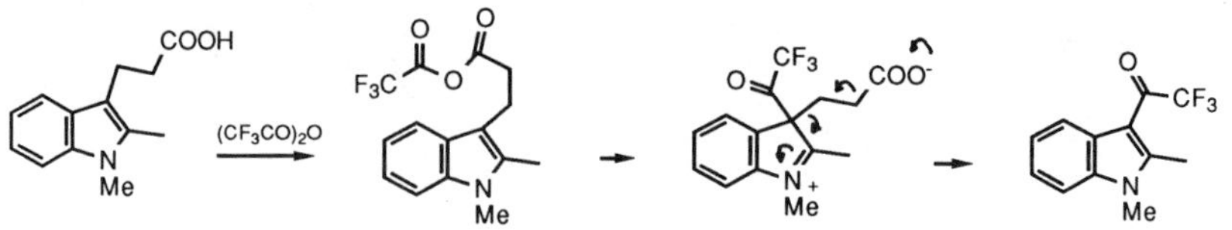

Irradiation of 5-aminotropolone in dilute mineral acid gives 2-amino-5-oxocyclopent-1-enylacetic acid [S. Seto, 1969]. Hydrate of a bicyclic ketone is expected to fragment in view of strain.

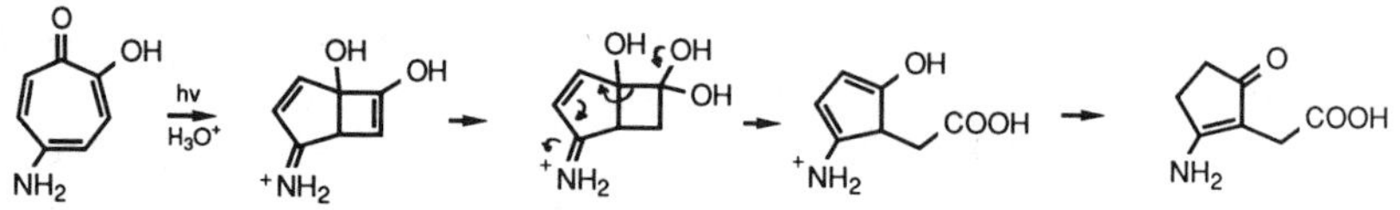

A remarkable rearrangement occurs after cycloaddition of 2,4-dimethyl-2-cyclobutenone to 1-(diethylamino)propyne [Ficini, 1977]. Formation of a tricyclic zwitterion preceding the fragmentation would undoubtedly be facilitated by the proximity of the enamine double bond to the carbonyl group.

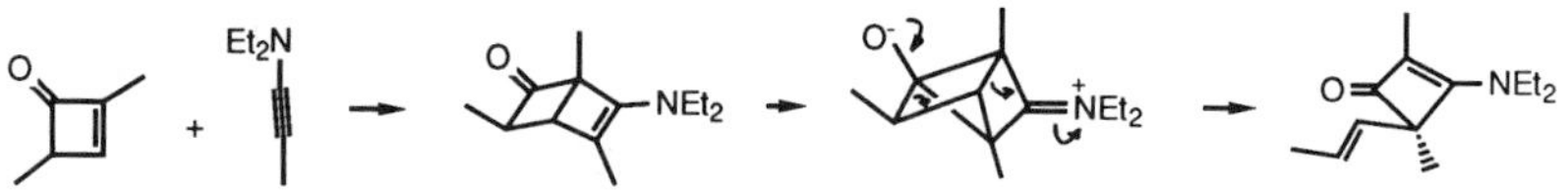

4-(Dimethylamino)benzaldehyde is a good one-carbon transferring agent. Its reaction with an organometallic reagent and then an arenediazonium salt constitutes a method of aldehyde synthesis [Stiles, 1960b]. Modification of the reaction sequence (before reaction with the diazonium salt) permits synthesis of ketones.

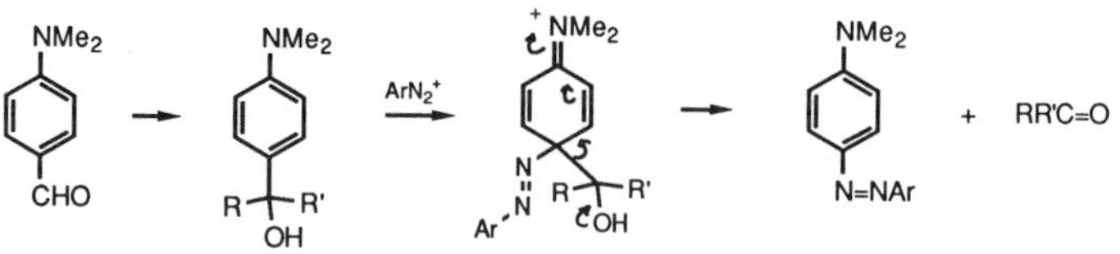

The reductive removal of functional groups including vinyl, 1-hydroxyalkyl, 1-methoxyethyl, and formyl on the periphery of hemins and protohemins by heating with polyphenolic compounds [Schumm, 1928, 1929; Bonnett, 1977] has been rationalized in terms of Friedel–Crafts reaction and fragmentation of the protonated pyrrole.

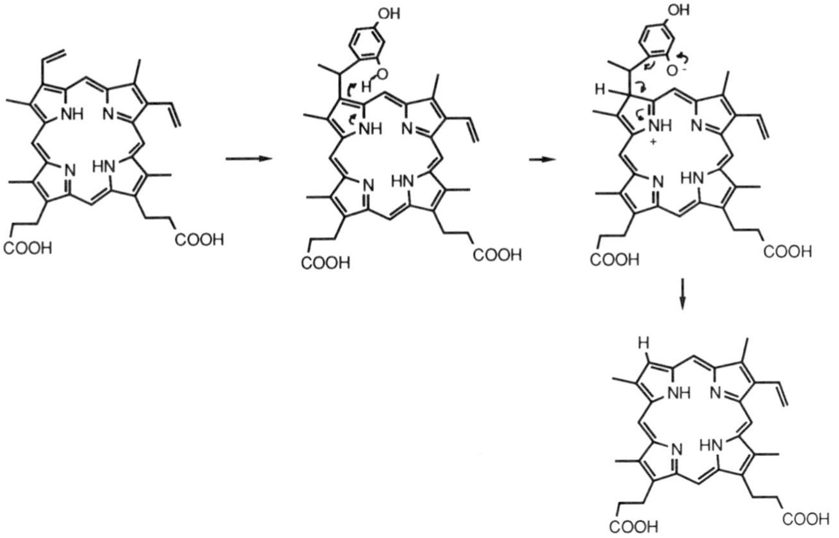

### 9.2.2. N4C=O Arrays

A classical example of this fragmentation is that which intervenes in the facile transformation of thebaine into thebenine [Gulland, 1923]. The reaction was initiated by opening of the dihydrobenzofuran and alkyl migration leading to a dienone with a **d4a=d** circuit. Hydrolysis of the iminium ion arising from the fragmentation followed by an intramolecular Friedel–Crafts reaction and aromatization complete the reaction.

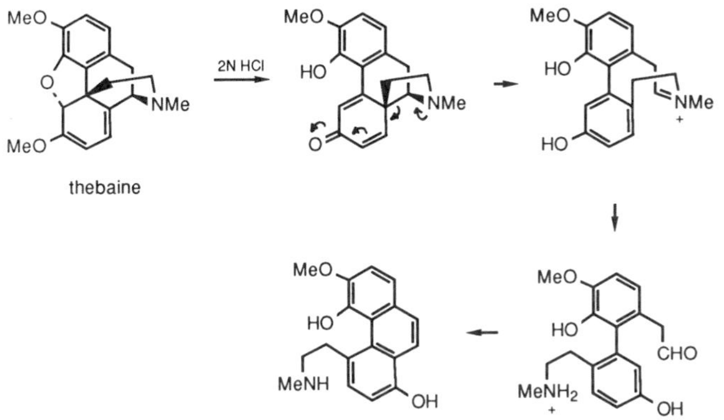

thebaine

The fragmentation step has circumstantial evidence in the biomimetic preparation of protostephanine [Battersby, 1968], and the reductive photolysis of thebaine [Theuns, 1984].

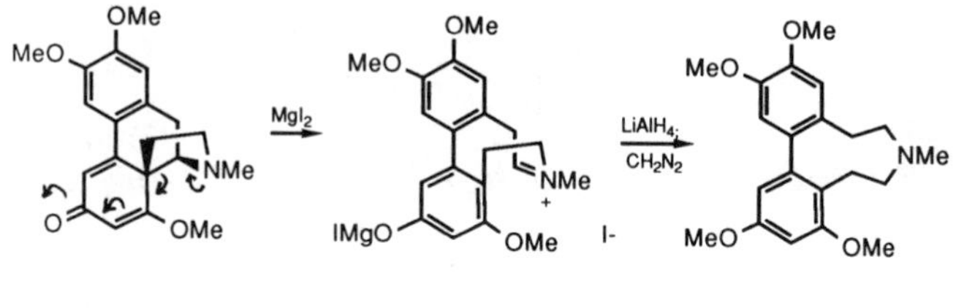

protostephanine

An analogous situation prevails in the anodic coupling of *N*-acyl-4-benzyl-1,2,3,4-tetrahydroisoquinolines which leads to phenanthrene products [Majeed, 1985]. Interestingly, the tetracyclic species obtained from corresponding isochroman-3-ones undergoes rearrangement only (without fragmentation).

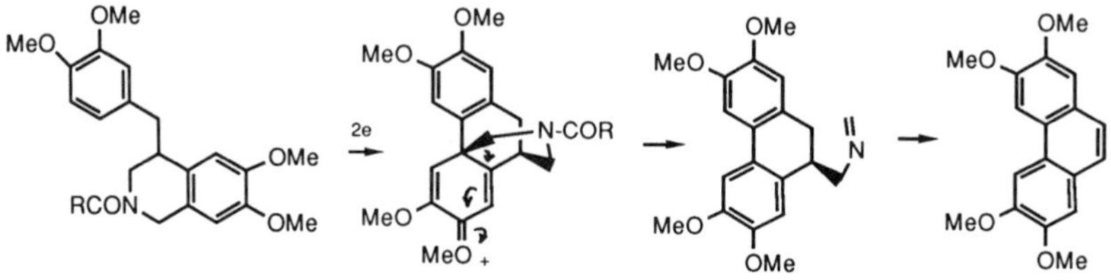

Phenolic oxidative coupling of 1-phenethyltetrahydroisoquinolines is believed to be the crucial step in the biosynthesis of homoerythrina alkaloids also. The biosynthetic schemes also entail fragmentation and recyclization. In vitro experiments [Kametani, 1970a; Kupchan, 1978] fully support these conjectures and controlled fragmentation of certain spirotetracyclic derivatives has been demonstrated [Marino, 1976].

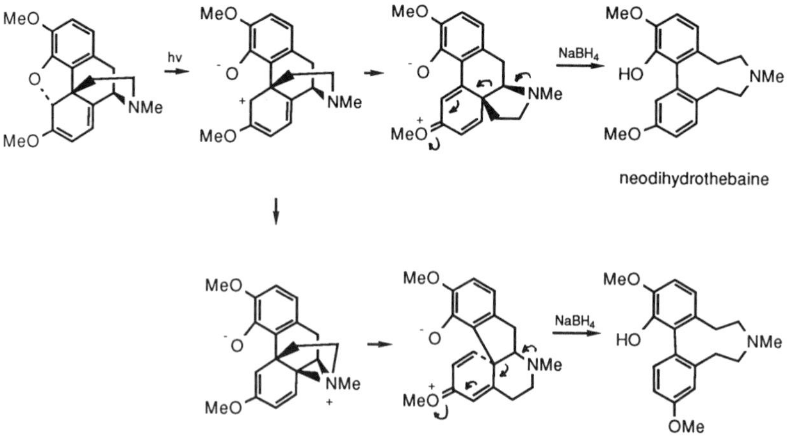

neodihydrothebaine

bractazonine

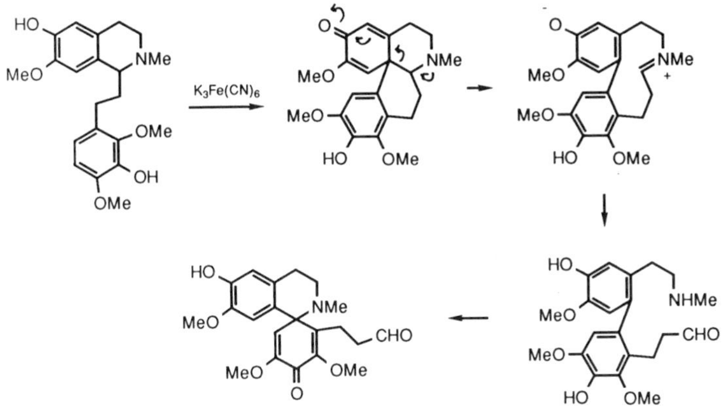

A potential synthetic intermediate for the morphine alkaloids interconverts with a bridged biaryl system by way of acid-catalyzed fragmentation and base-promited ring closure [Weller, 1986].

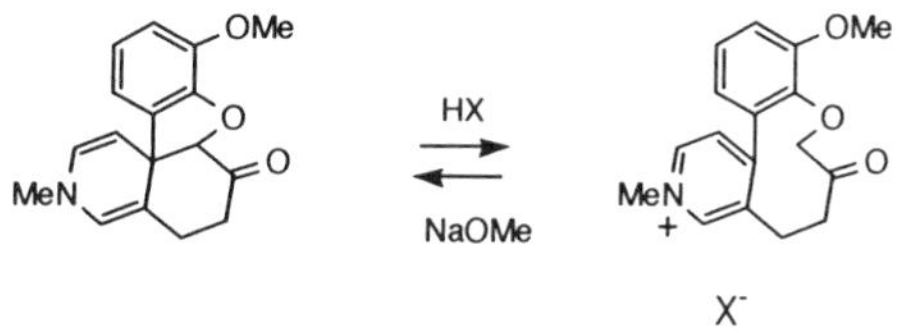

The photolysis product of *N*-carbomethoxynorcodeinone is a bridged biphenyl [Schultz, 1991] which arises from fragmentation of a spiroannulated cyclohexadienone intermediate.

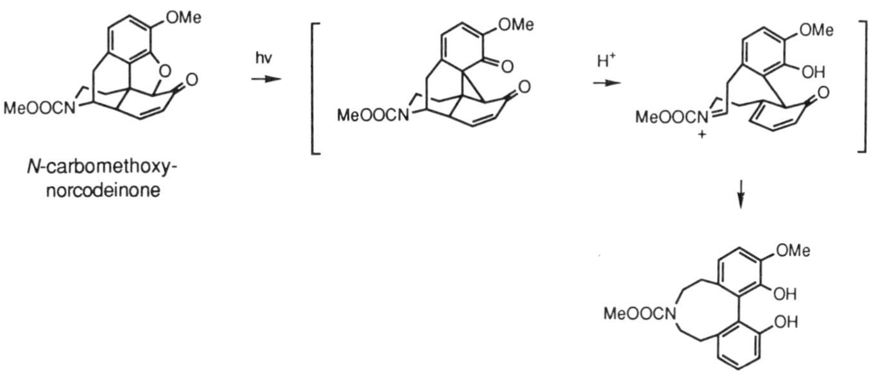

Lead tetraacetate oxidation of laudanine affords an *o*-quinone ether acetate which could be transformed into *N*-methyllaurotetanine and a 1-hydroxytetrahydroisoquinoline [Hara, 1984]. The latter compound is derivable from a fragmentation reaction.

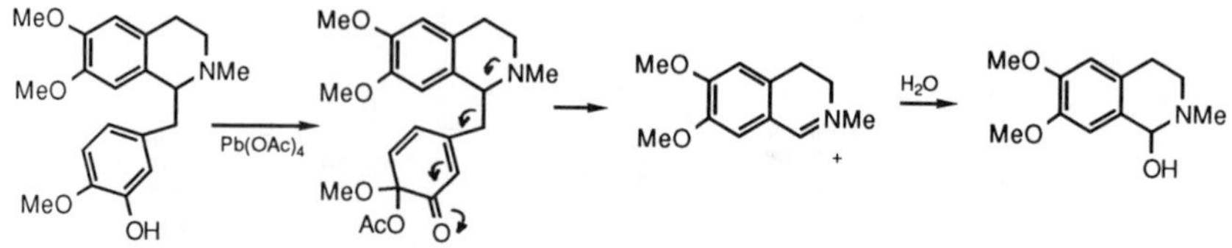

An acetoxycyclohexadienone obtained from a simpler isoquinoline has been found to undergo reductive isomerization on exposure to $BF_3$ in refluxing methanol [Hara, 1988]. The heterocycle suffers cleavage to return the molecule into an aromatic state, and subsequent Mannich condensation results in a trioxygenated tetrahydroisoquinoline.

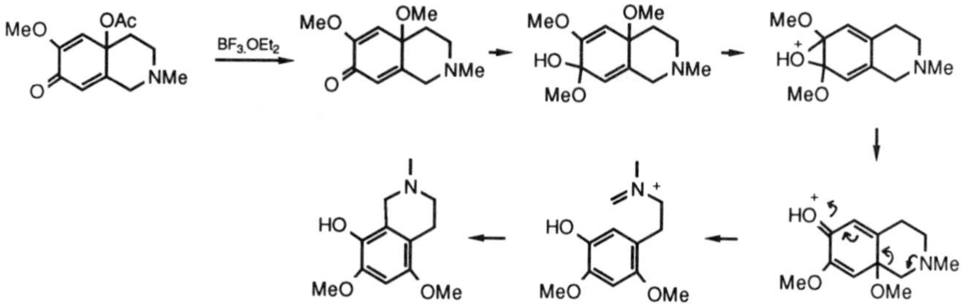

The 4-arylisoquinoline ring system is accessible from rearrangement of 3-aryl-4-hydroxy-1,2,3,4-tetrahydroisoquinolines under dehydrogenation conditions [Cushman, 1985]. Spirocyclic species are the pivotal intermediates.

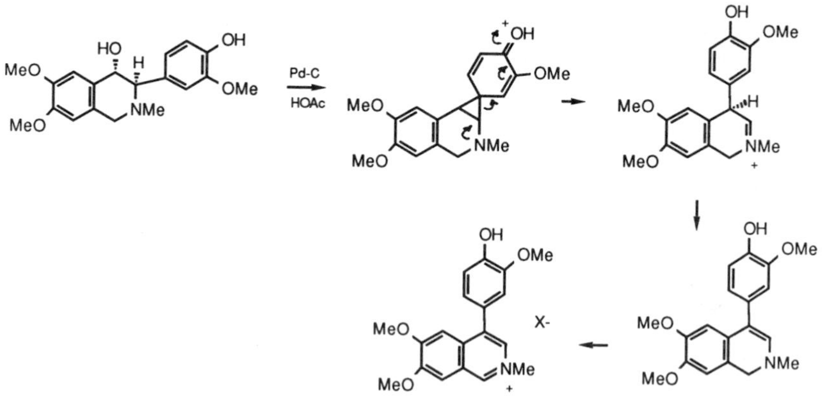

An elaboration of the ervatamine skeleton [Y. Langlois, 1975] consists of a Reformatskii reaction of a keto ester to introduce the missing elements of the C-ring. Unfortunately, fragmentation emerged as the more favorable reaction.

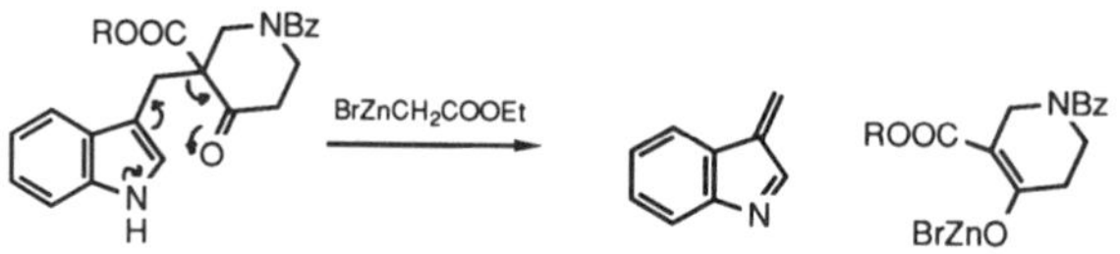

In a synthetic study toward vinoxine [Bosch, 1986] hydrogenation of a 4-alkylidene-1,4-dihydropyridine was carried out. However, the major product does not contain the pyridine moiety. The undesirable fragmentation may be suppressed by using acetic acid as solvent. The propensity for aromatization is the cause of the fragmentation and it had been reported [Wenkert, 1976] in connection with another synthetic work.

An assembly of dihydrocorynantheine from a tetracyclic ketone [Barczai-Beke, 1976] requires correction of the cyanoacetic ester chain stereochemistry. A successful equilibration of a dehydro derivative is via a retro-vinylogous Mannich reaction.

dihydrocorynantheine

The malonaldehydic ester pendant of 4,21-didehydrogeissoschizine is easily excised [Kan-Fan, 1980]. This reaction lends credence to the biosynthetic consideration that flavopereirine is thus originated.

Heating of (−)-tabersonine in acetic acid gave allocatharanthine as the major product [Muquet, 1972]. Apparently the chanoiminium species is rapidly isomerized of to a conjugate azabutadiene before recyclization.

(-)-tabersonine

Another pathway is open to $\Delta^{18}$-tabersonine along which the total obliteration of the C-ring generates a precursor of andranginine [Andriamialisoa, 1975].

An apparent C-20 epimerization of 19,20-dihydroakuammicine [Scott, 1974] can be viewed as involving a tricyclic dihydropyridinium intermediate.

An indoloazepine is pivotal of an outstanding approach to aspidosperma alkaloids such as pandoline [Kuehne, 1980] and vincadifformine [Kuehne, 1978]. Access of this heterocycle by reaction of the chloroindolenine derived from a tetrahydro-$\gamma$-carboline with thalliomalonate [Kuehne, 1979] is interesting as an orderly sequence of rearrangement, fragmentation, and Mannich reaction occurs.

The akuammiline to akuammicine transformation [Olivier, 1965] involves the loss of an acetoxymethyl group, formation of a cyclopropane intermediate, a fragmentation which is driven by the ester group and the basic nitrogen, and the closure of the C-ring.

akuammiline

A popular two-carbon expansion of cyclanones is by reaction of the corresponding enamines with a propiolic ester or acetylenedicarboxylic ester, and subsequent hydrolysis/decarboxylation. The isomerization of the primary adducts is an electrocyclic reversion [Boerth, 1975; Visser, 1982] although there must be some electronic aid from the donor and acceptor groups. A few applications of this reaction in synthesis may be mentioned, and they are: velleral, vellerolactone, and pyrovellerolactone [Froborg, 1978], (−)-steganone [E.R. Larson, 1982], the carbocyclic framework of the ophiobolins [Dauben, 1977a], and portion of the $C_{19}$-diterpene alkaloids [VanBeek, 1986].

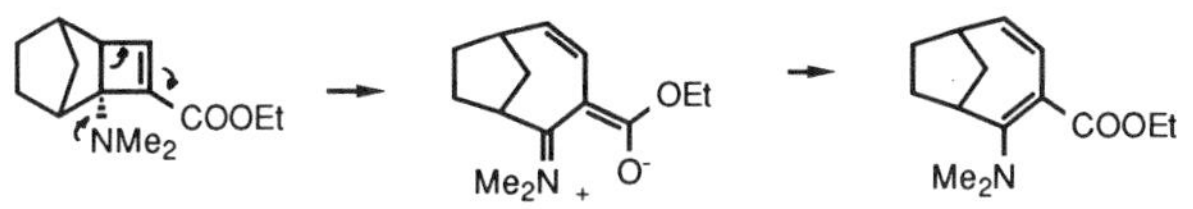

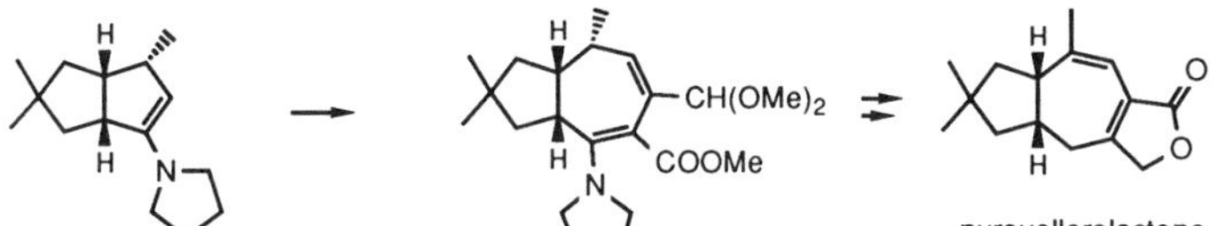

pyrovellerolactone

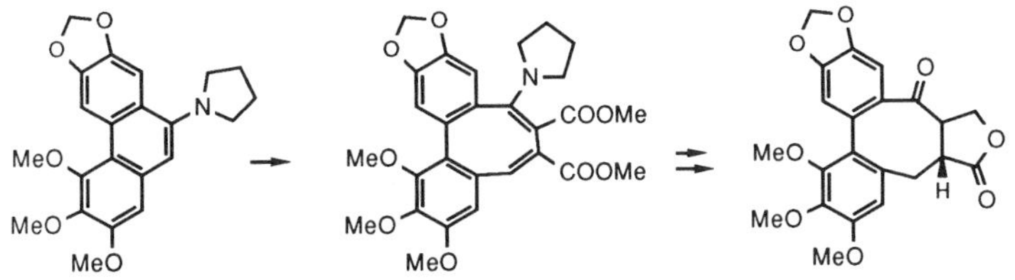

steganone

Indole is an enamine and its ring expansion has been studied [Acheson, 1972]. The incorporation of a second molecule of indole by a Michael reaction is not surprising, and actually it has an intramolecular analogy [McRae, 1982]. This is the formation of a dihydropentalene derivative from 1,4-dipyrrolidino-1,3-cyclohexadiene.

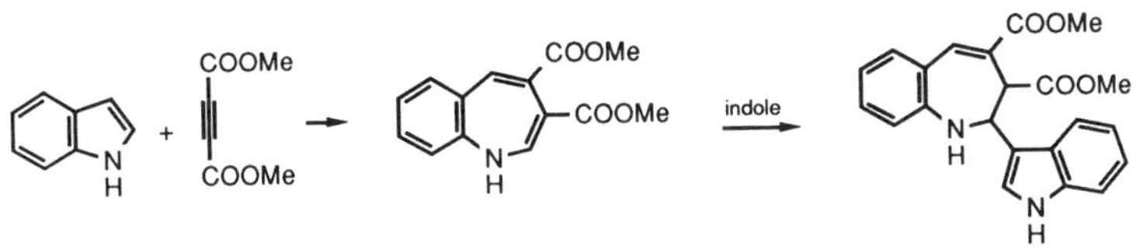

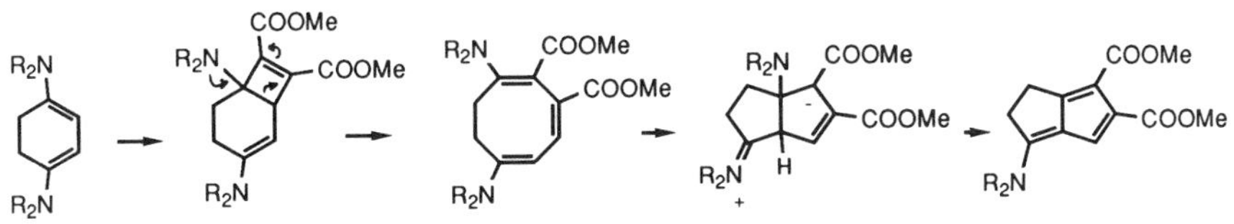

The formation of seven-membered heterocycles by thermolysis of the cycloadduct of *N*,*N*-dimethylaminoallene and acetylenic esters [Klop, 1983] can be rationalized by a reaction sequence involving ring opening, proton transfer, and cyclization.

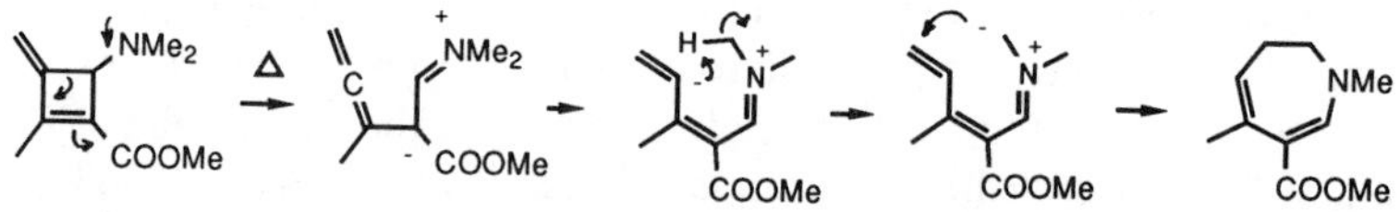

Irradiation of the nitrite esters derived from photoadducts of 2,2-dimethyl-3(2*H*)-furanone and alkenes induces two consecutive fragmentations leading to 1,5-ketoaldehydes [S,W. Baldwin, 1982a]. An excellent application of this transformation to synthesis of occidentalol has been achieved [S.W. Baldwin, 1982d].

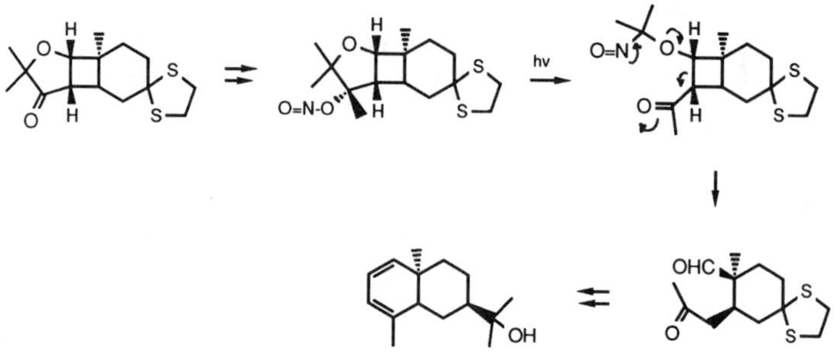

occidentalol

Reaction of a bromopropellanone with methylamine proceeds via $S_N2$ displacement and fragmentation [Yamaguchi, 1969]. It should be noted that propellanones which contain an acceptor group in place of the bromine atom adopt an alternative mode of fragmentation.

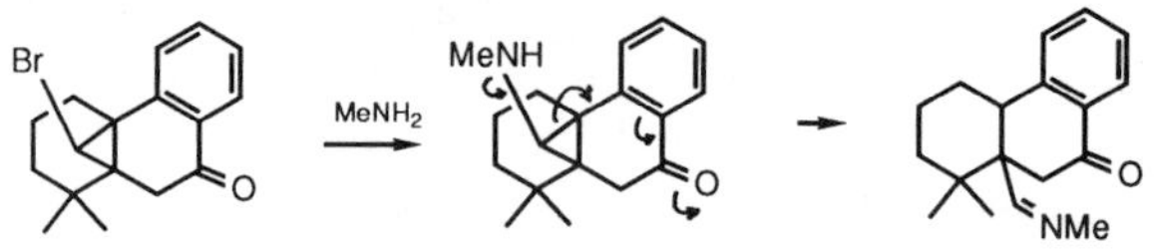

Zwitterions are formed by a thermal fragmentation of *β*-(aziridin-2-yl)acrylic esters [Hudlicky, 1986]. Reclosure of the ionic intermediates result in pyrrolidine-3-carboxylic esters.

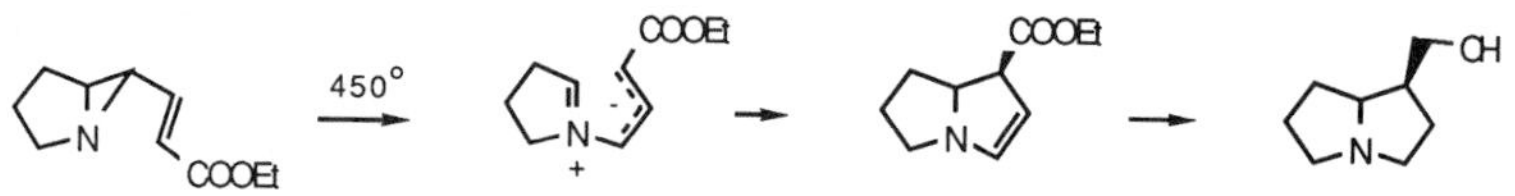

isoretronecanol

Formation of bicyclo[3.3.1]non-1-ene by thermolysis of a sulfoximine [Kim, 1975] is presumed to occur via its dissociation to a diazene and a concerted elimination of dinitrogen and carbon dioxide from the latter species.

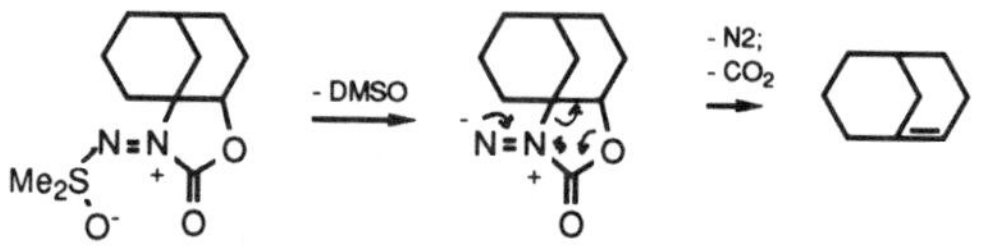

### 9.2.3. N4C=N Arrays

Acid hydrolysis of the morpholino nitrile derivative of 4-(dimethylamino)-benzaldehyde gives a by-product identified as bis-(4-dimethylamino)phenylacetonitrile [Bennett, 1970]. Quinoid intermediates are likely involved.

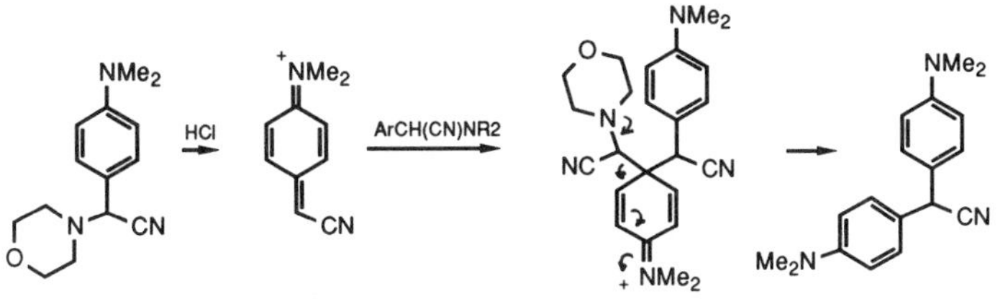

A study toward synthesis of sesbanine [Wanner, 1982] requires hydrolysis of an imine. However, fragmentation of the tetrahedral intermediate was found to complicate this route owing to the presence of the pyridinium moiety and the imide carbonyl groups.

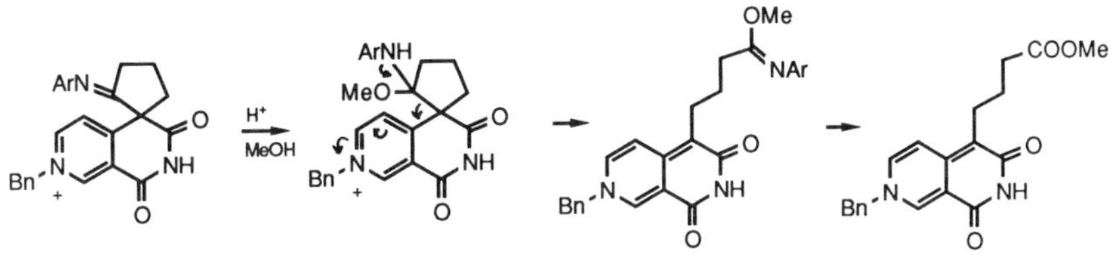

Chimonanthine has been isomerized to calycanthine by heating with acetic acid [Hall, 1967]. The bisindolenine tautomer probably intervenes in this transformation and in the formation of *N*-benzoyl-*N*-methyltryptamine from the benzoylation of calycanthine. The formation of a carboline on heating calycanthine with phthalic anhydride [Marion, 1938] may be similarly formulated.

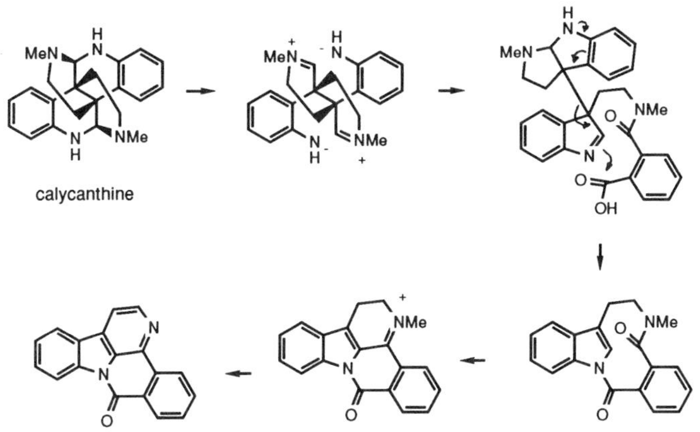

calycanthine

In a landmark synthesis of reserpine [Woodward, 1958] the closure of ring-C by the Bischler–Napieralski cyclization/reduction sequence generated the 3$\alpha$

isomer. In order to invert this center a $\gamma$-lactone bridging the E-ring was created. Treatment of the lactone with pivalic acid triggered C-2/C-3 bond cleavage (after protonation at C-2) and reclosure to the $3\beta$-H isomer in which the bulky indole moiety is in an equatorial orientation.

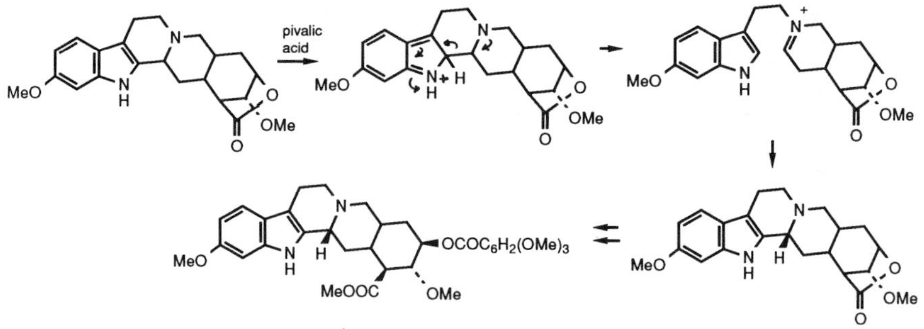

pivalic acid

reserpine

An entry into the aspidosperma alkaloid is associated with an acid-catalyzed cyclization of a diazoketone within the framework of a tetrahydro-$\beta$-carboline [Takano, 1978c, 1979b]. The initial product suffers a fragmentation that allows for the formation of the desired intermediate. A similar transformation could be effected by a Pummerer rearrangement [Genin, 1987], and for an aza analog of desethylaspidospermidine, a mild Lewis acid treatment of an *O*-acetylhydroxamate [Demuynck, 1989].

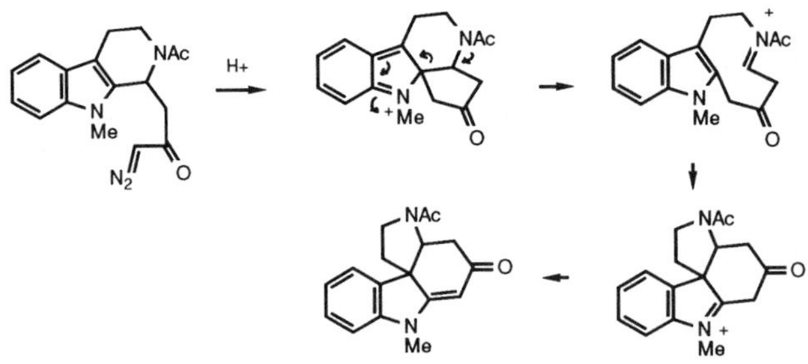

An extensive structural change results from the reduction of picraline [Britten, 1967] with zinc in hydrochloric acid. The original C-2/C-3 bond was cleaved during the reaction.

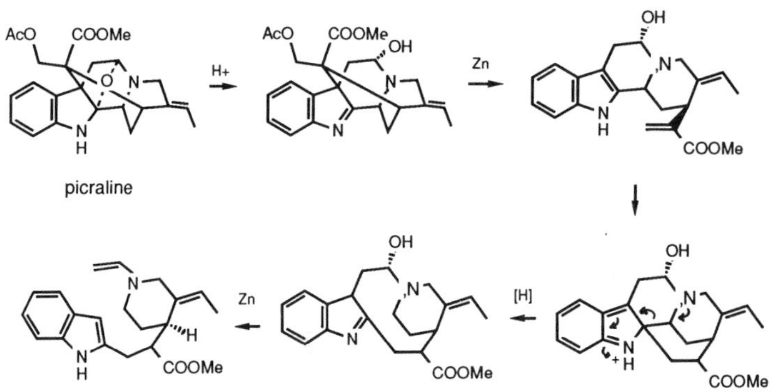

picraline

The seven-membered ring homolog of *N*-methyl-3,4-dihydrocarbazole reacts with an azidoformate to give a spirocyclic product [Bahadur, 1979]. Ring contraction of the initial adduct might have proceeded by a 1,2-migration or via fragmentation of a tetracyclic intermediate.

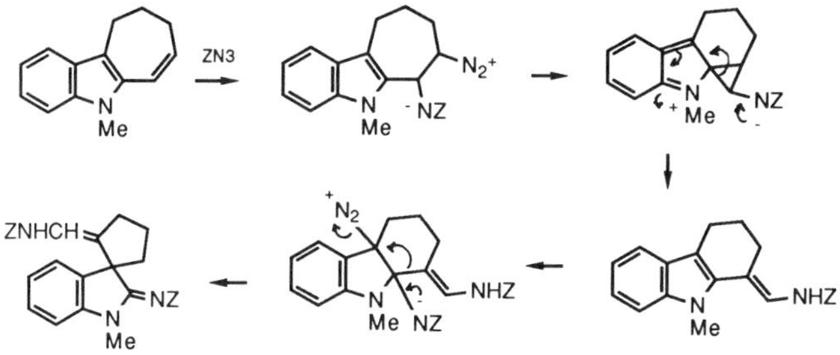

### 9.2.4. Other d4a=d Fragmentations

Under the Wolff–Kishner reduction conditions an indane-2-one in which each of the α-positions is spiroannulated with a cyclopropane ring has been found to undergo deletion of ethylene in addition to deoxygenation [Lemieux, 1989]. Apparently one of the cyclopropane rings suffered cleavage by attack of a hydrazine molecule, and the product eliminated diimide and ethylene. A normal reduction then followed.

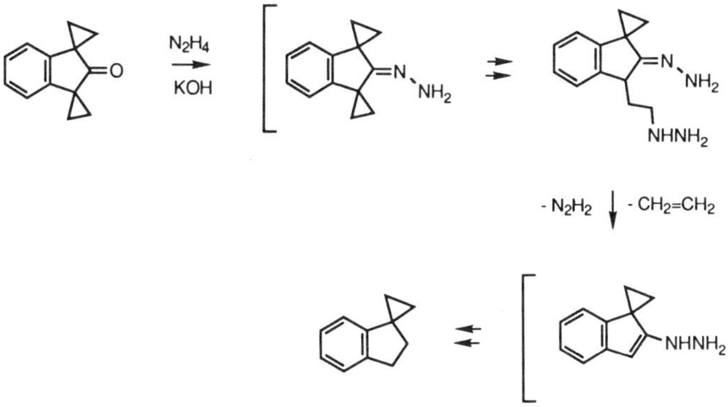

spiro[cyclopropane-1,1'-indan]

1-(*o*-Nitrobenzyl)isoquinolinium salts split into two moieties on borohydride reduction [Neumeyer, 1968]. It is interesting to note that normal reduction occurred when 3,4-dihydro-*N*-methylisoquinolinium analogs were subjected to the same conditions. Apparently the fragmentation urged by the nitro group is possible when attended by the regeneration of an aromatic isoquinolinium species.

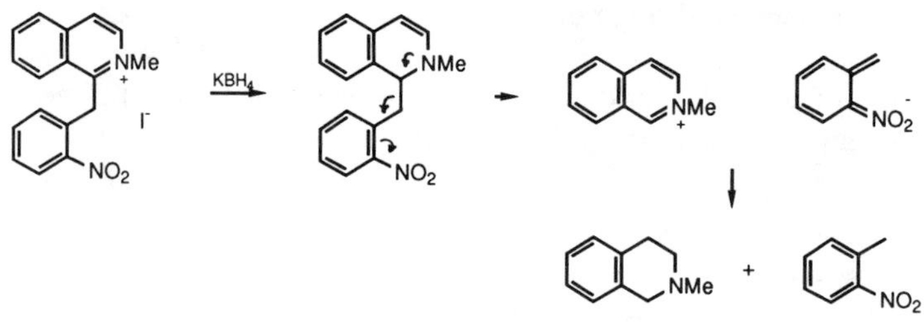

The major nitrosation product of camphor-3-glyoxylic acid is a hydroxamic acid [Hatfield, 1967]. The following mechanism portrays its mode of formation.

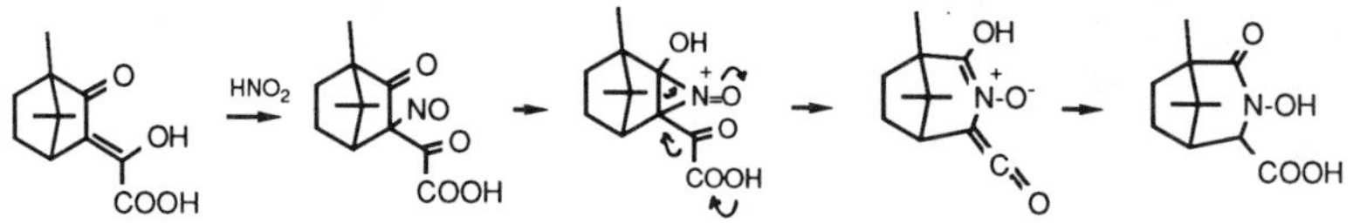

Thermal rearrangement of 1,3-dichloro-6-ethoxycarbonyl-2-thiabicyclo-[3.1.0]hex-3-ene leads to ethyl 2,4-dichloro-5-hydroxy-6-methylbenzoate [Gillespie, 1981]. A fragmentation is followed by a series of skeletal reorganizations.

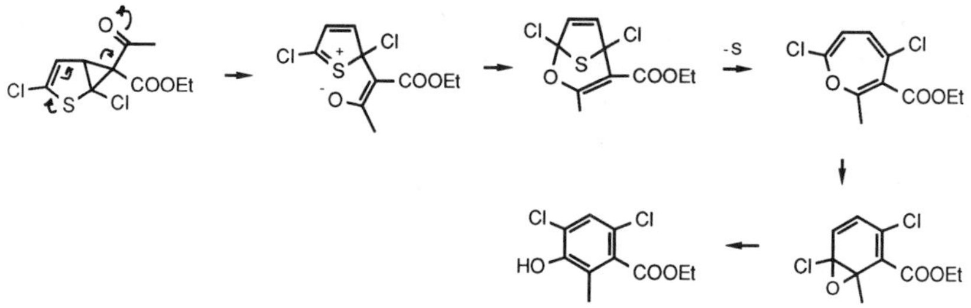

Ring expansion of α-(*n*-butylthiomethylene)cyclobutanones can be effected by converting the ketone group into the *O*-trimethylsilyl cyanohydrin [Byers, 1985].

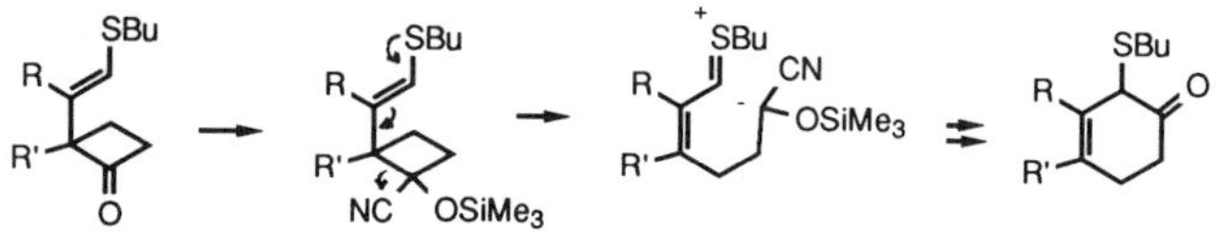

## 9.3. d4d SYSTEMS

The **d4d** system is disjoint and therefore not fragmentable unless contrapolarization (e.g. by redox manipulation) or rearrangement is performed.

Electrolysis of γ-hydroxy carboxylic acids generates products containing an olefin and a carbonyl group [Corey, 1957, 1959]. Ketals of γ-ketocarboxylic acids behave analogously, and this latter version has been applied to a synthesis of malyngolide [Wuts, 1986].

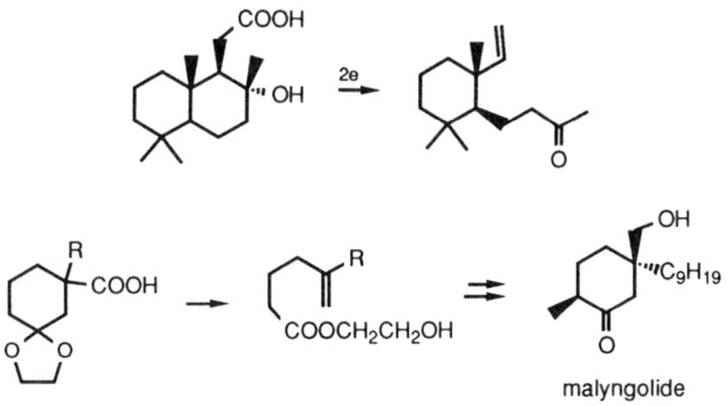

malyngolide

Fremy's salt mediates oxidation via radical mechanisms. The degradation of a pentacyclic aporphine lactam by this reagent [Castedo, 1982] is interesting and applicable to a synthesis of lysicamine. The last stage of the transformation involves decarboxylation-decarbonylation.

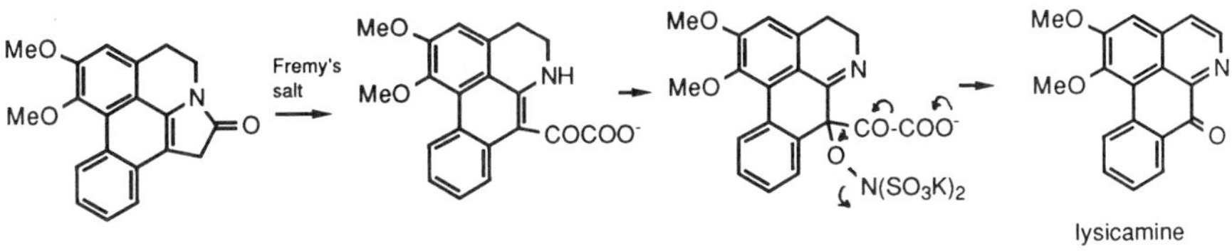

lysicamine

2-Hydroxy-2,5-cyclohexadienones react with superoxide anion to give lactols with loss of carbon monoxide [Frimer, 1982].

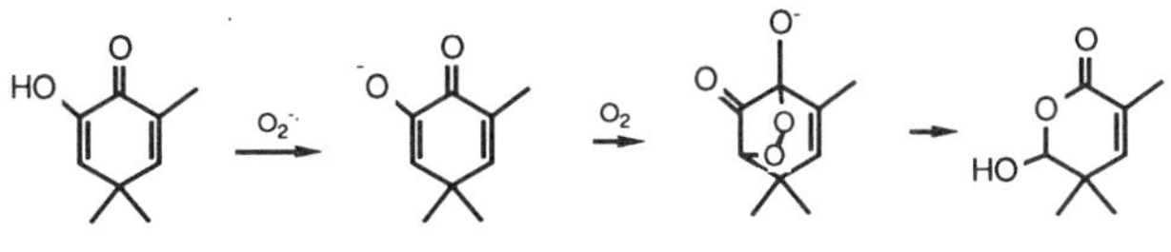

Decarbonylation of α-ketocarboxylic acids with iodosobenzene [Moriarty, 1981] may involve cyclic intermediates.

1,3-Diacyl compounds undergo addition with hydrogen peroxide and the adducts may fragment with or without alkyl migration [Payne, 1961]. The earliest report of this general transformation is that of Mannich [1941], and an application of this process is found in an approach to *trans*-chrysanthemic acid [Ho, 1983a] from 3-carene. β-Enaminoketones behave similarly [Zavialov, 1964].

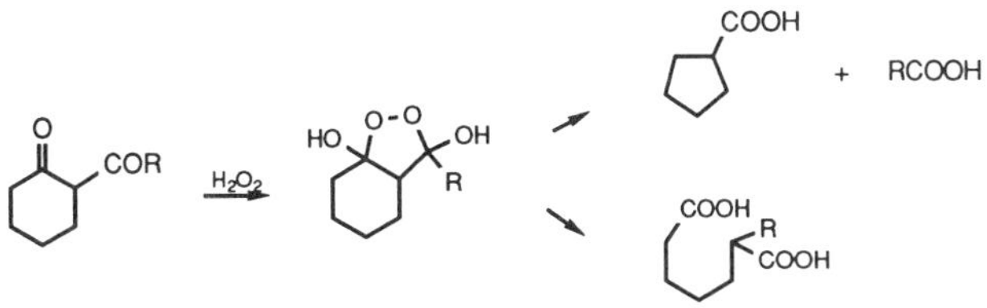

## 9.4. d5d SYSTEMS

### 9.4.1. O5X Arrays

Fragmentation of these systems usually leads to carbonyl compounds. For example, the photochemical reaction of 4-methyl-4-trichloromethyl-2,5-cyclohexadienone in methanol gives rise to a cyclopentenone [Schuster, 1971] via trapping of a bicyclic zwitterion by the solvent and chloride ion expulsion from the resulting enolate. On the other hand, the dichloromethyl cyclohexadienone appears to pursue a different reaction course [van Tamelen, 1962].

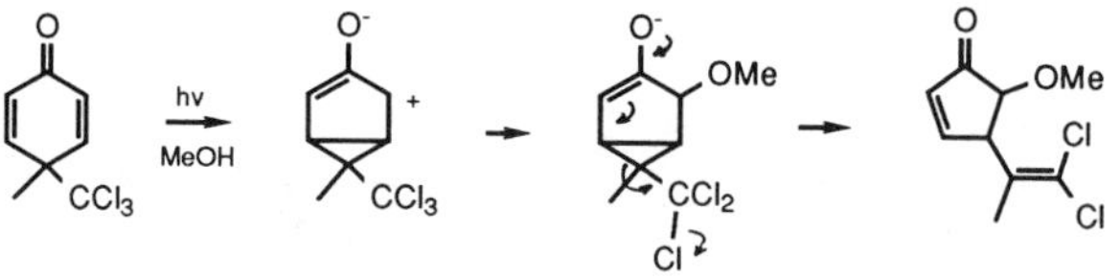

Reduction of a bischloromethyltricyclo[2.1.0.0$^{2,5}$]pentanone is attended by ring opening [Dowd, 1982]. The alcohol(ate) possesses two identical **d5d** fragmentation passages through which a large portion of angular strain of the molecule may be removed.

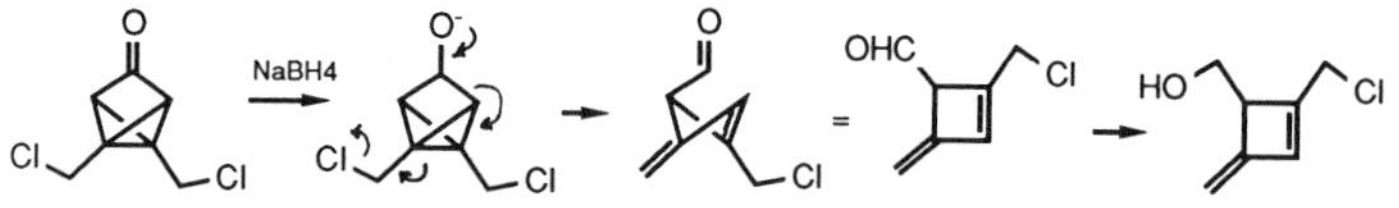

The absence of a dienone-phenol rearrangement by 5-benzoyloxy-5,6,7,8-tetrahydro-1,4a-dimethyl-2(4a*H*)-naphthalenone on heating with acid [Banerjee, 1985] and the formation of a tetrasubstituted benzene serve to indicate a fragmentation process to prevail.

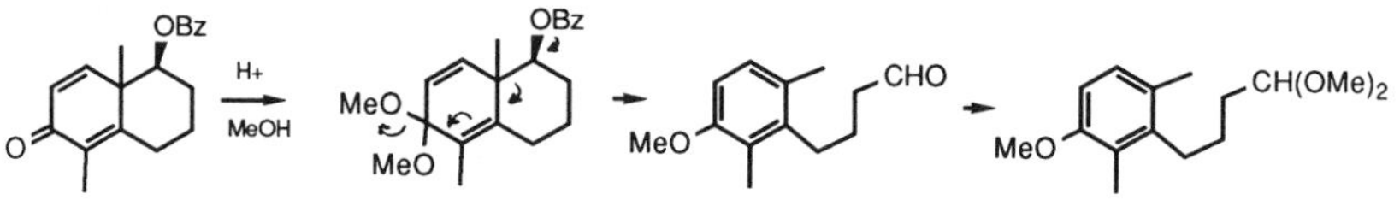

A chemical analogy for the last step of the microbial degradation of androst-4-ene-3,17-dione [Dodson, 1961] also involves the **d5d** fragmentation [Lythgoe, 1980].

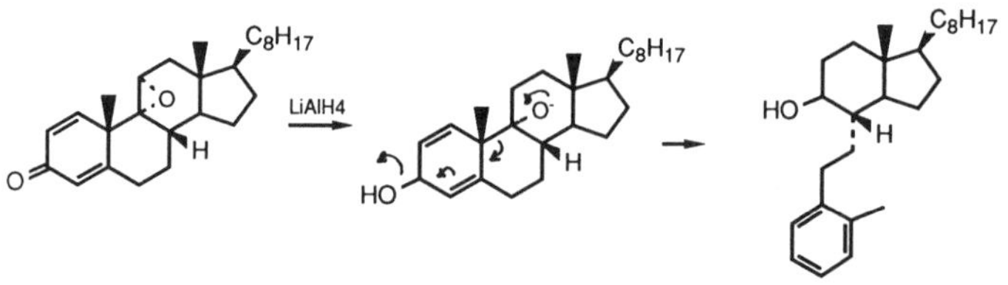

The Hofmann degradation of tazettine leading to a biphenyl [Ikeda, 1956] is apparently mechanistically related, with respect to the fragmentation step.

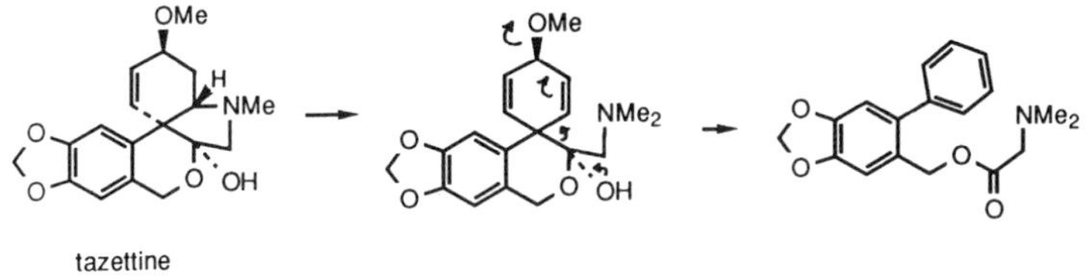

tazettine

The formation of a furyl ketone by acid treatment of *trans*-5,6-dihydroxy-5,6-dihydro-$\beta$-ionone [Skorianetz, 1973] is the result of chain-ring tautomerization and vinylogous 1,3-diol fragmentation. Simpler examples are also known [Skorianetz, 1975].

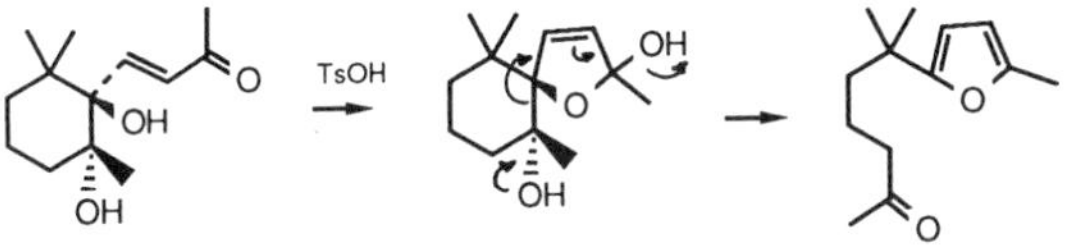

Reoxidation of the triol derived from alpinenone does not reconstitute the natural terpene [Itokawa, 1987]. Instead, hemiacetalization and subsequent fragmentation occur.

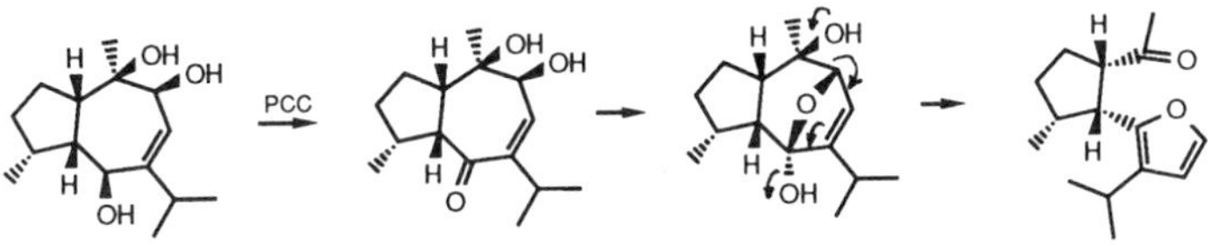

An acid-catalyzed product of dihydrocorymine has an interesting structure in that the lactam carbonyl is in the hydrate form, presumably due to steric inhibition of its interaction with the nitrogen atom and the angular strain of the bridge. It is degraded by a retro-Michael reaction with base [Massiot, 1983].

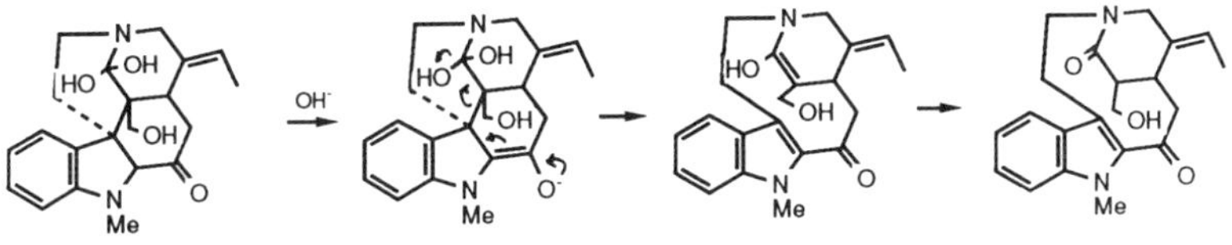

The xanthyl cation is aromatic. It is thus not surprising that 2-(9-xanthyl)-ethanol fragmented on exposure to acid [Deno, 1965]. Furthermore, the dehydration of 9-xanthylacetic acid involves a Friedel–Crafts acylation followed by fragmentation.

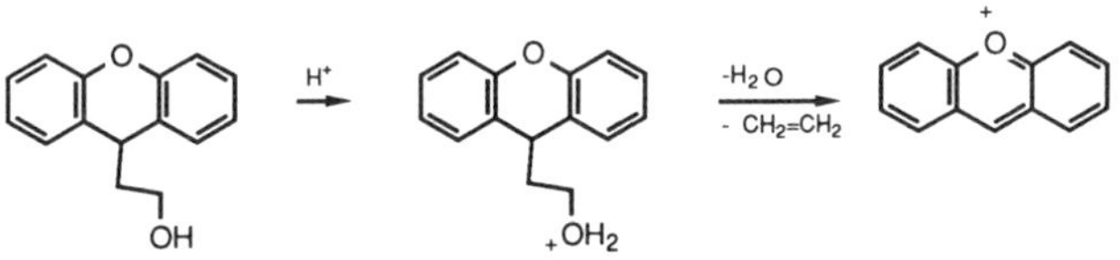

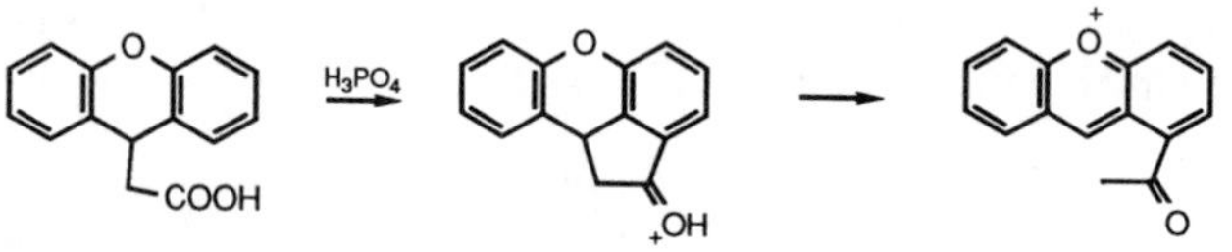

Ethanologous donor-assisted Beckmann fragmentations have been reported. In one case it represents an undesirable reaction in connection with synthesis of perhydrohistrionicotoxin [Ibuka, 1981], in another example it was designed as a step to generate the nine-membered ring of byssochlamic acid [Stork, 1972].

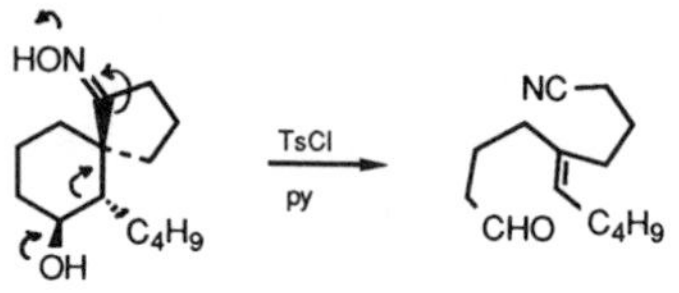

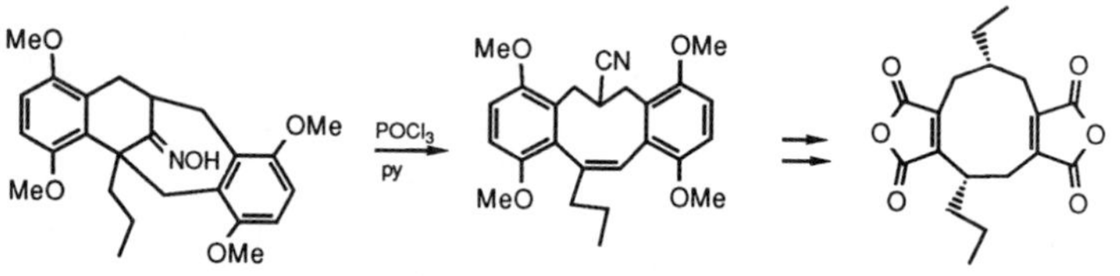

byssochlamic acid

Two amino-masking devices are 5-benzisoxazolylmethoxylcarbonyl [Kemp, 1975a] and 4-(dihydroxyboryl)benzyloxylcarbonyl groups [Kemp, 1975b]. Deprotection of the carbamates can be achieved by hydrogenolysis in the former instance and treatment with slightly basic hydrogen peroxide in the latter case. Generation of the phenoxide and hence a **d5d** system is the key.

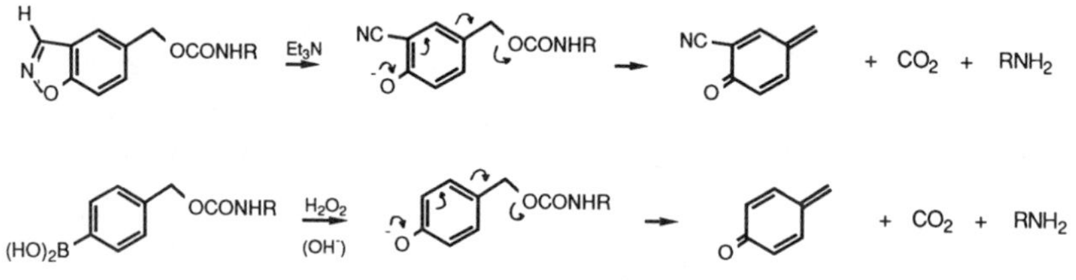

Protection of carboxylic acids as their 3′,5′-dimethoxybenzoin esters [Sheehan, 1971] accrues an advantage of mild and selective conditions (photolysis) for their cleavage. A *m*-methoxy group helps expulsion of the carboxylate ion as shown.

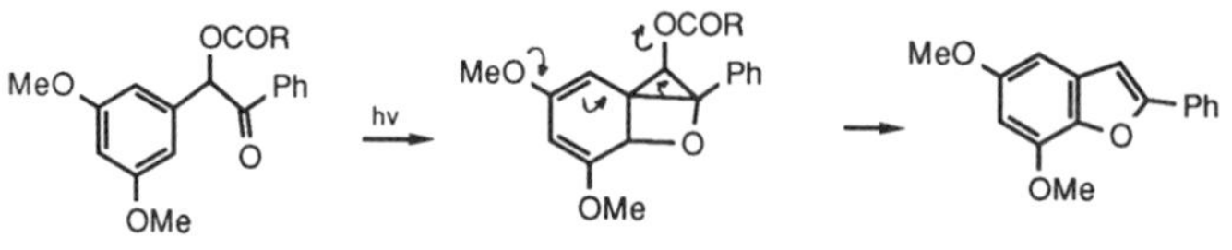

Concomitant decarboxylation and dehydration of prephenic acid [Plieninger, 1962] is facile because the product is aromatic. The extensive degradation of noranisatin seco acid to form dihydrocoumarins [K. Yamada, 1968] might also have involved a similar process.

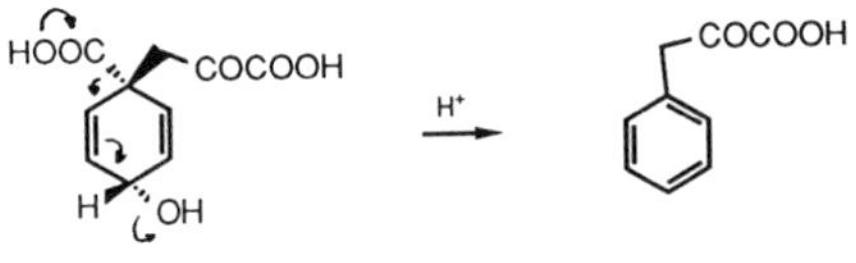

prephenic acid

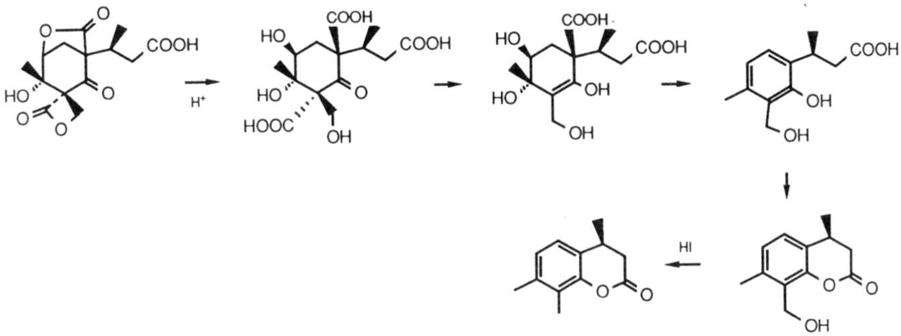

The double defunctionalization is a vinylogous reaction of that described for $\beta$-hydroxy carboxylic acids. Another example is the formation of 4-hydroxyindanone from the Diels–Alder adduct of 2-furoic acid and cyclopentadienone in alkaline solution [Gavina, 1981].

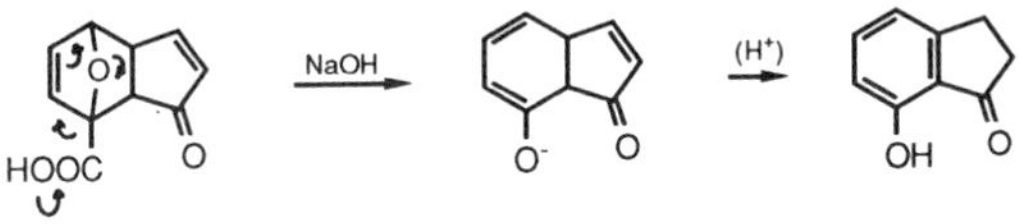

Oxidation of santonic acid with alkaline hydrogen peroxide affords aposantonic acid [Woodward, 1948; Hortmann, 1972]. Decarboxylation accompanies fragmentation of a hydroperoxide.

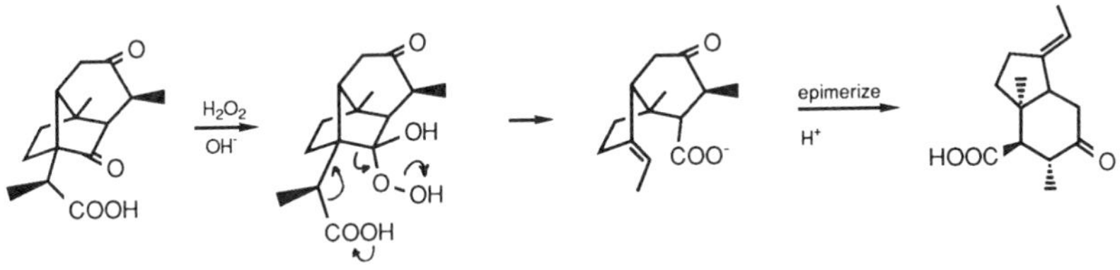

In a photochemical approach to synthesis of grandisol [Cargill, 1975] an allylic alcohol was subjected to ozonation. In situ fragmentation of the ozonide resulted in a keto acid.

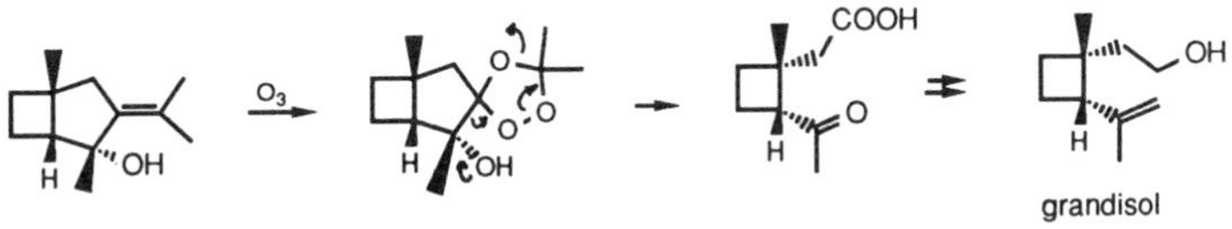

grandisol

Simultaneous cleavage of two C—C bonds and formation of a C=C bond by lead(IV) acetate oxidation of a 1,4-diketone hydrate constitutes one of the key transformations in a synthesis of triquinacene [Woodward, 1964].

HO O OH $\xrightarrow{Pb(OAc)_4}$ HO O O Pb(OAc)$_3$ → O O O ⇉ triqunacene

The reaction of diphenylcyclopropenone with most enamines does not proceed by C—C bond insertion. Amides are generated by way of cyclic intermediates [Bilinski, 1972]. The ylides formally derived from a 1,3-dipolar cycloaddition dismutate into the azabicyclo[3.1.0]hexenols which allow for ring cleavage to occur.

Borohydride reduction of *endo*-1,3,7,7-tetrachloro-2-azabicyclo[2.2.1]hept-5-yl acetate in ethanol results in a pyrrole derivative [Jung, 1980b]. Dehydrochlorination and imine reduction follow the fragmentation of the cyclopentane moiety.

Thebaine reacts with acetic anhydride to furnish many products, consistently among them is 3,6-dimethoxy-4,5-epoxyphenanthrene [Allen, 1984]. *N*-Acetylation instigates the loss of the piperidine moiety (cf. pentacyclic adduct formation from thebaine with an alkyl propiolate in polar solvents [K. Hayakawa, 1981]).

thebaine

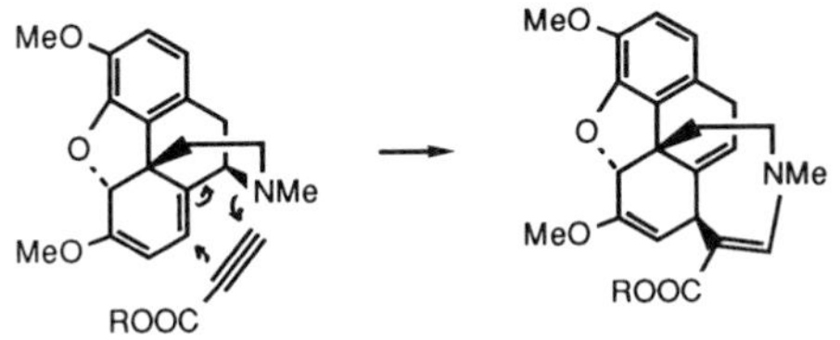

A number of syntheses using aziridinium intermediates actually involve their fragmentation prior to C—C bond formation [Kametani, 1973; 1982].

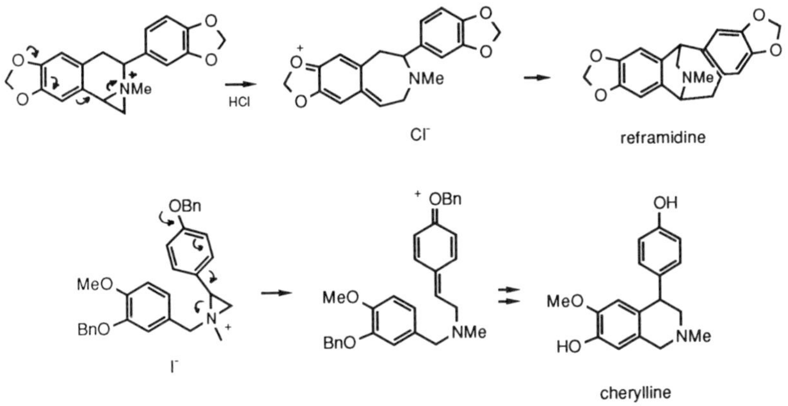

reframidine

cherylline

During a synthesis of ipecac alkaloids from norcamphor [Takano, 1982a] it was observed that equilibration at C-11b occurred while reducing the lactam carbonyl. The configurational change could have taken place in a B/C-seco intermediate.

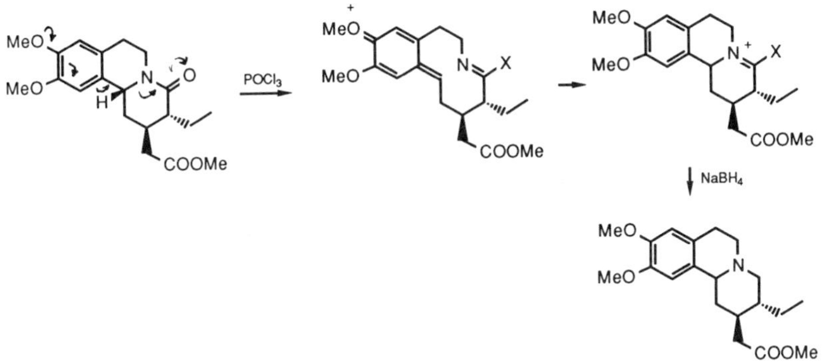

Cleavage of the intracyclic C—N bond of an isoquinoline annexed to a ten-membered ring [von Strandtmann, 1968] proceeds by a similar mechanism.

Benzazepines are rearrangement products of *N*-(4-nitrobenzyl)- and *N*-phenacyltetrahydroisoquinolinium salts in the presence of potassium *t*-butoxide [S. Smith, 1984]. Apparently quinonemethides are involved.

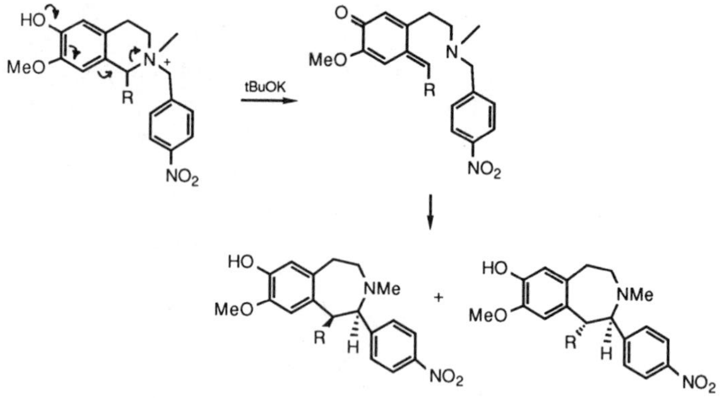

An attempt to dehydrate an aldoxime derived from forskolin 1,9-carbonate with 1,1′-carbonyldiimidazole [Saksena, 1985a] has led to a dihydro-$\gamma$-pyranone. This transformation is not unusual as the ethereal oxygen is in a perfect position to participate.

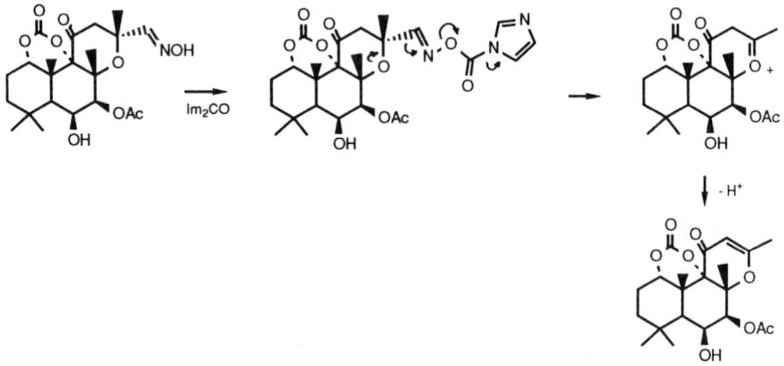

A mechanism for the dealkylation of trisubstituted olefins by nitrous acid has been suggested [Corey, 1987] as shown. Two nitration steps followed by Nef reaction of the allylic nitro group and subsequent tautomerization yield an isoxazole *N*-oxide. *O*-Nitrosation then sets up a fragmentation.

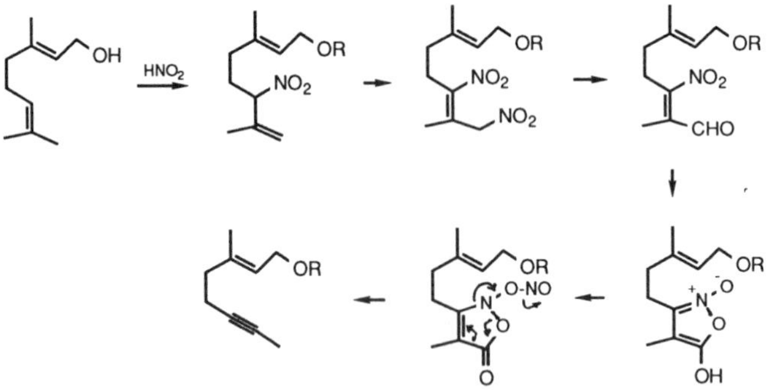

The Eschenmoser–Tanabe fragmentation is probably the most famous "named" fragmentation besides the Grob fragmentation. This widely applicable method for C—C bond cleavage of enecarbonyl compounds via the epoxy intermediates is also characterized by transposition of the functional groups. To the original protocol [J. Schreiber, 1967; Tanabe, 1967] of using tosylhydrazide has been added many other alternative reagents, including *N*-aminoaziridines [Felix, 1972], 2,4-dinitrobenzenesulfonylhydrazide [Corey, 1975], and mesitylenesulfonylhydrazide [C.B. Reese, 1981], which serve to expand the scope of the fragmentation. It is also possible to induce fragmentation of tosylhydrazones of unsaturated ketones by epoxidation [Sunthankar, 1972].

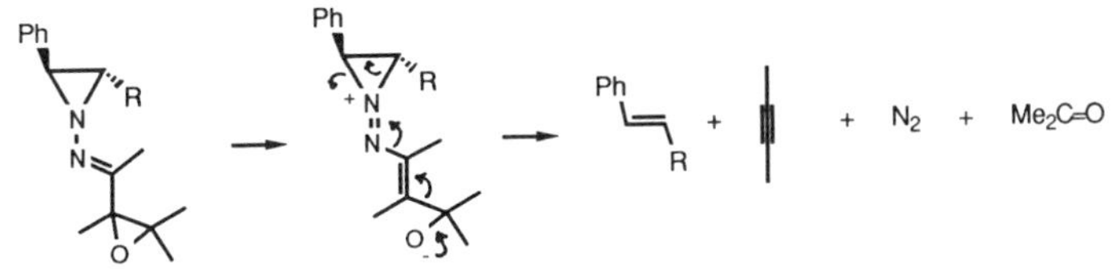

The decomposition of epoxy diazirenes [Borrevang, 1968] to generate the same products is of little preparative value, while fragmentation of epoxy ketoximes [P. Wieland, 1967] may fill a certain need. The gross mechanistic perspective for all these variants is the same.

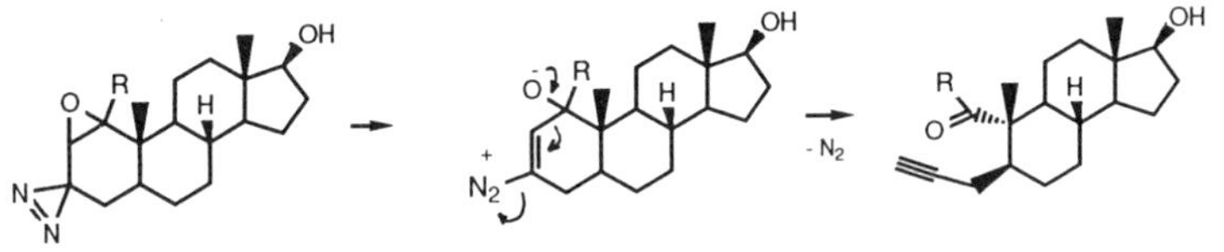

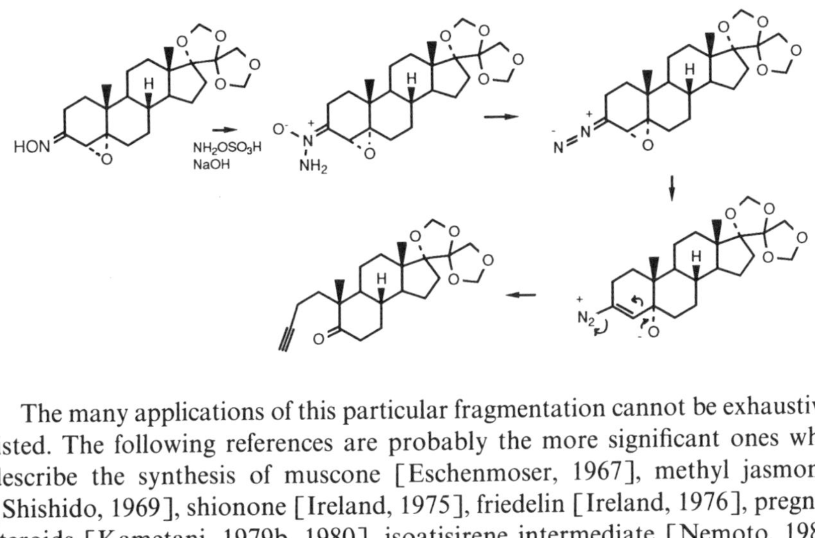

The many applications of this particular fragmentation cannot be exhaustively listed. The following references are probably the more significant ones which describe the synthesis of muscone [Eschenmoser, 1967], methyl jasmonate [Shishido, 1969], shionone [Ireland, 1975], friedelin [Ireland, 1976], pregnane steroids [Kametani, 1979b, 1980], isoatisirene intermediate [Nemoto, 1985], and medium-sized ring systems [Lange, 1974; Batzold, 1976].

An alternative method for the fragmentation above is via *N*-bromination of the unsaturated tosylhydrazones in the presence of an alcohol [Fehr, 1979].

### 9.4.2. N5X Arrays

Many fragmentations belonging to this category are found in indole derivatives. For example, 3,3'methylene-(2,2'-dimethyl)bisindole splits up on reaction with 4-dimethylaminobenzaldehyde in an acidic solution [Biswas, 1969].

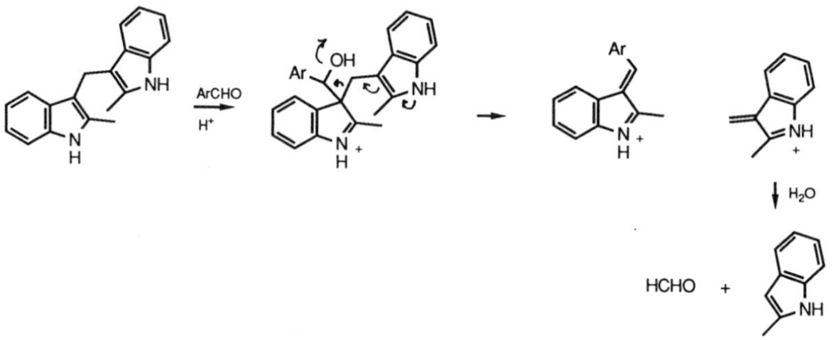

The modified Polonovski reaction using trifluoroacetic anhydride as reagent on tabernaemontanine [Husson, 1973; Mangeney, 1978] affords a seco species which is capable of recyclization with a more stable enamine, resulting in a new tetracyclic skeleton characteristic of ervatamine.

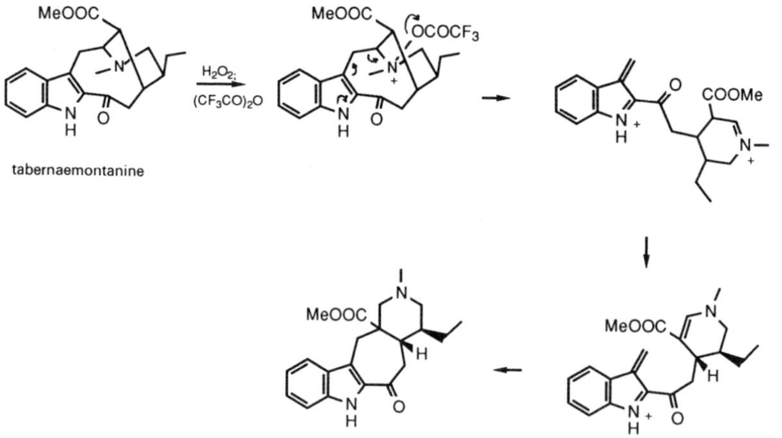

tabernaemontanine

The formation of vallesamine from stemmadenine $N_b$-oxide [Scott, 1978] is biomimetic. The recyclization step may be arrested, as in the reaction of *N*-anhydrovinblastine $N_b$-oxide [Mangeney, 1979]. In the case of catharanthine lactone *N*-oxide, the presence of vindoline causes an aminal formation [Honma, 1977; Kutney, 1977] from the fragmented species.

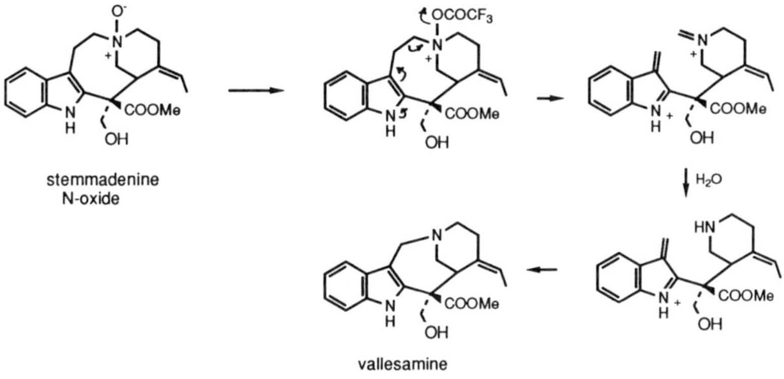
stemmadenine
N-oxide
OCOCF3
COOMe
OH
H2O
vallesamine

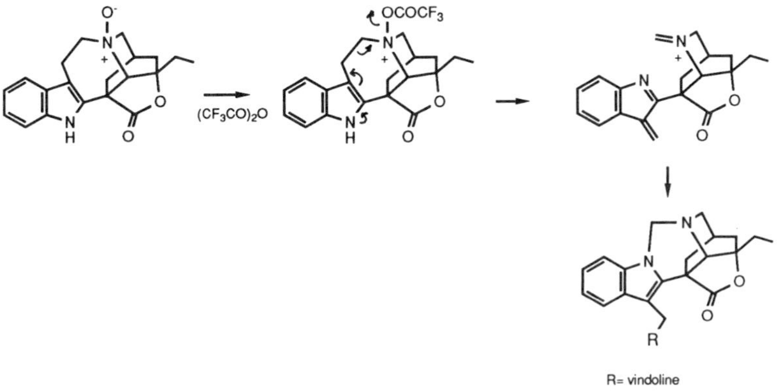
(CF3CO)2O
R= vindoline

The solvolysis study of chlorotabersonine which aimed at synthesis of meloscine-type substances [Hofheinz, 1976] showed the formation of an indolo[2,3-*b*]quinoline derivative. Double fragmentation, hydrolysis, and Diels–Alder reaction would be the steps of the reaction.

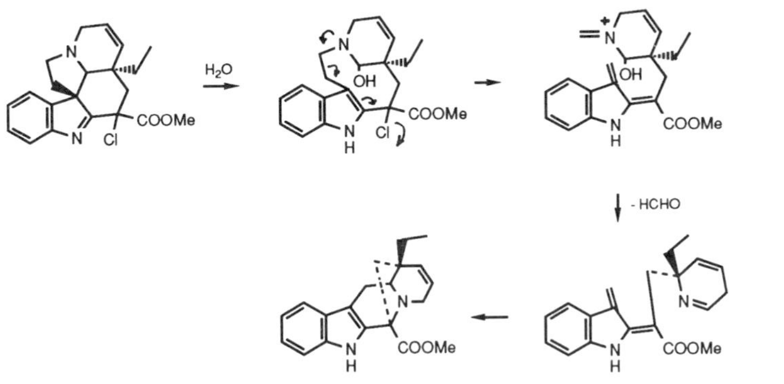
H2O
COOMe
- HCHO

5-(Tricyanoethenyl)-1,2,4-thiadiazoles undergo disproportionation in ethanol [Franz, 1976]. A reasonable mechanism is shown here.

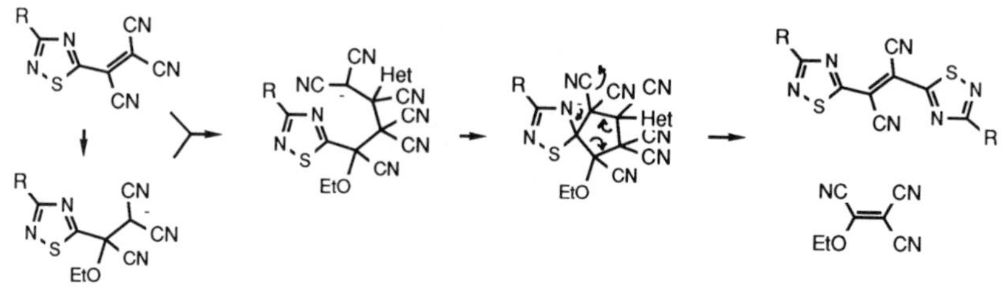

In the determination of the absolute configuration of macroline [Neukomm, 1981] a correlation with a degradation product of (+)-ajmaline was made. $N_b$-Methyl-21-deoxydihydroajmaline reacts with lead tetraacetate to give an indole, fragmentation of the cyclopentanol being affected.

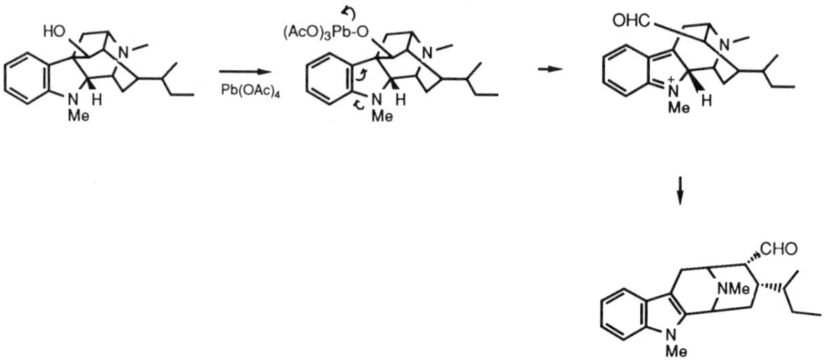

Cleavage of tetrahydro-$\beta$-carbolines is facilitated by proper activation at the $N_b$ atom [Magnus, 1989b]. A well-established strategy for acquiring quebrachamine-type compounds is based on this fragmentation [Takano, 1980a]. Ironically, the intervention of the fragmentation during removal of lactam carbonyl [Takano, 1979a] prevented the completion of an alternative route to quebrachamine, the fragmented intermediate being afforded an opportunity to undergo a Diels–Alder reaction.

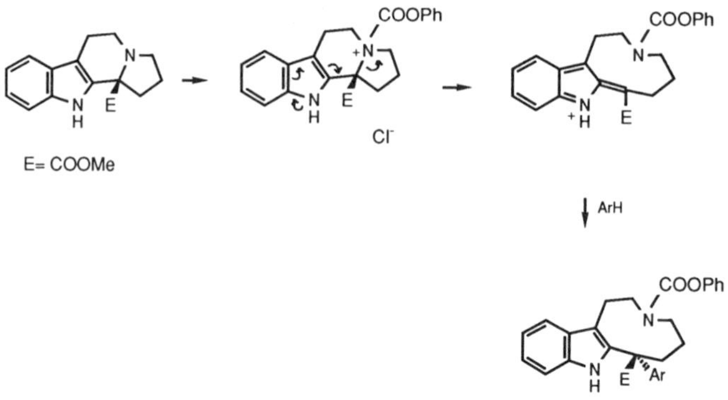

E= COOMe

Two interesting minor products from a Bischler–Napieralski cyclization of *N*-acetyltryptamine in the presence of indole are 2-methyltryptamine and *tris*(3-indolyl)methane [Frost, 1989], they being formed from the major reaction product by way of a cleavage-recombination series of reactions.

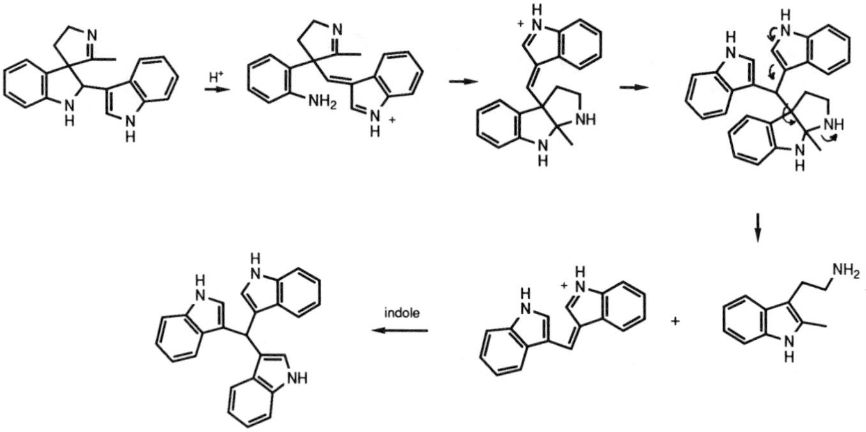

The loss of a $CH_2CH_2NH$ segment from a tryptamine derivative accompanies its annulation [Takano, 1979e]. The final step of C—C bond cleavage is favored by aromatization.

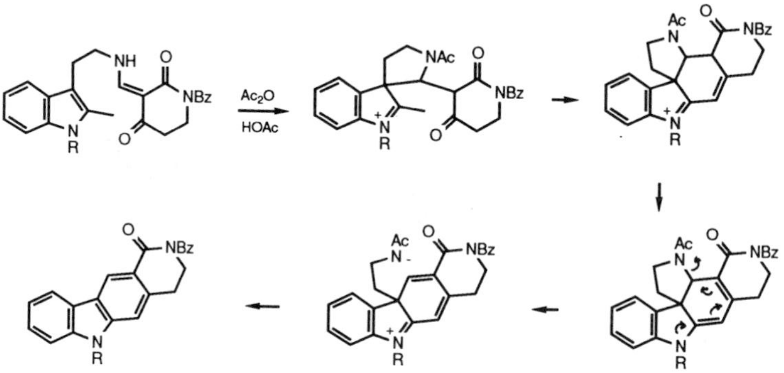

The action of hydroxylamine on diazepinones results in *N*-hydroxypyrrolones [J.A. Moore, 1959]. The reagent adds across the C=N and C=O bonds to afford bridged intermediates which decompose with elimination of the element of $CH_2$=N(R)—NH.

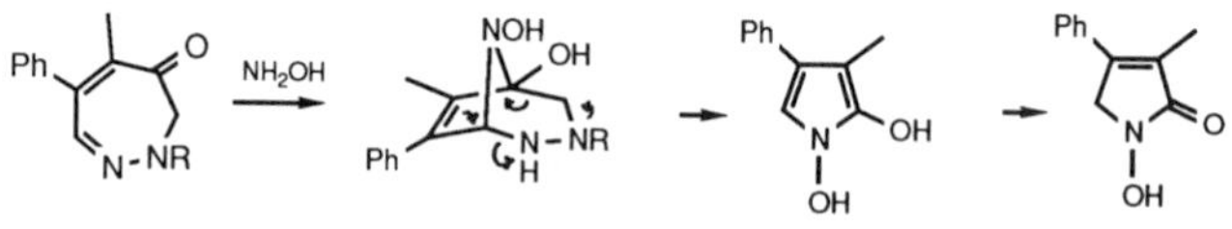

The Schmidt reaction products from 1,3-bishomocubanone monoketal include, in addition to the lactam, two interesting esters of rearranged skeletons [Mehta, 1980]. Both compounds are derivable from fragmentation with participation of the ketal oxygen and/or bromine when the latter is present at a bridgehead.

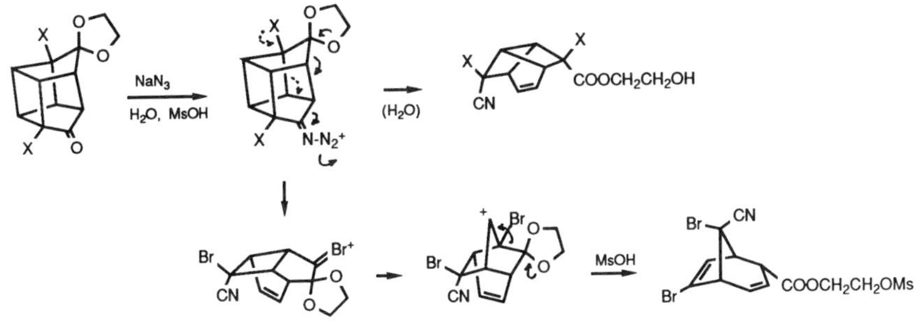

Wolff–Kishner reduction of *trans*-π-bromocamphor yields limonene [Gustafson, 1965]. The decomposition can be formulated as a **d5d** fragmentation.

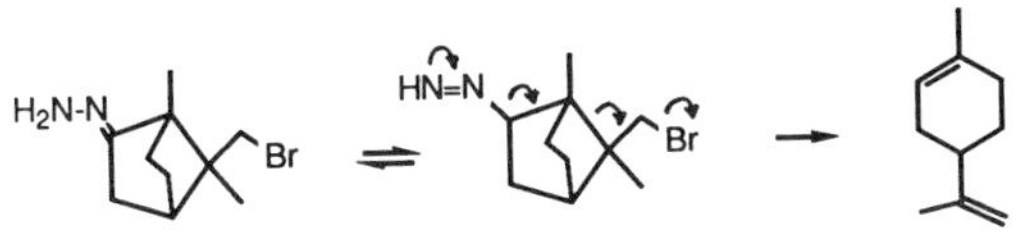

During a synthesis of strychnine [Woodward, 1963a] a Dieckmann condensation was attempted to construct ring-C. However, an $N_b$-tosyl group which is five atoms away from the anionic center rendered the molecule fragmentable.

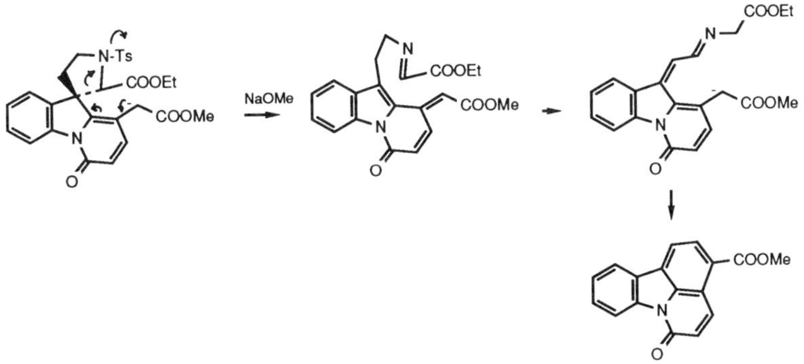

## 9.5. d6a=d SYSTEMS

Alcoholysis of *C*-acyl Meldrum's acid represents a convenient method for synthesis of β-keto esters [Oikawa, 1978].

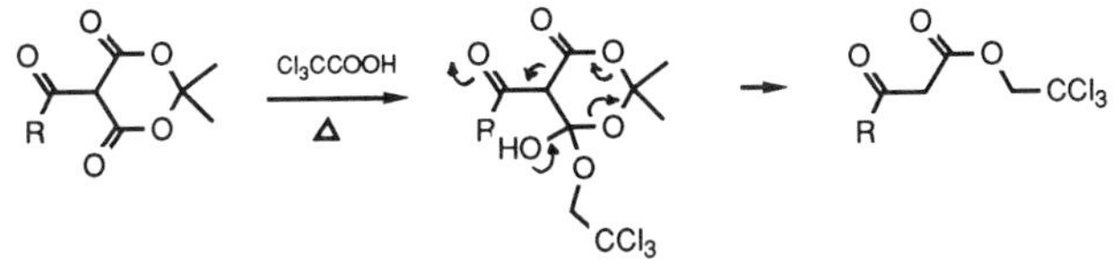

An unexpected transformation took place in the reaction of the diketone obtained from cyclopentadiene and 2,5-dihalo-1,4-benzoquinone by thermal and photocycloadditions with ethyl diazoacetate [Marchand, 1988]. There were two consecutive fragmentation processes succeeding the ring enlargement.

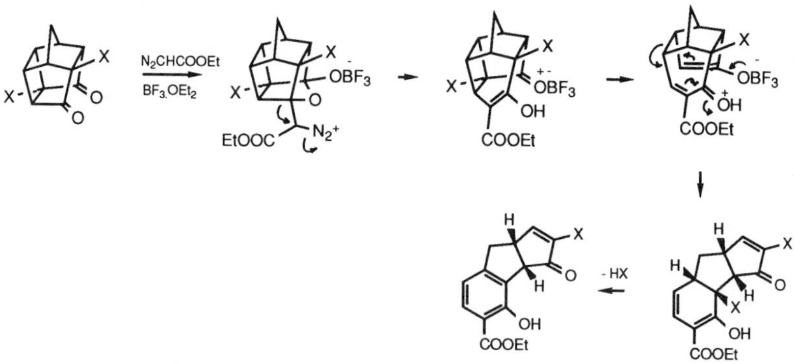

Triphenylphosphine converts the epidiepoxide generated from photooxygenation of 2,3-homotropone into 3,5-cyclooctadien-1,2-dione [Ito, 1975]. The reduction of O—O linkage and elimination of triphenylphosphine oxide results in a homotropolone which tautomerizes (fragments).

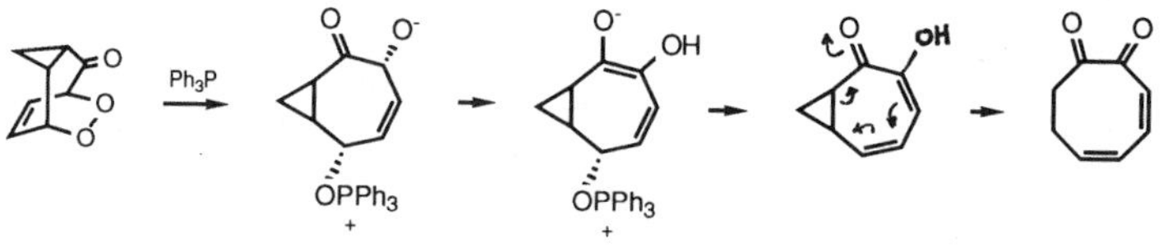

The prototype of a B-homophyllocladane system is accessible via intramolecular carbene addition and acid-catalyzed fragmentation [Ghatak, 1977]. The participation of an oxy substituent aromatic substituent in the cleavage of an allylic C—C bond is also involved in the rearrangement grisadienediones to depsidones [Sala, 1978].

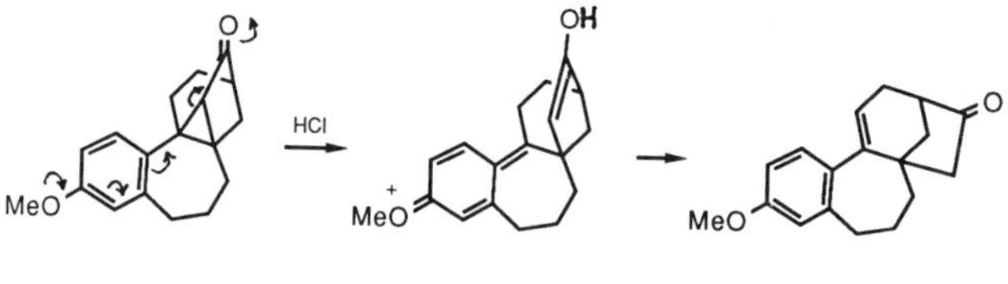

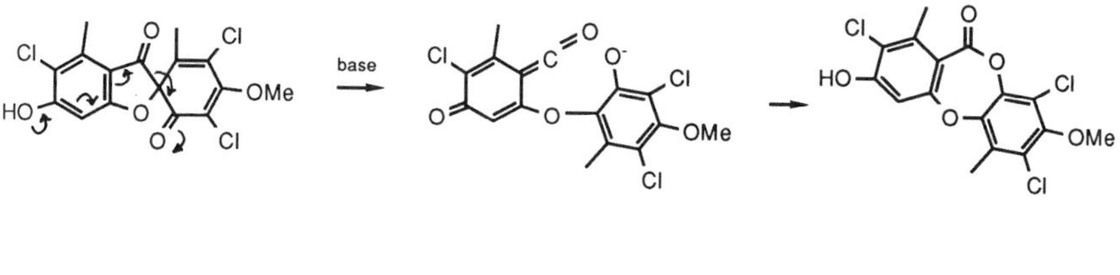

Oxidation of methyldopa ethyl ester gives rise to a quinonemethide which aromatizes by a formal ester group migration. The formation of a cyclopropanone hemiacetal intermediate seems reasonable [Cheng, 1982].

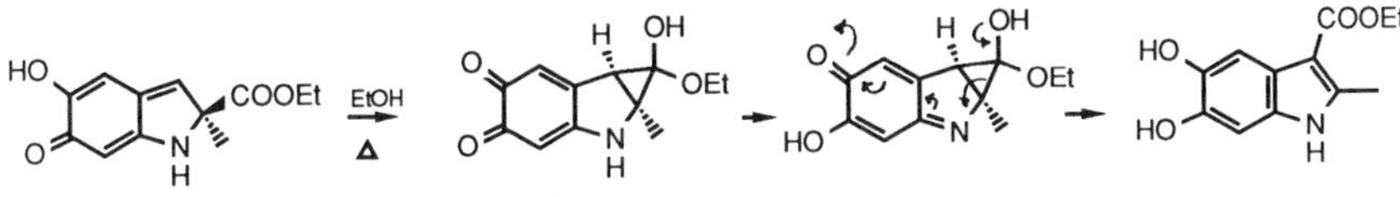

An unusual ring fission of a hydrophenanthrenone [Durani, 1983] indicates an important role of the methoxy substituent on the aromatic ring in directing the reaction course.

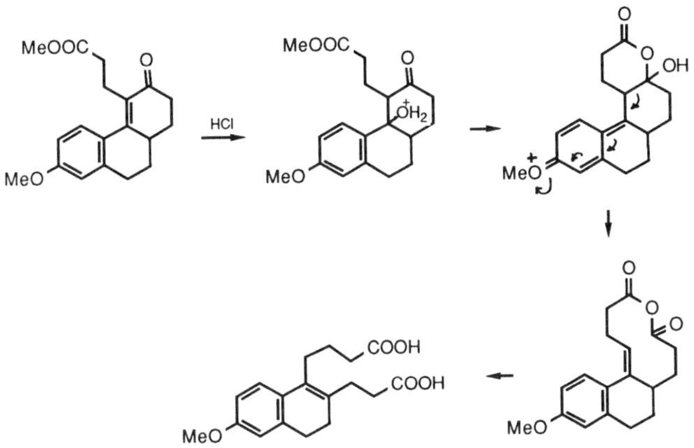

The AB-1 dimer of acronycine is a precursor of another dimer AB-2 [Funayama, 1989]. The rearrangement to the latter follows a fragmentation–reattachment route.

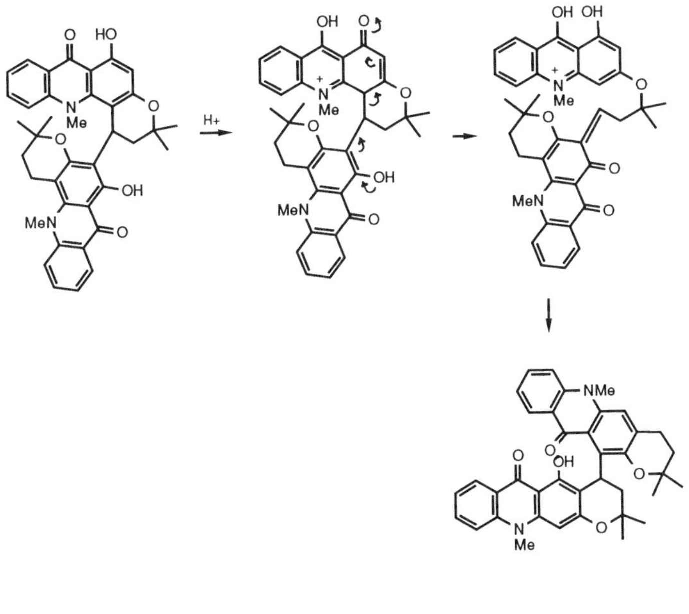

The thermal decomposition of bisanthraquinone pigments such as rugulosin [Sankawa, 1968] must be initiated by retro-aldol reactions that sever two of the interconnecting C—C bonds. Aromatization provides the driving force for the dehydrative fragmentation which results in two anthraquinone moieties.

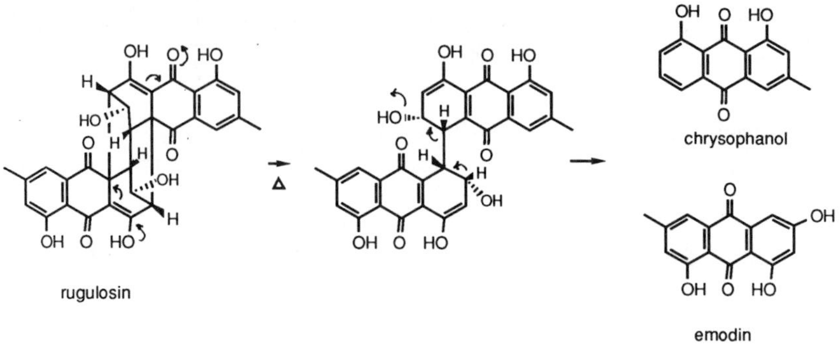

rugulosin

chrysophanol

emodin

A peroxidase-catalyzed oxidation of 2′,4,4′-trihydroxychalcone is a quinol ether called chalaurenol [Begly, 1982]. A fragmentation mechanism for its formation has been suggested.

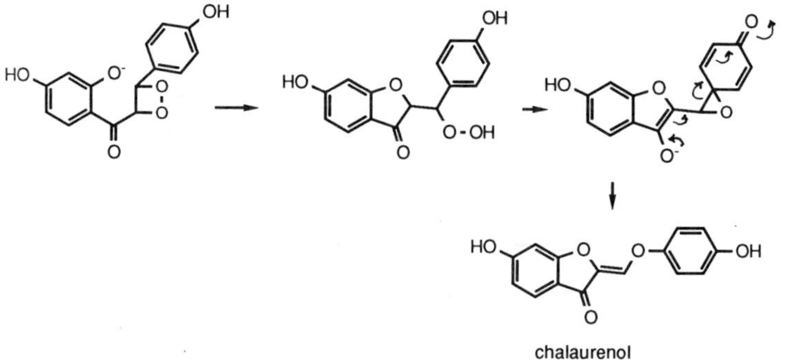

chalaurenol

Ethyl 2-cyano-3-(4-hydroxyphenyl)acrylate is degraded by hydrogen peroxide to furnish hydroquinone [Igarashi, 1967]. Apparently a spirocyclic epoxide formation supersedes the normal reaction.

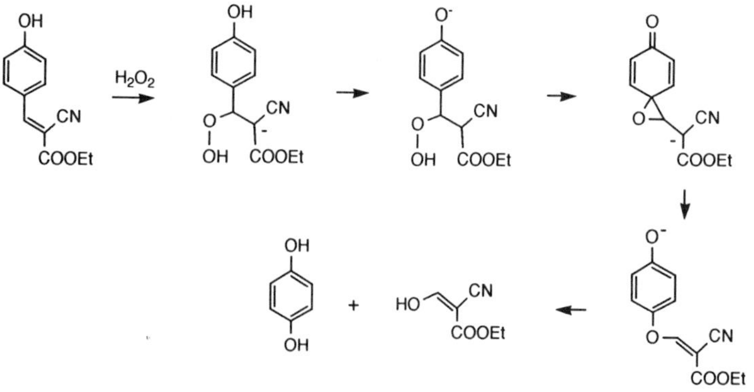

(1*S*,2*S*)-1-Hydroxy-2-[(*S*)-valylamino]cyclobutane-1-acetic acid may target towards crucial pyridoxal enzymes such as cystathionine-γ-synthetase of bacteria via in situ hydrolysis to give the aminocyclobutane. Condensation and fragmentation of the Schiff base would lead to inhibition [Adlington, 1983].

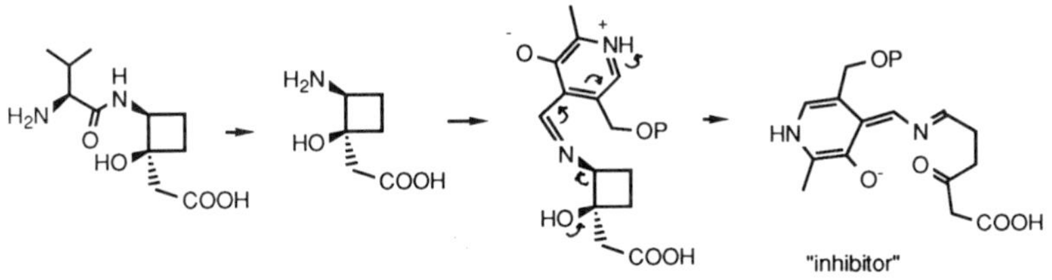

"inhibitor"

(21*H*,24*H*)-Bilin-1,19-diones have been cleaved by thiobarbituric acid, yielding two pyrromethenone fragments [Manitto, 1980]. An adduct emerges when protonation of the pyrrole ring is suppressed by a base. The protonation is prerequisite of the fragmentation.

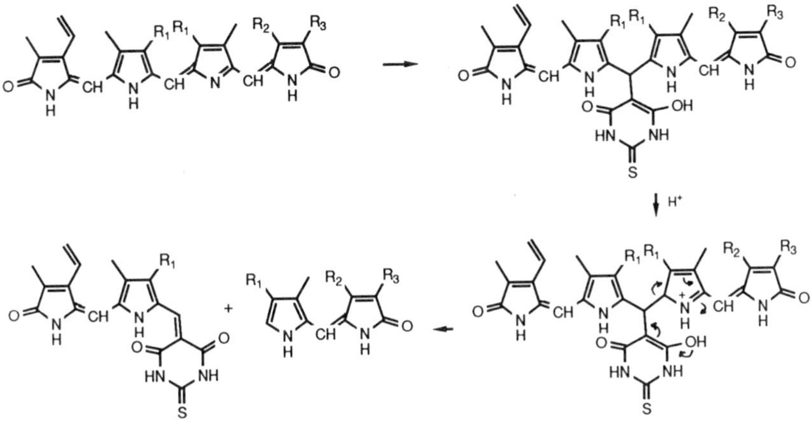

Isomerization of (−)-sophoramine to (+)-isosophoramine [Okuda, 1963] involves a transannular fragmentation-reclosure sequence.

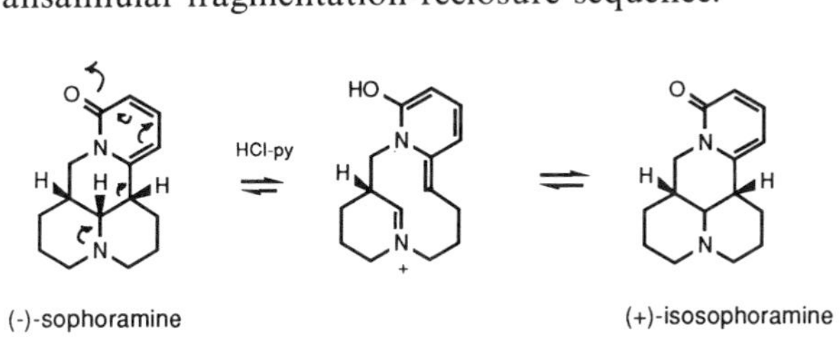

(-)-sophoramine

(+)-isosophoramine

## 9.6. d6d SYSTEMS

Fragmentation of the **d6d** array usually involve oxidoreduction because of the disjoint arrangement of the polar substituents. Two examples are given here.

The sidechain degradation of tryptophan derivatives and indole-3-acetic acid has been accomplished by using oxidizing agents including iodobenzene diacetate [Moriarty, 1985].

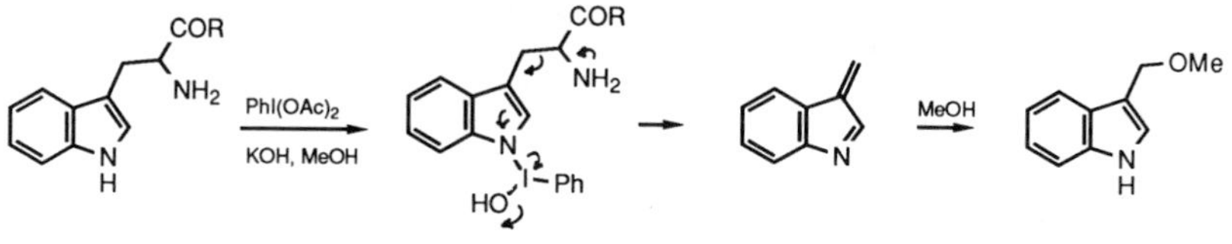

Nitro compounds afford nitrenes on heating with organic phosphites through stepwise deoxygenation. An unusual reaction course is followed by 1-(2′-nitrophenethyl)-tetrahydroisoquinolines because the amino nitrogen participates in the expulsion of triethyl phosphate from the first adducts. The spirocyclic intermediates generate aromatic nitrenes by elimination of ethylene [Kametani, 1975].

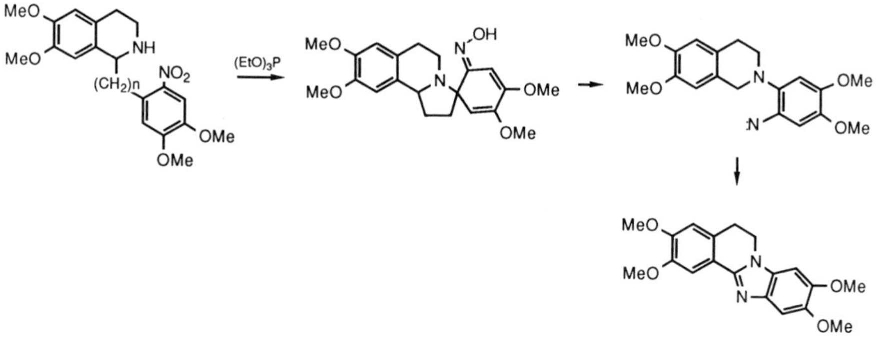

## 9.7. d7d SYSTEMS

An apparent transannular dehydrobromination of 6-bromo-7-oxototarol occurs when it is treated with sodium bicarbonate in dimethyl sulfoxide [Cambie, 1968]. The participation of the phenolic oxygen atom is critical.

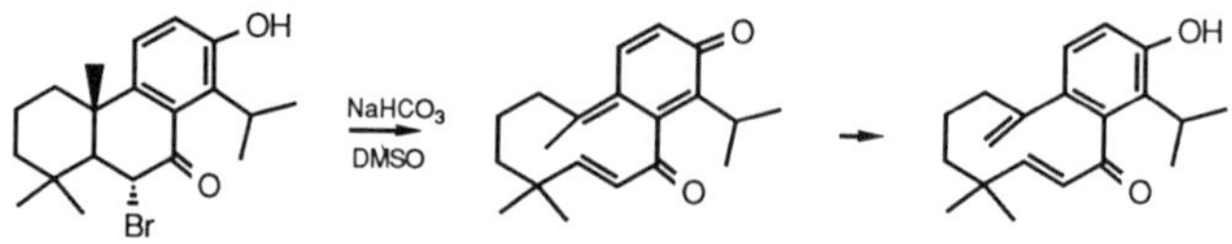

The diketones arisen from Swern-type oxidation of 7-halobicyclo[4.1.0]-heptane-2,3- or 3,4-diols undergo dehydrohalogenation in situ [Banwell, 1985] because of the presence of a **d5d** or **d7d** circuit in the norcaradiene tautomers.

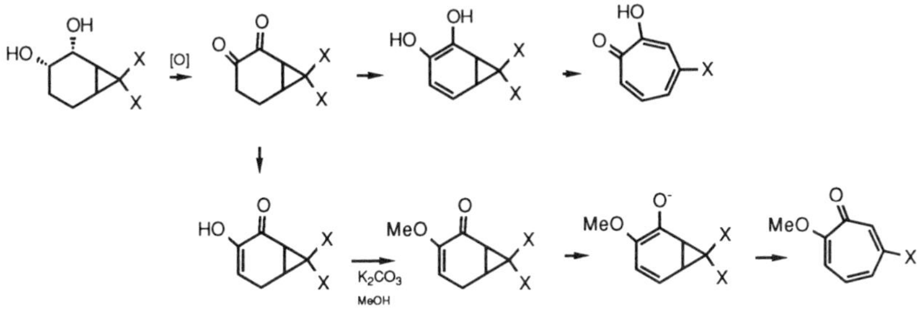

Magallanesine is probably an artifact originating from reaction of berberine with chloroform and oxygen followed by rearrangement [Valencia, 1985]. If the amidic nitrogen did not participate in the dechlorination step then one of the oxygen substituents must have, and the fragmentation is appropriately a **d7d** version.

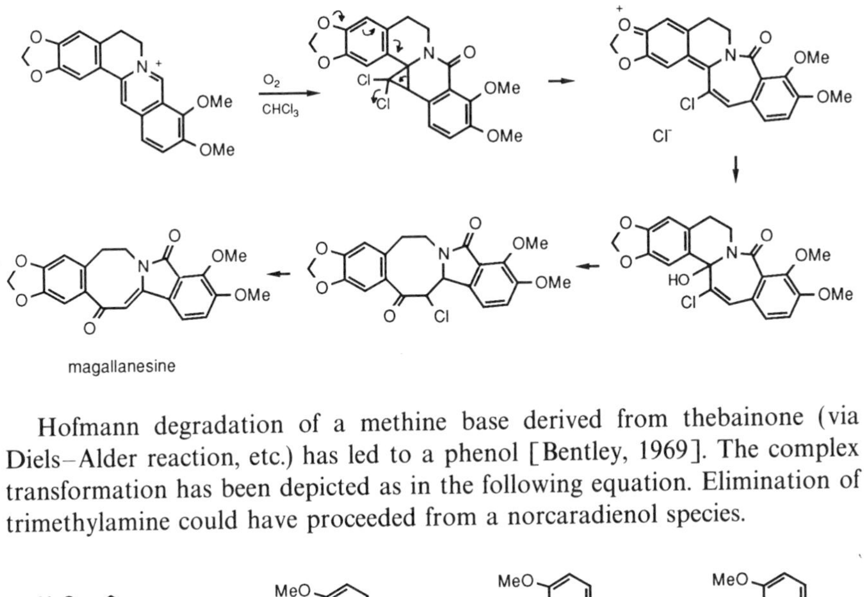

magallanesine

Hofmann degradation of a methine base derived from thebainone (via Diels–Alder reaction, etc.) has led to a phenol [Bentley, 1969]. The complex transformation has been depicted as in the following equation. Elimination of trimethylamine could have proceeded from a norcaradienol species.

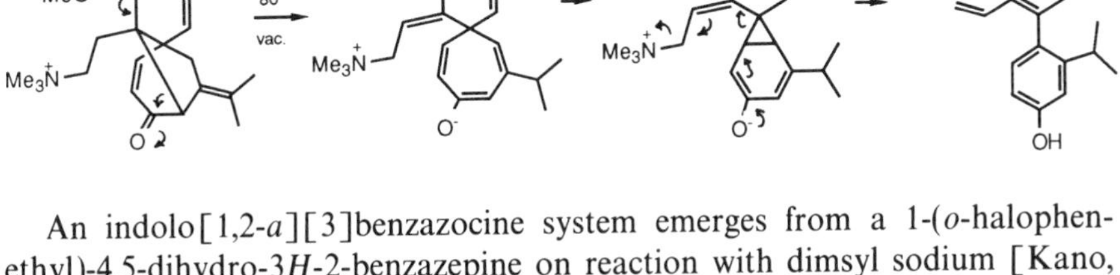

An indolo[1,2-*a*][3]benzazocine system emerges from a 1-(*o*-halophenethyl)-4,5-dihydro-3*H*-2-benzazepine on reaction with dimsyl sodium [Kano, 1978]. This transformation may be considered as involving benzyne interception by the nitrogen atom, followed by nucleophilic attack of the dimsyl anion on the iminium species prior to a fragmentative reorganization.

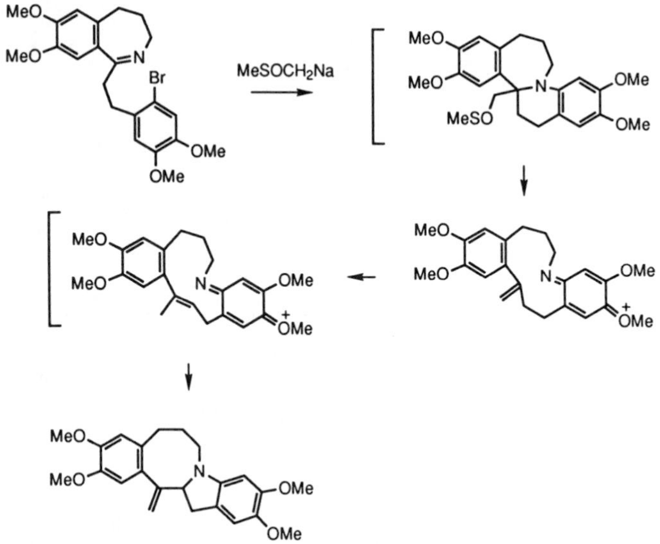

The major product from Beckmann rearrangement of a dibenzosuberone oxime with phosphorus pentachloride is a biphenyl [Forbes, 1968a]. Cleavage of the C—C bond is greatly assisted by the *p*-methoxy substituent.

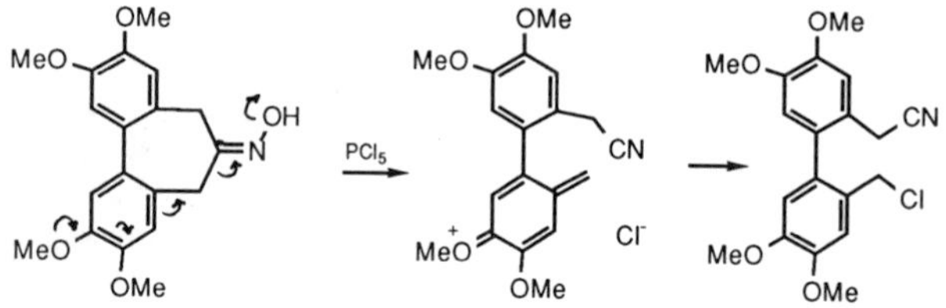

A crucial step in a synthesis of steganone [Magnus, 1984] is the solvolytic generation of the eight-membered ring from a cyclopropylcarbinol. This process was assisted by an oxygen substituent in the *para*-position of an aromatic ring. Formally, the reaction is a **d7d** fragmentation.

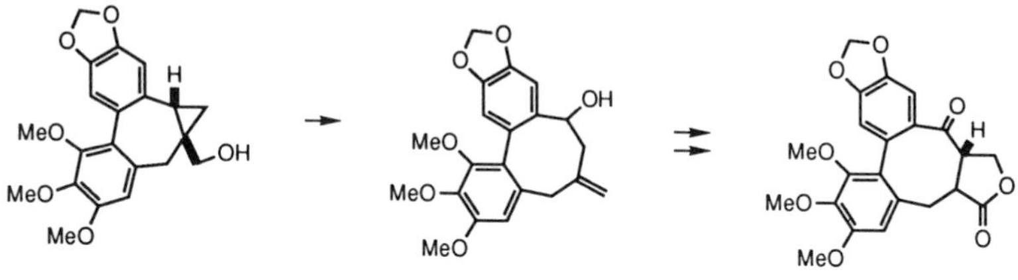

steganone

## 9.8. d8a=d SYSTEMS

The reductive cleavage of an aromatic moiety from a vinblastine model [Magnus, 1987b] has considerable bearings on the mode of action of the bisindole anticancer drugs. The fragmented species is susceptible to attack by nucleophiles which would restore aromaticity of the system.

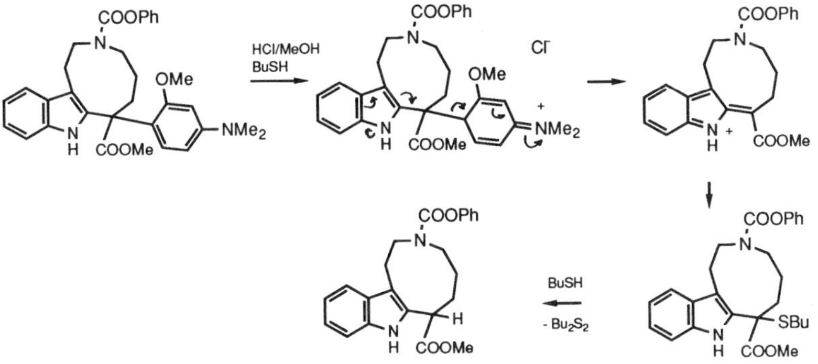

The remarkable cyclization with inversion of ring-D during biosynthesis of uroporphyrinogen-III from hydroxymethylbilane may involve a spirocyclic intermediate. Support for this mechanistic proposal has come from reaction of a linear analog with acid [Hawker, 1987]. Although four products have been obtained, by far the major (62%) tripyrrole has a structure which could be accommodated only by the fragmentation–recombination scheme. Results from experiments with mixture of α-free pyrromethanes [Jackson, 1989] are also consistent with such a route.

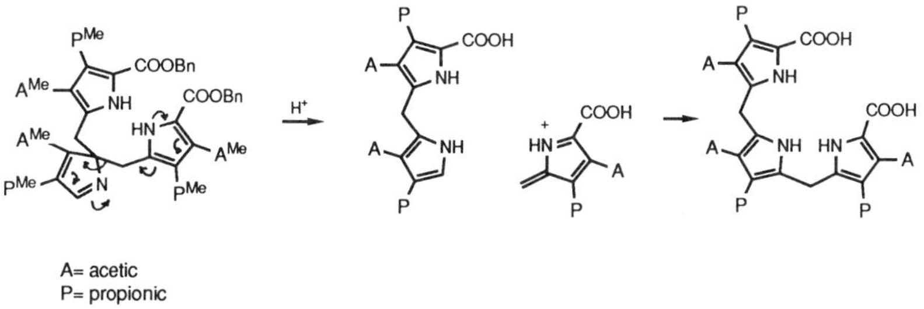

A= acetic
P= propionic

# REFERENCES


Abadir, B.J., Cook, J.W., Loudon, J.D., Steel, D.K.V. (1953) *J. Chem. Soc.* 2350.

Abramovitch, R.A., Cue, B.W., Jr. (1973) *J. Org. Chem.* **38:** 173.

Acheson, R.M., Bridson, J.N., Cameron, T.S. (1972) *J. Chem. Soc. Perkin Trans. I* 968.

Adam, G. (1967) *Liebigs Ann. Chem.* **709**: 191.

Adam, G., Voight, D., Schreiber, K. (1971) *Tetrahedron* **27**: 2181.

Adam, W., Rodriguez, A. (1981) *Tetrahedron Lett.* **22**: 3505.

Adams, J., Leblanc, Y., Rokach, J. (1984a) *Tetrahedron Lett.* **25**: 1227.

Adams, J., Rokach, J. (1984b) *Tetrahedron Lett.* **25**: 35.

Adams, J., Belley, M. (1986) *J. Org. Chem.* **51**: 3878.

Adcock, W., Coope, J., Shiner, V.J. (1990) *J. Org. Chem.* **55**: 1411.

Adderley, C.J.R., Baddeley, G.V., Hewgill, F.R. (1967) *Tetrahedron* **23**: 4143.

Adlington, R.M., Baldwin, J.E., Jones, R.H., Murphy, J.A., Parisi, M.F. (1983) *Chem. Commun.* 1479.

Agathocleous, D., Cox, G., Page, M.I. (1985) *Tetrahedron Lett.* **27**: 1631.

Agosta, W.C., Herron, D.K. (1969) *J. Org. Chem.* **34**: 2782.

Agosta, W.C., Foster, A.M. (1971) *Chem. Commun.* **34**: 433.

Ahmad, A., Spenser, I.D. (1960) *Can. J. Chem.* **38**: 1625.

Ahmad, F.B.H., Bruce, J.M., Khalafy, J., Sabetian, K. (1981) *Chem. Commun.* 169.

Ahmad, S.A., Rigby, W., Taylor, R.B. (1966) *J. Chem. Soc.* (*C*) 772.

Ahmed, Z., Cava, M.P. (1983) *J. Am. Chem. Soc.* **105**: 682.

Aimi, N., Inaba, I., Watanabe, M., Shibata, S. (1969) *Tetrahedron* **25**: 1825.

Ainley, A.D., Robinson, R. (1937) *J. Chem. Soc.* 369.

Akpuaka, M.U., Beddoes, R.L., Bruce, J.M., Fitzjohn, S., Millis, O.S. (1982) *Chem. Commun.* 686.

Albert, R.M., Butler, G.B. (1977) *J. Org. Chem.* **42**: 674.

Alder, R.W., Colclough, D., Mowlam, R.W. (1991) *Tetrahedron Lett.* **32**: 7755.

Allen, A.C., Cooper, D.A., Moore, J.M., Teer, C.B. (1984) *J. Org. Chem.* **49**: 3462.

Allred, E.L., Smith, R.L. (1966) *J. Org. Chem.* **31**: 3498.

Allred, E.L., Flynn, C.R. (1975) *J. Am. Chem. Soc.* **97**: 614.

Almansa, C., Carceller, E., Moyano, A., Serratosa, F. (1986) *Tetrahedron* **42**: 3637.

Alonso, M.E. (1987) *Art of problem solving in organic chemistry*, Wiley, NY, 225.

Ananthanarayan, T.P., Gallagher, T., Magnus, P. (1982) *Chem. Commun.* 709.

Ananthanarayan, T.P., Magnus, P., Norman, A.W. (1983) *Chem. Commun.* 1096.

Anciaux, A., Eman, A., Dumont, W., Krief, A. (1975) *Tetrahedron Lett.* 1617.

Ando, M., Tajima, K., Takase, K. (1978) *Chem. Lett.* 617.

Andriamialisoa, R.Z., Diatta, L., Rasoanaivo, P., Langlois, N., Potier, P. (1975) *Tetrahedron* **31**: 2347.

Andriantsiferana, M., Besselievre, R., Riche, C., Husson, H.-P. (1977) *Tetrahedron Lett.* 2587.

Aneja, R., Pelletier, S.W. (1964) *Tetrahedron Lett.* 669.

Anschütz, R. (1880) *Ber.* **13**: 1541.

Aono, M., Terao, Y., Achiwa, K. (1986) *Heterocycles* **24**: 313.

ApSimon, J.W., Buccini, J.A., Chau, A.S.Y. (1974) *Tetrahedron Lett.* 539.

Arata, Y., Tanaka, K., Yoshifuji, S., Kanemoto, S. (1979) *Chem. Pharm. Bull.* **27**: 981.

Arcus, C.L., Prydal, B.S. (1954) *J. Chem. Soc.* 4018.

Ardill, H., Grigg, R., Sridharan, V., Malone, J. (1987) *Chem. Commun.* 1296.

Arigoni, D., Barton, D.H.R., Corey, E.J., Jeger, O., Caglioti, L., Dev, S., Ferrini, P.G., Glazier, E.R., Melera, A., Pradhan, S.K., Schaffner, K., Sternhell, S. Templeton, J.E., Tobinaga, S. (1960) *Experientia* **16**: 41.

Arigoni, D. (1962) *Gazz. Chim. Ital.* **92**: 884.

Arlt, D., Jautelat, M., Lantzsch, R. (1981) *Angew. Chem. Intern. Ed.* **20**: 703.

Aschan, O. (1894) *Ber.* **27**: 3504.

Ashcroft, A.E., Davies, D.T., Sutherland, J.K. (1984) *Tetrahedron* **40**: 4579.

Atkinson, R.S., Miller, J.E. (1977) *Chem. Commun.* 35.

Atta-ur-Rahman, Bashir, M. (1983) *Heterocycles* **20**: 59.

Atta-ur-Rahman, Malik, S., Albert, K. (1986) *Z. Naturforsch.* **B41**: 386.

Attwood, M.R., Brown, B.R., Lisseter, S.G., Torrero, C.L., Weaver, P.M. (1984) *Chem. Commun.* 177.

Autry, R.L., Scullard, P.W. (1968) *J. Am. Chem. Soc.* **90**: 4917.

Ayer, W.A., Ward, D.E., Browne, L.J., Delbaere, L.T.J., Hoyano, Y. (1981) *Can. J. Chem.* **59**: 26665.

Babler, J.H., Moormann, A.E. (1976) *J. Org. Chem.* **41**: 1477.

Bac, N.V., Langlois, Y. (1982) *J. Am. Chem. Soc.* **104**: 7666.

Bach, R.D., Tubergen, M.W., Klix, R.C. (1986) *Tetrahedron Lett.* **27**: 3565.

Bacos, D., Celerier, J.P., Lhommet, G. (1987) *Tetrahedron Lett.* **28**: 2353.

Baer, H.H., Kovar, J. (1971a) *Can. J. Chem.* **49**: 1940.

Baer, H.H., Rank, W. (1971b) *Can. J. Chem.* **49**: 3197.

Baeyer, A., Drewsen, V. (1882) *Ber.* **15**: 2856.

Bagli, J.F., Bogri, T. (1972) *J. Org. Chem.* **37**: 2132.

Bahadur, G.A., Bailey, A.S., Costello, G., Scott, P.W. (1979) *J. Chem. Soc. Perkin Trans. I* 2154.

Bailey, A.S., Holton, A.G., Seager, J.F. (1972) *J. Chem. Soc. Perkin Trans. I* 1003.

Bailey, A.S., Vandrevala, M.H. (1980) *J. Chem. Soc. Perkin Trans. I* 1512.

Bailey, A.S., Ellis, J.H., Peach, J.M., Pearman, M.L. (1983) *J. Chem. Soc. Perkin Trans. I* 2425.

Baisted, D.J., Whitehurst, J.S. (1965) *J. Chem. Soc.* 2340.

Bakale, R.P., Scialdone, M.A., Johnson, C.R. (1990) *J. Am. Chem. Soc.* **112**: 6729.

Baker, R., Mason, T.J., Salter, J.C. (1970) *Chem. Commun.* 509.

Balasubramanian, K.K., Selvaraj, S. (1980) *Synthesis* 138.

Baldwin, J.E., Ganno, J.E. (1969) *Tetrahedron Lett.* 1101.

Baldwin, S.W., Page, E.H., Jr. (1972) *Chem. Commun.* 1337.

Baldwin, S.W., Crimmins, M.T. (1980) *J. Am. Chem. Soc.* **102**: 1198.

Baldwin, S.W. (1981) in (A. Padwa, ed.) *Organic photochemistry*, Chap. 2.

Baldwin, S.W., Blomquist, H.R. (1982a) *J. Am. Chem. Soc.* **104**: 4990.

Baldwin, S.W., Crimmins, M.T. (1982b) *J. Am. Chem. Soc.* **104**: 1132.

Baldwin, S.W., Fredericks, J.E. (1982c) *Tetrahedron Lett.* **23**: 1235.

Baldwin, S.W., Landmesser, N.G. (1982d) *Tetrahedron Lett.* **23**: 4443.

Baldwin, S.W., Martin, G.F., Nunn, D.S. (1985) *J. Org. Chem.* **50**: 5720.

Baldwin, S.W., McIver, J.M. (1987) *J. Org. Chem.* **52**: 320.

Ballantine, J.A., Alam, M., Fishhlock, G.W. (1977) *J. Chem. Soc. Perkin Trans. I* 1781.

Ban, Y., Yoshida, K., Goto, J., Oishi, T., Takeda, E. (1983) *Tetrahedron* **39**: 3657.

Banerjee, A.K., Hurtado, H.E. (1985) *Tetrahedron* **41**: 3029.

Bang, L., Guest, I.G., Ourisson, G. (1972) *Tetrahedron Lett.* 2089.

Bang, L., Ourisson, G. (1973) *Tetrahedron* **29**: 2097.

Banholzer, K., Schmid, M. (1956) *Helv. Chim. Acta* **39**: 548.

Banwell, M.G. (1983) *Chem. Commun.* 1453.

Banwell, M.G., Onrust, R. (1985) *Tetrahedron Lett.* **26**: 4543.

Baraldi, P.G., Barco, A., Benetti, S., Pollini, G.P., Zanirato, V. (1984) *Tetrahedron Lett.* **25**: 4291.

Barczai-Beke, M., Dornyei, G., Toth, G., Tamas, J., Czantay, C. (1976) *Tetrahedron* **32**: 1153.

Barker, A.J., Pattenden, G. (1983) *J. Chem. Soc. Perkin Trans. I* 1901.

Barneis, Z.J., Carr, J.D., Warnet, R.J., Wheeler, D.M.S. (1968) *Tetrahedron* **24**: 5053.

Bartlett, M.F., Dickel, D.F., Taylor, W.I. (1958) *J. Am. Chem. Soc.* **80**: 126.

Bartlett, M.F., Sklar, R., Taylor, W.I., Schlittler, E., Amai, R.L.S., Beak, P., Bringi, N.V., Wenkert, E. (1962) *J. Am. Chem. Soc.* **84**: 622.

Bartlett, P.D., Ando, T. (1970) *J. Am. Chem. Soc.* **92**: 7518.

Barton, D.H.R., Brooks, C.J.W., Holness, N.J. (1951) *J. Chem. Soc.* 278.

Barton, D.H.R., Holness, N.J., Overton, K.H., Rosenfelder, W.J. (1952) *J. Chem. Soc.* 3751.

Barton, D.H.R., deMayo, P. (1956) *J. Chem. Soc.* 142.

Barton, D.H.R., Cheung, H.T., Cross, A.D., Jackman, L.M., Martin-Smith, M. (1961) *J. Chem. Soc.* 5061.

Barton, D.H.R., Sutherland, J.K. (1965) *J. Chem. Soc.* 1769.

Barton, D.H.R., Werstiuk, N.H. (1968) *J. Chem. Soc.* (*C*) 148.

Barton, D.H.R., Garbers, C.F., Giacopello, D., Harvey, R.G., Lessard, J., Taylor, D.R. (1969) *J. Chem. Soc.* (*C*) 1050.

Barton, D.H.R., Bhatnagar, N.Y., Blazejewski, J.-C., Charpiot, B., Finet, J.P., Lester, D.J., Motherwell, W.B., Papoula, M.T.B., Stanforth, S.P. (1985) *J. Chem. Soc. Perkin Trans. I* 2657.

Basyne, D.W., Nicol, A.J., Tennant, G. (1975) *Chem. Commun.* 782.

Bates, H.A., Rapoport, H. (1979) *J. Am. Chem. Soc.* **101**: 1259.

Battersby, A.R., Bhatnagar, A.K., Hackett, P., Thornber, C.W., Staunton, J. (1968) *Chem. Commun.* 1214.

Batzold, F.H., Robinson, C.H. (1976) *J. Org. Chem.* **41**: 313.

Baudouy, R., Crabbe, P., Greene, A.E., LeDrian, C., Orr, A.F. (1977) *Tetrahedron Lett.* 2973.

Baxter, A.D., Roberts, S.M., Scheinmann, F., Wakefield, B.J., Newton, R.F. (1983) *Chem. Commun.* 932.

Baxter, I., Cameron, D.W., Sanders, J.K.M., Titman, R.B. (1972) *J. Chem. Soc. Perkin Trans. I* 2046.

Bayne, D.W., Nicol, A.J., Tennant, G. (1975) *Chem. Commun.* 782.

Beart, J.E., Lilley, T.H., Haslam, E. (1985) *J. Chem. Soc. Perkin Trans. I* 1429.

Becker, D., Harel, Z., Birnbaum, D. (1975) *Chem. Commun.* 377.

Becker, D., Birnbaum, D. (1980) *J. Org. Chem.* **45**: 570.

Beckmann, S., Geiger, H. (1965) *Chem. Ber.* **98**: 516.

Beecham, A.F. (1957) *J. Am. Chem. Soc.* **79**: 3257, 3262.

Beereboom, J.J. (1963) *J. Am. Chem. Soc.* **85**: 3525.

Begly, M.J., Crombie, L., London, M., Savin, J., Whiting, D.A. (1982) *Chem. Commun.* 1319.

Bégué, J.P., Charpentier-Morize, M., Pardo, C., Sansoulet, J. (1978) *Tetrahedron* **34**: 293.

Bennett, D.J., Kirby, G.W., Moss, V.A. (1970) *J. Chem. Soc.* (*C*) 2049.

Bennett, P., Donnelly, J.A., Fox, M.J. (1979) *J. Chem. Soc. Perkin Trans. I* 2990.

Bentley, K.W., Robinson, R. (1952) *J. Chem. Soc.* 947.

Bentley, K.W., Dominguez, J., Ringe, J.P. (1957) *J. Org. Chem.* **22**: 409, 418.

Bentley, K.W., Ball, J.C. (1958) *J. Org. Chem.* **23**: 1720.

Bentley, K.W., Crocker, H.P., Walser, R., Fulmor, W., Morton, G.O. (1969) *J. Chem. Soc.* (*C*) 2225.

Bentrude, W.G., Johnson, W.D., Khan, W.A., Witt, E.R. (1972) *J. Org. Chem.* **37**: 631.

Bergman, R.G., Rajadhyaksha, V.J. (1970) *J. Am. Chem. Soc.* **92**: 2163.

Beringer, F.M., Galton, S.A. (1963) *J. Org. Chem.* **28**: 3250.

Berkowitz, W.F. (1972) *J. Org. Chem.* **37**: 341.

Berkowitz, W.F., Grenetz, S.C. (1976) *J. Org. Chem.* **41**: 10.

Berkowitz, W.F., Perumattam, J., Amarasekara, A. (1985) *Tetrahedron Lett.* **26**: 3665.

Bernhard, H.O., Reed, J.N., Snieckus, V. (1977) *J. Org. Chem.* **42**: 1093.
Bertrand, M., Dulcere, J.P., Gil, G. (1980) *Tetrahedron Lett.* **21**: 1945.
Besselièvre, R., Husson, H.-P. (1976) *Tetrahedron Lett.* 1873.
Bettolo, R.M., Tagliatesta, P., Lupi, A., Bravetti, D. (1983) *Helv. Chim. Acta* **66**: 760.
Bhandari, K.S., Snieckus, V. (1971) *Synthesis* 327.
Bhat, V., Cookson, R.C. (1981) *Chem. Commun.* 1123.
Biellmann, J.F., Goeldner, M.P. (1971) *Tetrahedron* **27**: 1789.
Bienz, S., Guggisberg, A., Walchli, R., Hesse, M. (1988a) *Helv. Chim. Acta* **71**: 1708.
Bienz, S., Hesse, M. (1988b) *Helv. Chim. Acta* **71**: 1704.
Biggi, G., deHoog, A.J., Del Cima, F., Pietra, F. (1973) *J. Am. Chem. Soc.* **95**: 7108.
Bilinski, V., Steinfels, M.A., Dreiding, A.S. (1972) *Helv. Chim. Acta* **55**: 1075.
Billups, W.E., Blakeney, A.J. (1976) *J. Am. Chem. Soc.* **98**: 7817.
Birch, A.J., Lahey, F.N. (1953) *Aust. J. Chem.* **6**: 379.
Birch, A.H., Brown, J.N., Stansfield, F. (1964) *J. Chem. Soc.* 5343.
Birch, A.J., Hill, J.S. (1966) *J. Chem. Soc.* (*C*) 419.
Birch, A.J., Keeton, R. (1968) *J. Chem. Soc.* (*C*) 109.
Birch, A.J., Macdonald, P.L., Powell, V.H. (1969) *Tetrahedron Lett.* 351.
Birch, A.M., Pattenden, G. (1983) *J. Chem. Soc. Perkin Trans. I* 1913.
Bird, C.W., Brown, A.L., Chan, C.C., Lewis, A. (1989) *Tetrahedron Lett.* **30**: 6223.
Birladeanu, L., Harris, D.L., Winstein, S. (1970) *J. Am. Chem. Soc.* **92**: 6387.
Biswas, K.M., Jackson, A.H. (1969) *Tetrahedron* **25**: 227.
Blanco, L., Slougui, N., Rousseau, G., Conia, J.-M. (1981) *Tetrahedron Lett.* **22**: 645.
Blatt, A.H., Hawkins, W.L. (1936) *J. Am. Chem. Soc.* **58**: 81.
Blattel, R.A., Yates, P. (1972a) *Tetrahedron Lett.* 1069.
Blattel, R.A., Yates, P. (1972b) *Tetrahedron Lett.* 1073.
Block, E., Lafitte, J.-A., Eswarakrishnan, V. (1986) *J. Org. Chem.* **51**: 3428.
Boch, M., Korth, T., Nelke, J.M., Pike, D., Radunz, H., Winterfeldt, E. (1972) *Chem. Ber.* **105**: 2126.
Bodendorf, K., Mayer, R. (1965) *Chem. Ber.* **98**: 3554.
Boeckman, R.K., Jr., Arvanitis, A., Voss, M.E. (1989) *J. Am. Chem. Soc.* **111**: 2737.
Böhm, K., Gössinger, E., Müller, R. (1989) *Tetrahedron* **45**: 1391.
Boerth, D.W., Van Catledge, F.A. (1975) *J. Org. Chem.* **40**: 3319.
Böttger, D., Welsel, P. (1983) *Tetrahedron Lett.* **24**: 5201.
Bohlmann, F., Rufer, C. (1964) *Chem. Ber.* **97**: 1770.
Bohlmann, F., Schumann, D. (1965) *Chem. Ber.* **98**: 3133.
Boivin, J., Elkaim, L., Ferro, P.G., Zard, S.Z. (1991) *Tetrahedron Lett.* **32**: 5321.
Boland, W., Jaenicke, L. (1978) *Chem. Ber.* **111**: 3262.
Bond, R.F., Boeyens, J.C.A., Holzapfel, C.W., Steyns, P.S. (1979) *J. Chem. Soc. Perkin Trans. I* 1751.
Bondon, D., Pietrasanta, Y., Pucci, B. (1976) *Tetrahedron* **32**: 2401.
Bonnett, R., Campion-Smith, I.H., Page, A.J. (1977) *J. Chem. Soc. Perkin Trans. I* 68.
Bordwell, F.G., Scamehorn, R.G., Knipe, A.C. (1970) *J. Am. Chem. Soc.* **92**: 2172.
Borrevang, P., Hjort, J., Rapala, R.T., Edie, R. (1968) *Tetrahedron Lett.* 4905.

Borsche, W. (1912) *Liebigs Ann. Chem.* **390**: 1.

Bosch, J., Domingo, A., Linares, A. (1983a) *J. Org. Chem.* **48**: 1075.

Bosch, J., Moral, M., Rubiralta, M. (1983b) *Heterocycles* **20**: 509.

Bosch, J., Bennasar, M.-L., Zulaica, E. (1986) *J. Org. Chem.* **51**: 2289.

Bosshard, H.H., Mory, R., Schmid, M., Zollinger, H. (1959) *Helv. Chim. Acta* **42**: 1653.

Botta, M., Castelli, S., Gamacora, A. (1985) *Tetrahedron* **41**: 2913.

Bottini, A.T., Grob, C.A., Schumacher, E., Zergenyi, J. (1966) *Helv. Chim. Acta* **49**: 2516.

Bramwell, A.F., Crombie, L., Knight, M.H. (1965) *Chem. Commun.* 1265.

Bremner, J.B., Jaturonrusmee, W., Engelhardt, L.M., White, A.H. (1989) *Tetrahedron Lett.* **30**: 3213.

Brenneisen, P., Grob, C.A., Jackson, R.A., Ohta, M. (1965) *Helv. Chim. Acta* **48**: 146.

Brieger, G., Hachey, D.L., Ciaramitaro, D. (1969) *J. Org. Chem.* **34**: 220.

Brinker, U.H., Wüster, H., Maas, G. (1985) *Chem. Commun.* 1812.

Britten, A.Z., Joule, J.A., Smith, G.F. (1967) *Tetrahedron* **23**: 1971.

Brockmann, H., Lenk, W. (1959) *Chem. Ber.* **92**: 1880.

Brook, P.R., Kitson, D.E. (1978) *Chem. Commun.* 87.

Brown, D.S., Foster, R.W.G., Marples, B.A., Mason, K.G. (1980) *Tetrahedron Lett.* **21**: 5057.

Brown, E., Dhal, R., Robin, J.-P. (1979) *Tetrahedron Lett.* 733.

Brown, H.C., Prasad, J.V.N.V. (1985) *J. Org. Chem.* **50**: 3002.

Brown, H.L., Buchanan, G.L., O'Donnell, J. (1979) *J. Chem. Soc. Perkin Trans. I* 1740.

Brown, J.M., Cresp, T.M., Mander, L.N. (1977) *J. Org. Chem.* **42**: 3984.

Brown, R.T., Blackstock, W.P., Wingfield, M. (1984) *Tetrahedron* **25**: 1831.

Brownbridge, P., Fleming, I., Pearce, A., Warren, S. (1976) *Chem. Commun.* 751.

Bruce, J.M., Creed, D., Dawes, K. (1971) *J. Chem. Soc. (C)* 2244.

Bryant, L.R.B., Coyle, J.D., Challiner, J.F., Haws, E.J. (1984) *Tetrahedron Lett.* **25**: 1087.

Bryant, R.J., McDonald, E. (1975) *Tetrahedron Lett.* 3841.

Buchanan, G.L., Maxwell, C., Henderson, W. (1965) *Tetrahedron* **21**: 3273.

Buchanan, G.L., Ferguson, G., Lawson, A.M., Pollard, D.R. (1966a) *Tetrahedron Lett.* 530.

Buchanan, G.L., McLay, G.W. (1966b) *Tetrahedron* **22**: 1521.

Buchanan, G.L., Young, G.A.R. (1973) *J. Chem. Soc. Perkin Trans. I* 2404.

Buchanan, G.L., Kean, N.B., Taylor, R. (1975) *Tetrahedron* **31**: 1583.

Buchanan, J.G.S.C., Davis, B.R. (1967) *J. Chem. Soc. (C)* 1340.

Buchanan, J.G.S.C., Crump, D.R., Davis, B.R. (1979) *J. Chem. Soc. Perkin Trans. I* 2826.

Buckley, D., Gibson, M.S. (1956) *J. Chem. Soc.* 3242.

Büchi, G., Coffen, D.L., Kocsis, K., Sonnet, P.E., Ziegler, F.E. (1965) *J. Am. Chem. Soc.* **87**: 2073.

Büchi, G., Wüest, H. (1968) *Tetrahedron* **24**: 2049.

Büchi, G., Wüest, H. (1969) *J. Org. Chem.* **34**: 857.

Büchi, G., Kulsa, P., Ogasawara, K., Rosati, R.L. (1970) *J. Am. Chem. Soc.* **92**: 999.

Büchi, G., Carlson, J.A., Powell, J.E., Tietze, L.-F. (1973a) *J. Am. Chem. Soc.* **95**: 540.

Büchi, G., Hochstrasser, U., Pawlak, W. (1973b) *J. Org. Chem.* **38**: 4348.

Buggle, K., Power, J. (1980) *J. Chem. Soc. Perkin Trans. I* 1070.

Burckhardt, U., Grob, C.A., Kiefer, H.R. (1967) *Helv. Chim. Acta* **50**: 231.

Burgstahler, A.W., Walker, D.E., Jr., Kuebrich, J.P., Schowen, R.L. (1972) *J. Org. Chem.* **37**: 1272.

Burka, L.T., Bowen, R.M., Wilson, B.J., Harris, T.M. (1974) *J. Org. Chem.* **39**: 3241.

Burks, J.E., Jr., Crandall, J.K. (1984) *J. Org. Chem.* **49**: 4663.

Burrows, E.P., Szafraniec, L.L. (1986) *J. Org. Chem.* **51**: 4706.

Buyle, R. (1964) *Helv. Chim. Acta* **47**: 2449.

Byers, J.H., Spencer, T.A. (1985) *Tetrahedron Lett.* **26**: 713.

Caine, D., Frobese, A.A. (1977) *Tetrahedron Lett.* 3107.

Caine, D., Crews, E., Salvino, J.M. (1983) *Tetrahedron Lett.* **24**: 2083.

Caine, D., McCloskey, C.J., VanDerveer, D. (1985) *J. Org. Chem.* **50**: 175.

Caine, D., McCloskey, C.J., Atwood, J.L., Bott, S.G., Zhang, H.-M., VanDerveer, D. (1987a) *J. Org. Chem.* **52**: 1280.

Caine, D., Stanhope, B. (1987b) *Tetrahedron* **43**: 5545.

Cambie, R.C., Gallagher, R.T. (1968) *Tetrahedron* **24**: 4631.

Cambie, R.C., Crump, D.R., Duve, R.N. (1969) *Aust. J. Chem.* **22**: 1975.

Campbell, A.C., Maidment, M.S., Pick, J.H., Stevenson, D.F.M., Woods, G.F. (1978) *J. Chem. Soc. Perkin Trans. I* 163.

Campbell, S.F., Constantino, M.G., Brocksom, T.G., Petragnani, N. (1975) *Synth. Commun.* **5**: 353.

Cargill, R.L., Wright, B.W. (1975) *J. Org. Chem.* **40**: 120.

Carman, R.M., Brecknell, D.J., Deeth, H.C. (1966) *Tetrahedron Lett.* 4387.

Carruthers, W., Qureshi, M.I. (1973) *J. Chem. Soc. Perkin Trans. I* 51.

Carruthers, W., Orridge, A. (1977) *J. Chem. Soc. Perkin Trans. I* 2411.

Cartier, D., Ouahrani, M., Levy, J. (1989) *Tetrahedron Lett.* **30**: 1951.

Casadevall, E., Pouet, Y. (1978) *Tetrahedron* **34**: 1921.

Casals, P.-F., Ferard, J., Ropert, R. (1976) *Tetrahedron Lett.* 3077.

Castedo, L., Saá, C., Saá, J.M., Suau, R. (1982) *J. Org. Chem.* **47**: 513.

Castro, J.L., Castedo, L., Riguera, R. (1985) *Tetrahedron Lett.* **26**: 1561.

Cattanach, C.J., Cohen, A., Heath-Brown, B. (1968) *J. Chem. Soc. (C)* 1235.

Cauwberghs, S.G., DeClercq, P.J. (1988) *Tetrahedron Lett.* **29**: 6501.

Cava, M.P., Glamkowski, E.J., Weintraub, P.M. (1966) *J. Org. Chem.* **31**: 2755.

Cavill, G.W.K., Hall, C.D. (1967) *Tetrahedron* **23**: 1119.

Cech, J., Santavy, F. (1949) *Coll. Czech. Chem. Commun.* **14**: 532.

Cervantes, H., Khac, D.D., Fétizon, M., Guir, F., Beloeil, J.-C., Lallemand, J.-Y., Prangé, T. (1986) *Tetrahedron* **42**: 3491.

Chakraborty, D.P., Chowdhury, B.K. (1968) *J. Org. Chem.* **33**: 1265.

Challand, B.D., Kronis, G., Lange, G.L., deMayo, P. (1967) *Chem. Commun.* 704.

Challand, B.D., deMayo, P. (1968) *Chem. Commun.* 982.

Chalmers, A.M., Baker, A.J. (1974) *Tetrahedron Lett.* 4529.

Chamberlain, P., Rooney, A.E. (1979) *Tetrahedron Lett.* 383.

Chan, T.H., Massuda, D. (1977) *J. Am. Chem. Soc.* **99**: 936.

Chandrasekaran, S., Venkataramani, P.S., Srinivasan, K.G., Swaminathan, S. (1962) *Tetrahedron Lett.* 729.

Chass, D.A., Buddhasukh, D., Magnus, P.D. (1978) *J. Org. Chem.* **43**: 1750.

Cheng, A.C., Shulgin, A.T., Castagnoli, N. (1982) *J. Org. Chem.* **47**: 5258.

Chin-you, N., MacLean, D.B., Prakash, A., Calvo, C. (1971) *Can. J. Chem.* **49**: 3240.

Chow, K., Moore, H.W. (1987) *Tetrahedron Lett.* **28**: 5013.

Chu, A., Mander, L.N. (1988) *Tetrahedron Lett.* **29**: 2727.

Clark, R.D., Untch, K.G. (1979) *J. Org. Chem.* **44**: 248, 253.

Clarke, C., Fleming, I., Fortunak, J.M.D., Gallagher, P.T., Honan, M.C., Mann, A., Nubling, C.O., Raithby, P.R., Wolff, J.J. (1988) *Tetrahedron* **44**: 3931.

Clarke, T., Stewart, J.D., Ganem, B. (1987) *Tetrahedron Lett.* **28**: 6253.

Clauder, O., Gesztes, K., Szasz, K. (1962) *Tetrahedron Lett.* 1147.

Clausen, T.P., Keller, J.W., Reichardt, P.B. (1990) *Tetrahedron Lett.* **31**: 4537.

Closs, G.L., Schwartz, G.M. (1961) *J. Org. Chem.* **26**: 2609.

Coates, R.M., Sandefur, L.O., Smillie, R.D. (1975) *J. Am. Chem. Soc.* **97**: 1619.

Coates, R.M., Muskopf, J.W., Senter, P.A. (1985) *J. Org. Chem.* **50**: 3541.

Cocker, J.D., Halsall, T.G. (1957) *J. Chem. Soc.* 3441.

Cocker, W., Lauder, H.S.J., Shannon, P.V.R. (1974) *J. Chem. Soc. Perkin Trans. I* 194.

Cocker, W., Grayson, D.H. (1975) *J. Chem. Soc. Perkin Trans. I* 1347.

Cohen, T., Yu, L.-C., Daniewski, W.M. (1985) *J. Org. Chem.* **50**: 4596.

Coke, J.L., Williams, H.J., Natarajan, S. (1977) *J. Org. Chem.* **42**: 2380.

Collado, I.G., Macias, F.A., Massanet, G.M., Luis, F.R. (1986) *Tetrahedron* **42**: 3611.

Colvin, E.W., Raphael, R.A., Roberts, J.S. (1971) *Chem. Commun.* 858.

Colvin, E.W., Malchenko, S., Raphael, R.A., Roberts, J.S. (1973) *J. Chem. Soc. Perkin Trans. I* 1989.

Comins, D.L., Abdullah, A.H., Smith, R.K. (1983) *Tetrahedron Lett.* **24**: 2711.

Conant, J.B., Coyne, B.B. (1922) *J. Am. Chem. Soc.* **44**: 2530.

Confalone, P., Pizzolato, G. (1981) *J. Am. Chem. Soc.* **103**: 4251.

Connolly, J.D., McCrindle, R., Overton, K.H. (1968) *Tetrahedron* **24**: 1489.

Connolly, J.D., Henderson, R., McCrindle, R., Overton, K.H., Bhacca, N.S. (1973) *J. Chem. Soc. Perkin Trans. I* 865.

Connolly, P.J., Heathcock, C.H. (1985) *J. Org. Chem.* **50**: 4135.

Conover, L.H. (1956) *Chem. Soc. Spec. Publ.* **5**: 48.

Conrow, K. (1966) *J. Org. Chem.* **31**: 1050.

Cooper, J.D., Vitullo, V.P., Whalen, D.L. (1971) *J. Am. Chem. Soc.* **93**: 6293.

Cope, A.C., Graham, E.S., Marshall, D.J. (1954) *J. Am. Chem. Soc.* **76**: 6159.

Corbella, A., Jommi, G., Rindone, B., Scolastico, C. (1969) *Tetrahedron* **25**: 4835.

Corey, E.J., Sauers, R.R., Swann, S. Jr. (1957) *J. Am. Chem. Soc.* **79**: 5826.

Corey, E.J., Sauers, R.R. (1959) *J. Am. Chem. Soc.* **81**: 1743.

Corey, E.J., Mitra, R.B., Uda, H. (1964a) *J. Am. Chem. Soc.* **86**: 485.

Corey, E.J., Ohno, M., Mitra, R.B., Vatachencherry, P.A. (1964b) *J. Am. Chem. Soc.* **86**: 478.

Corey, E.J., Ruden, R.A. (1973) *J. Org. Chem.* **38**: 834.

Corey, E.J., Ulrich, P. (1975) *Tetrahedron Lett.* 3685.

Corey, E.J., Trybulski, E.J., Suggs, J.W. (1976a) *Tetrahedron Lett.* 4577.

Corey, E.J., Wollenberg, R.H. (1976b) *Tetrahedron Lett.* 4705.

Corey, E.J., Pearce, H. (1979) *J. Am. Chem. Soc.* **101**: 5841.

Corey, E.J., De, B. (1984) *J. Am. Chem. Soc.* **106**: 2735.

Corey, E.J., Seibel, W.L., Kappos, J.C. (1987) *Tetrahedron Lett.* **28**: 4921.

Corona, T., Crotti, P., Macchia, F., Ferretti, M. (1984) *J. Org. Chem.* **49**: 377.

Coulson, C.A., O'Leary, B., Mallon, R.B. (1978) *"Hückel Theory for Organic Chemists"*, Academic Press, p. 132–133.

Coxon, J.M., Hartshorn, M.P., Kirk, D.N. (1967) *Tetrahedron* **23**: 3511.

Coxon, J.M., Garland, R.P., Hartshorn, M.P., Lane, G.A. (1968) *Chem. Commun.* 1506.

Coxon, J.M., Hartshorn, M.P., Kirk, D.N. (1969) *Tetrahedron* **25**: 2603.

Crabb, T.A., Wilkinson, J.R. (1977) *J. Chem. Soc. Perkin Trans. I* 953.

Crabbé, P., Cervantes, A., Cruz, A., Gleazzi, E., Iriarte, J., Velvarde, E. (1973) *J. Am. Chem. Soc.* **95**: 6653.

Crabtree, E.V., Poziomek, E.J. (1967) *J. Org. Chem.* **32**: 1231.

Cramer, F., Scheit, K.H., Baldauf, H.J. (1962) *Chem. Ber.* **95**: 1657.

Crandall, J.K., Machleder, W.H. (1968) *J. Am. Chem. Soc.* **90**: 7292, 7346.

Crandall, J.K., Conover, W.W. (1978) *J. Org. Chem.* **43**: 3533.

Creary, X., Rollin, A.J. (1977) *J. Org. Chem.* **42**: 4231.

Criegee, R., Kraft, L., Rank, B. (1933) *Liebigs Ann. Chem.* **507**: 159.

Cross, B.E., Grove, J.F., MacMillan, J., Mulholland, T.P.C. (1958) *J. Chem. Soc.* 2520.

Cross, B.E., Grove, J.F., Morrison, A. (1961) *J. Chem. Soc.* 2498.

Cross, B.E., Hanson, J.R., Briggs, L.H., Cambie, R.C., Rutledge, P.S. (1963) *Proc. Chem. Soc.* 17.

Cusack, N.J., Davis, B.R. (1965) *J. Org. Chem.* **30**: 2062.

Cushman, M., Choong, T.C., Valko, J.T., Koleck, M.P. (1980) *J. Org. Chem.* **45**: 5067.

Cushman, M., Mohan, P. (1985) *Tetrahedron Lett.* **26**: 4563.

Dane, E., Heiss, R., Schafer, H. (1959) *Angew. Chem.* **71**: 339.

d'Angelo, J., Ortuno, R.-M., Brault, J.-F., Ficini, J., Riche, C., Bouchaudy, J.-F. (1983) *Tetrahedron Lett.* **24**: 1489.

Danishefsky, S., Tsai, M.Y., Dynak, J. (1975) *Tetrahedron* 239.

D'Arcy, R., Grob, C.A., Kaffenberger, T., Krasnobajew, V. (1966) *Helv. Chim. Acta* **49**: 185.

Dastur, K.P. (1974) *J. Am. Chem. Soc.* **96**: 2605.

Dauben, W.G., McFarland, J.W. (1960) *J. Am. Chem. Soc.* **82**: 4245.

Dauben, W.G., Hart, D.J. (1977a) *J. Org. Chem.* **42**: 922.

Dauben, W.G., Hart, D.J. (1977b) *J. Org. Chem.* **42**: 3787.

Dauben, W.G., Bunce, R.A. (1983) *J. Org. Chem.* **48**: 4642.

Dave, V., Warnhoff, E.W. (1976) *Tetrahedron Lett.* 4695.

Davis, B.R., Woodgate, P.W. (1966) *J. Chem. Soc.* (*C*) 2006.

Dean, F.M., Evans, C.A., Francis, T., Robertson, A. (1957) *J. Chem. Soc.* 1577.

Dean, R.T., Rapoport, H. (1978) *J. Org. Chem.* **43**: 4183.

DeGraaf, S.A.G., Pandit, U.K. (1973) *Tetrahedron* **29**: 4263; (1974) *Tetrahedron* **30**: 1115.

DeKimpe, N., Schamp, N. (1974) *Tetrahedron Lett.* 3779.

DeKimpe, N., Verhe, R., DeBuyck, L., Schamp, N. (1985) *Tetrahedron Lett.* **26**: 2709.

Delepine, M. (1895) *Compt. rend.* **120**: 501; (1897) *Compt. rend.* **124**: 292.

deMayo, P. (1957) *Perfumery Essent. Oil Rec.* **48**: 18.

deMayo, P., Takashita, J., Sattar, A.B.M.A. (1962) *Proc. Chem. Soc.* 119.

deMayo, P. (1964) *Pure Appl. Chem.* **9**: 597.

deMeijere, A. (1974) *Tetrahedron Lett.* 1845, 1849.

Demuth, M. (1984) *Chimia* **38**: 257.

Demuth, M., Wietfeld, B., Pandey, B., Schaffner, K. (1985) *Angew. Chem. Inter. Ed.* **24**: 763.

Demuth, M., Palomer, A., Sluma, H.-D., Dey, A.K., Kruger, C., Tsay, Y.-H. (1986) *Angew. Chem.* **98**: 1093.

Demuynck, L., Cherest, M., Lusinchi, X., Thal, C. (1989) *Tetrahedron Lett.* **30**: 719.

Dennis, W.H., Hull, L.A., Rosenblatt, D.H. (1967) *J. Org. Chem.* **32**: 3783.

Deno, N.C., Sacher, E. (1965) *J. Am. Chem. Soc.* **87**: 5120.

Desai, R.D. (1932) *J. Chem. Soc.* 1097.

DeSelms, R.D. (1969) *Tetrahedron Lett.* 1179.

Deslongchamps, P., Lafontaine, J., Ruest, L., Soucy, P. (1977) *Can. J. Chem.* **55**: 4117.

Deslongchamps, P. (1984) *Aldrichimica Acta* **17**: 59.

Desmaele, D., Ficini, J., Guingant, A., Kahn, P. (1983a) *Tetrahedron Lett.* **24**: 3079.

Desmaele, D., Ficini, J., Guingant, A., Touzin, A.M. (1983b) *Tetrahedron Lett.* **24**: 3083.

deWaal, H.L., Prinsloo, B.J., Arndt, R.R. (1966) *Tetrahedron Lett.* 6169.

Deyrup, J.A., Vestling, M.M., Hagan, W.V., Yun, H.Y. (1969) *Tetrahedron* **25**: 1467.

Dhanak, D., Kuroda, R., Reese, C.B. (1987) *Tetrahedron Lett.* **28**: 1827.

Diatta, L., Andriamialisoa, R.Z., Langlois, N., Potier, P. (1976) *Tetrahedron* **32**: 2839.

Diekmann, J. (1963) *J. Org. Chem.* **28**: 2880.

Disanayaka, B.W., Weedon, A.C. (1985) *Chem. Commun.* 1282.

Djuric, S., Sarkar, T., Magnus, P. (1980) *J. Am. Chem. Soc.* **102**: 6885.

Dodson, R.M., Muir, R.D. (1961) *J. Am. Chem. Soc.* **83**: 4627.

Doecke, C.W., Klein, G., Paquette, L.A. (1978) *J. Am. Chem. Soc.* **100**: 1596.

Doering, W.v.E., Haines, R.M. (1954) *J. Am. Chem. Soc.* **76**: 482.

Dolby, L.J., Iwamoto, R.H. (1965) *J. Org. Chem.* **30**: 2420.

Donnelly, J.A., Fox, M.J., Sharma, T.C. (1979a) *Tetrahedron* **35**: 875.

Donnelly, J.A., Fox, M.J., Sharma, T.C. (1979b) *Tetrahedron* **35**: 1987.

Döpke, W., Flentje, H., Jeffs, P.W. (1968) *Tetrahedron* **24**: 4459.

Dorsey, D.A., King, S.M., Moore, H.W. (1986) *J. Org. Chem.* **51**: 2814.

Dowd, P., Schappert, R., Garner, P. (1982) *Tetrahedron Lett.* **23**: 7.

Dowd, P., Kaufman, C., Kaufman, P., Paik, Y.H. (1985a) *Tetrahedron Lett.* **26**: 2279.

Dowd, P., Kaufman, C., Paik, Y.H. (1985b) *Tetrahedron Lett.* **26**: 2283.

Duc, D.K.M., Fetizon, M., Hanna, I., Olesker, A. (1981) *Tetrahedron Lett.* **22**: 3847.

Duddeck, H., Feuerhelm, H.-T., Snatzke, G. (1979) *Tetrahedron Lett.* **829.**

Dufraisse, C. (1936) *Bull. Soc. Chim. Fr.* [5]**3**: 1880.

Dumont, W., Bayet, P., Krief, A. (1974) *Angew. Chem. Intern. Ed.* **13**: 804.

Dumont, W., Krief, A. (1976) *Angew. Chem. Intern. Ed.* **15**: 161.

Durani, S., Kapil, R.S. (1983) *J. Chem. Soc. Perkin Trans. I* 211.

Duthaler, R.O., Stingelin-Schmid, R.S., Ganter, C. (1976) *Helv. Chim. Acta* **59**: 307.

Dystra, H.B., Lewis, J.F., Boord, C.E. (1930) *J. Am. Chem. Soc.* **52**: 3396; and later papers.

East, D.S., Ishikawa, K., McMurry, T.B.H. (1973) *J. Chem. Soc. Perkin Trans. I* 2563.

Eaton, P.E. (1964) *Tetrahedron Lett.* 3695.

Eaton, P.E., Mueller, R.H., Carlson, G.R., Cullison, D.A., Cooper, G.F., Chou, T.-C., Krebs, E.P. (1977) *J. Am. Chem. Soc.* **99**: 2751.

Eaton, P.E., Andrews, G.D., Krebs, E.P., Kunai, A. (1979) *J. Org. Chem.* **44**: 2824.

Ebnother, A., Niklaus, P., Suess, R. (1969) *Helv. Chim. Acta* **52**: 629.

Edwards, O.E., Ferrari, C. (1964) *Can. J. Chem.* **42**: 172.

Edwards, O.E. (1965) *Chem. Commun.* 318.

Edwards, P.N., Smith, G.F. (1961) *J. Chem. Soc.* 1458.

Eilerman, R.G., Willis, B.J. (1981) *Chem. Commun.* 30.

Eisch, J.J., Dua, S.K., Behrooz, M. (1985) *J. Org. Chem.* **50**: 3674.

Elliott, W.W., Hammick, D.L. (1951) *J. Chem. Soc.* 3402.

Emmer, G., Graf, W. (1981) *Helv. Chim. Acta* **64**: 1398.

Enderle, H.G. (1967) Dissert. Univ. Basel.

Engler, T.A., Combrink, K.D., Reddy, J.P. (1989) *Chem. Commun.* 454.

English, J., Brutcher, F.V. (1952) *J. Am. Chem. Soc.* **74**: 4279.

English, J., Bliss, A.D. (1956) *J. Am. Chem. Soc.* **78**: 4057.

Eschenmoser, A., Furst, A. (1951) *Experientia* **7**: 209.

Eschenmoser, A., Felix, D., Ohloff, G. (1967) *Helv. Chim. Acta* **50**: 708.

Evans, E.H., Hewson, A.T., March, L.A., Nowell, I.W. (1982) *Chem. Commun.* 1342.

Exon, C., Nobbs, M., Magnus, P. (1981) *Tetrahedron* **37**: 4515.

Fang, F.G., Maier, M.E., Danishefsky, S.J., Schulte, G. (1990) *J. Org. Chem.* **55**: 831.

Farges, M.G. (1962) *Bull. Soc. Chim. Fr.* 1266.

Farina, F., Paredes, M.C., Valderrama, J.A. (1990) *J. Chem. Soc. Perkin Trans. I* 2345.

Fayos, I.J., Clardy, J., Dolby, L.J., Farnham, T. (1977) *J. Org. Chem.* **42**: 1349.

Fehr, C., Ohloff, G., Büchi, G. (1979) *Helv. Chim. Acta* **62**: 2655.

Feldman, P.L., Rapoport, H. (1987) *J. Am. Chem. Soc.* **109**: 1603.

Felix, D., Schreiber, J., Ohloff, G., Eschenmoser, A. (1971) *Helv. Chim. Acta* **54**: 2896.

Felix, D., Müller, R.K., Horn, U., Joos, R., Schreiber, J., Eschenmoser, A. (1972) *Helv. Chim. Acta* **55**: 1276.

Ferreira, M.A., King, T.J., Ali, S., Thomson, R.H. (1980) *J. Chem. Soc. Perkin Trans. I* 249.

Ferris, A.F. (1960) *J. Org. Chem.* **25**: 12.

Feuer, H., Anderson, R.S. (1961) *J. Am. Chem. Soc.* **83**: 2960.

Ficini, J., Touzin, A.M. (1972) *Tetrahedron Lett.* 2097.

Ficini, J., d'Angelo, J., Noire, J. (1974) *J. Am. Chem. Soc.* **96**: 1213.

Ficini, J., d'Angelo, J. (1976) *Tetrahedron Lett.* 687.

Ficini, J., Falou, S., d'Angelo, J. (1977) *Tetrahedron Lett.* 1931.

Ficini, J., Guingant, A., d'Angelo, J. (1979) *J. Am. Chem. Soc.* **101**: 1318.

Ficini, J., Guingant, A., d'Angelo, J., Stork, G. (1983) *Tetrahedron Lett.* **24**: 907.

Fickes, G.N., Kemp, K.C. (1973) *Chem. Commun.* 84.

Findlay, J.A., Desai, D.N., Lonergan, G.C., White, P.S. (1980) *Can. J. Chem.* **58**: 2827.

Fischer, H.P., Grob, C.A. (1968) *Helv. Chim. Acta* **51**, 153.

Fischer, W. (1974) Dissert. Univ. Basel.

Fisher, W., Grob, C.A. (1975) *Tetrahedron Lett.* 3547.

Fisher, W., Grob, C.A., von Sprecher, G. (1979) *Tetrahedron Lett.* 473.

Fittig, R., Worringer, L. (1885) *Liebigs Ann. Chem.* **227**: 1.

Flecker, P., Schomburg, D., Thies, P.W., Winterfeldt, E. (1984a) *Tetrahedron* **40**: 4843.

Flecker, P., Winterfeldt, E. (1984b) *Tetrahedron* **40**: 4853.

Fleming, I., Hatanaka, N., Ohta, H., Simamura, O., Yoshida, M. (1972) *Chem. Commun.* 242.

Fleming, I., Michael, J.P. (1978) *Chem. Commun.* 245.

Föhlisch, B., Herrscher, O. (1985) *Tetrahedron* **41**: 1979.

Forbes, E.J., Warrell, D.C., Fry, W.J. (1967) *J. Chem. Soc.* 1693.

Forbes, E.J., Gray, C.J. (1968a) *Tetrahedron* **24**: 2795.

Forbes, E.J., Gregory, M.J., Warrell, D.C. (1968b) *J. Chem. Soc.* (*C*) 1969.

Francisco, C.G., Freire, R., Hernandez, R., Salazar, J.A., Suarez, E., Cortes, M. (1983) *J. Chem. Soc. Perkin Trans. I* 2757.

Franck, B., Teetz, V. *Angew. Chem. Intern. Ed.* **10**: 411.

Franck, R.W., Rizzi, G.P., Johnson, W.S. (1964) *Steroids* **4**: 463.

Franz, J.E., Howe, R.K., Pearl, H.K. (1976) *J. Org. Chem.* **41**: 620.

Frater, G. (1976) *Helv. Chim. Acta* **59**: 164.

Freeman, J.P., Duchamp, D.J., Chidester, C.G., Slomp, G., Szmuszkovicz, J., Raban, M. (1982) *J. Am. Chem. Soc.* **104**: 1380.

Freeman, P.K., Johnson, R.C. (1969) *J. Org. Chem.* **34**: 1751.

Friary, R., Seidl, V. (1986) *J. Org. Chem.* **51**: 3214.

Fried, J., Klingsberg, A. (1953) *J. Am. Chem. Soc.* **75**: 4929.

Frimer, A.A., Gilinsky-Sharon, P., Aljadeff, G. (1982) *Tetrahedron Lett.* 1301.

Froborg, J., Magnusson, G. (1978) *J. Am. Chem. Soc.* **100**: 6728.

Frost, J.R., Gaudilliere, B.R.P., Kauffmann, E., Loyaux, D., Normand, N., Petry, G., Poirier, P., Wenkert, E., Wick, A. (1989) *Heterocycles* **28**: 175.

Fuhrer, H., Ganguly, A.K., Gopinath, K.W., Govindachari, T.R., Nagarajan, K., Pai, B.R., Parthasarathy, P.C. (1970) *Tetrahedron* **26**: 2371.

Fuji, K., Ito, N., Uchida, I., Fujita, E. (1985) *Chem. Pharm. Bull.* **33**: 1034.

Fujita, E., Fujita, T., Fuji, K., Ito, N. (1966) *Tetrahedron* **22**: 3423.

Fujita, E., Fujita, T., Nagao, Y. (1969a) *Tetrahedron* **25**: 3717.

Fujita, E., Fujita, T., Shibuya, M. (1969b) *Tetrahedron* **25**: 2517.

Fukumoto, K., Chihiro, M., Ihara, M., Kametani, T., Honda, T. (1983) *J. Chem. Soc. Perkin Trans. I* 2567.

Funayama, S., Cordell, G.A. (1989) *Heterocycles* **29**: 815.

Funk, R.L., Horcher, L.H.M., III, Daggett, J.U., Hansen, M.M. (1983) *J. Org. Chem.* **48**: 2632.

Gallagher, T., Magnus, P., Huffman, J.C. (1982) *J. Am. Chem. Soc.* **104**: 1140.

Gallagher, T., Magnus, P.D. (1983) *J. Am. Chem. Soc.* **105**: 2086.

Gambacorta, A., Nicoletti, R., Forcellese, M.L. (1971) *Tetrahedron* **27**: 985.

Gammill, R.B., Gold, P.M., Mizsak, S.A. (1980) *J. Am. Chem. Soc.* **102**: 3095.

Ganguli, M., Burka, L.T., Harris, T.M. (1984) *J. Org. Chem.* **49**: 3762.

Garst, M.E., Roberts, V.A., Prussin, C. (1982) *J. Org. Chem.* **47**: 3969.

Gassman, P.G., Marshall, J.L., Macmillan, J.G. (1973) *J. Am. Chem. Soc.* **95**: 6319.

Gaviña, F., Costero, A.M., Gil, P., Palazon, B., Luis, S.V. (1981) *J. Am. Chem. Soc.* **103**: 1797.

Gebreyesus, T., Djerassi, C. (1972) *J. Chem. Soc. Perkin Trans. I* 849.

Geetha, P., Narasimhan, K., Swaminathan, S. (1979) *Tetrahedron Lett.* 565.

Geisel, M., Grob, C.A., Wohl, R.A. (1969) *Helv. Chim. Acta* **52**: 2206.

Genin, D., Andriamialisoa, R.Z., Langlois, N., Langlois, Y. (1987) *Heterocycles* **26**: 377.

Genot, A., Florent, J.-C., Monneret, C. (1987) *J. Org. Chem.* **52**: 1057.

Gerlach, H. (1977) *Helv. Chim. Acta* **60**: 3039.

Ghatak, U.R., Ray, J.K. (1977) *J. Chem. Soc. Perkin Trans. I* 518.

Ghosez, L., Montaigne, R., Mollet, P. (1966) *Tetrahedron Lett.* 135.

Gibbons, E.G. (1982) *J. Am. Chem. Soc.* **104**: 1767.

Gibson, C.S., Simonsen, J.L. (1925) *J. Chem. Soc.* **127**: 1294.

Gillespie, R.J., Murray-Rust, J., Murray-Rust, P., Porter, A.E.A. (1981) *Tetrahedron* **37**: 743.

Girotra, N.N., Wendler, N.L. (1979) *Tetrahedron Lett.* 4793.

Gleiter, R., Stohrer, W.D., Hoffmann, R. (1972) *Helv. Chim. Acta* **55**: 893.

Goldschmidt, Z., Finkel, D. (1983) *J. Chem. Soc. Perkin Trans. I* 45.

Goldstein, M.J., Johnson, M.W., Taylor, R.T. (1982) *Tetrahedron Lett.* **23**: 3331.

Goldstein, S., Vannes, P., Honge, C., Frisque-Hesbain, A.M., Wiaux-Zamar, C., Ghosez, L., Germain, G., Declercq, J.P., Van Meersche, M., Arrieta, J.M. (1981) *J. Am. Chem. Soc.* **103**: 4616.

Gompper, R., Schwarzensteiner, M.-L. (1983) *Angew. Chem.* **95**: 553.

Gompper, R., Noth, H., Rattay, W., Schwarzensteiner, M.-L., Spes, P., Wagner, H.-U. (1987) *Angew. Chem.* **99**: 1071.

Gordon, E.M., Pluscec, J., Ondetti, M.A. (1981) *Tetrahedron Lett.* **22**: 1871.

Gorman, M., Neuss, N. (1963) *144th ACS Nat. Meet.* 38M.

Goto, T., Kishi, Y., Takahashi, S., Hirata, Y. (1965) *Tetrahedron* **21**: 2059.

Govindachari, T.R., Nagarajan, K., Parthasarathy, P.C. (1969) *Chem. Commun.* 823.

Gray, R.W., Dreiding, A.S. (1977) *Helv. Chim. Acta* **60**: 1969.

Grayson, D.H., Wilson, J.R.H. (1984) *Chem. Commun.* 1695.

Greenhill, J.V., Mohamed, M.I. (1979) *J. Chem. Soc. Perkin Trans. I* 1411.

Greenlee, M.L. (1981) *J. Am. Chem. Soc.* **103**: 2425.

Grimm, E.L., Reissig, H.-U. (1985) *J. Org. Chem.* **50**: 242.

Grob, C.A., Baumann, W. (1955) *Helv. Chim. Acta* **38**: 594.

Grob, C.A. (1957) *Experientia* **13**: 126.

Grob, C.A., Hoegerle, R.M., Ohta, M. (1962) *Helv. Chim. Acta* **45**: 1823.

Grob, C.A., Csapilla, J., Cseh, G. (1964a) *Helv. Chim. Acta* **47**: 1590.

Grob, C.A., Schwarz, W. (1964b) *Helv. Chim. Acta* **47**: 1870.

Grob, C.A., Kiefer, H.R., Lutz, H.J., Wilkens, H.J. (1967a) *Helv. Chim. Acta* **50**: 416.

Grob, C.A., Schiess, P.W. (1967b) *Angew. Chem. Intern. Ed.* **6**: 1.

Grob, C.A., von Tschammer, H. (1968) *Helv. Chim. Acta* **51**: 1083.

Grob, C.A. (1969) *Angew. Chem. Intern. Ed.* **8**: 535.

Grob, C.A., Kostka, K., Kuhnen, F. (1970) *Helv. Chim. Acta* **53**: 608

Grob, C.A., Unger, F.M., Weiler, E.D., Weiss, A. (1972) *Helv. Chim. Acta* **55**: 501.

Grob, C.A., Kunz, W., Marbet, P.R. (1975a) *Tetrahedron Lett.* 2613.

Grob, C.A., Schmitz, B., Sutter, A., Weber, A.H. (1975b) *Tetrahedron Lett.* 3551.

Gulland, J.M., Robinson, R. (1923) *J. Chem. Soc.* **123**: 980

Gustafson, D.H., Erman, W.F. (1965) *J. Org. Chem.* **30**: 1665.

Guthrie, R.W., Valenta, Z., Wiesner, K. (1966) *Tetrahedron Lett.* 4645.

Guyot, M., Molho, D. (1973) *Tetrahedron Lett.* 3433.

Habermehl, G., Andres, H., Miyashita, K., Witkop, B., Daly, J.W. (1976) *Liebigs Ann. Chem.* 1577.

Hajicek, J., Gretler, R., Hesse, M. unpubl. results, quoted in Stach (1988).

Hajicek, J., Trojanek, J. (1981) *Tetrahedron Lett.* **22**: 2927.

Hall, E.S., McCapra, F., Scott, A.I. (1967) *Tetrahedron* **23**: 4131.

Halsall, T.G., McHale, P.J., Mendez, A.M. (1978) *J. Chem. Soc. Perkin Trans. I* 1606.

Hamilton, R.J., Mander, L.N., Sethi, S.P. (1986) *Tetrahedron* **42**: 2881.

Hamon, D.P.G., Taylor, G.F. (1975a) *Tetrahedron Lett.* 1623.

Hamon, D.P.G., Taylor, G.F., Young, R.N. (1975b) *Synthesis* 428.

Han, W.C., Takahashi, K., Cook, J.M., Weiss, U., Silverton, J.V. (1982) *J. Am. Chem. Soc.* **104**: 318.

Hanson, J.R., Raines, D., Wadsworth, H. (1978) *J. Chem. Soc. Perkin Trans. I* 743.

Hara, H., Hashimoto, F., Hoshino, O., Umezawa, B. (1984) *Tetrahedron Lett.* **25**: 3615.

Hara, H., Akiba, T., Miyashita, T., Hoshino, O., Umezawa, B. (1988) *Tetrahedron Lett.* **29**: 3815.

Hara, S., Taguchi, H., Yamamoto, H., Nozaki, H. (1975) *Tetrahedron Lett.* 1545.

Harapanhalli, R.S. (1988) *Liebigs Ann. Chem.* 1009.

Hardegger, E., Schellenbaum, M., Huwyler, R., Zust, A. (1957) *Helv. Chim. Acta* **40**: 1816.

Harley-Mason, J., Kaplan, M. (1965) *Chem. Commun.* 298.

Hart, H., Cipriani, R.A. (1962) *J. Am. Chem. Soc.* **84**: 3697.

Hart, H., Chen, S.-M., Lee, S., Ward, D.L., Kung, W.-J.H. (1980) *J. Org. Chem.* **45**: 2091.

Hartshorn, M.P., Kirk, D.N., Wallis, A.F.A. (1964) *J. Chem. Soc.* 5494.

Hasek, R.H., Clark, R.D., Chaudet, J.H. (1961) *J. Org. Chem.* **26**: 3130.

Hassner, A., Dillon, J., Onan, K.D. (1986) *J. Org. Chem.* **51**: 3315.

Hatfield, L.D., Huntsman, W.D. (1967) *J. Org. Chem.* **32**: 1800.

Hauptmann, H., Mader, M. (1977) *Tetrahedron Lett.* 3151.

Hawker, C.J., Stark, W.M., Battersby, A.R. (1987) *Chem. Commun.* 1313.

Hawkins, R.G.E., Large, R. (1973) *J. Chem. Soc. Perkin Trans. I* 2169.

Hayakawa, K., Motohiro, S., Fujii, I., Kanematsu, K. (1981) *J. Am. Chem. Soc.* **103**: 4605.

Hayakawa, K., Ueyama, K., Kanematsu, K. (1985) *J. Org. Chem.* **50**: 1963.

Heaney, H., Hollinshead, J.H., Sharma, R.P. (1978) *Tetrahedron Lett.* 67.

Heathcock, C.H. (1966) *J. Am. Chem. Soc.* **88**: 4110.

Heathcock, C.H., Badger, R.A. (1972) *J. Org. Chem.* **37**: 234.

Heathcock, C.H., Tice, C.M., Gemroth, T.C. (1982) *J. Am. Chem. Soc.* **104**: 6081.

Heine, H.-G., Hartmann, W. (1981) *Synthesis* **706.**

Heinz, U., Adams, E., Klintz, R., Welzel, P. (1990) *Tetrahedron* **46**: 4217.

Hendrickson, J.B., Boeckman, R.K., Jr. (1971) *J. Am. Chem. Soc.* **93**: 1307.

Hershenson, F.M., Swenton, L., Prodan, K.A. (1980) *Tetrahedron Lett.* **21**: 2617.

Herz, W. (1957) *J. Am. Chem. Soc.* **79**: 5011.

Herz, W. (1962) *J. Am. Chem. Soc.* **84**: 3857.

Herz, W., Nair, M.G., Prakash, D. (1975) *J. Org. Chem.* **40**: 1017.

Heyes, R.G., Robertson, A. (1935) *J. Chem. Soc.* **681.**

Hikino, H., Kohama, T., Takemoto, T. (1969) *Tetrahedron* **25**: 1037.

Hill, R.K (1962) *J. Org. Chem.* **27**: 29.

Hill, R.K., Ledford, N.D. (1975) *J. Am. Chem. Soc.* **97**: 666.

Hinshaw, W.B., Levy, J., LeMen, J. (1971) *Tetrahedron Lett.* **995.**

Hirama, M., Ito, S. (1977) *Tetrahedron Lett.* 627.

Hirao, T., Fujihara, Y., Kurokawa, K., Oshiro, Y., Agawa, T. (1986) *J. Org. Chem.* **51**: 2830.

Hirata, T., Suga, T. (1971) *Bull. Chem. Soc. Jpn* **44**: 2833.

Hiyama, T., Mishima, T., Kitatani, K., Nozaki, H. (1974) *Tetrahedron Lett.* 3297.

Hiyama, T., Kanakura, A., Yamamoto, H., Nozaki, H. (1978) *Tetrahedron Lett.* 3051.

Ho, T.-L., Wong, C.M. (1973a) *Chem. Commun.* 224; (1973b) *Synth. Commun.* **3**: 145.

Ho, T.-L. (1977) *Hard and soft acids and bases principle in organic chemistry*, Academic Press, NY.

Ho, T.-L., Din, Z.U. (1982a) *Synth. Commun.* **12**: 257.

Ho, T.-L., Liu, S.-H. (1982b) *Synth. Commun.* **12**: 995.

Ho, T.-L. (1983a) *Synth. Commun.* **13**: 761.

Ho, T.-L., Stark, C.J. (1983b) *Liebigs Ann. Chem.* 1446.

Ho, T.-L. (1988) *Rev. Chem. Interm.* **9**: 117.

Ho, T.-L. (1989) *Res. Chem. Interm.* **11**: 157.

Ho, T.-L. (1991) *Polarity control for synthesis*, Wiley, NY.

Hobson, J.D., Malpass, J.R. (1967) *J. Chem. Soc.* (*C*) 1645.

Hodges, R., White, E.P., Shannon, J.S. (1964) *Tetrahedron Lett.* 371.

Hoffman, R.V., Orphanides, G.G., Shechter, H. (1978) *J. Am. Chem. Soc.* **100**: 7927.

Hoffmann, R.W. (1965) *Chem. Ber.* **98**: 222.

Hoffmann, R.W., Lotze, M., Reiffen, M., Steinbach, K. (1981) *Liebigs Ann. Chem.* 581.

Hoffmann, R.W., Ditrich, K. (1989) *Liebigs Ann. Chem.* 23.

Hofheinz, W., Schönholzer, P., Bernauer, K. (1976) *Helv. Chim. Acta* **59**: 1213.

Hogeveen, H., Nusse, B.J. (1978) *J. Am. Chem. Soc.* **100**: 3110.

Holland, J.M., Jones, D.W. (1970) *J. Chem. Soc.* (*C*) 530.

Hollenstein, S., Laube, T. (1990) *Angew. Chem. Int. Ed. Engl.* **29**: 188.

Holton, R.A., Kennedy, R.M. (1984) *Tetrahedron Lett.* **25**: 4455.

Holton, R.A., Juo, R.R., Kim, H.B., Williams, A.D., Harusawa, S., Lowenthal, R.E., Yogai, S. (1988) *J. Am. Chem. Soc.* **110**: 6558.

Honma, Y., Ban, Y. (1977) *Heterocycles* **6**: 291.

Hora, J., Toube, T.P., Weedon, B.C.L. (1970) *J. Chem. Soc.* 241.

Hori, K., Nomura, K., Mori, S., Yoshii, E. (1989) *Chem. Commun.* 712.

Horner, L., Weber, K.H., Dürckheimer, W. (1961) *Chem. Ber.* **94**: 2881.

Horowitz, R.M., Ullyot, G.E. (1952) *J. Am. Chem. Soc.* **74**: 587.

Hortmann, A.G. (1970) *Tetrahedron Lett.* **52**: 99.

Hortmann, A.G., Daniel, D.S. (1972) *J. Org. Chem.* **37**: 4446.

Houminer, Y. (1985) *J. Org. Chem.* **50**: 786.

House, H.O., Gannon, W.F. (1958) *J. Org. Chem.* **23**: 879.

House, H.O., Bryant, W.M. (1966) *J. Org. Chem.* **31**: 2755.

House, H.O., Blankley, C.J. (1968) *J. Org. Chem.* **33**: 47.

House, H.O., Melillo, D.G., Sauter, F.J. (1973) *J. Org. Chem.* 38: 741.

House, H.O., Lee, T.V. (1979) *J. Org. Chem.* **44**: 2819.

Howard, G.A., Slater, C.A. (1957) *J. Chem. Soc.* 1924.

Hoye, T.R., Kurth, M.J. (1980) *J. Org. Chem.* **45**: 3549.

Huber, U.A., Dreiding, A.S. (1970) *Helv. Chim. Acta* **53**: 495.

Hudlicky, T., Seoane, G., Frazier, J.O., Kwart, L.D., Tiedje, M., Beal, C., Seoane, A. (1986) *J. Am. Chem. Soc.* **108**: 3755.

Hudlicky, T., Fleming, A., Lovelace, T.C. (1989a) *Tetrahedron* **45**: 3021.

Hudlicky, T., Fleming, A., Radesca, L. (1989b) *J. Am. Chem. Soc.* **111**: 6691.

Hugel, G., Gourdier, B., Lévy, J., LeMen, J. (1974) *Tetrahedron Lett.* 1597.

Hugel, G., Laronze, J.-Y., Laronze, J., Lévy, J. (1981a) *Heterocycles* **16**: 581.

Hugel, G., Massiot, G., Lévy, J., LeMen, J. (1981b) *Tetrahedron* **37**: 1369.

Huisgen, R., Kolbeck, W. (1965) *Tetrahedron Lett.* 783.

Hunter, N.R., Green, N.A., McKinnon, D.M. (1980) *Tetrahedron Lett.* **21**: 4589.

Hurst, J.J., Whitham, G.H. (1960) *J. Chem. Soc.* 2864.

Husson, A., Langlois, Y., Riche, C., Husson, H.-P. (1973) *Tetrahedron* **29**: 3095.

Hutchinson, J.H., Money, T. (1985) *Can. J. Chem.* **63**: 3182; (1987) *Can. J. Chem.* **65**: 1.

Ibuka, T., Mori, Y., Inubushi, Y. (1976) *Tetrahedron Lett.* 3169.

Ibuka, T., Mitsui, Y., Hayashi, K., Minakata, H., Inubushi, Y. (1981) *Tetrahedron Lett.* **22**: 4425.

Igarashi, M., Midorikawa, H. (1967) *J. Org. Chem.* **32**: 3399.

Ikeda, M., Matsugashita, S., Tamura, Y. (1977) *J. Chem. Soc. Perkin Trans. I* 1770.

Ikeda, M., Ohno, K., Takahashi, M., Homma, K., Uchino, T., Tamura, Y. (1983) *Heterocycles* **20**: 1005.

Ikeda, T., Taylor, W.I., Tsuda, Y., Uyeo, S., Yajima, H. (1956) *J. Chem. Soc.* 4749.

Immer, H., Bagli, J.F. (1968) *J. Org. Chem.* **33**: 2457.

Incze, M., Soti, F., Kardos-Balogh, Z., Szantay, C. (1985) *Heterocycles* **23**: 2843.

Ingold, C.K. (1922a) *J. Chem. Soc.* **121**: 1143.

Ingold, C.K., Perren, E.A., Thorpe, J.F. (1922b) *J. Chem. Soc.* **121**: 1765.

Inokuchi, T., Takagishi, S., Akahoshi, F., Torii, S. (1987) *Chem. Lett.* 1553.

Iorio, M.A. (1984) *Heterocycles* **22**: 2207.

Ireland, R.E., Kowalski, C.J., Tilley, J.W., Walba, D.M. (1975) *J. Org. Chem.* **40**: 990.

Ireland, R.E., Walba, D.M. (1976) *Tetrahedron Lett.* 1071.

Ireland, R.E., Dow, W.S., Godfrey, J.D., Thaisrivongs, S. (1984) *J. Org. Chem.* **49**: 1001.

Ireland, R.E., Häbich, D., Norbeck, D.W. (1985) *J. Am. Chem. Soc.* **107**: 3271.

Ishikawa, K., McMurry, T.B.H. (1973) *J. Chem. Soc. Perkin Trans. I* 914.

Ishizumi, K., Inaba, S., Yamamoto, H. (1973) *J. Org. Chem.* **38**: 2617.

Isoe, S., Hyeon, S.B., Ichikawa, H., Katsumura, S., Sakan, T. (1968) *Tetrahedron Lett.* 5561.

Ito, Y., Oda, M., Kitahara, Y. (1975) *Tetrahedron Lett.* 239.

Ito, Y., Saegusa, T. (1977) *J. Org. Chem.* **42**: 2326.

Ito, Y., Nakajo, E., Nakatsuka, M., Saegusa, T. (1983) *Tetrahedron Lett.* **24**: 2881.

Itokawa, H., Morita, H., Watanabe, K., Takase, A., Iitaka, Y. (1984) *Chem. Lett.* 1687.

Itokawa, H., Morita, H., Osawa, K., Watanabe, K., Iitaka, Y. (1987) *Chem. Pharm. Bull.* **35**: 2849.

Ivanov, P.M., Osawa, E., Klunder, A.J.H., Zwanenburg, B. (1985) *Tetrahedron* **41**: 975.

Iwata, C., Yamada, M., Shinoo, Y., Kobayashi, K., Okada, H. (1977) *Chem. Commun.* 888.

Iwata, C., Yamada, M., Shinoo, Y. (1979) *Chem. Pharm. Bull.* **27**: 274.

Iwata, C., Fusaka, T., Fujiwara, T., Tomita, K., Yamada, M. (1981) *Chem. Commun.* 463.

Iwata, C., Akiyama, T., Miyashita, K. (1987) *Chem. Ind.* 294.

Iwata, C., Takemoto, Y., Doi, M., Imanishi, T. (1988) *J. Org. Chem.* **53**: 1623.

Jackson, A.H., Lertwanawatana, W., Pandey, R.K., Rao, K.R.N. (1989) *J. Chem. Soc. Perkin Trans. I* 374.

Jackson, L.B., Waring, A.J. (1985) *Chem. Commun.* 857.

Jaecklin, A.P., Skrabal, P., Zollinger, H. (1971) *Helv. Chim. Acta* **54**: 2870.

Janculey, J., Nerdel, F., Frank, D., Barth, G. (1967) *Chem. Ber.* **100**: 715.

Jansson, K., Ahlfors, S., Frejd, T., Kihlberg, J., Magnusson, G., Dahmen, J., Noori, G., Stenvall, K. (1988) *J. Org. Chem.* **53**: 5629.

Jaroszewski, J.W., Ettlinger, M.G. (1982) *J. Org. Chem.* **47.**
3974.

Jayaraman, S., Thambidurai, S., Srinivasan, J., Rathinasababapathy, R., Rajagopalan, K. (1989) *Tetrahedron* **45**: 1845.

Johnson, P.Y. (1972) *Tetrahedron Lett.* 1991.

Johnson, P.Y., Berman, M. (1975) *J. Org. Chem.* **40**: 3046.

Jones, E.C.S., Kenner, J. (1930) *J. Chem. Soc.* 919.

Jones, R.L., Rees, C.W. (1969) *J. Chem. Soc.* (*C*) 2249, 2251.

Jones, S., Kennewell, P.D., Westwood, R., Sammes, P.G. (1990) *Chem. Commun.* 497.

Jones, V.K., Jones, L.B. (1970) *Tetrahedron Lett.* 3171.

Joule, J.A., Smith, G.F. (1962) *J. Chem. Soc.* 312.

Julia, M., Bagot, J., Siffert, O. (1973) *Bull. Soc. Chim. Fr. II* 1424.

Julia, S., Linstrumelle, G. (1966) *Bull. Soc. Chim. Fr.* 3490.

Jung, M.E., Pan, Y.-G. (1980a) *Tetrahedron Lett.* **21**: 3127.

Jung, M.E., Shapiro, J.J. (1980b) *J. Am. Chem. Soc.* **102**: 7862.

Jung, M.E., Miller, S.J. (1981) *J. Am. Chem. Soc.* **103**: 1984.

Jung, M.E., Lowe, J.A., III, Lyster, M.A., Node, M., Pfluger, R.W., Brown, R.W. (1984) *Tetrahedron* **40**: 4751.

Kaczmarek, R., Blechert, S. (1986) *Tetrahedron Lett.* **27**: 2845.

Kaffenberger, T. (1966) Dissert. Univ. Basel.

Kaiya, T., Shirai, N., Sakakibara, J. (1979) *Chem. Commun.* 431.

Kaiya, T., Shirai, N., Sakakibara, J. (1981) *Chem. Commun.* 22.

Kalyanasundaram, M., Rajagopalan, K., Swaminathan, S. (1980) *Tetrahedron Lett.* **21**: 4391.

Kametani, T., Fukumoto, K., Kawatsu, M., Fujihara, M. (1970a) *J. Chem. Soc.* (*C*) 922; (1970b) *J. Chem. Soc.* (*C*) 2209.

Kametani, T., Hirata, S., Ogasawara, K. (1973) *J. Chem. Soc. Perkin Trans. I* 1466.

Kametani, T., Fujimoto, Y., Mizushima, M. (1975) *Heterocycles* **3**: 619.

Kametani, T., Matsumoto, H., Satoh, Y., Nemoto, H., Fukumoto, K. (1977) *J. Chem. Soc. Perkin Trans. I* 376.

Kametani, T., Matsumoto, H., Nemoto, H., Fukumoto, K. (1978) *J. Am. Chem. Soc.* **100**: 6218.

Kametani, T., Ohsawa, T., Ihara, M. (1979a) *Heterocycles* **12**: 913.

Kametani, T., Suzuki, K., Nemoto, H. (1979b) *Chem. Commun.* 1127; (1980) *Tetrahedron Lett.* 1469; (1981a) *J. Am. Chem. Soc.* **103**: 2890.

Kametani, T., Honda, T., Matsumoto, H., Fukumoto, K. (1981b) *J. Chem. Soc. Perkin Trans. I* 1383.

Kamenati, T., Higashiyama, K., Honda, T., Otomasu, H. (1982) *J. Chem. Soc. Perkin Trans. I* 2935.

Kametani, T., Kanaya, N., Nakayama, A., Mochizuki, T., Yokohama, S., Honda, T. (1986) *J. Org. Chem.* **51**: 624.

Kametani, T., Nishimura, M., Higurashi, K., Suzuki, Y., Tsubuki, M., Honda, T. (1987) *J. Org. Chem.* **52**: 5233.

Kan-Fan, C., Husson, H.-P. (1980) *Tetrahedron Lett.* **21**: 4265.

Kano, S., Yokomatsu, T. (1978) *Tetrahedron Lett.* 1209.

Kaplan, M.L. (1967) *J. Org. Chem.* **32**: 2346.

Karrer, P., Kebrle, J., Thakkar, H.M. (1950) *Helv. Chim. Acta* **33**: 1711.

Katagiri, N., Sato, M., Saikawa, S., Sakamoto, T., Muto, M., Kaneko, C. (1985) *Chem. Commun.* 189.

Katagiri, N., Akatsuka, H., Haneda, T., Kaneko, C., Sera, A. (1988) *J. Org. Chem.* **53**: 5464.

Kato, M., Kurihara, H., Yoshikoshi, A. (1979) *J. Chem. Soc. Perkin Trans. I* 2740.

Kato, M., Tooyama, Y., Yoshikoshi, A. (1989) *Chem. Lett.* 61.

Kato, T., Sato, M., Kitagawa, Y. (1978) *Chem. Pharm. Bull.* **26**: 632.

Kato, T., Sato, M., Kitagawa, Y. (1978b) *J. Chem. Soc. Perkin Trans. I* 352.

Katsumura, S., Isoe, S. (1982) *Helv. Chim. Acta* **65**: 1927.

Kaufmann, T., Echsler, K.J., Hamsen, A., Kriegsmann, R., Steinseifer, F., Vahrenhorst, A. (1978) *Tetrahedron Lett.* 4391.

Kawazu, K., Mitsui, T. (1966) *Tetrahedron Lett.* 3519.

Keana, J.F.W., Heo, G.S., Gaughan, G.T. (1985) *J. Org. Chem.* **50**: 2346.

Kelly, R.B., Zamecnik, J. (1970) *Chem. Commun.* 1102.

Kelly, R.B., Zamecnik, J., Beckett, B.A. (1971) *Chem. Commun.* 479.

Kelly, R.B., Eber, J., Hung, H.-K. (1973) *Can. J. Chem.* **51**: 2534.

Kelly, T.R., Echavarrenn, A., Whiting, A., Weibel, F.R., Miki, Y. (1986) *Tetrahedron Lett.* **27**: 6049.

Kemp, D.S., Hoyng, C.F. (1975a) *Tetrahedron Lett.* 4625.

Kemp, D.S., Roberts, D.C. (1975b) *Tetrahedron Lett.* 4629.

Kende, A.S. (1956) *Chem. Ind.*, 1053.

Kende, A.S., Kaldor, I. (1989) *Tetrahedron Lett.* **30**: 7329.

Kent, R.H., Anselme, J.-P. (1968) *Can. J. Chem.* **46**: 2322.

Khandelwal, Y., Moraes, G., deSouza, N.J., Fehlhaber, H.W., Paulus, E.F. (1986) *Tetrahedron Lett.* **27**: 6249.

Kiang, A.K., Loh, S.K., Demanczyk, M., Gemenden, C.W., Papariello, G.J., Taylor, W.I. (1966) *Tetrahedron* **22**: 3293.

Kienzle, F., Mergelsberg, I., Stadlwieser, J., Arnold, W. (1985) *Helv. Chim. Acta* **68**: 1133.

Kierstead, R.W., Linstead, R.P., Weedon, B.C.L. (1952) *J. Chem. Soc.* 3616.

Kiguchi, T., Naito, T., Ninomiya, I. (1987) *Heterocycles* **26**: 1747.

Kiguchi, T., Kuninobu, N., Naito, T., Ninomiya, I. (1989) *Heterocycles* **28**: 19.

Kim, M., White, J.D. (1975) *J. Am. Chem. Soc.* **97**: 451.

Kimbrough, R.D., Jr., Askins, R.W. (1967) *J. Org. Chem.* **32**: 3683.

Kinast, G., Tietze, L.-F. (1976) *Chem. Ber.* **109**: 3626.

Kinoshita, T., Tatara, S., Sankawa, U. (1985) *Chem. Pharm. Bull.* **33**: 1770.

Kiseleva, N.I., Baskakov, Y.A., Faddeeva, M.I., Putsykin, Y.G. (1981) *Zh. Org. Khim.* **17**: 343.

Kitahara, Y. (1956) *Sci. Rep. Tohoku Univ.* I **39**: 250.

Klatte, F., Rosentreter, U., Winterfeldt, E. (1977) *Angew. Chem. Intern. Ed.* **16**: 878.

Klehr, M., Schäfer, H.J. (1975) *Angew. Chem. Intern. Ed.* **14**: 247.

Klein, K.P., Demmin, T.R., Oxenrider, B.C., Rogic, M.M., Tetenbaum, M.T. (1979) *J. Org. Chem.* **44**: 275.

Klop, W., Brandsma, L. (1983) *Chem. Commun.* 988.

Klunder, A.J.H., Ariaans, G.J.A., Zwanenburg, B. (1984) *Tetrahedron Lett.* **Lett. 25**: 5457.

Klunder, A.J.H., deValk, W.C.G.M., Verlaak, J.M.J., Schellekens, J.W.M., Noordik, J.H., Parthasarathi, V., Zwannenburg, B. (1985) *Tetrahedron* **41**: 963.

Klyne, W., Prelog, V. (1960) *Experientia* **16**: 521.
Knox, J.R., Slobbe, J. (1971) *Tetrahedron Lett.* 2149.
Kobayashi, H., Himeshima, Y., Sonoda, T. (1983) *Chem. Lett.* 1211.
Kobayashi, S., Tezuka, T., Ando, W. (1979) *Chem. Commun.* 508.
Kobayashi, Y., Taguchi, T., Mamada, M., Shimizu, H., Murahashi, H. (1979) *Chem. Pharm. Bull.* **27**: 3123.
Kocienski, P., Todd, M. (1983) *J. Chem. Soc. Perkin Trans. I* 1777.
Kodama, M., Takahashi, T., Kurihara, T., Ito, S. (1980) *Tetrahedron Lett.* **21**: 2811.
Koelsch, C.F., Byers, D.J. (1940) *J. Am. Chem. Soc.* **62**: 560.
Koelsch, C.F., Wawzonek, S. (1943) *J. Am. Chem. Soc.* **65**: 755.
Kögl, F., Becker, H., Detzel, A., deVoss, G. (1928) *Liebigs Ann. Chem.* **465**: 211.
Kohen, F. (1966) *Chem. Ind.* 1378.
Kohler, E.P., Richtmyer, N.K., Hester, W.F. (1931) *J. Am. Chem. Soc.* **53**: 205.
Kojima, T., Inouye, Y., Kakisawa, H. (1985) *Chem. Lett.* 323.
Kouno, I., Irie, H., Kawan, N. (1984) *J. Chem. Soc. Perkin Trans. I* 2511.
Koyama, H., Okawara, H., Kobayashi, S., Ohno, M. (1985) *Tetrahedron Lett.* **26**: 2685.
Kozikowski, A.P., Floyd, W.C. (1978) *Tetrahedron Lett.* 19.
Kozikowski, A.P., Ames, A. (1981) *J. Am. Chem. Soc.* **103**: 3923.
Kozikowski, A.P., Chen., C., Ball, R.G. (1990) *Tetrahedron Lett.* **31**: 5869.
Krapcho, A.P., Peters, R.C.H. (1968) *Chem. Commun.* 1615.
Krapcho, A.P., Horn, D.E., Rao, D.R., Abegaz, B. (1972) *J. Org. Chem.* **37**: 1575.
Krapcho, A.P., Jahngren, E.G.E. (1974) *J. Org. Chem.* **39**: 1322, 1650.
Krapcho, A.P. (1982) *Synthesis* 805, 893.
Kraus, G.A., Molina, M.T., Walling, J.A. (1987a) *J. Org. Chem.* **52**: 1273.
Kraus, G.A., Thomas, P.J., Hon, Y.-S. (1987b) *Chem. Commun.* 1849.
Kraus, W., Chassin, C., Chassin, R. (1969) *Tetrahedron* **25**: 3681.
Kraus, W., Chassin, C. (1970) *Liebigs Ann. Chem.* **735**: 198.
Krepski, L.R., Hassner, A. (1978) *J. Org. Chem.* **43**: 3173.
Kricheldorf, H.R., Schwarz, G., Kaschig, J. (1977) *Angew. Chem. Intern. Ed.* **16**: 550.
Krief, A., Surleraux, D., Frauenrath, H. (1988) *Tetrahedron Lett.* **29**: 6157.
Krook, M.A., Miller, M.J., Eigenbrot, C. (1985) *Chem. Commun.* 1265.
Krow, G.R., Reilly, J. (1972) *Tetrahedron Lett.* 3129.
Ksander, G.M., McMurry, J.E., Johnson, M. (1977) *J. Org. Chem.* **42**: 1180.
Kuehne, M.E., Roland, D.M., Hafter, R. (1978) *J. Org. Chem.* **43**: 3705.
Kuehne, M.E., Matsko, T.H., Bohnert, J.C., Kirkemo, C.L. (1979) *J. Org. Chem.* **44**: 1063.
Kuehne, M.E., Kirkemo, C.L., Matsko, T.H., Bohnert, J.C. (1980) *J. Org. Chem.* **45**: 3259.
Kuehne, M.E., Bohnert, J.C. (1981) *J. Org. Chem.* **46**: 3443.
Kuehne, M.E., Earley, W.G. (1983a) *Tetrahedron* **39**: 3707; (1983b) *Tetrahedron* **39**: 3917.
Kuehne, M.E., Podhorez, D.E. (1985) *J. Org. Chem.* **50**: 924.
Kuhn, R., Teller, E. (1968) *Liebigs Ann. Chem.* **715**: 106.

Kulinkovich, O.G., Tishchenko, I.G., Roslik, N.A. (1982) *Synthesis* 931.

Kupchan, S.M., Dhingra, O.P., Kim, C.K., Kameswaran, V. (1978) *J. Org. Chem.* **43**: 2521.

Kurosawa, Y., Takada, A. (1981) *Chem. Pharm. Bull.* **29**: 2871.

Kurosawa, Y., Nemoto, Y., Sakakura, A., Ogura, M., Takada, A. (1984) *Chem. Pharm. Bull.* **32**: 3366.

Kurth, M.J., Rodriguez, M.J. (1989) *Tetrahedron* **45**: 6963.

Kutney, J.P., Joshua, A.V., Liao, P.-h. (1977) *Heterocycles* **6**: 297.

Kuwajima, I., Urabe, H. (1982) *J. Am. Chem. Soc.* **104**: 6831.

Kwantes, P.M., Schmitz, R.F., Boutkan, C., Klumpp, G.W. (1978) *Tetrahedron Lett.* 3237.

LaForge, F.B., Smith, L.E. (1929) *J. Am. Chem. Soc.* **51**: 2574; (1930) *J. Am. Chem. Soc.* **52**: 1080.

Lai, J.T., Westfahl, J.C. (1980) *J. Org. Chem.* **45**: 1513.

Lal, K., Salomon, R.G. (1989) *J. Org. Chem.* **54**: 2628.

Landmesser, N.G. (1980) Ph.D. Thesis, Duke Univ.

Lange, G.L., deMayo, P. (1967) *Chem. Commun.* 705.

Lange, G.L., Hall, T.W. (1974) *J. Org. Chem.* **34**: 3819.

Langlois, N., Andriamialisoa, R.Z. (1979) *J. Org. Chem.* **44**: 2468.

Langlois, Y., Potier, P. (1975) *Tetrahedron* **31**: 423.

Lantos, I., Bhattacharjee, D., Eggleston, D.S. (1986) *J. Org. Chem.* **51**: 4147.

Lapworth, A. (1898) *J. Chem. Soc.* **73**: 495.

Larsen, S.D., Monti, S.A. (1977) *J. Am. Chem. Soc.* **99**: 8015.

Larsen, S.D., Monti, S.A. (1979) *Synth. Commun.* **9**: 141.

Larson, E.R., Raphael, R.A. (1982) *J. Chem. Soc. Perkin Trans. I* 521.

Lavie, D., Kashman, Y., Glotter, E., Danieli, N. (1966) *J. Chem. Soc. (C)* 1757.

Lavie, D., Levy, E.C., Rosito, C., Zelnik, R. (1970) *Tetrahedron* **26**: 219.

LeBel, N.A., Caprathe, B.W. (1985) *J. Org. Chem.* **50**: 3938.

Leboff, A., Carbonnelle, A.-C., Alazard, J.-P., Thal, C., Kende, A.S. (1987) *Tetrahedron Lett.* **28**: 4163.

Lee, J., Tang, J., Snyder, J.K. (1987) *Tetrahedron Lett.* **28**: 3427.

Lee, S.Y., Niwa, M., Snider, B.B. (1988) *J. Org. Chem.* **53**: 2356.

Lee-Ruff, E., Turro, N.J., Amice, P., Conia, J.-M. (1969) *Can. J. Chem.* **47**: 2797.

Leete, E. (1978) *Chem. Commun.* 1055.

Leffingwell, J.C. (1970) *Tetrahedron Lett.* 1653.

LeGoff, E. (1962) *J. Am. Chem. Soc.* **84**: 3786.

Lemieux, R.P., Beak, P. (1989) *Tetrahedron Lett.* **30**: 1353.

Leonard, N.J., Gelfand, S. (1955) *J. Am. Chem. Soc.* **77**: 3272.

Leriverend, P., Conia, J.-M. (1966) *Bull. Soc. Chim. Fr.* 121.

Lévy, J., Pierron, C., Lukacs, G., Massiot, G., LeMen, J. (1976) *Tetrahedron Lett.* 669.

Lewars, E., Morrison, G. (1977) *Tetrahedron Lett.* 501.

Lewin, G., Poisson, J., Lamotte-Brasseur, J. (1982) *Tetrahedron* **38**: 3291.

Lewis, K.G., Mulquiney, C.E. (1977) *Tetrahedron* **33**: 463.

Lillie, T.J., Musgrave, O.C., Skoyles, D. (1976) *J. Chem. Soc. Perkin Trans. I* 2155.

Lind, H., Winkler, T. (1980) *Tetrahedron Lett.* **21**: 119.
Little, R.D., Dawson, J.R. (1978) *J. Am. Chem. Soc.* **100**: 4607.
Liu, H.-J., Valenta, Z., Wilson, J.S., Yu, T.T.J. (1969) *Can. J. Chem.* **47**: 509.
Liu, H.-J., Valenta, Z., Yu, T.T.J. (1970) *Chem. Commun.* 1116.
Liu, H.-J., Lee, S.P. (1977) *Tetrahedron Lett.* 3699.
Liu, H.-J., Chan, W.H. (1979) *Can. J. Chem.* **57**: 708.
Liu, H.-J., Dieck-Abularach, T. (1982) *Tetrahedron Lett.* **23**: 295.
Liu, H.J., Llinas-Brunet, M. (1988) *Can. J. Chem.* **66**: 528.
Liu, T.M.H., Reamer, R.A., Grabowski, E.J.J. (1979) *J. Chem. Soc. Perkin Trans. I* 42.
Lockhart, R.W., Kitadani, M., Einstein, F.W.B., Chow, Y.L. (1978) *Can. J. Chem.* **56**: 2897.
Lorenzi-Riatsch, A., Nakashita, Y., Hesse, M. (1981) *Helv. Chim. Acta* **64**: 1854; (1984) *Helv. Chim. Acta* **67**: 249.
Lowe, J.N., Ingraham, L.L. (1970) *J. Am. Chem. Soc.* **93**: 3801.
Lown, J.W., Akhtar, M.H. (1972) *Can. J. Chem.* **50**: 2236.
Lukacs, G., Cloarec, L., Luschini, X. (1970) *Tetrahedron Lett.* 89.
Lunkwitz, K., Pritzkow, W., Schmid, G. (1968) *J. Prakt. Chem.* **37**: 319.
Lythgoe, B., Roberts, D.A. (1980) *J. Chem. Soc. Perkin Trans. I* 892.
Macdonald, T.L. (1978a) *Tetrahedron Lett.* 4201.
Macdonald, T.L. (1978b) *J. Org. Chem.* **43**: 3621.
Mackay, C., Waigh, R.D. (1984) *Heterocycles* **22**: 687.
MacMillan, J., Pryce, R.J. (1967) *J. Chem. Soc.* (*C*) 740.
MacMillan, J., Vanstone, A.E., Yeboah, S.K. (1972) *J. Chem. Soc. Perkin Trans. I* 2898.
Madge, N.C., Holmes, A.B. (1980) *Chem. Commun.* 956.
Magid, R.M., Grayson, C.R., Cowsar, D.R. (1968) *Tetrahedron Lett.* 4819.
Magnus, P., Schultz, J., Gallagher, T. (1984) *Chem. Commun.* 1179.
Magnus, P., Pappalardo, P., Southwell, I. (1986) *Tetrahedron* **42**: 3215.
Magnus, P., Cairns, P.M., Moursounidis, J. (1987a) *J. Am. Chem. Soc.* **109**: 2469.
Magnus, P., Ladlow, M., Elliott, J. (1987b) *J. Am. Chem. Soc.* **109.**
7929.
Magnus, P., Matthews, I.R., Schultz, J., Waditschatka, R., Huffman, J.C. (1988) *J. Org. Chem.* **53**: 5772.
Magnus, P., Katoh, T., Matthews, I.R., Huffman, J.C. (1989a) *J. Am. Chem. Soc.* **111**: 6707.
Magnus, P., Ladlow, M., Elliott, J., Kim, C.S. (1989b) *Chem. Commun.* 518.
Magnus, P., Mugrage, B. (1990) *J. Am. Chem. Soc.* **110**: 462.
Mahajan, J.R., Ferreira, G.A.L., Araujo, H.C. (1972) *Chem. Commun.* 1078.
Mahajan, J.R. (1976) *Synthesis* 110.
Mahajan, J.R., DeAraujo, H.C. (1981) *Synthesis* 49.
Majeed, A.J., Patel, P.J., Sainsbury, M. (1985) *J. Chem. Soc. Perkin Trans. I* 1195.
Maksimov, V.I. (1965) *Tetrahedron* **21**: 687.
Mandal, A.N., Bhattacharya, S., Raychaudhuri, S.R., Chatterjee, A. (1988) *J. Chem. Res.* (*S*) 366.
Mander, L.N., Ritchie, E., Taylor, W.C. (1967) *Aust. J. Chem.* **20**: 981.

Mangeney, P. (1978) *Tetrahedron* **34**: 1359.

Mangeney, P., Andriamialisoa, R.Z., Lallemand, J.-Y., Langlois, N., Langlois, Y., Potier, P. (1979) *Tetrahedron* **35**: 2175.

Manitto, P., Monti, D. (1980) *Chem. Commun.* 178.

Mann, J., Wilde, P.D., Finch, M.W. (1985) *Chem. Commun.* 1543.

Mannich, C. (1941) *Ber.* **74B**: 1007.

Mannien, K., Krieger, H. (1967) *Tetrahedron Lett.* 2071.

Marchand, A.P., Chou, T.-c. (1973) *J. Chem. Soc. Perkin Trans. I* 1948.

Marchand, A.P., Chou, T.-c., Ekstrand, J.D., van der Helm, D. (1976) *J. Org. Chem.* **41**: 1438.

Marchand, A.P., Reddy, G.M., Watson, W.H., Nagl, A. (1988) *J. Org. Chem.* **53**: 5969.

Marchand, A.P., Vidyasagar, V. (1989) *197th ACS Nat. Meet.* ORGN 209.

Marchand, A.P., Reddy, S.P., Rajapaksa, D., Ren, C.-t., Watson, W.H., Kashyap, R.P. (1990) *J. Org. Chem.* **55**: 3493.

Marino, J.P., Samanen, J.M. (1976) *J. Org. Chem.* **41**: 179.

Marino, J.P., de la Praddilla, R.F., Laborde, E. (1984) *J. Org. Chem.* **49**: 5279.

Marino, J.P., Laborde, E. (1985) *J. Am. Chem. Soc.* **107**: 734.

Marino, J.P., Silveira, C., Comasseto, J., Petragnani, N. (1987) *J. Org. Chem.* **52**: 4139.

Marino, J.P., Long, J.K. (1988) *J. Am. Chem. Soc.* **110**: 7916.

Marion, L., Manske, R.H.F. (1938) *Can. J. Res.* **B16**: 432.

Marquez, V.E., Rao, K.V.B., Silverton, J.V., Kelley, J.A. (1984) *J. Org. Chem.* **49**: 912.

Marschall, H., (1972) *Chem. Ber.* **105**: 541.

Marschall, H., Vogel, F. (1974) *Chem. Ber.* **107**: 2176.

Marshall, J.A., Scanio, C.J.V. (1965) *J. Org. Chem.* **30**: 3019.

Marshall, J.A., Scanio, C.J.V., Iburg, W.J. (1967) *J. Org. Chem.* **32**: 3750.

Marshall, J.A., Bundy, G.L., Fanta, W.I. (1968a) *J. Org. Chem.* **33**: 3913.

Marshall, J.A., Partridge, J.J. (1968b) *J. Org. Chem.* **33**: 4090.

Marshall, J.A., Babler, J.H. (1969a) *J. Org. Chem.* **34**: 4186.

Marshall, J.A., Brady, S.F. (1969b) *Tetrahedron Lett.* 1387.

Marshall, J.A., Arrington, J.P. (1971a) *J. Org. Chem.* **36**: 214.

Marshall, J.A., Belletire, J. (1971b) *Tetrahedron Lett.* 871.

Marshall, J.A., Greene, A.E. (1971c) *J. Org. Chem.* **36**: 2035.

Marshall, J.A., Ruth, J.A. (1974a) *J. Org. Chem.* **39**: 1971.

Marshall, J.A., Seitz, D.E. (1974b) *J. Org. Chem.* **39**: 1814.

Marshall, J.A., Seitz, D.E. (1975a) *J. Org. Chem.* **40**: 534.

Marshall, J.A., Snyder, W.R. (1975b) *J. Org. Chem.* **40**: 1656.

Martin, J., Parker, W., Raphael, R.A. (1964) *J. Chem. Soc.* 289.

Martin, J., Parker, W., Stewart, T., Stevenson, J.R. (1972) *J. Chem. Soc. Perkin Trans. I* 1760.

Martin, P. (1973) Dissert. Univ. Basel.

Massiot, G., Zeches, M., Thepenier, P., Jacquier, M.J., LeMen-Olivier, L., Delande, C. (1982) *Chem. Commun.* 768.

Massiot, G., LeVaud, C., Vercauteren, J., LeMen-Olivier, L., Levy, J. (1983) *Helv. Chim. Acta* **66**: 2414.

Mastalerz, H. (1984) *J. Org. Chem.* **49**: 4092.
Masuoka, N., Kamikawa, T. (1976) *Tetrahedron Lett.* 1691.
Matoba, K., Noda, H., Yamazaki, T. (1987) *Chem. Pharm. Bull.* **35**: 2627.
Matsuo, K., Ohta, M., Tanaka, K. (1985) *Chem. Pharm. Bull.* **33**: 4063.
Matsuura, T., Ogura, K. (1968) *Tetrahedron* **24**: 6167.
Matz, J.R., Cohen, T. (1981) *Tetrahedron Lett.* **22**: 2459.
Mauperin, P., Levy, J., LeMen, J. (1971) *Tetrahedron Lett.* 999.
McAndrew, B.A. (1979) *J. Chem. Soc. Perkin Trans. I* 1837.
McCurry, P.M. (1971) *Tetrahedron Lett.* 1845.
McElvain, S.M., Weyna, P.L. (1959) *J. Am. Chem. Soc.* **81**: 2579.
McKennis, J.S., Brener, L., Ward, J.S., Pettit, R. (1971) *J. Am. Chem. Soc.* **93**: 4957.
McMorris, T.C., Anchel, M. (1965) *J. Am. Chem. Soc.* **87**: 1594.
McMurry, J.E., Isser, S.J. (1972) *J. Am. Chem. Soc.* **94**: 7132.
McRae, J., Moss, V.A., Raphael, R.A. (1982) *Tetrahedron* **38**: 2097.
Mahta, G., Kapoor, S.K. (1974) *J. Org. Chem.* **39**: 2618.
Mehta, G., Suri, S.C. (1980) *Tetrahedron Lett.* **21**: 1093.
Mehta, G., Rao, K.S., Bhadbhade, M.M., Venkatesan, K. (1981) *Chem. Commun.* 755.
Mehta, G., Rao, K.S. (1983) *Tetrahedron Lett.* **24**: 809.
Mehta, G., Rao, K.S. (1985) *J. Org. Chem.* **50**: 5537.
Meinwald, J., Emerman, S.L. (1956) *J. Am. Chem. Soc.* **78**: 5087.
Meinwald, J., Chapman, O.L. (1959) *J. Am. Chem. Soc.* **81**: 5800.
Meinwald, J., Labana, S.S., Chadha, M.S. (1963) *J. Am. Chem. Soc.* **85**: 582.
Menicagli, R., Malanga, C., Lardici, L., Pecunioso, A. (1983) *J. Chem. Res.* (*S*) 124.
Menicagli, R., Malanga, C., Lardici, L. (1985) *J. Chem. Res.* (*S*) 20.
Meyers, C.Y., Kolb, V.M. (1978) *J. Org. Chem.* **43**: 1985.
Michne, W.F. (1976) *J. Org. Chem.* **41**: 894.
Milenkov, B., Guggisberg, A., Hesse, M. (1987) *Tetrahedron Lett.* **28**: 315.
Miller, J.A., Ullah, G.M. (1989) *J. Chem. Soc. Perkin Trans. I* 633.
Miller, L.S., Grohmann, K., Dannenberg, J.J., Todaro, L. (1981) *J. Am. Chem. Soc.* **103**: 6249.
Miller, R.B., Reichenbach, T. (1974) *Tetrahedron Lett.* 543.
Minato, H., Matsumoto, M., Katayama, T. (1973) *J. Chem. Soc. Perkin Trans. I* 1819.
Mincione, E., Bovicelli, P. (1988) *J. Chem. Res.* (*S*) 370.
Miyahara, K., Kumamoto, F., Kawasaki, T. (1980) *Tetrahedron Lett.* **21**: 83.
Miyano, M., Dorn, C.R. (1972) *J. Org. Chem.* **37**: 259.
Miyano, S., Abe, N., Mibu, N., Takeda, K., Sumoto, K. (1983) *Synthesis* 401.
Miyashita, K., Kumamoto, F., Kawasaki, T. (1980) *Tetrahedron Lett.* **21**: 83.
Miyashita, M., Yoshikoshi, A. (1974) *J. Am. Chem. Soc.* **96**: 1917.
Miyashita, M., Hoshino, M., Yoshikoshi, A. (1988) *Tetrahedron Lett.* **29**: 347.
Mohr, P., Tamm, C. (1987) *Tetrahedron Lett.* **28**: 391, 395.
Molines, H., Nguyen, T., Wakselman, C. (1985) *Synthesis* 754.
Mongrain, M., Lafontaine, J., Bélanger, A., Deslongchamps, P. (1970) *Can. J. Chem.* **48**: 3273.

Moniot, J.L., el Rahman, A., Shamma, M. (1977) *Tetrahedron Lett.* 3787.

Monti, S.A., White, G.L. (1975) *J. Org. Chem.* **40**: 215.

Moon, M.P., Komin, A.P., Wolfe, J.F., Morris, G.F. (1983) *J. Org. Chem.* **48**: 2392.

Moore, H.W. (1979) *Acc. Chem. Res.* **12**: 125.

Moore, J.A., Binkert, J. (1959) *J. Am. Chem. Soc.* **81**: 6029.

Moore, R.E., Rapoport, H. (1973) *J. Org. Chem.* **38**: 215.

Mori, K., Nakahara, Y., Matsui, M. (1970) *Tetrahedron Lett.* 2411.

Moriarty, R.M., Gupta, S.C., Hu, H., Berenschot, D.R., White, K.B. (1981) *J. Am. Chem. Soc.* **103**: 686.

Moriarty, R.M., Sultana, M. (1985) *J. Am. Chem. Soc.* **107**: 4559.

Moriarty, R.M., Vaid, R.K., Ravikumar, V.T., Vaid, B.K., Hopkins, T.E. (1988) *Tetrahedron* **44**: 1603.

Moriconi, E.J., Creegan, F.J. (1966) *J. Org. Chem.* **31**: 2091.

Morita, Y., Savaskan, S., Jaeggi, K.A., Hesse, M., Renner, U., Schmid, H. (1975) *Helv. Chim. Acta* **58**: 211.

Morita, Y., Hesse, M., Renner, U., Schmid, H. (1976) *Helv. Chim. Acta* **59**: 532.

Morzycki, J.W., Mudz, J. (1989) *Heterocycles* **28**: 75.

Moss, R.A., Zdrojewski, T. (1991) *Tetrahedron Lett.* **32**: 5667.

Mukaiyama, T. (1976) *Angew. Chem.* **88**: 111.

Mukaiyama, T., Murakami, M. (1987) *Synthesis* 1043.

Mukharji, P.C., DasGupta, T.K. (1969) *Tetrahedron* **25**: 5275.

Mukherjee-Müller, G., Gilgen, P., Zsindely, J., Schmid, H. (1977) *Helv. Chim. Acta* **60**: 1758.

Mulder, A.J. (1980) *Ger. Offen.* 2945812.

Müller, E.P., Jeger, O. (1975) *Helv. Chim. Acta* **58**: 2173.

Mulzer, J., Pointner, A., Chucholowski, A., Bruntrup, G. (1979) *Chem. Commun.* 52.

Muquet, M., Kunesch, N., Poisson, J. (1972) *Tetrahedron* **28**: 1363.

Murai, A., Takatsuru, H., Masamune, T. (1980) *Bull. Chem. Soc. Jpn* **53**: 1049.

Murai, A., Sato, S., Masamune, T. (1981) *Tetrahedron Lett.* 1033.

Murakami, D., Tokura, N. (1958) *Bull. Chem. Soc. Jpn* **31**: 1044.

Murakami, Y., Yokoyama, Y., Okuyama, N. (1983) *Tetrahedron Lett.* **24**: 2189.

Murakami, Y., Yokoyama, Y., Aoki, C., Miyagi, C., Watanabe, T., Ohmoto, T. (1987) *Heterocycles* **26**: 875.

Murray, R.K., Jr., Ford, T.M. (1977) *J. Org. Chem.* **42**: 1806.

Muskopf, J.W., Coates, R.M. (1985) *J. Org. Chem.* **50**: 69.

Musser, A.K., Fuchs, P.L. (1982) *J. Org. Chem.* **47**: 3121.

Muxfeldt, H., Haas, G., Hardtmann, G., Kathawala, F., Moobery, J.B., Vedejs, E. (1979) *J. Am. Chem. Soc.* **101**: 689.

Nagai, Y., Irie, A., Uno, H., Minami, S. (1979) *Chem. Pharm. Bull.* **27**: 1922.

Nagaoka, H., Ohsawa, K., Takata, T., Yamada, Y. (1984) *Tetrahedron Lett.* **25**: 5389.

Nagaoka, H., Kobayashi, K., Matsui, T., Yamada, Y. (1987) *Tetrahedron Lett.* **28**: 2021.

Nagaoka, H., Kobayashi, K., Yamada, Y. (1988) *Tetrahedron Lett.* **29**: 5945.

Nagaoka, H., Shimano, M., Yamada, Y. (1989) *Tetrahedron Lett.* **30**: 971.

Nagata, W., Sugasawa, T., Narisada, M., Wakabayashi, T., Hayase, Y. (1963) *J. Am. Chem. Soc.* **85**: 2342; (1967) *J. Am. Chem. Soc.* **89**: 1499.

Nagata, W., Narisada, M., Wakabayashi, T. (1968) *Chem. Pharm. Bull.* **16**: 875.

Nagumo, S., Suemune, H., Sakai, K. (1988) *Tetrahedron Lett.* **29**: 6927.

Nagumo, S., Suemune, H., Sakai, K. (1991) *Tetrahedron Lett.* **32**: 5585.

Nair, V., Chamberlain, S.D. (1985) *J. Org. Chem.* **50**: 5069.

Naito, T., Iida, N., Ninomiya, I. (1981) *Chem. Commun.* 44.

Naito, T., Kaneko, C. (1983) *Chem. Pharm. Bull.* **31**: 366.

Nakanishi, K., Ohashi, M., Tada, M., Yamada, Y. (1965) *Tetrahedron* **21**: 1231.

Nakano, T., Haces, A., Martin, A., Rojas, A. (1981) *J. Chem. Soc. Perkin Trans. I* 2075.

Nakano, T., Spinelli, A.C., Martin, A., Usubillaga, A., McPhail, A.T., Onan, K.D. (1985) *J. Chem. Soc. Perkin Trans. I* 1693.

Nakatani, K., Isoe, S. (1985) *Tetrahedron Lett.* **26**: 2209.

Narisada, M., Watanabe, F., Nagata, W. (1971) *Tetrahedron Lett.* 3681.

Naruto, S., Nishimura, H., Kaneko, H. (1975) *Chem. Pharm. Bull.* **23**: 1276.

Natsume, M., Utsunomiya, I. (1984) *Chem. Pharm. Bull.* **32**: 2477.

Natsume, M., Utsonomiya, I., Yamaguchi, K., Sakai, S. (1985) *Tetrahedron* **41**: 2115.

Nefedov, O., Nowizkaja, N., Iwaschenko, A. (1967) *Liebigs Ann. Chem.* **707**: 217.

Nemoto, H., Hashimoto, M., Kurobe, H., Fukumoto, K., Kametani, T. (1985) *J. Chem. Soc. Perkin Trans. I* 927.

Nerdel, F., Goetz, H., Wolff, M. (1960) *Liebigs Ann. Chem.* **632**: 65.

Nerdel, F., Frank, D., Lengert, H.J. (1965) *Chem. Ber.* **98**: 728.

Nerdel, F., Frank, D., Marschall, H. (1967) *Chem. Ber.* **100**: 720.

Nerdel, F., Barth, G., Frank, D., Weyerstahl, P. (1969a) *Chem. Ber.* **102**: 407.

Nerdel, F., Dahl, H., Weyerstahl, P. (1969) *Tetrahedron Lett.* 809.

Nerdel, F., Frank, D., Metasch, W., Gerner, K., Marschall, H. (1970) *Tetrahedron* **26**: 1589.

Neukomm, G., Kletzhändler, E., Hesse, M. (1981) *Helv. Chim. Acta* **64**: 90.

Neumeyer, J.L., McCarthy, M., Weinhardt, K.K., Levins, P.L. (1968) *J. Org. Chem.* **33**: 2890.

Newman, M.S., Magerlein, B.J. (1968) *Org. React.* **5**: 413.

Nicholson, J.R., Singh, G., McCullough, K.J., Wightman, R.H. (1989) *Tetrahedron* **45**: 889.

Nickon, A., Iwadare, T., McGuire, F.J., Mahajan, J.R., Narang, S.A., Umezawa, B. (1970) *J. Am. Chem. Soc.* **92**: 1688.

Nicoletti, R., Forcellese, M.L. (1968) *Tetrahedron* **24**: 6519.

Nishiyama, H., Sakuta, K., Osaka, N., Itoh, K. (1983) *Tetrahedron Lett.* **24**: 4021.

Nishiyama, H., Matsumoto, M., Arai, H., Sakaguchi, H., Itoh, K. (1986) *Tetrahedron Lett.* **27**: 1599.

Nishiyama, H., Sakuta, K., Osaka, N., Arai, H., Matsumoto, M., Itoh, K. (1988) *Tetrahedron* **44**: 2413.

Node, M., Kawabata, T., Ueda, M., Fujimoto, M., Fuji, K., Fujita, E. (1982) *Tetrahedron Lett.* **23**: 4047.

Noguchi, S., Morita, K. (1963) *Chem. Pharm. Bull.* **10**: 1235.

Noyce, D.S. (1950) *J. Am. Chem. Soc.* **72**: 924.

Nozoe, T., Kitahara, Y., Doi, K. (1951) *J. Am. Chem. Soc.* **73**: 1895.

Numan, H., Wynberg, H. (1978) *J. Org. Chem.* **43**: 2232.

Nwaji, M.N., Onyiriuka, O.S. (1974) *Tetrahedron Lett.* 2255.

Ochiai, M., Arimoto, M., Fujita, E. (1981) *Chem. Commun.* 460.

Ochiai, M., Iwaki, S., Ukita, T., Nagao, Y. (1987) *Chem. Lett.* 133.

Oda, K., Ohnuma, T., Ban, Y. (1984) *J. Org. Chem.* **49**: 953.

Ogawa, T., Tomisawa, K., Sota, K. (1988) *Heterocycles* **27**: 2815.

Ogino, T., Kubota, T., Manaka, K. (1976) *Chem. Lett.* 323.

Ohkata, K., Kubo, Y., Tamura, A., Hanafusa, T. (1975) *Chem. Lett.* 859.

Ohloff, G., Becker, J., Schulte-Elte, K.H. (1967) *Helv. Chim. Acta* **50**: 705.

Ohno, M. (1963) *Tetrahedron Lett.* 1753.

Ohno, M., Okamoto, M., Kawabe, N., Umezawa, H., Yakeuchi, T., Iinuma, H., Takahashi, S. (1971) *J. Am. Chem. Soc.* **93**: 1285.

Ohno, M., Ido, M., Eguchi, S. (1988) *Chem. Commun.* 1530.

Ohnuma, T., Sekine, Y., Ban, Y. (1979) *Heterocycles* **12**: 151.

Ohnuma, T., Tabe, M., Shiiya, K., Ban, Y. (1983b) *Tetrahedron Lett.* **24**: 4249.

Ohnuma, T., Nagasaki, M., Tabe, M., Ban, Y. (1983a) *Tetrahedron Lett.* **24**: 4253.

Ohshiro, Y., Ishida, M., Shibata, J., Minami, T., Agawa, T. (1982) *Chem. Lett.* 587.

Ohtaka, H., Morisaki, M., Ikekawa, N. (1973) *J. Org. Chem.* **38**: 1688.

Oikawa, Y., Sugano, K., Yonemitsu, O. (1978) *J. Org. Chem.* **43**: 2087.

Okamoto, Y., Senokuchi, K., Kanematsu, K. (1984) *Chem. Pharm. Bull.* **32**: 4593; (1985) *Chem. Pharm. Bull.* **33**: 3074.

Okuda, S., Kamata, H., Tsuda, K. (1963) *Chem. Pharm. Bull.* **11**: 1349.

Olivier, L., Levy, J., LeMen, J., Janot, M.M., Budzikiewicz, H., Djerassi, C. (1965) *Bull. Soc. Chim. Fr.* 868.

Olofson, R.A., Zimmerman, D.M. (1967) *J. Am. Chem. Soc.* **89**: 5057.

Ono, N., Miyake, H., Kaji, A. (1984) *J. Org. Chem.* **49**: 4997.

Openshaw, H.T., Whittaker, N. (1963) *J. Chem. Soc.* 1449, 1461.

Opitz, G., Zimmermann, F. (1963) *Liebigs Ann. Chem.* **662**: 178.

Oppolzer, W., Hauth, H., Pfäffli, P., Wenger, R. (1977) *Helv. Chim. Acta* **60**: 1801.

Oppolzer, W., Bird, T.G.C. (1979) *Helv. Chim. Acta* **62**: 1199.

Oppolzer, W., Burford, S.C. (1980a) *Helv. Chim. Acta* **63**: 788.

Oppolzer, W., Wylie, R.D. (1980b) *Helv. Chim. Acta* **62**: 1198.

Oppolzer, W., Zutterman, F., Bättig, K. (1983) *Helv. Chim. Acta* **66**: 522.

Oppolzer, W., Godel, T. (1984) *Helv. Chim. Acta* **67**: 1154.

O'Shea, M.G., Kitching, W. (1989) *Tetrahedron* **45**: 1177.

Overman, L.E., Jacobsen, E.J. (1982) *J. Am. Chem. Soc.* **104**: 7225.

Overman, L.E., Ricca, D.J. (1992) in Trost, B.M., Fleming, I. (ed.) *Comprehensive organic synthesis*, Vol. 2 Pergamon, Oxford.

Ozaki, S., Nishiguchi, S., Masui, M. (1984) *Chem. Pharm. Bull.* **32**: 2609.

Padwa, A., Crumrine, D., Hartman, R., Layton, R. (1967) *J. Am. Chem. Soc.* **89**: 4435.

Padwa, A., Wisnieff, T.J., Walsh, E.J. (1986) *J. Org. Chem.* **51**: 5036.

Paquette, L.A., Rosen, M. (1966) *Tetrahedron Lett.* **311.**

Paquette, L.A., Lang, S.A., Short, M.R., Parkinson, B., Clardy, J. (1972) *Tetrahedron Lett.* 3141.

Paquette, L.A. (1973) *MTP Int. Rev. Sci. Org. Chem. Ser.* 1, 5, 127.

Paquette, L.A., Davis, R.F., James, D.R. (1974) *Tetrahedron Lett.* 1615.

Paquette, L.A., Snow, R.A., Muthard, J.L., Cynkowski, T. (1978) *J. Am. Chem. Soc.* **100**: 1600.

Paquette, L.A., Wyvratt, M.J., Schallner, O., Muthard, J.L., Begley, W.J., Blankenship, R.M., Balogh, D. (1979) *J. Org. Chem.* **44**: 3616.

Paquette, L.A., Roberts, R.A., Drtina, G.J. (1984) *J. Am. Chem. Soc.* **106**: 6690.

Paquette, L.A., Maynard, G.D., Ra, C.S., Hoppe, M. (1989) *J. Org. Chem.* **54**: 1408.

Parham, W.E., Bolon, D.A., Schweizer, E.E. (1961) *J. Am. Chem. Soc.* **83**: 603.

Parham, W.E., Sperley, R.J. (1967) *J. Org. Chem.* **32**: 926.

Parham, W.E., Dooley, J.F. (1968) *J. Org. Chem.* **33**: 1476.

Parham, W.E., Dooley, J.F., Meilahn, M.K., Greidanus, J.W. (1969) *J. Org. Chem.* **34**: 1474.

Partridge, J.J., Chadha, N.K., Uskokovic, M.R. (1973) *J. Am. Chem. Soc.* **95**: 532.

Pattenden, G., Teague, S.J. (1984) *Tetrahedron Lett.* **25**: 3021.

Paulsen, H., Stoye, D. (1966) *Chem. Ber.* **99**: 908.

Payne, G. (1961) *J. Org. Chem.* **26**: 4793.

Pearlman, B.A. (1979) *J. Am. Chem. Soc.* **101**: 6404.

Pelletier, S.W., Mody, N.V., Finer-Moore, J., Ateya, A.-M.M., Schramm, L.C. (1981) *Chem. Commun.* 327.

Pelletier, S.W., Glinski, J.A., Varughese, K.I., Maddry, J., Mody, N.V. (1983) *Heterocycles* **20**: 413.

Pfenninger, J., Graf, W. (1980) *Helv. Chim. Acta* **63**: 1562.

Pfoertner, K.-H., Meister, W., Oberhänsli, W.E., Schönholzer, P., Vetter, W. (1975) *Helv. Chim. Acta* **58**: 846.

Phillips, R.R. (1959) *Org. React.* **10**: 143.

Piccard, J. (1873) *Ber.* **6**: 884.

Piers, E., Abeysekera, B.F., Herbert, D.J., Suckling, I.D. (1982) *Chem. Commun.* 404.

Pillay, K.S., Lockhart, R.N., Tezuka, T., Chow, Y.L. (1974) *Chem. Commun.* 80.

Plesek, J. (1956) *Chem. Listry* **50**: 1854.

Plieninger, H. (1962) *Angew. Chem. Intern. Ed.* **1**: 367.

Pohmakotr, M., Pisutjaroenpong, S. (1985) *Tetrahedron Lett.* **26**: 3613.

Posner, G.H., Asirvatham, E. (1986) *Tetrahedron Lett.* **27**: 663.

Posner, G.H., Asirvatham, E., Webb, K.S., Jew, S.S. (1987) *Tetrahedron Lett.* **28**: 5071.

Prelog, V., Wieland, P. (1944) *Helv. Chim. Acta* **27**: 1127.

Prinzbach, H., Berger, H., Lüttringhaus, H. (1965) *Angew. Chem. Intern. Ed.*, **4**: 435.

Pritzkow, W., Pritz, U. (1966) *J. Prakt. Chem.* **33**: 235.

Pritzkow, W., Rösler, W. (1967) *Liebigs Ann. Chem.* **703**: 66.

Prout, F.S. (1973) *J. Org. Chem.* **38**: 399.

Qian, L., Ji, R. (1989) *Tetrahedron Lett.* **30**: 2089.

Quallich, G.J., Schlessinger, R.H. (1979) *J. Am. Chem. Soc.* **101**: 7627.

Quilico, A., Freri, M. (1928) *Gazz. Chim. Ital.* **58**: 380.

Quirion, J.-C., Grierson, D.S., Royer, J., Husson, H.-P. (1988) *Tetrahedron* **29**: 3311.

Radulescu, D., Gheorgiu, G. (1927) *Ber.* **59**: 186.

Raju, N., Rajagopalan, K., Swaminathan, S., Shoolery, J.N. (1980) *Tetrahedron Lett.* **21**: 1577.

Rama Rao, A.V., Yadav, J.S., Reddy, B., Mehendale, A.R. (1984) *Tetrahedron* **40**: 4643.

Ranu, B.C., Sarkar, D.C. (1988) *Chem. Commun.* 245.

Rao, C.S.S., Kumar, G., Rajagopalan, K., Swaminathan, S. (1982) *Tetrahedron* **38**: 2195.

Ratcliffe, A.H., Smith, G.F., Smith, G.N. (1973) *Tetrahedron Lett.* 5179.

Raverty, W.D., Thomson, R.H., King, T.J. (1977) *J. Chem. Soc. Perkin Trans. I* 1204.

Read, G., Ruiz, V.M. (1973) *J. Chem. Soc. Perkin Trans. I* 235, 1223.

Reese, C.B., Stebles, M.R.D. (1972) *Chem. Commun.* 1231.

Reese, C.B., Sanders, H.P. (1981) *Synthesis* 276.

Reinecke, M.G., Kray, L.R., Francis, R.F. (1972) *J. Org. Chem.* **37**: 3489.

Reissig, H.-U., Böhm, I. (1983) *Tetrahedron Lett.* **24**: 715.

Reissig, H.-U. (1985) *Tetrahedron Lett.* **26**: 3943.

Reusch, W., Mattison, P. (1967) *Tetrahedron* **23**: 1953.

Reydellet, V., Helquist, P. (1989) *Tetrahedron Lett.* **30**: 6837.

Rice, K.C., Weiss, U., Akiyama, T., Highet, R.J., Lee, T., Silverton, J.V. (1975) *Tetrahedron Lett.* 3767.

Richou, O., Vaillancourt, V., Faulkner, D.J., Albizati, K.F. (1989) *J. Org. Chem.* **54**: 4729.

Risinger, G.E., Haag, W.G., III (1973) *J. Org. Chem.* **38**: 3646.

Roberts, J.C., Thompson, D.J. (1971) *J. Chem. Soc.* (*C*) 3493.

Roberts, M.R., Schlessinger, R.H. (1979) *J. Am. Chem. Soc.* **101**: 7626.

Roberts, M.R., Schlessinger, R.H. (1981) *J. Am. Chem. Soc.* **103**: 724.

Robinson, R. (1958) *Bull. Soc. Chim. Fr.* 125.

Rogic, M.M., Vitrone, J., Swedloff, M.D. (1975) *J. Am. Chem. Soc.* **97**: 3848.

Romo, J., Romo de Vivar, A. (1957) *J. Am. Chem. Soc.* **79**: 1118.

Roush, W.R., D'Ambra, T.E. (1983) *J. Am. Chem. Soc.* **105**: 1059.

Roush, W.R., Russo-Rodriguez, S., (1987) *J. Org. Chem.* **52**: 598.

Rowe, F.M., Levin, E., Burns, A.C., Davis, J.S.H., Tepper, W. (1926) *J. Chem. Soc.* 690.

Rowley, M., Tsukamoto, M., Kishi, Y. (1989) *J. Am. Chem. Soc.* **111**: 2735.

Rubin, M.B., Gutman, A.L., Inbar, S. (1979a) *Tetrahedron Lett.* 889.

Rubin, M.B., Inbar, S. (1979b) *Tetrahedron Lett.* 5021.

Rubiralta, M., Diez, A., Balet, A., Bosch, J. (1987) *Tetrahedron* **43**: 3021.

Rüchardt, C., Schwarzer, H. (1966) *Chem. Ber.* **99**: 1878.

Ruggeri, R.B., McClure, K.F., Heathcock, C.H. (1989) *J. Am. Chem. Soc.* **111**: 1530.

Ruzicka, L., Jeger, O., Winter, M. *Helv. Chim. Acta* **26**: 265.

Saá, C., Guitian, E., Castedo, L., Suau, R., Saá, J.M. (1986) *J. Org. Chem.* **51**: 2781.

Saha, B., Roy, S.C., Satyanarayana, G.O.S.V., Ghatak, U.R., Seal, A., Ray, S., Bandyopadhyay, R., Ghosh, M., Das, B., Basak, B.S. (1985) *J. Chem. Soc. Perkin Trans. I* 505.

Sakai, T., Miyata, K., Ishikawa, M., Takeda, A. (1985) *Tetrahedron Lett.* **26**: 4727.

Sakan, T., Isoe, S., Hyeon, S.B. (1967) *Tetrahedron Lett.* 1623.

Sakkers, P.J.D., Vankan, J.M.J., Klunder, A.J.H., Zwanenburg, B. (1979) *Tetrahedron Lett.* 897.

Saksena, A.K., McPhail, A.T., Onan, K.D. (1981) *Tetrahedron Lett.* **22**: 2067.

Saksena, A.K., Green, M.J., Shue, H.-J., Wong, J.K. (1985a) *Chem. Commun.* 1748.

Saksena, A.K., Green, M.J., Shue, H.-J., Wong, J.K., McPhail, A.T. (1985b) *Tetrahedron Lett.* **26**: 551.

Sala, T., Sargent, M.V. (1978) *Chem. Commun.* 1043.

Sammes, M.P., Maini, P.N., Katritzky, A.R. (1984) *Chem. Commun.* 354.

Sampath, V., Schore, N.E. (1983) *J. Org. Chem.* **48**: 4882.

Sampath, V., Lund, E.C., Knudsen, M.J., Olmstead, M.M., Schore, N.E. (1987) *J. Org. Chem.* **52**: 3595.

Sankawa, U., Seo, S., Kobayashi, N., Ogihara, Y., Shibata, S. (1968) *Tetrahedron Lett.* 5557.

Sano, T., Hirose, M., Horiguchi, Y., Takayanagi, H., Ogura, H., Tsuda, Y. (1987) *Chem. Pharm. Bull.* **35**: 4730.

Sasaki, T., Eguchi, S., Ishii, T. (1970) *J. Org. Chem.* **35**: 249.

Sasaki, T., Eguchi, S., Toru, T. (1969) *J. Am. Chem. Soc.* **91**: 3390; (1971) *J. Org. Chem.* **36**: 3460.

Sasaki, T., Kanematsu, K., Murata, M. (1973) *Tetrahedron* **29**: 529.

Sato, M., Ebine, S., Tsunetsugu, J. (1977) *Tetrahedron Lett.* 855.

Sato, M., Ogasawara, H., Sekiguchi, K., Kaneko, C. (1984) *Heterocycles* **22**: 2563.

Sato, T., Kikuchi, T., Sootome, N., Murayama, E. (1985) *Tetrahedron Lett.* **26**: 2205.

Sato, T., Watanabe, M., Murayama, E. (1986) *Tetrahedron Lett.* **27**: 1621.

Sato, Y., Ikekawa, N. (1961) *J. Org. Chem.* **26**: 5058.

Satterthwait, A.C., Westheimer, F.H. (1978) *J. Am. Chem. Soc.* **100**: 3197.

Sayigh, A.A.R., Tilley, J.N., Ulrich, H. (1964) *J. Org. Chem.* **29**: 3344.

Scheffer, J.R., Wong, Y.-F. (1980) *J. Am. Chem. Soc.* **102**: 6604.

Schell, F.M., Cook, P.M. (1984) *J. Org. Chem.* **49**: 4067.

Schiehser, G.A., White, J.D. (1980) *J. Org. Chem.* **45**: 1864.

Schirmann, P.J., Matthews, R.S., Dittmer, D.C. (1983) *J. Org. Chem.* **48**: 4426.

Schlecht, M.F., Kim, H. (1985) *Tetrahedron Lett.* **26**: 127.

Schlessinger, R.H., Nugent, R.A. (1982) *J. Am. Chem. Soc.* **104**: 1116.

Schmidt, H., Ried, W. (1969) *Tetrahedron Lett.* 2431.

Schmitz, B. (1974) Dissert. Univ. Basel.

Schneider, W.P., Axen, U., Lincoln, F.H., Pike, J.E., Thompson, J.L. (1969) *J. Am. Chem. Soc.* **91**: 5372.

Schneider, W.P., Morge, R.A. (1978) *J. Org. Chem.* **43**: 759.

Schönberg, A., Moubasher, R. (1949) *J. Chem. Soc.* 212.

Schöpf, C. (1927) *Liebigs Ann. Chem.* **452**: 411.

Schöpf, C., Benz, G., Dursch, F., Burkhardt, W., Rokohl, R. (1972) *Liebigs Ann. Chem.* **755**: 86.

Schofield, J.A., Todd, A. (1961) *J. Chem. Soc.* 2316.

Scholz, D. (1981) *Chem. Ber.* **114**: 909.

Schore, N.E., Najdi, S.D. (1987) *J. Org. Chem.* **52**: 5296.

Schreiber, J., Leimgruber, W., Pesaro, M., Schudel, P., Threlfall, T., Eschenmoser, A. (1961) *Helv. Chim. Acta* **44**: 540.

Schreiber, J., Felix, D., Eschenmoser, A., Winter, M., Gautschi, F., Schulte-Elte, K.H., Sundt, E., Ohloff, G., Kalvoda, J., Kaufmann, H., Wieland, P., Anner, G. (1967) *Helv. Chim. Acta* **50**: 2101.

Schug, K., Guengerich, C.P. (1979) *J. Am. Chem. Soc.* **101**: 235.

Schultz, A.G., Staib, R.R., Eng, K.K. (1987) *J. Org. Chem.* **52**: 2968.

Schultz, A.G., Graves, D.M., Jacobson, R.R., Tham, F.S. (1991) *Tetrahedron Lett.* **32**: 7499.

Schumann, D., Schmid, H. (1963) *Helv. Chim. Acta* **46**: 1996.

Schumm, O. (1928) *Z. Physiol. Chem.* **176**: 122; **178**: 1.

Schuster, D.I., Liu, K. (1971) *J. Am. Chem. Soc.* **93**: 6711.

Schwartz, J.S., Applegate, H.E. (1967a) *J. Org. Chem.* **32**: 1241.

Schwartz, J.S., Applegate, H.E., Bouchard, J.L., Weisenborn, F.L. (1967b) *J. Org. Chem.* **32**: 1238.

Schweizer, E.E., Parham, W.E. (1960) *J. Am. Chem. Soc.* **82**: 4085.

Scott, A.I., Yeh, C.-L. (1974) *J. Am. Chem. Soc.* **96**: 2273.

Scott, A.I., Yeh, C.-L., Greenslade, D. (1978) *Chem. Commun.* 947.

Scott, A.I., Williams, H.J., Stolowich, N.J., Kasruso, P., Gonzalez, M.D., Muller, G., Hlineny, K., Savvidis, E., Schneider, E., Traub-Eberhard, U., Wirth, G. (1989) *J. Am. Chem. Soc.* **111**: 1897.

Seebach, D., Peletis, N. (1969) *Angew. Chem. Intern. Ed.* **8**: 450.

Seebach, D., Beck, A.K. (1974) *Angew. Chem. Intern. Ed.* **13**: 806.

Seebach, D., Beck, A.K. (1975) *Chem. Ber.* **108**: 314.

Seitz, D.E., Zapata, A. (1981) *Synthesis* **557.**

Sekiya, M., Ohashi, Y., Terao, Y., Ito, K. (1976) *Chem. Pharm. Bull.* **24**: 369.

Semmelhack, M.F., Tomesch, J.C. (1977) *J. Org. Chem.* **42**: 2657.

Seto, H., Hirokawa, S., Fujimoto, Y., Tatsuno, T. (1983) *Chem. Lett.* 989.

Seto, H., Sakaguchi, M., Fujimoto, Y., Tatsuno, T., Yoshioka, H. (1985a) *Chem. Pharm. Bull.* **33**: 412.

Seto, H., Tsunoda, S., Ikeda, H., Fujimoto, Y., Tatsuno, T., Yoshioka, H. (1985) *Chem. Pharm. Bull.* **33**: 2594.

Seto, J., Fujimoto, Y., Tatsuno, T., Yoshioka, H. (1985) *Synth. Commun.* **15**: 1217.

Seto, S., Sugiyama, H., Takanaka, S., Watanabe, H. (1969) *J. Chem. Soc.* (*C*) 1625.

Seyferth, D., Dertouzos, H., Suzuki, R., Mui, J.Y.-P. (1967) *J. Org. Chem.* **32**: 2980.

Seyferth, D., Hopper, S.P., Darragh, K.V. (1969) *J. Am. Chem. Soc.* **91**: 6536.

Sezaki, M., Kondo, S., Maeda, K., Umezawa, H., Ohno, M. (1970) *Tetrahedron* **26**: 5171.

Shea, K.J., Fruscella, W.M., England, W.P. (1987) *Tetrahedron Lett.* **28**: 5623.

Shamma, M., Moniot, J.L., Smeltz, L.A., Shores, W.A., Töke, L. (1977) *Tetrahedron* **33**: 2907.

Sheehan, J.C., Wilson, R.M., Oxford, A.W. (1971) *J. Am. Chem. Soc.* **93**: 7222.

Shibuya, M., Jaisli, F., Eschenmoser, A. (1979) *Angew. Chem. Intern. Ed.* **18**: 636.

Shimada, J., Hashimoto, K., Kim, B.H., Nakamura, E., Kuwajima, I. (1984) *J. Am. Chem. Soc.* **106**: 1759.

Shishido, K., Koizumi, S., Utimoto, K. (1969) *J. Org. Chem.* **34**: 2661.

Shishido, K., Matsuura, T., Fukumoto, K., Kametani, T. (1983) *Chem. Pharm. Bull.* **31**: 57.

Shimizu, M., Ando, R., Kuwajima, I. (1984) *J. Org. Chem.* **49**: 1230.

Siegfried, T.H. (1975) Dissert. Univ. Basel.

Silverman, R.B. (1981) *J. Org. Chem.* **46**: 4789.

Simpson, T.J., Pemberton, A.D. (1989) *Tetrahedron* **45**: 2451.

Skorianetz, W., Ohloff, G. (1973) *Helv. Chim. Acta* **56**: 2025; (1975) *Helv. Chim. Acta* **58**: 771.

Slawjanow, A. (1907) *J. Russ. Phys. Chem. Soc.* **39**: 140.

Sliwa, H., Ouattara, L. (1987) *Heterocycles* **26**: 3065.

Slougui, N., Rousseau, G., Conia, J.-M. (1982) *Synthesis* 58.

Slougui, N., Rousseau, G. (1985) *Tetrahedron* **41**: 2643.

Slougui, N., Rousseau, G. (1987) *Tetrahedron Lett.* **28**: 1651.

Smalberger, T.M., Vleggaar, R., Weber, J.C. (1975) *Tetrahedron* **31**: 2297.

Smith, A.B., Toder, B.H., Branca, S.J. (1976) *J. Am. Chem. Soc.* **98**: 7456.

Smith, D.C. (1957) *J. Chem. Soc.* 2690.

Smith, E.M., Shapiro, E.L., Teutsch, G., Weber, L., Herzog, H.L., McPhail, A.T., Tschang, P.-S.W., Meinwald, J. (1974) *Tetrahedron Lett.* 3519.

Smith, K.M., Langry, K.C. (1983a) *J. Org. Chem.* **48**: 500.

Smith, K.M., Miura, M., Tabba, H.D. (1983b) *J. Org. Chem.* **48**: 4779.

Smith, L.I., Rogier, E.R. (1951) *J. Am. Chem. Soc.* **73**: 3837.

Smith, R., Livinghouse, T. (1983) *J. Org. Chem.* **48**: 1554.

Smith, S., Jr., Elango, V., Shamma, M. (1984) *J. Org. Chem.* **49**: 581.

Smula, V., Cundasawmy, Holland, H.L., MacLean, D.B. (1973) *Can. J. Chem.* **51**: 3287.

Snatzke, G., Langen, H., Himmelreich, J. (1971) *Liebigs Ann. Chem.* **744**: 142.

Sobti, R.R., Dev, S. (1970) *Tetrahedron* **26**: 649.

Sommer, K. (1970) *Z. Anorg. Allgem. Chem.* **377**: 273.

Souchet, M., Guilhem, J., LeGoffic, F. (1987) *Tetrahedron Lett.* **28**: 2371.

Spencer, T.A., Schmiegel, K.K., Williamson, K.L. (1963) *J. Am. Chem. Soc.* **85**: 3785.

Spry, D.O., Bhala, A.R., Spitzner, W.A., Jones, N.D., Swartzendruber, J.K. (1984) *Tetrahedron Lett.* **25**: 2531.

Spurr, P.R., Hamon, D.P.G. (1983) *J. Am. Chem. Soc.* **105**: 4734.

Squattrito, P.J., Williard, P.G. (1989a) *Tetrahedron Lett.* **30**: 5211.

Stach, H., Hesse, M. (1988) *Tetrahedron* **44**: 1573.

Sternbach, D., Shibuya, M., Jaisli, F., Bonetti, M., Eschenomoser, A. (1979) *Angew. Chem. Intern. Ed.* **18**: 634.

Stevens, C.L., Mukherjee, T.K., Traynelis, V.J. (1956) *J. Am. Chem. Soc.* **78**: 2264.

Stevens, K.L., Lundin, R., Davis, D.L. (1975) *Tetrahedron* **31**: 2749.

Stevens, R.V., Lee, A.W.M. (1982) *Chem. Commun.* 103.

Stevens, R.V., Lawrence, D.S. (1985) *Tetrahedron* **41**: 93.

Stiles, M., Miller, R.G. (1960a) *J. Am. Chem. Soc.* **82**: 3802.

Stiles, M., Sisti, A.J. (1960b) *J. Org. Chem.* **25**: 1691.

Still, W.C., Tsai, M.-Y. (1980) *J. Am. Chem. Soc.* **102**: 3654.

Stork, G., Breslow, R. (1953) *J. Am. Chem. Soc.* **75**: 3292.

Stork, G., Borowitz, I. (1960) *J. Am. Chem. Soc.* **82**: 4307.

Stork, G., Dolfini, J.E. (1963) *J. Am. Chem. Soc.* **85**: 2872.

Stork, G., Tabak, J.M., Blount, J. (1972) *J. Am. Chem. Soc.* **94**: 4735.

Stork, G., Macdonald, T.L. (1975) *J. Am. Chem. Soc.* **97**: 1264.

Streef, J.W., van der Plas, H.C., Wei, Y.Y., Declercq, J.P., Van Meersche, M. (1987) *Heterocycles* **26**: 685.

Sugawara, T., Kuwajima, I. (1985) *Tetrahedron Lett.* **26**: 5571.

Sunthankar, S.V., Mehendale, S.D. (1972) *Tetrahedron Lett.* 2481.

Suzuki, M., Kowata, Kurosawa, E. (1980) *Tetrahedron Lett.* **36**: 1551.

Suzuki, T., Sato, E., Unno, K., Kametani, T. (1985a) *Heterocycles* **23**: 835; (1985b) *Heterocycles* **23**: 839.

Swaminathan, S., Srinivasan, K.G., Venkataramani, P.S. (1970) *Tetrahedron* **26**: 1453.

Swan, G.A. (1950) *J. Chem. Soc.* 1534.

Swindell, C.S., deSolms, J., Springer, J.P. (1984) *Tetrahedron Lett.* **25**: 3797, 3801.

Szantay, C., Keve, T., Bölcskei, H., Megyeri, G., Gacs-Baitz, E. (1985) *Heterocycles* **23**: 1885.

Taga, J. (1964) *Chem. Pharm. Bull.* **12**: 389.

Tagawa, H., Kubo, S. (1984) *Chem. Pharm. Bull.* **32**: 3407.

Taguchi, T., Takigawa, T., Tawara, Y., Morikawa, T., Kobayashi, Y. (1984) *Tetrahedron Lett.* **25**: 5689.

Takahashi, T., Nomura, K., Satoh, J. (1983) *Chem. Commun.* 1441.

Takano, S., Hatakeyama, S., Ogasawara, K. (1977) *Chem. Commun.* 68.

Takano, S., Hatakeyama, S., Ogasawara, K. (1978a) *Tetrahedron Lett.* 2519.

Takano, Sasaki, M., Kanno, H., Shishido, K., Ogasawara, K. (1978b) *J. Org. Chem.* **43**: 4169.

Takano, S., Shishido, K., Sato, M., Yuta, K., Ogasawara, K. (1978c) *Chem. Commun.* 943.

Takano, S., Hatakeyama, S., Ogasawara, K. (1979a) *J. Am. Chem. Soc.* **101**: 6414.

Takano, S., Shishido, K., Matsuzaka, J., Sato, M., Ogasawara, K. (1979b) *Heterocycles* **13**: 307.

Takano, S., Takahashi, M., Hatakeyama, S., Ogasawara, K. (1979c) *Chem. Commun.* **556;** (1979d) *Heterocycles* **12**: 765.

Takano, S., Yuta, K., Hatakeyama, S., Ogasawara, K. (1979e) *Tetrahedron Lett.* 369.

Takano, S., Chiba, K., Yonaga, M., Ogasawara, K. (1980a) *Chem. Commun.* 616.

Takano, S., Masuda, K., Ogasawara, K. (1980) *Chem. Commun.* 887.

Takano, S., Murakata, C., Ogasawara, K. (1980c) *Heterocycles* **14**: 1301.

Takano, S., Yonaga, M., Yamada, S., Hatakeyama, S., Ogawara, K. (1981) *Heterocycles* **15**: 309.

Takano, S., Hatakeyama, S., Takahashi, Y., Ogasawara, K. (1982a) *Heterocycles* **17**: 263.

Takano, S., Morimoto, M., Masuda, K., Ogasawara, K. (1982b) *Chem. Pharm. Bull.* **30**: 4238.

Takano, S., Sato, N., Otaki, S., Ogasawara, K. (1987) *Heterocycles* **25**: 69.

Takano, S., Kijima, A., Sugihara, T., Satoh, S., Ogasawara, K. (1989) *Chem. Lett.* 87.

Takaoka D. (1979) *J. Chem. Soc. Perkin Trans. I* 2711.

Takashita, H., Sato, T., Murai, T., Ito, S. (1969) *Tetrahedron Lett.* 3095.

Takeda, K., Horibe, I., Minato, H. (1973) *J. Chem. Soc. Perkin Trans. I* 2212.

Takeshita, H., Hatsui, T., Kato, N., Masuda, T., Tagoshi, H. (1982) *Chem. Lett.* 1153.

Takeuchi, K., Yoshida, M. (1989) *J. Org. Chem.* **54**: 3772.

Takeuchi, N., Kasawa, T., Ikeda, R., Shimizu, K., Hatakeyama, K., Aida, Y., Kaneko, Y., Tobinaga, S. (1984) *Chem. Pharm. Bull.* **32**: 2249.

Tamura, Y., Kita, Y., Ishibashi, H., Ikeda, M. (1972) *Tetrahedron Lett.* 1977.

Tamura, Y., Ishibashi, H., Hirai, M., Kita, Y., Ikeda, M. (1975) *J. Org. Chem.* **40**: 2702.

Tamura, Y., Morita, I., Tsubouchi, H., Ikeda, M. (1982) *Heterocycles* **17**: 163.

Tanabe, M., Crowe, D.F., Dehn, R.L., Detre, G. (1967) *Tetrahedron Lett.* 3739.

Tanaka, M., Suemune, H., Sakai, K. (1988) *Tetrahedron Lett.* **29**: 1733.

Tarbell, D.S., Carman, R.M., Chapman, D.D., Cremer, S.E., Cross, A.D, Huffman, K.R., Kunstmann, M., McCorkindale, N.J., McNally, J.G., Rosowsky, A., Varino, F.H.L., West, R.L. (1961) *J. Am. Chem. Soc.* **83**: 3096.

Tatsuta, K., Akimoto, K., Kinoshita, M. (1979) *J. Am. Chem. Soc.* **101**: 6116.

Taylor, M.D., Smith, III, A.B. (1983) *Tetrahedron Lett.* **24**: 1867.

ter Borg, A.P., Razenberg, E., Kloosterziel, H. (1965) *Rec. Trav. Chim. Pays-Bas* **84**: 1230.

Ternansky, R.J., Balogh, D.W., Paquette, L.A. (1982) *J. Am. Chem. Soc.* **104**: 4503.

Theuns, H.G., LaVos, G.F., tenNoever de Brauw, M.C., Salemink, C.A. (1984) *Tetrahedron Lett.* **25**: 4161.

Thiellier, H.P.M., Koomen, G.-J., Pandit, U.K. (1975) *Heterocycles* **3**: 707.

Thompson, H.W., Wong, J.K. (1985) *J. Org. Chem.* **50**: 404.

Tietze, L.-F. (1974) *Chem. Ber.* **107**: 2499.

Tietze, L.-F., Kinast, G., Uzar, H.C. (1979) *Angew. Chem. Intern. Ed.* **18**: 541.

Tietze, L.-F., Reichert, U. (1980) *Angew. Chem.* **92**: 832.

Tobe, Y., Kishida, T., Yamashita, T., Kakiuchi, K., Odaira, Y. (1985) *Chem. Lett.* 1437.

Toda, T., Saito, K., Mukai, T. (1972) *Tetrahedron Lett.* 1981.

Tolstikov, G.A., Schultz, E.E., Spirikhin, L.V. (1986) *Tetrahedron* **42**: 591.

Torii, S., Inokuchi, T., Takahashi, N. (1978) *J. Org. Chem.* **43**: 5020.

Torii, S., Inokuchi, T., Oi, R. (1982) *J. Org. Chem.* **47**: 47.

Toussaint, O., Capdevielle, P., Maumy, M. (1984) *Tetrahedron Lett.* **25**: 3819.

Traynelis, V.J., Schield, J.A., Lindley, W.A., MacDowell, D.W.H. (1978) *J. Org. Chem.* **43**: 3379.

Tresca, J.P., Fourrey, J.L., Polonsky, J., Wenkert, E. (1973) *Tetrahedron Lett.* 895.

Trost, B.M., Bogdonowicz, M. (1972) *J. Am. Chem. Soc.* **94**: 4777.

Trost, B.M., Preckel, M. (1973) *J. Am. Chem. Soc.* **95**: 7862.

Trost, B.M., Hiroi, K., Holy, N. (1975a) *J. Am. Chem. Soc.* **97**: 5873.

Trost, B.M., Keeley, D.E. (1975b) *J. Org. Chem.* **40**: 2013.

Trost, B.M., Preckel, M., Leichter, L.M. (1975c) *J. Am. Chem. Soc.* **97**: 2224.

Trost, B.M., Frazee, W.J. (1977a) *J. Am. Chem. Soc.* **99**: 6124.

Trost, B.M., Tamaru, Y. (1977b) *J. Am. Chem. Soc.* **99**: 3101.

Trost, B.M., Bogdanowicz, M.J., Frazee, W.J., Salzmann, T.N. (1978a) *J. Am. Chem. Soc.* **100**: 5512.

Trost, B.M., Ochiai, M., McDougal, P.G. (1987b) *J. Am. Chem. Soc.* **100**: 7103.

Trost, B.M., Jungheim, L.N. (1980a) *J. Am. Chem. Soc.* **102**: 7910.

Trost, B.M., Vincent, J.E. (1980b) *J. Am. Chem. Soc.* **102**: 5680.

Trost, B.M., Hiemstra, H. (1982) *J. Am. Chem. Soc.* **104**: 886.

Truce, W.E., Shepherd, J.P. (1977) *J. Am. Chem. Soc.* **99**: 6453.

Tsai, T.Y.R., Tsai, C.S.J., Sy, W.W., Shanbhag, M.N., Liu, W.C., Lee, S.F., Wiesner, K. (1977) *Heterocycles* **7**: 217.

Tsankova, E. (1982) *Tetrahedron Lett.* **23**: 771.

Tsuda, Y., Isobe, K., Sano, T., Morimoto, A. (1975) *Chem. Pharm. Bull.* **23**: 98.

Tsuda, Y., Hosoi, S., Ohshima, T., Kaneuchi, S., Murata, M., Kiuchi, F., Yoda, J., Sano, T. (1985) *Chem. Pharm. Bull.* **33**: 3574.

Tsuda, Y., Kiuchi, F., Umeda, I., Iwasa, E., Sakai, Y. (1986) *Chem. Pharm. Bull.* **34**: 3614.

Tsuji, J., Nisar, M., Minami, I. (1986) *Tetrahedron Lett.* **27**: 2483.

Tsunetsugu, J., Asai, M., Hiruma, S., Kurata, Y., Mori, A., Ono, K., Uchiyama, H., Sato, M., Ebine, S. (1983) *J. Chem. Soc. Perkin Trans. I* 285.

Tsuzuki, K., Hashimoto, H., Shirahama, H., Matsumoto, T. (1977c) *Chem. Lett.* 1469.

Uff, B.C., Powell, M.J., Curran, A.C.W. (1980) *Chem. Commun.* 1059.

Ullman, E.F. (1959) *J. Am. Chem. Soc.* **81**: 5386.

Umehara, M., Hishida, S., Okumoto, S., Ohba, S., Ito, M., Saito, Y. (1987) *Bull. Chem. Soc. Jpn* **60**: 4474.

Underwood, G.R., Ramamoorthy, B. (1970) *Tetrahedron Lett.* 4125.

Uno, H., Kurokawa, M. (1978) *Chem. Pharm. Bull.* **26**: 312.

Urry, W.H., Szecsi, P., Ikoku, C., Moore, D.W. (1964) *J. Am. Chem. Soc.* **86**: 2224.

Utaka, M., Fujita, Y., Takeda, A. (1982) *Chem. Lett.* 1607.

Utaka, M., Nakatani, M., Takeda, A. (1985) *Tetrahedron* **41**: 2163.

Utaka, M., Kuriki, H., Sasaki, T., Takeda, A. (1986a) *J. Org. Chem.* **51**: 935.

Utaka, M., Nakatani, M., Takeda, A. (1986b) *J. Org. Chem.* **51**: 1140.

Valencia, E., Freyer, A.J., Shamma, M., Fajardo, V. (1984) *Tetrahedron Lett.* **25**: 599.

Valencia, E., Fajardo, V., Freyer, A., Shamma, M. (1985) *Tetrahedron Lett.* **26**: 993.

Valenta, Z., Papadopoulos, S., Podesva, C. (1961) *Tetrahedron* **15**: 100.

Van Beek, G., van der Baan, J.L., Klumpp, G.W., Bickelhapt, F. (1986) *Tetrahedron* **42**: 5111.

Vander Zwan, M.C., Hartner, F.W., Reamer, R.A., Tull, R. (1978) *J. Org. Chem.* **43**: 509.

van Tamelen, E.E., Kirk, K., Brieger, G. (1962) *Tetrahedron Lett.* 939.

van Tamelen, E.E., Haarstad, V.B., Orvis, R.L. (1968) *Tetrahedron* **24**: 687.

van Tamelen, E.E., Oliver, L.K. (1970) *J. Am. Chem. Soc.* **92**: 2136.

Vedejs, E., Martinez, G.R. (1980) *J. Am. Chem. Soc.* **102**: 7993.

Vedejs, E., Larsen, S.D. (1984) *J. Am. Chem. Soc.* **106**: 3030.

Verhe, R., DeBuyck, L., De Kimpe, N., Kudesia, V.P., Schamp, N. (1977) *J. Org. Chem.* **42**: 1256.

Visser, G.W., Verboom, W., Reinhoudt, D.N., Harkema, S., van Hummel, G.J. (1982) *J. Am. Chem. Soc.* **104**: 6842.

von Strandtmann, M., Puchalski, C., Shavel, J. (1968) *J. Org. Chem.* **33**: 4010, 4015.

Waddell, T.G. (1985) *Tetrahedron Lett.* **26**: 6277.

Waddell, T.G., Ross, P.A. (1987) *J. Org. Chem.* **52**: 4802.

Wagner, R.B., Moore, J.A. (1950) *J. Am. Chem. Soc.* **72**: 974.

Wakamatsu, T., Akasaka, K., Ban, Y. (1977) *Tetrahedron Lett.* 2755.

Wälchli, R., Bienz, S., Hesse, M. (1985) *Helv. Chim. Acta* **68**: 484.

Wall, R.T. (1970) *Tetrahedron* **26**: 2107.

Wallach, O. (1918) *Ann. Chem.* **414**: 271.

Wanag, G. (1937) *Ber.* **70**: 274.

Wang, F., Liang, X. (1986) *Tetrahedron* **42**: 265.

Wanner, M.J., Koomen, G.J., Pandit, U.K. (1982) *Tetrahedron* **38**: 2741.

Waring, A.J., Zaidi, J.H. (1985) *J. Chem. Soc. Perkin Trans. I* 631.

Warner, P.M., Palmer, R.F., Lu, S.-L. (1977) *J. Am. Chem. Soc.* **99**: 3773.

Warner, P.M., Chen, B.-L., Wada, E. (1981) *J. Org. Chem.* **46**: 4795.

Washburne, S.S., Chawla, R.R. (1971) *J. Organomet. Chem.* **31**: C20.

Wasserman, H.H., Ives, J.L. (1976) *J. Am. Chem. Soc.* **98**: 7868.

Wasserman, H.H., Han, W.T., Schaus, J.M., Faller, J.W. (1984) *Tetrahedron Lett.* **25**: 3111.

Wasserman, H.H., Ives, J.L. (1985a) *J. Org. Chem.* **50**: 3573.

Wasserman, H.H., Pickett, J.E. (1985b) *Tetrahedron* **41**: 2155.

Watanabe, Y., Takeda, T., Aubo, K., Ueno, Y., Toru, T. (1992) *Chem. Lett.* 159.

Watthey, J.W.H., Desai, M. (1982) *J. Org. Chem.* **47**: 1755.

Weigele, M., Tengi, J.P., DeBernardo, S., Czajkowski, R., Leimgruber, W. (1976) *J. Org. Chem.* **41**: 388.

Weinreb, S.M., McMorris, T.C., Anchel, M. (1971) *Tetrahedron Lett.* 3489.

Weinreb, S.M., Demko, D.M., Lessen, T.A., Demers, J.P. (1986) *Tetrahedron Lett.* **27**: 2099.

Weitz, E., Scheffer, A. (1921) *Ber.* **54**: 2327.

Weller, D.D., Runyan, M.T. (1986) *Tetrahedron Lett.* **27**: 4829.

Wender, P.A., Howbert, J.J. (1982) *Tetrahedron Lett.* **23**: 3983.

Wender, P.A., Fisher, K. (1986) *Tetrahedron Lett.* **27**: 1857.

Wender, P.A. (1989) *Pure Appl. Chem.* **61**: 469.

Wender, P.A., von Geldern, T.W., Levine, B.H. (1988) *J. Am. Chem. Soc.* **110**: 4858.

Wender, P.A., Manly, C.J. (1990) *J. Am. Chem. Soc.* **112**: 8579.

Wendler, N.L., Taub, D., Kuo, H. (1960) *J. Am. Chem. Soc.* **82**: 5701.

Wenkert, E., Kariv, E. (1965) *Chem. Commun.* 570.

Wenkert, E., Haviv, F., Zeitlin, A. (1969) *J. Am. Chem. Soc.* **91**: 2299.

Wenkert, E., Bakuzis, P., Baumgarten, R.J., Doddrell, D., Jeffs, P.W., Leicht, C.L., Mueller, R.A., Yoshikoshi, A. (1970) *J. Am. Chem. Soc.* **92**: 1617.

Wenkert, E., Bakuzis, P., Baumgarten, R.J., Leicht, C.L., Schenk, H.P. (1971) *J. Am. Chem. Soc.* **93**: 3208.

Wenkert, E., Bakuzis, P., Baumgarten, R.J., Leicht, C.L., Schenk, H.P. (1971) *J. Am. Chem. Soc.* **93**: 3208.

Wenkert, E., Chang, C.-J., Chawla, H.P.S., Cochran, D.W., Hagaman, E.W., King, J.C., Orito, K. (1976) *J. Am. Chem. Soc.* **98**: 3645.

Wenkert, E., Buckwalter, B.L., Craveiro, A.A., Sanchez, E.L., Sathe, S.S. (1978a) *J. Am. Chem. Soc.* **100**: 1267.

Wenkert, E., Hudlicky, T., Showalter, H.D.H. (1987b) *J. Am. Chem. Soc.* **100**: 4893.

Wenkert, E., Alonso, M.E., Buckwalter, B.L., Sanchez, E.L. (1983) *J. Am. Chem. Soc.* **105**: 2021.

Wenkert, E., Greenberg, R.S., Raju, M.S. (1985) *J. Org. Chem.* **50**: 4681.

Westley, J.W., Evans, R.H., Jr., Williams, T., Stempel, A. (1970) *Chem. Commun.* 71.

Weyler, W., Jr., Pearce, D., Moore, H.W. (1973) *J. Am. Chem. Soc.* **95**: 2603.

Weyler, W., Jr., Duncan, W.G., Moore, H.W. (1975) *J. Am. Chem. Soc.* **97**: 6187.

Wharton, P.S., Bohlen, D.H. (1961) *J. Org. Chem.* **26**: 3615.

Wharton, P.S., Hiegel, G.A., Coombs, R.V. (1963) *J. Org. Chem.* **28**: 3217.

Wharton, P.S., Hiegel, G.A. (1965) *J. Org. Chem.* **30**: 3254.

Wharton, P.S., Sundin, C.E., Johnson, D.W., Kluender, H.C. (1972) *J. Org. Chem.* **37**: 34.

White, J.D., Torii, S., Nogami, J. (1974) *Tetrahedron Lett.* 2879.

Whitmore, F.C. (1948) *Chem. Eng. News* **26**: 668.

Wieland, H., Kajiro, K. (1933) *Liebigs Ann. Chem.* **506**: 60.

Wieland, P., Kaufmann, H., Eschenmoser, A. (1967) *Helv. Chim. Acta* **50**: 2108.

Wiesner, K., Valenta, Z., Ayer, W.A., Fowler, L.R., Francis, J.E. (1958) *Tetrahedron* **4**: 87.

Wiesner, K., Götz, M., Simmons, D.L., Fowler, L.R., Bachelor, F.W., Brown, R.F.C., Büchi, G. (1959) *Tetrahedron Lett.* 15.

Wiesner, K., Bickelhaupt, F., Babin, D.R., Götz, M. (1960) *Tetrahedron* **9**: 254.

Wiesner, K., Götz, M., Simmons, D.L., Fowler, L.R. (1963) *Coll. Czech. Chem. Commun.* **28**: 2462.

Wiesner, K., Musil, V., Wiesner, K.J. (1968a) *Tetrahedron Lett.* 5643.

Wiesner, K., Uyeo, S., Philipp A., Valenta, Z. (1968b) *Tetrahedron Lett.* 6279.

Weisner, K., Ho, P.T., Chang, D., Lam, Y.K., Pan, C.S.J., Ren, W.Y. (1973) *Can. J. Chem.* **51**: 3978.

Wilcox, C.F., Jr., Nealy, D.L. (1963) *J. Org. Chem.* **28**: 3450.

Wilder, P., Hsieh, W.-C. (1975) *J. Org. Chem.* **40**: 717.

Williamson, M.P., Williams, D.H. (1981) *J. Am. Chem. Soc.* **103**: 6580.

Wilson, N.D.V., Jackson, A., Gaskell, A.J., Joule, J.A. (1968) *Chem. Commun.* 584.

Wilson, S.R., Zucker, P.A. (1988) *J. Org. Chem.* **53**: 4682.

Wilton, J.H., Doskotch, R.W. (1983) *J. Org. Chem.* **48**: 4251.

Winkler, J.D., Hershberger, P.M., Springer, J.P. (1986a) *Tetrahedron Lett.* **27**: 5177.

Winkler, J.D., Hey, J.P. (1986b) *J. Am. Chem. Soc.* **108**: 6425.

Winkler, J.D., Hey, J.P., Darling, S.D. (1986c) *Tetrahedron Lett.* **27**: 5959.

Winkler, J.D., Hey, J.P., Williard, P.G. (1986d) *J. Am. Chem. Soc.* **108**: 6425.

Winkler, J.D., Henegar, K.E. (1987a) *J. Am. Chem. Soc.* **109**: 2850.

Winkler, J.D., Hey, J.P., Hannon, F.J. (1987b) *Heterocycles* **25**: 55.

Winkler, J.D., Muller, C.L., Scott, R.D. (1988) *J. Am. Chem. Soc.* **110**: 4831.

Winkler, J.D., Hershberger, P.M. (1989a) *J. Am. Chem. Soc.* **111**: 4852.

Winkler, J.D., Muller, C.L., Hey, J.P., Ogilvie, R.J., Haddad, N., Squattrito, P.J., Williard, P.G. (1989b) *Tetrahedron Lett.* **30**: 5211.

Winkler, J.D., Scott, R.D., Williard, P.G. (1990) *J. Am. Chem. Soc.* **112**: 8971.

Wiseman, J.R., Vanderbilt, J.J. (1978) *J. Am. Chem. Soc.* **100**: 7730.

Witkop, B., Patrick, J.B. (1951) *J. Am. Chem. Soc.* **73**: 2188.

Wittig, G., Frommeld, H.D. (1964) *Chem. Ber.* **97**: 3548.

Wolff, S., Agosta, W.C. (1984) *J. Am. Chem. Soc.* **106**: 2363.

Wolinsky, J., Hutchins, R.O., Gibson, T.W. *J. Org. Chem.* **33**: 407.

Woodward, R.B., Doering, W.v.E. (1945) *J. Am. Chem. Soc.* **67**: 860.

Woodward, R.B., Brutschy, F.J., Baer, H. (1948) *J. Am. Chem. Soc.* **70**: 4216.

Woodward, R.B., Bader, F.E., Bickel, H., Frey, A.J., Kierstead, R.W. (1958) *Tetrahedron* **2**: 1.

Woodward, R.B. (1960) *Angew. Chem.* **72**: 651.

Woodward, R.B., Cava, M.P., Ollis, W.D., Hunger, A., Daeniker, H.U., Schenker, K. (1963a) *Tetrahedron* **19**: 247.

Woodward, R.B., Yates, P. (1963b) *J. Am. Chem. Soc.* **85**: 551, 553.

Woodward, R.B., Fukunaga, T., Kelly, R.C. (1964) *J. Am. Chem. Soc.* **86**: 3162.

Woodward, R.B. (1966) *Science* **153**: 487.

Woodward, R.B. (1968) *Pure Appl. Chem.* **17**: 519.

Woodward, R.B. (1971) *Pure Appl. Chem.* **25**: 283.

Woodward, R.B. (1973) *Pure Appl. Chem.* **33**: 145.

Wrister, J., Brener, L., Pettit, R. (1970) *J. Am. Chem. Soc.* **92**: 7499.

Wuts, P.G.M., Cheng, M.-C. (1986) *J. Org. Chem.* **51**: 2844.

Xie, Z.F., Suemune, H., Sakai, K. (1988) *Chem. Commun.* 612.

Yadav, J.S., Patil, D.G., Krishna, R.R., Chawla, H.P.S., Dev, S. (1982) *Tetrahedron* **38**: 1003.

Yamada, K., Takada, S., Nakamura, S., Hirata, Y. (1968) *Tetrahedron* **24**: 199.

Yamaguchi, H., Okamoto, T. (1969) *Chem. Pharm. Bull.* **17**: 214.

Yamakawa, K., Sakaguchi, R., Nakamura, T., Watanabe, K. (1976) *Chem. Lett.* 991.

Yamamoto, Y., Uyeo, S., Ueda, K. (1964) *Chem. Pharm. Bull.* **12**: 386.

Yamamura, S., Sasaki, K., Hirata, Y., Chen, Y.-P., Hsu, H.-Y. (1973) *Tetrahedron Lett.* 4877.

Yamazaki, T., Matoba, K., Itooka, T., Chintani, M., Momose, T., Muraoka, O. (1987) *Chem. Pharm. Bull.* **35**: 3453.

Yates, B.L., Quijano, J. (1969) *J. Org. Chem.* **34**: 2506.

Yates, P., Stout, G.H. (1958) *J. Am. Chem. Soc.* **80**: 1691.

Yates, P., Anderson, C.D. (1963) *J. Am. Chem. Soc.* **85**: 2937.

Yates, P., Crawford, R.J. (1966) *J. Am. Chem. Soc.* **88**: 1561.

Yates, P., Abrams, G.D., Betts, M.J., Goldstein, S. (1971) *Can. J. Chem.* **49**: 2850.

Yokoe, I., Nakamura, K., Higuchi, K., Shirataki, Y., Komatsu, M., Miyamae, H. (1985) *Heterocycles* **23**: 291.

Yoneda, S., Kawase, T., Yoshida, Z. (1978) *J. Org. Chem.* **43**: 1980.

Yonemitsu, O., Witkop, B., Karle, I.L. (1967) *J. Am. Chem. Soc.* **89**: 1039.

Yoshida, J., Isoe, S. (1987) *Chem. Lett.* 631.

Yoshida, J., Murata, T., Isoe, S. (1988) *J. Organomet. Chem.* **345**: C23.

Yoshioka, M., Hoshino, M. (1971) *Tetrahedron Lett.* 2413.

Young, J.C., Blackwell, B.A., ApSimon, J.W. (1986) *Tetrahedron Lett.* **27**: 1019.

Zagorski, M.G., Salomon, R.G. (1980) *J. Am. Chem. Soc.* **102**: 2501.

Zajac, W.W., Jr., Dampawan, P., Ditrow, D. (1986) *J. Org. Chem.* **51**: 2617.

Zavialov, S.I., Vinogradova, L.P., Kondratieva, G.V. (1964) *Tetrahedron* **20**: 2745.

Zbiral, E., Nestler, G. (1970a) *Tetrahedron* **26**: 2945.

Zbiral, E., Nestler, G., Kischa, K. (1970b) *Tetrahedron* **26**: 1427.

Ziegler, F.E., Reid, G.R., Studt, W.L., Wender, P.A. (1977) *J. Org. Chem.* **42**: 1991.

Ziegler, F.E., Jaynes, B.H. (1987) *Tetrahedron Lett.* **28**: 2339.

Ziegler, F.E., Sobolov, S.B. (1990) *J. Am. Chem. Soc.* **112**: 2749.

Zoch, H.-G., Szeimies, G., Romer, R., Germain, G., Declercq, J.-P. (1983) *Chem. Ber.* **116**: 2285.

Zurflüh, R., Wall, E.N., Siddall, J.B., Edwards, J.A. (1968) *J. Am. Chem. Soc.* **90**: 6224.

# INDEX